Immunoreceptor Signaling

A subject collection from *Cold Spring Harbor Perspectives in Biology*

ALSO FROM COLD SPRING HARBOR LABORATORY PRESS

OTHER SUBJECT COLLECTIONS FROM COLD SPRING HARBOR PERSPECTIVES IN BIOLOGY

Cell–Cell Junctions

Cell Biology of Bacteria

Generation and Interpretation of Morphogen Gradients

NF-κB: A Network Hub Controlling Immunity, Inflammation, and Cancer

Symmetry Breaking in Biology

The Origins of Life

The p53 Family

RELATED LABORATORY MANUALS

Molecular Cloning: A Laboratory Manual, Third Edition

Transcriptional Regulation in Eukaryotes: Concepts, Strategies, and Techniques, Second Edition

Protein–Protein Interactions: A Molecular Cloning Manual, Second Edition

OTHER RELATED TITLES

The TGF-β Family

Immunoreceptor Signaling

A subject collection from *Cold Spring Harbor Perspectives in Biology*

EDITED BY

Lawrence E. Samelson

Andrey Shaw

COLD SPRING HARBOR LABORATORY PRESS
Cold Spring Harbor, New York • www.cshlpress.com

Immunoreceptor Signaling

A Subject Collection from *Cold Spring Harbor Perspectives in Biology*
Articles online at www.cshperspectives.org

Printed in the United States of America

Publisher	John Inglis
Acquisition Editor	Richard Sever
Project Manager	Joan Ebert
Production Editor	Kaaren Kockenmeister
Cover Designer	Denise Weiss

Front cover artwork: The drawing is an artist's rendition of the interaction between the antigen-presenting cell (depicted in brown) and the T lymphocyte (depicted in blue), during T-cell activation. The artist has attempted to convey the variety of transmembrane receptors and ligands on each cell type that participates in this process. (Image courtesy of Dr. Philip Stork, The Vollum Institute, Oregon Health & Sciences University.)

Library of Congress Cataloging-in-Publication Data

Immunoreceptor signaling : a subject collection from Cold Spring Harbor Perspectives in Biology / edited by Lawrence E. Samelson, Andrey S. Shaw.
p. cm.
Includes bibliographical references and index.
ISBN 978-0-87969-896-6 (hardcover : alk. paper)
1. Immune response--Regulation. 2. Immunocompetent cells. 3. Cellular signal transduction. 4. Cell receptors. I. Samelson, L. E. II. Shaw, Andrey S. III. Title.

QR186.I475 2010
616.07'95--dc22

2010038473

10 9 8 7 6 5 4 3 2 1

All World Wide Web addresses are accurate to the best of our knowledge at the time of printing.

All Cold Spring Harbor Laboratory Press publications may be ordered directly from Cold Spring Harbor Laboratory Press, 500 Sunnyside Blvd., Woodbury, New York 11797-2924. Phone: 1-800-843-4388 in Continental U.S. and Canada. All other locations: (516) 422-4100. FAX: (516) 422-4097. E-mail: cshpress@cshl.edu. For a complete catalog of Cold Spring Harbor Laboratory Press publications, visit our website at http://www.cshlpress.com/.

Contents

Preface

IMMUNORECEPTOR SIGNALING AS A TOPIC OF RESEARCH has a history of about 30 years. Though the effects of plant lectins on lymphocyte activation were studied back in the 1960s, the modern era began once the lymphocyte antigen receptors were isolated and characterized. After a number of false starts, the T-cell antigen receptor (TCR) was finally identified and its multi-subunit character described in the 1980s. Though surface immunoglobulin was well known to be the recognition component of the B-cell antigen receptor (BCR), the signaling components of this receptor as well as the family of Fc receptors were also identified during this period.

Since then and with the involvement of increasing numbers of researchers, the field has expanded and has flourished due to the power of modern multidisciplinary research. Cellular immunology approaches demonstrated how receptor engagement affects lymphocyte activation, cytokine production, and elaboration of cell-surface molecules. Biochemical approaches not only resulted in isolation of the receptors but also an increasing list of downstream signaling molecules. The latter include protein tyrosine and serine kinases as well as a growing list of their substrates, all central to immunoreceptor signaling. The analysis of complexes of signaling molecules and the protein–protein interactions that make up these complexes remains an active area of research.

The molecular biology revolution that affected all of biomedical research over the same timespan also helped to transform immunoreceptor signaling research. Modern molecular biological tools allowed for the identification and characterization of the receptors and downstream signaling molecules. Genetic approaches such as the generation of mutants and their expression then allowed for the role of these molecules to be defined in cells and in engineered mice. Recently, the application of cell biologic and modern imaging techniques has triggered a new wave of discoveries. The application of these methods led to the discovery of synapses formed during cell–cell contact that are involved in signaling. Microscopic studies now track interactions of single signaling molecules within cells in real time. With the increasing acquisition of large quantities of data, computational approaches are being used to help understand the complexity of these signaling pathways.

Our hope here is to familiarize the reader with the latest discoveries in this field. We also hope that they convey a sense of the excitement about the incredible progress achieved over the last 20 years. The chapters cover a wide range of topics and, in aggregate, our hope is that they provide a comprehensive sense of the current state of the field.

We begin with discussion of TCR and BCR structure and function. Downstream of both of these receptors and receptors of the innate immune system, discussed next, are the protein tyrosine kinases (PTKs) of the Syk family. We continue with a chapter reviewing one of these PTKs, the central TCR-coupled kinase, ZAP-70, followed by chapters focusing on two of the proximal substrates of this kinase, LAT and SLP-76. Both chapters cover the topic of protein–protein interactions and signaling complexes. Members of the Tec family of PTKs, which are also activated following immunoreceptor engagement and are in turn responsible for activation of a variety of signaling enzymes, are the subject of the next chapter.

Engagement of antigen or immunoglobulin receptors alone does not result in full cellular activation. A number of additional receptors are required and this topic is reviewed in chapters on CD28 co-stimulation and the SLAM and SAP families of molecules. The activation of immunoreceptors leads to the activation of serine kinases and changes in lipid metabolism and ultimately in activation of transcription factors such as NF-κB, as well as to changes in the cytoskeleton; chapters on these topics follow. We conclude with two chapters with a more global perspective. The first describes formation of the immune synapse of activated lymphocytes. Many of the molecules reviewed in the preceding chapters localize to this structure. The final chapter brings a systems approach to the study of the immense complexity of immunoreceptor signaling.

We hope that this collection brings together many of the major topics and approaches in this field. Perhaps this overview will aid in stimulating the next many years of fruitful research. We very much appreciate the efforts of several individuals at Cold Spring Harbor Laboratory Press. Richard Sever initiated the project and provided enthusiastic support for it. Kaaren Kockenmeister supervised production of the chapters and Joan Ebert guided and prodded editors and authors alike. Finally, we are grateful to our colleagues whose work has led the field and whose authorship is responsible for *any* success of this volume.

LARRY SAMELSON
ANDREY SHAW

Immunoreceptor Signaling

A subject collection from *Cold Spring Harbor Perspectives in Biology*

Structural Biology of the T-Cell Receptor: Insights into Receptor Assembly, Ligand Recognition, and Initiation of Signaling

Kai W. Wucherpfennig[1,2,3], Etienne Gagnon[1], Melissa J. Call[1], Eric S. Huseby[4], and Matthew E. Call[5]

[1]Department of Cancer Immunology and AIDS, Dana-Farber Cancer Institute, Boston, Massachusetts 02115

[2]Department of Neurology, Harvard Medical School, Boston, Massachusetts 02115

[3]Program in Immunology, Harvard Medical School, Boston, Massachusetts 02115

[4]Department of Pathology, University of Massachusetts Medical School, Worchester, Massachusetts 01655

[5]Department of Biological Chemistry and Molecular Pharmacology, Harvard Medical School, Boston, Massachusetts 02115

Correspondence: kai_wucherpfennig@dfci.harvard.edu

The T-cell receptor (TCR)-CD3 complex serves as a central paradigm for general principles of receptor assembly, ligand recognition, and signaling in the immune system. There is no other receptor system that matches the diversity of both receptor and ligand components. The recent expansion of the immunological structural database is beginning to identify key principles of MHC and peptide recognition. The multicomponent assembly of the TCR complex illustrates general principles used by many receptors in the immune system, which rely on basic and acidic transmembrane residues to guide assembly. The intrinsic binding of the cytoplasmic domains of the CD3ε and ζ chains to the inner leaflet of the plasma membrane represents a novel mechanism for control of receptor activation: Insertion of critical CD3ε tyrosines into the hydrophobic membrane core prevents their phosphorylation before receptor engagement.

LIGAND BINDING BY THE EXTRACELLULAR DOMAINS OF T-CELL RECEPTORS: STRUCTURAL BASIS OF αβ AND γδ TCR BINDING TO A DIVERSE GROUP OF LIGANDS

The structures of many T-cell receptors (TCRs) in their ligand-bound state have now been determined, allowing some of the general rules of antigen recognition to be distilled. Though the best studied interaction is the binding of αβTCR with peptide-MHC complexes, it is highly instructive to compare the prototypical binding to peptide-MHC with the recognition of lipid antigen-CD1 complexes and engagement of nonclassical MHC class I molecules by γδTCRs. These comparisons show that the recognition strategies for these

 doi: 10.1101/cshperspect.a005140
Cite this article as *Cold Spring Harb Perspect Biol* 2010;2:a005140

three classes of TCR ligands are strikingly different (Fig. 1).

Many structures of human and mouse complexes have shown a typical binding mode, which maximizes αβTCR interaction with the MHC-bound peptide (Garboczi et al. 1996; Garcia et al. 1996; Rudolph et al. 2006). This binding mode places the hypervariable loops of the TCR, the CDR3α and CDR3β loops, over the center of the bound peptide (Fig. 1A,D). Two other TCR loops, CDR1α and CDR1β, frequently also contribute to peptide recognition by binding over amino-terminal and carboxy-terminal peptide segments, respectively. Most contacts to the MHC molecule are typically made by the germline-encoded CDR1 and CDR2 loops of both TCRα and β, but the CDR3 loops can also contribute to MHC binding. Thus, six TCR loops can be involved in peptide and MHC recognition and usually provide substantial specificity for both components (Fig. 1A,D).

αβTCRs not only recognize MHC-bound peptides, but also CD1-bound lipid antigens. CD1 molecules have a similar overall fold as MHC class I molecules, but their groove is substantially wider and more hydrophobic to

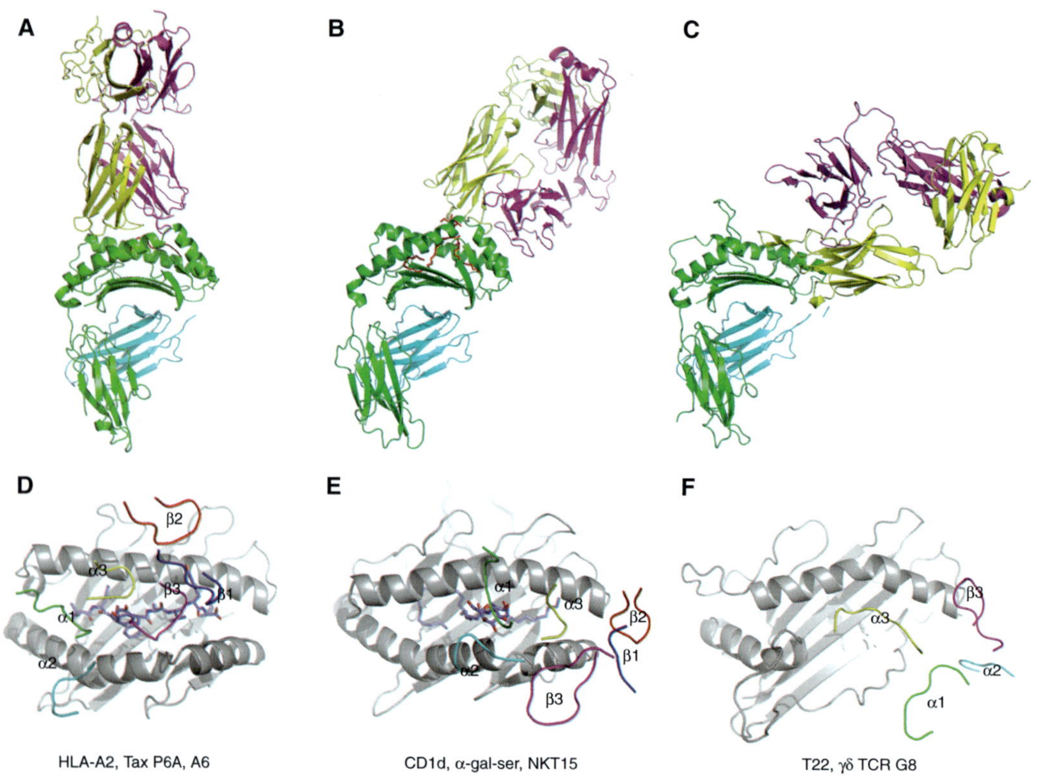

Figure 1. TCR structures with different classes of ligands show distinct binding solutions. (*A–C*). The docking topologies are compared for the αβTCR A6 bound to a peptide-MHC class I complex (HLA-A2 with Tax P6A peptide; PDB entry 1QRN) (*A*), the αβTCR NKT15 bound to the complex of CD1d and the glycolipid α-galactosylceramide (PDB entry 2PO6) (*B*), and the γδTCR G8 bound to the nonclassical MHC class Ib molecule T22 (PDB entry 1YPZ) (*C*). (*D–F*) The placement of the TCR loops is compared for the same complexes. (*D*) The A6 TCR uses all six loops to contact the MHC molecule and the bound peptide. The four germline-encoded loops (α1, α2, β1, and β2) contribute to MHC binding, while the two CDR3 loops are positioned over the peptide (α3 and β3). (*E*) The NKT15 TCR contacts the CD1d-bound glycolipid only through germline elements encoded in the invariant TCRα chain (α1 and α3 loops). (*F*) The γδ TCR G8 inserts the CDR3δ chain loop into the hydrophobic groove of T22; other TCR loops do not appear to be essential for T22 binding.

accommodate a diverse group of lipids, glycolipids, and lipopeptides (Barral and Brenner 2007). NK T cells recognize lipid antigens bound to CD1d, with α-galactosylceramide being the best characterized lipid antigen (Bendelac et al. 2007). In the structure of a human NK TCR, the TCR binds over the extreme end of the CD1d binding cleft (Borg et al. 2007). Furthermore, the TCR is positioned approximately parallel to the long axis of the CD1d binding groove, in contrast to the typical diagonal footprint for most αβTCRs on peptide-MHC (Fig. 1 B,E). NK T cells use a limited TCR repertoire for CD1d-lipid binding, in particular an invariant TCRα chain and a specific Jα segment (Vα24-Jα18) (Bendelac et al. 2007). In the structure only the glycosyl head group of the lipid antigen is exposed and it is contacted solely by germline-encoded loops, CDR1α (encoded by Vα24) and CDR3α (encoded by Jα18) (Fig. 1E). Only germline-encoded segments thus confer specificity for the glycolipid antigen by this NK TCR (Borg et al. 2007). This is in striking contrast to the central role of the hypervariable CDR3α and β loops in discriminating between MHC-bound peptides.

The ligands for most γδ T cells remain unknown, but in mice the closely related nonclassical MHC class Ib molecules T10 and T22 serve as ligands for a sizable population of γδ T cells in nonimmunized mice (Shin et al. 2005). The structure of the G8 γδTCR bound to T22 shows a binding mode that is entirely different from αβTCR engagement of peptide-MHC (Adams et al. 2005). The G8 TCR binds T22 sideways at a highly tilted angle (Fig. 1C), contrasting with the essentially parallel alignment of the long axes of the αβTCR and peptide-MHC. T22 does not have a bound peptide and the segment that corresponds to the amino-terminal part of the α2 helix of classical MHC class I molecules is actually unwound, exposing the hydrophobic groove. The CDR3 loop of TCRδ inserts into this hydrophobic groove and contributes a substantial fraction of the buried surface (Fig. 1F). Interestingly, transfer of this CDR3δ loop into an αβ TCR resulted in T22 binding (Adams et al. 2008). Thus, a single loop of this γδTCR appears to be sufficient for T22 binding.

The overall binding mode thus differs greatly for all three classes of ligands: αβTCRs typically bind in a diagonal orientation over most of the MHC-bound peptide, whereas the NK TCR sits over only part of the CD1d binding groove parallel to the long axis of the groove. The most extreme case is the γδTCR that binds sideways to T22 and inserts one loop into the exposed T22 groove.

Structural Insights Into the Mechanism of MHC Restriction

There has been much debate on the fascinating problem of why αβTCRs are "MHC restricted," and the structural database is now large enough to extract key principles. The overall binding mode of most crystallized αβTCRs with peptide-MHC is similar, even though there is substantial variation in the binding angle relative to the long axis of the MHC molecule (Rudolph et al. 2006; Garcia et al. 2009). Importantly, the general location of the TCR chains on the peptide-MHC ligand is similar in all studied structures: The TCRβ chain is positioned over the α1 helix of the MHC molecule and the TCRα chain over the other MHC helix (α2 in MHC class I, which corresponds to β1 in MHC class II). TCRs have a similar overall fold to antibody Fab segments, but if there were no inherent rules to TCR binding to peptide-MHC, the opposite placement (TCRα rather than TCRβ over the MHC α1 helix) should have been seen by now.

Genes of the MHC are extremely polymorphic. In man, there are >800 MHC class I alleles and >600 alleles of MHC class II (Robinson et al. 2009). If TCRs are prebiased to interact with MHC proteins, essential genetic elements have to be shared between all MHC proteins. It has been appreciated for some time that the overall structure of MHC proteins and their bound peptide is quite similar (Stern and Wiley 1994). Likely because of selective pressure by pathogens, the vast majority of allelic variation between MHC proteins is located within the peptide binding groove (Bjorkman et al. 1987). Conventional alignment of the α1 domain of MHC class I and II molecules failed to show substantial homology in solvent-exposed residues that could be contacted by the TCR. However, if the alignment is shifted by a single turn of an α-helix substantial homology is evident among solvent exposed residues available for interaction with TCRs (Huseby et al. 2005; Huseby et al. 2008). For example, when the surface exposed side chains of H2-K^b and I-A^b are compared, there is substantial conservation between these MHC class I and class II proteins (Table 1). Interestingly, these conserved amino acids are located within TCR binding footprints (Fig. 2). Thus, conserved/chemically similar amino acids on the helices of all MHC proteins may be the elements that the TCR V-domains are genetically encoded to bind.

The diagonal binding orientation on the peptide-MHC surface is similar among most crystallized αβ TCRs, but because of variation in this binding angle, it has not been possible to identify MHC residues contacted by all TCRs (Baker et al. 2001; Rudolph et al. 2006). Rather, the growing structural database suggests an alternative hypothesis: Individual V-genes or groups of related V-genes have coevolved with MHC, which has resulted in preferred contacts by these V-domains with MHC (Garcia et al. 2009). This hypothesis is supported by a substantial number of crystal structures involving six different Vβ8.2 TCRs and one Vβ8.1 TCR in complex with mouse I-A molecules. These structures show close convergence of the CDR1β and CDR2β contacts with the I-A α1 chain helix (Reinherz et al. 1999; Feng et al. 2007; Dai et al. 2008; Garcia et al. 2009). This Vβ8 interaction motif has been seen in structures with different I-A alleles, Vα segments, and peptides, and mutation of these TCRβ residues reduced or abolished activation by a panel of T cells. However, at least some V-gene segments can use more than one set of interactions to bind a given MHC molecule. For example, two Vβ2-containing TCRs with different peptide specificities show distinct docking footprints on the H-2K^b α1 helix (Reiser et al. 2000; Reiser et al. 2003). The hypervariable CDR3 loops of the TCR and

Table 1. Alignment of solvent accessible murine MHC class I and class II residues

MHC class II

Alpha 1 domain

	55	57	61	64	68	72	75
IA b, u	D	Q	Q	A	H	V	K
IA d, g7, k	E	Q	Q	A	H	I	K
IA s	D	Q	Q	A	Y	I	K
IA f	D	Q	Q	A	H	I	K
IA q	D	Q	Q	A	H	G	K
IE d, k	E	Q	A	A	A	V	E
IE u	E	Q	A	A	A	V	K

Beta 1 domain

	81	77	73	70	69	66	64
IA b, d	H	T	A	R	E	E	Q
IA g7, f, k, q	H	T	A	R	E	67-Y	Q
IA u	Y	T	A	R	E	67-Y	Q
IA s	H	T	A	Q	E	67-Y	N
IE b, k, s, u	H	T	A	Q	E	E	Q
IE d	H	T	A	D	E	E	Q

MHC class I

Alpha 1 domain

	58	61	62	65	69	72	76	79
Kb	E	E	R	Q	G	Q	V	R
Kd	E	E	E	Q	S	Q	V	R
Kk	E	E	R	Q	G	Q	V	R
Db	E	E	R	Q	G	Q	V	R
Dk	E	E	R	Q	G	Q	V	R
Ld, q	E	E	R	Q	G	Q	V	R

Alpha 2 domain

	166	163	158	155	154	152	150	149
Kb	E	T	A	R	E	E	A	Q
Kd	E	E	A	Y	E	D	A	Q
Kk	E	T	A	R	E	D	A	Q
Db	E	E	A	H	E	A	S	Q
Dk	E	T	A	R	E	A	A	Q
Ld, q	E	E	A	Y	E	A	A	Q

Helical residues of murine MHC class I and class II proteins were aligned based on a previously reported approach that takes a shift in the MHC class I and class II α1 helices into account (Huseby et al. 2005). Identical residues are highlighted in yellow, structurally similar residues in orange. In some MHC class II molecules, there is a deletion corresponding to one amino acid and residues 66/67 located in the same position of the β1 domain are indicated.

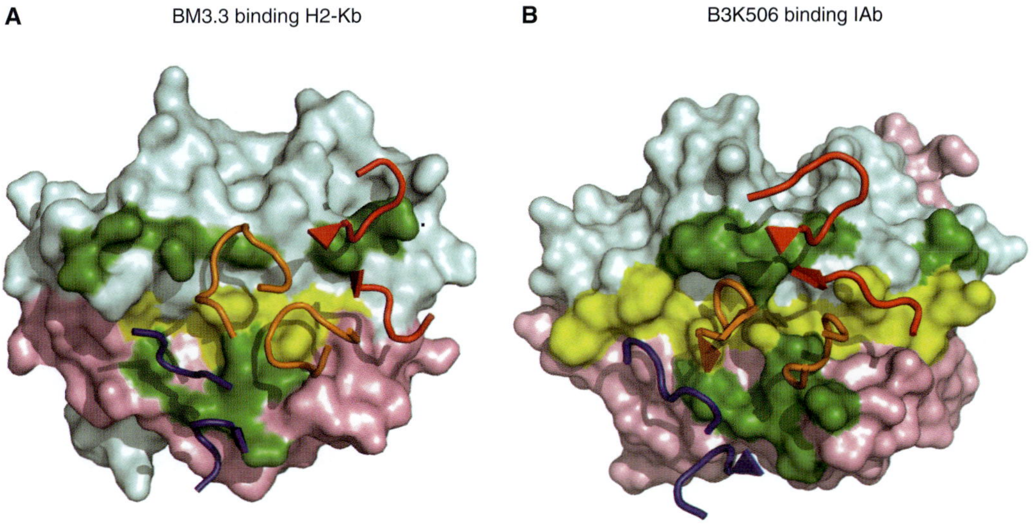

Figure 2. Conserved helical residues of MHC class I and class II proteins are often contacted by CDR1 and CDR2 residues of TCRs. Conserved MHC helical residues are highlighted in green for structures of H-2K^b (*A*) and I-A^b (*B*). The BM3.3 TCR (H-2K^b; PDB entry 1NAM) and the B3K506 TCR (I-A^b; PDB entry 3C5Z) contact these conserved helical MHC residues. TCR CDR1β and CDR2β are colored red, CDR3α and CDR3β orange, and CDR1α and CDR2α blue. Peptide residues are colored yellow. Colored cyan are the H2-K^b α1 and IA^b α1 domains, and magenta H2-K^b α2 and IA^b β1 domains.

the MHC-bound peptide are probably involved in determining which binding interactions with the MHC helices are possible and energetically most favorable.

The identification of the conserved Vβ8 interaction motif with I-A proteins provided an opportunity to examine if these germline-encoded TCR amino acids are required for T-cell development. Particularly prominent in this interaction motif are the contributions of the CDR2β loop residues Y46, Y48, and E54. In mice expressing single, rearranged TCRβ chains, individual mutation of these CDR2β residues to alanine substantially reduced development of the entire TCR repertoire (Scott-Browne et al. 2009). The phenotype was particularly strong for mutation of the two tyrosines (Y46 and Y48) in the CDR2β loop. One of these tyrosines (Y46) is conserved in another mouse Vβ segment (Vβ6) and thymic CD4 T cells were again significantly reduced when Y46 was mutated to alanine. Interestingly, the lowered Vβ affinity appeared to be compensated by selection of TCRα chains with higher affinity for MHC. These results show that thymic selection is controlled by germline-encoded MHC contact points.

This conclusion is supported by the finding that a substantial fraction of T cells from mice with limited negative selection are highly MHC and peptide cross-reactive (Huseby et al. 2005). Interestingly, some of these TCRs are positively selected on both MHC class I and class II, which shows that some T cells can recognize conserved features of the MHC class I and II helices. Structural comparison of highly specific and highly cross-reactive TCRs from these mice showed that the highly cross-reactive TCR had a much smaller concentrated interaction surface that was more hydrophobic (Dai et al. 2008). The interaction with the MHC became highly focused on a limited number of TCR amino acids, which included the CDR2β residues Y46, Y48, and E54 discussed above. The germline-coded TCR contacts with MHC can thus yield T cells with a high degree of MHC and peptide cross-reactivity, which are normally eliminated by negative selection (Huseby et al. 2008).

An alternative hypothesis to explain the diagonal binding mode of the TCR with peptide-MHC proposes that this geometry is driven by the requirement to accommodate the coreceptors CD4/CD8 or by a requirement for formation of TCR dimers/oligomers before initiation of signaling (Ding et al. 1998). No structural data are yet available to test the validity of these ideas. It is important to keep in mind that these hypotheses are not mutually exclusive. For example, the requirement for formation of a higher-order structure could have defined the structural boundaries for coevolution of TCR with MHC.

A common theme emerging from all of these structural studies is that particular germline elements are used in different ways for αβ or γδTCR recognition of peptide-MHC, lipid-CD1 complexes, or nonclassical MHC class I molecules. In the case of the G8 γδTCR, a single TCR loop appears to be sufficient for binding to the nonclassical MHC class Ib molecule T22 (Adams et al. 2005). In contrast, αβTCRs typically use four germline-encoded loops for binding to the MHC helices, which enables the hypervariable CDR3 loops to probe the contents of the peptide binding groove (Rudolph et al. 2006). The recognition mode is yet again different for αβ TCRs binding to CD1d-lipid complexes (Borg et al. 2007). In this specialized case, the germline encoded Vα and Jα segments of the invariant TCRα chain are used to bind to a CD1 embedded glycolipid antigen.

ASSEMBLY OF T-CELL RECEPTOR-CD3 COMPLEXES

How is an encounter with appropriate peptide-MHC or lipid-CD1 complexes actually "read" into a T cell? The mature TCR proteins carry no signaling motifs, but rather transmit signals via tyrosine-based ITAM motifs in the cytoplasmic portions of the TCR-associated CD3γε, CD3δε, and ζζ modules (Germain and Stefanova 1999; Samelson 2002). The precise mechanisms linking MHC binding outside the cell to early biochemical events inside the cell remain an area of vigorous investigation and significant controversy. Proposed triggering models range from receptor clustering and coreceptor recruitment to an array of conformational change models in which TCR proteins are proposed to physically impinge on the CD3 modules to transmit signals (Kuhns et al. 2006; Choudhuri and van der Merwe 2007). Our ability to distinguish among these different models depends critically on a solid understanding of the spatial organization of the complete TCR-CD3 complex. However, no high-resolution structure of an intact complex is yet available. As such, we must rely on three types of available experimental data to inform our working models: (1) measurements of the stoichiometric relationships among subunits, (2) available atomic-resolution structures of folded domains, and (3) biochemical data identifying the molecular surfaces involved in intersubunit contacts.

TCR-CD3 Stoichiometry

The αβTCR is noncovalently associated with three dimeric signaling modules: CD3δε, CD3γε, and ζζ (Call and Wucherpfennig 2005). On ligand binding by the TCR, cytoplasmic ITAM motifs in the CD3 and ζ chains become phosphorylated by the Src-family kinase Lck (Weiss and Littman 1994), constituting the earliest detectable biochemical consequence of TCR ligation. Early genetic, biochemical, and immunofluorescence-based experiments established that the intact TCR-CD3 complex contains two copies of CD3ε (Blumberg et al. 1990b; de la Hera et al. 1991), consistent with a model in which both CD3δε and CD3γε are incorporated into each αβTCR complex. Similar approaches have failed to detect more than one TCRα or TCRβ chain in intact TCR-CD3 complexes (Punt et al. 1994; Call et al. 2002), indicating that the receptor is monovalent with respect to the ligand-binding module. This model is supported by direct biochemical measurements of the TCR-CD3 stoichiometry in ER-assembled complexes, in which the composition was found to be TCRαβ:CD3δε:CD3γε:ζζ (Call et al. 2004).

The composition of the γδTCR and pre-TCR complexes differ from αβTCR. Mice deficient in CD3γ, CD3ε, or ζ subunits show a near-complete arrest of thymocyte development at the $CD4^{-}CD8^{-}$ (DN) stage, the point at which pre-TCR signaling is required for further maturation (reviewed in Dave 2009). However, in CD3δ-deficient mice, thymocyte development proceeded to the $CD4^+CD8^+$ (DP) stage, suggesting that pre-TCR function was intact, and the γδ T-cell compartment was unperturbed (Dave et al. 1997). Thus, whereas CD3γ, CD3ε, and ζ are required for all developmental stages and T-cell lineages, CD3δ may be dispensable for γδ and pre-TCR function in mice. A recent report has indeed shown that surface-expressed TCR-CD3 complexes on CD3δ-sufficient murine γδ T cells do not contain CD3δε modules, and quantitative flow cytometry measurements yielded a stoichiometry of TCRγδ:CD3γε(2):ζζ (Hayes and Love 2006). We note that this appears to be a difference between mouse and human γδTCR, which does incorporate CD3δε (Siegers et al. 2007). Both αβ and γδTCR thus require two CD3 modules, and based on our current understanding of TCR-CD3 assembly mechanisms (see the following discussion), the same is likely to be true for pre-TCR.

Mechanisms Directing TCR-CD3 Complex Assembly

Like the TCR, CD3δε and CD3γε heterodimers are formed through interactions between their extracellular Ig domains. The subunit interface is formed primarily through an extended contact along two β strands that terminate in the stalk regions connecting the Ig domains to the TM domains (Sun et al. 2001; Arnett et al. 2004; Kjer-Nielsen et al. 2004; Sun et al. 2004). In contrast, the disulfide-linked ζζ homodimer, having only a very small (nine amino acid) ectodomain, forms through contacts predominantly within the TM domain (Rutledge et al. 1992; Call et al. 2002; Call et al. 2006). A major remaining challenge has been to identify the molecular surfaces responsible for stabilizing contacts among these four dimeric modules.

Soluble ectodomain fragments of TCR and CD3 dimers show no measurable affinity for one another, and we now understand that the determinants for assembly are contained primarily within the TM and juxtamembrane regions. The TCR proteins have three basic amino acids within the TM domains: Two in TCRα and one in TCRβ. Similarly, CD3 and ζ proteins each contain a single aspartic or glutamic acid in their TM domains, creating a pair of acidic residues in each dimeric module. Cellular transfection studies established that at least some of these ionizable residues participate in assembly (Alcover et al. 1990; Blumberg et al. 1990a; Cosson et al. 1991; Manolios et al. 1991). A more comprehensive mutagenesis analysis using in vitro-assembled complexes revealed that each of the basic TM residues in the TCR specifically recruits one of the three signaling modules in contacts requiring *both* acidic TM residues (Call et al. 2002; Call et al. 2004). The picture that emerges from these studies is that of an ordered assembly process organized around three trimeric intramembrane interactions: CD3δε associates with TCRα, and CD3γε with TCRβ through the centrally placed lysine residues, and then the ζζ module associates with TCRα through an arginine residue in the upper third of the TM domain (Fig. 3A). Alanine substitution at any one of these nine positions prevents formation of a complete TCR-CD3 complex.

The identity and placement of the basic TM residues are conserved among αβTCR, γδTCR, and pre-TCR sequences, implying that the assembly mechanism described for αβTCR applies to all three complexes. This would account for the presence of two CD3 modules in γδTCR (and possibly pre-TCR) even when CD3δ is not available. Selectivity for specific TCR chains in the assembly process is likely determined by sequences in the extracellular portions of the CD3 proteins, because a chimeric CD3γ bearing the CD3δ TM domain can be incorporated into complete αβTCR-CD3 complexes and rescues surface TCR expression in CD3γ-deficient T cells (Wegener et al. 1995; Call et al. 2002). Other lines of evidence also point to important interactions among ectodomains. Mutations in the TCRα connecting peptide have adverse effects on assembly and signaling (Werlen et al. 2000), and a specific "CxxC" motif in the stalk regions of the CD3 proteins is required for association with TCR (Xu et al. 2006). The nine amino acid extracellular

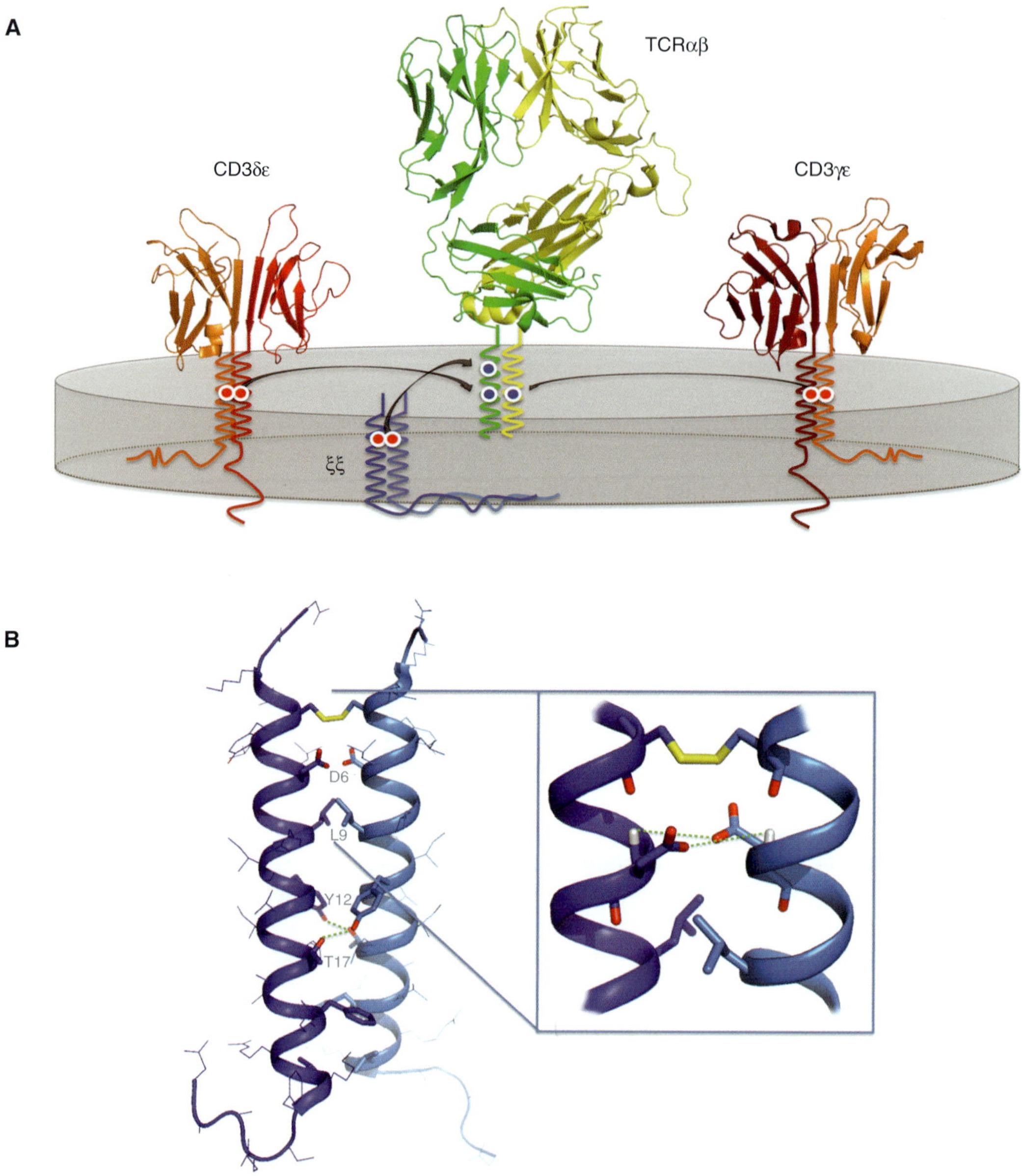

Figure 3. Assembly of the TCR-CD3 complex. (*A*) The TCR-CD3 complex is composed of four dimeric modules: TCRαβ (or γδ), CD3δε, CD3γε, and ζζ, which associate through intramembrane contacts to form the intact complex. Each dimeric signaling module associates with the TCR through a pair of acidic TM residues that bind a specific basic TM residue in the TCR. The result is an octameric complex with the stoichiometry as shown. Ribbon structures were generated from PDB entries 1MI5 (LC13 αβTCR), 1XMW (CD3δε), and 1SY6 (CD3γε). (*B*) NMR structure of the ζζ TM homodimer in detergent micelles (PBD entry 2HAC). The aspartic acid pair (D6-D6) that mediates assembly with TCR is located at the helix dimer interface and stabilized by interhelical H-bonds between side-chain oxygens and backbone amide protons (indicated by dotted lines). Further interface contacts include methyl and aromatic packing (such as L9-L9) and two tyrosine-threonine H-bonds (between Y12 and T17) that are critical for dimer formation.

stalks of the ζζ dimer also appear to play a role in association with TCR, because mutations in this region adversely affect assembly (Xu and Wucherpfennig, unpubl.). Likewise, mutations in loop regions in both TCRα and TCRβ constant domains have been implicated in receptor complex stability as well as signaling functions (Kuhns and Davis 2007; Beddoe et al. 2009).

Conservation of Membrane-based Receptor Complex Assembly

The ζζ dimer is one of four homodimeric modules known to provide the signaling capacity to many activating receptors in the immune system (Call and Wucherpfennig 2007). The Fc receptor γ (Fcγ) subunit associates with a subset of Fc receptors, natural killer (NK) cell receptors, and activating receptors expressed on osteoclasts and platelets. DAP10 and DAP12 associate with NK cell receptors (Lanier 2009). These four signaling proteins share three common features: They are all disulfide-linked homodimers, all bear an aspartic acid pair in the dimeric TM domains, and all carry tyrosine-based cytoplasmic motifs that are phosphorylated on receptor triggering.

With very few exceptions, the receptors that associate with these signaling modules have a basic residue in the TM domain that is required for assembly. In a series of biochemical studies that included a broad representation of receptor complexes, a set of common features was identified (Feng et al. 2005; Garrity et al. 2005; Feng et al. 2006; Feng et al. 2007). First, in every case investigated, the association required both aspartic acid residues; alanine substitution of only one in the pair invariably resulted in dramatic assembly defects. Second, the determinants for intramembrane assembly are contained almost entirely within this focused contact site, because the TM domains of receptors could be substituted with poly-leucine or poly-valine sequences without disrupting assembly, as long as the basic residue remained in its native position (Feng et al. 2006). Finally, extracellular and intracellular domains were not required for the intramembrane assembly, because short peptides encompassing TM and juxtamembrane sequences were sufficient to build receptor complexes.

Structural Basis for Intramembrane Assembly

Precise structural information is required to understand why this particular intramembrane arrangement is desirable. The only atomic-resolution structure to emerge so far is the solution NMR structure of the ζζ TM dimer (Call et al. 2006). Consistent with the requirement for a *pair* of acidic TM residues, the sidechains of the two aspartic acids are colocalized at the helix dimer interface (Fig. 3B). The close apposition of two acidic groups is stabilized by a hydrogen-bonding network that includes both intra- and interhelical hydrogen bonds. The specific geometry of the di-aspartate site seems to be critical for binding to the receptor, because even the chemically conservative substitution of glutamic acid can result in major assembly defects (Call et al. 2002; Call et al. 2006).

Functional Ramifications of Understanding Receptor Assembly

At the level of TCR-CD3 complex formation within the ER, the intramembrane assembly mechanism provides an important quality control checkpoint. The basic and acidic residues that organize and stabilize the complex also act as signals for degradation of subunits that remain unassembled (Bonifacino et al. 1990; Call and Wucherpfennig 2005; Call et al. 2006). Assembly and destruction are therefore directly competing processes that function together to ensure that only intact, functional receptor complexes reach the T-cell surface.

The sequestration of the most critical stabilizing contacts within the membrane may provide a requisite degree of conformational freedom allowing rearrangements among extracellular and intracellular domains to transmit signals across the plasma membrane. The highly focused and polar nature of these contacts (especially in the hydrophobic bilayer interior where competing ions are lacking) may even allow for reorientations of TM helices around fixed points to mechanically communicate conformational changes to the intracellular signaling domains (Engelman 2003). These are difficult hypotheses to test, and will require sophisticated new experimental approaches for monitoring molecular motions in living cells receiving activating signals.

LIPID BINDING BY THE CD3ε AND ζ CYTOPLASMIC DOMAINS

Each ITAM has two tyrosines and two aliphatic residues (YxxL/I$x_{6\text{-}12}$YxxL/I) and phosphorylation of both tyrosines by Lck or Fyn is required for binding of the tandem SH2 domains of ZAP-70 (Reth 1989; Weiss and Littman 1994; Hatada et al. 1995). Most textbooks show the cytoplasmic domains of the CD3 subunits as flexible chains in the cytosol, a rendering that suggests that they would be continuously accessible to tyrosine kinases. Is this model correct?

Lipid Binding of ITAMS

The inner leaflet of the plasma membrane has a negative charge because phosphatidylserine (PS) is almost exclusively localized to the inner leaflet in live cells. Phosphatidylserine is the most abundant anionic lipid in cellular membrane (representing ~20% of inner leaflet lipids), and its asymmetric distribution is maintained by an ATP-dependent lipid flippase referred to as aminophospholipid translocase (Devaux 1991; Fridriksson et al. 1999; Devaux et al. 2008). This lipid asymmetry is lost when cells become apoptotic, enabling detection of such cells by labeling with the PS binding protein annexin V. Other anionic lipids present at significantly lower densities contribute to the negative charge, including phosphatidylglycerol, phosphatidic acid, and a variety of phosphoinositides in different phosphorylation states (Fridriksson et al. 1999).

The cytoplasmic domains of both CD3ε and ζ chains have a net positive charge, which raised the question whether they could bind to the inner leaflet of the plasma membrane through electrostatic interactions. Using in vitro assays with synthetic lipid vesicles, the ζ chain cytoplasmic domain was shown to bind to vesicles with a net negative charge, but not to vesicles lacking such a charge (Aivazian and Stern 2000). Binding resulted in a substantial increase in α-helical content detected by circular dichroism measurements, and these authors proposed that the three ITAMs fold into a α-helical structure upon lipid binding. Binding to synthetic lipid vesicles was later also shown for the CD3ε cytoplasmic domain as well as the cytoplasmic domain of Fcγ, which serves as a signaling module for activating Fc receptors (Sigalov et al. 2006; Xu et al. 2008; Deford-Watts et al. 2009). CD3ε has a higher net positive charge than ζ and bound acidic lipid vesicles with higher affinity. Mutation of two clusters of basic residues in the amino-terminal part of the CD3ε cytoplasmic domain abrogated lipid binding, confirming the importance of electrostatic interactions (Xu et al. 2008). Little or no binding was detected for the cytoplasmic domains of the CD3γ and CD3δ chains in these in vitro assays, which lack a net positive charge, although it remains possible that they bind cooperatively with ζ and CD3ε to cellular membranes.

It was critical to establish that such lipid binding actually occurs in cells because all binding studies had been performed in vitro with synthetic lipid vesicle preparations. A cellular fluorescence resonance energy transfer (FRET) assay was developed to directly address this question (Xu et al. 2008). In a fully extended conformation, the 57 amino acid cytoplasmic domain of CD3ε is at a substantial distance from the plasma membrane (~200 Å). This distance is outside of the range for FRET because energy transfer from donor to acceptor fluorophore is highly distance dependent, with the upper limit being ~100 Å (Kenworthy 2001). Efficient FRET between a fluorescent protein attached to the carboxy-terminus of CD3ε and a fluorescent dye in the plasma membrane is therefore only expected in the lipid-bound state of CD3ε, but not when the cytoplasmic domain has dissociated from the membrane. The validity of this FRET-based approach was tested with constructs using flexible linkers of increasing length (3, 25, and 50 amino acids).

A high FRET signal was observed with the wild-type CD3ε cytoplasmic domain, but not with a mutant that failed to show binding in the in vitro assay described above (Sigalov et al. 2006; Xu et al. 2008).

Structure of the Membrane-bound ITAM

Determination of the structure of the CD3ε cytoplasmic domain required a lipid bilayer surface sufficiently large in size to enable proper binding of the cytoplasmic domain. Detergent micelles with the appropriate lipid headgroups would not be suitable because of their high curvature. The solution to this problem was to use bicelles, which represent flat discs formed using a mixture of short- and long-chain lipids. The long-chain lipids form the bilayer core of such bicelles, whereas the short chain lipids associate at the rim. Bicelles have a size similar to small proteins and thus tumble quickly enough for solution NMR studies (Prosser et al. 2006). This approach enabled determination of the structure of the CD3ε cytoplasmic domain in a lipid-bound state (Xu et al. 2008). A surprising finding was that the cytoplasmic domain was actually inserted into the bilayer, with the peptide backbone being located at the interface between the hydrophilic headgroup region and the hydrophobic acyl chain layer. This overall position of the backbone enabled insertion of all hydrophobic side chains into the hydrophobic core of the bilayer. The structure of the ITAM itself shows insertion of all four key residues, the two tyrosines, and the two aliphatic residues, into the acyl chain region of the bilayer (Fig. 4). Helical structure was confined to short segments centered on the tyrosines, and circular dichroism experiments confirmed the presence of a small amount of α-helical structure that was lost when the tyrosines and aliphatic residues of the ITAM were mutated. Interestingly, the amino-terminal part of the CD3ε cytoplasmic domain is rather flexible, indicating that the ITAM is connected to the TM domain through a membrane-bound flexible linker (Xu et al. 2008). Functional studies showed that lipid binding by CD3ε or ζ prevented phosphorylation by Lck (Aivazian and Stern 2000; Xu et al. 2008). This means that dissociation of the ITAMs from the membrane is required as one of the initial events in TCR activation.

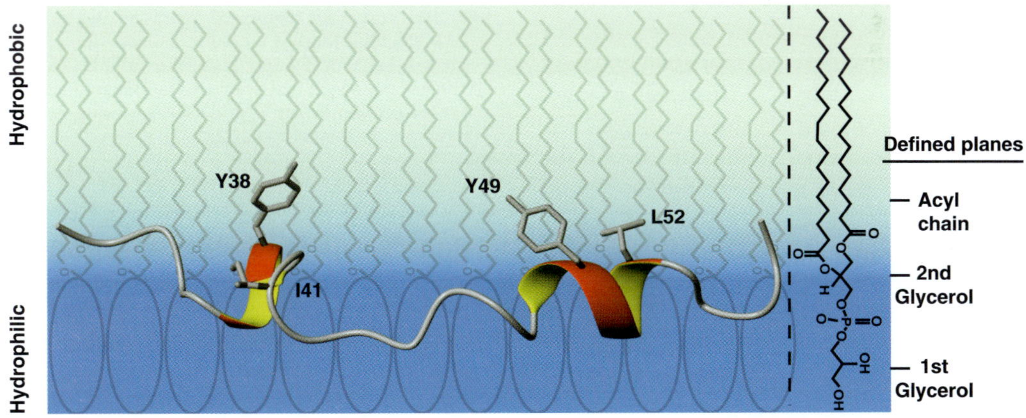

Figure 4. Structure of the CD3ε ITAM in the lipid-bound state. The two tyrosines (Y38 and Y49) and the two aliphatic residues (I41 and L52) of the ITAM protrude into the hydrophobic acyl chain region of the lipid bilayer. The peptide backbone of the ITAM resides primarily at the interface between the hydrated lipid headgroup region and the hydrophobic bilayer interior. The hydrophobic layer of the bilayer is shaded light blue and the hydrated lipid headgroup region in dark blue. The POPG structure graphic to the right shows the location of the hydrophilic headgroup and the hydrophobic acyl chains of the lipid (PDB entry 2K4F).

Possible Mechanisms for ITAM Release from the Membrane

The mechanisms resulting in dissociation of the CD3ε and ζ cytoplasmic domains from the membrane on TCR triggering remain unknown. It is unlikely that subtle conformational change in the extracellular domains of the TCR-CD3 complex during peptide-MHC binding are sufficient due to the flexible nature of the N-terminal part of the cytoplasmic domain (Aivazian and Stern 2000; Xu et al. 2008). Other mechanisms must therefore account for the change in tyrosine accessibility during TCR triggering.

It has been proposed that the continued movement of T cells across antigen presenting cells results in a mechanical force on the TCR-peptide-MHC recognition unit (Ma et al. 2008; Kim et al. 2009). Such a mechanical force could possibly result in dissociation of the ITAM from the membrane but would have to be strong or sustained enough to act through the flexible amino-terminal part of the cytoplasmic domain on the ITAM. A second possibility is that clustering of TCR-CD3 complexes in early microclusters at the immunological synapse results in competition among the cytoplasmic domains for membrane surface (Aivazian and Stern 2000). TCR microclusters can form before initiation of signaling because they are observed even in the presence of a Src kinase inhibitor that blocks Lck activity (Campi et al. 2005). A third hypothesis is that TCR clustering changes the lipid environment in the vicinity of the clustered TCRs, reducing the affinity of the cytoplasmic domains for the membrane. Lipid binding is highly sensitive to the density of negatively charged phospholipids due to the essential role of basic residues in the cytoplasmic domains for lipid binding.

CONCLUDING REMARKS

Structural data are now available not only for the extracellular domains, but also some of the transmembrane and cytoplasmic domains of the TCR-CD3 complex. These structural studies create a solid framework for understanding key functional aspects of the TCR-CD3 complex and many other activating receptors in the immune system.

ACKNOWLEDGMENTS

This work was supported by grants to K.W.W. to the National Institutes of Health (RO1 AI054520 and AI064177). E.S.H. was supported in part by Beckman Young Investigator and Searle Scholar awards.

REFERENCES

Adams EJ, Chien YH, Garcia KC. 2005. Structure of a γδ T cell receptor in complex with the nonclassical MHC T22. *Science* **308:** 227–231.

Adams EJ, Strop P, Shin S, Chien YH, Garcia KC. 2008. An autonomous CDR3δ is sufficient for recognition of the nonclassical MHC class I molecules T10 and T22 by γδ T cells. *Nat Immunol* **9:** 777–784.

Aivazian D, Stern LJ. 2000. Phosphorylation of T cell receptor zeta is regulated by a lipid dependent folding transition. *Nat Struct Biol* **7:** 1023–1026.

Alcover A, Mariuzza RA, Ermonval M, Acuto O. 1990. Lysine 271 in the transmembrane domain of the T-cell antigen receptor β chain is necessary for its assembly with the CD3 complex but not for α/β dimerization. *J Biol Chem* **265:** 4131–4135.

Arnett KL, Harrison SC, Wiley DC. 2004. Crystal structure of a human CD3-ϵ/δ dimer in complex with a UCHT1 single-chain antibody fragment. *Proc Natl Acad Sci* **101:** 16268–16273.

Baker BM, Turner RV, Gagnon SJ, Wiley DC, Biddison WE. 2001. Identification of a crucial energetic footprint on the α1 helix of human histocompatibility leukocyte antigen (HLA)-A2 that provides functional interactions for recognition by tax peptide/HLA-A2-specific T cell receptors. *J Exp Med* **193:** 551–562.

Barral DC, Brenner MB. 2007. CD1 antigen presentation: How it works. *Nat Rev Immunol* **7:** 929–941.

Beddoe T, Chen Z, Clements CS, Ely LK, Bushell SR, Vivian JP, Kjer-Nielsen L, Pang SS, Dunstone MA, Liu YC, et al. 2009. Antigen ligation triggers a conformational change within the constant domain of the αβ T cell receptor. *Immunity* **30:** 777–788.

Bendelac A, Savage PB, Teyton L. 2007. The biology of NKT cells. *Annu Rev Immunol* **25:** 297–336.

Bjorkman PJ, Saper MA, Samraoui B, Bennett WS, Strominger JL, Wiley DC. 1987. The foreign antigen binding site and T cell recognition regions of class I histocompatibility antigens. *Nature* **329:** 512–518.

Blumberg RS, Alarcon B, Sancho J, McDermott FV, Lopez P, Breitmeyer J, Terhorst C. 1990a. Assembly and function of the T cell antigen receptor. Requirement of either the lysine or arginine residues in the transmembrane region of the α chain. *J Biol Chem* **265:** 14036–14043.

Blumberg RS, Ley S, Sancho J, Lonberg N, Lacy E, McDermott F, Schad V, Greenstein JL, Terhorst C. 1990b. Structure of the T-cell antigen receptor: Evidence for two CD3 ϵ subunits in the T-cell receptor-CD3 complex. *Proc Natl Acad Sci* **87:** 7220–7224.

Bonifacino JS, Cosson P, Klausner RD. 1990. Colocalized transmembrane determinants for ER degradation and subunit assembly explain the intracellular fate of TCR chains. *Cell* **63:** 503–513.

Borg NA, Wun KS, Kjer-Nielsen L, Wilce MC, Pellicci DG, Koh R, Besra GS, Bharadwaj M, Godfrey DI, McCluskey J, et al. 2007. CD1d-lipid-antigen recognition by the semi-invariant NKT T-cell receptor. *Nature* **448:** 44–49.

Call ME, Wucherpfennig KW. 2005. The T cell receptor: Critical role of the membrane environment in receptor assembly and function. *Annu Rev Immunol* **23:** 101–125.

Call ME, Wucherpfennig KW. 2007. Common themes in the assembly and architecture of activating immune receptors. *Nat Rev Immunol* **7:** 841–850.

Call ME, Pyrdol J, Wucherpfennig KW. 2004. Stoichiometry of the T-cell receptor-CD3 complex and key intermediates assembled in the endoplasmic reticulum. *Embo J* **23:** 2348–2357.

Call ME, Pyrdol J, Wiedmann M, Wucherpfennig KW. 2002. The organizing principle in the formation of the T cell receptor-CD3 complex. *Cell* **111:** 967–979.

Call ME, Schnell JR, Xu C, Lutz RA, Chou JJ, Wucherpfennig KW. 2006. The structure of the zetazeta transmembrane dimer reveals features essential for its assembly with the T cell receptor. *Cell* **127:** 355–368.

Campi G, Varma R, Dustin ML. 2005. Actin and agonist MHC-peptide complex-dependent T cell receptor microclusters as scaffolds for signaling. *J Exp Med* **202:** 1031–1036.

Choudhuri K, van der Merwe PA. 2007. Molecular mechanisms involved in T cell receptor triggering. *Semin Immunol* **19:** 255–261.

Cosson P, Lankford SP, Bonifacino JS, Klausner RD. 1991. Membrane protein association by potential intramembrane charge pairs. *Nature* **351:** 414–416.

Dai S, Huseby ES, Rubtsova K, Scott-Browne J, Crawford F, Macdonald WA, Marrack P, Kappler JW. 2008. Crossreactive T Cells spotlight the germline rules for αβ T cell-receptor interactions with MHC molecules. *Immunity* **28:** 324–334.

Dave VP. 2009. Hierarchical role of CD3 chains in thymocyte development. *Immunol Rev* **232:** 22–33.

Dave VP, Cao Z, Browne C, Alarcon B, Fernandez-Miguel G, Lafaille J, de la Hera A, Tonegawa S, Kappes DJ. 1997. CD3 δ deficiency arrests development of the α β but not the γ δ T cell lineage. *Embo J* **16:** 1360–1370.

de la Hera A, Muller U, Olsson C, Isaaz S, Tunnacliffe A. 1991. Structure of the T cell antigen receptor (TCR): Two CD3 ϵ subunits in a functional TCR/CD3 complex. *J Exp Med* **173:** 7–17.

Deford-Watts LM, Tassin TC, Becker AM, Medeiros JJ, Albanesi JP, Love PE, Wulfing C, van Oers NS. 2009. The cytoplasmic tail of the T cell receptor CD3 ε subunit contains a phospholipid-binding motif that regulates T cell functions. *J Immunol* **183:** 1055–1064.

Devaux PF. 1991. Static and dynamic lipid asymmetry in cell membranes. *Biochemistry* **30:** 1163–1173.

Devaux PF, Herrmann A, Ohlwein N, Kozlov MM. 2008. How lipid flippases can modulate membrane structure. *Biochim Biophys Acta* **1778:** 1591–1600.

Ding YH, Smith KJ, Garboczi DN, Utz U, Biddison WE, Wiley DC. 1998. Two human T cell receptors bind in a similar diagonal mode to the HLA-A2/Tax peptide complex using different TCR amino acids. *Immunity* **8:** 403–411.

Engelman DM. 2003. Electrostatic fasteners hold the T cell receptor-CD3 complex together. *Mol Cell* **11:** 5–6.

Feng D, Bond CJ, Ely LK, Maynard J, Garcia KC. 2007. Structural evidence for a germline-encoded T cell receptor-major histocompatibility complex interaction 'codon'. *Nat Immunol* **8:** 975–983.

Feng J, Call ME, Wucherpfennig KW. 2006. The assembly of diverse immune receptors is focused on a polar membrane-embedded interaction site. *PLoS Biol* **4:** e142.

Feng J, Garrity D, Call ME, Moffett H, Wucherpfennig KW. 2005. Convergence on a distinctive assembly mechanism by unrelated families of activating immune receptors. *Immunity* **22:** 427–438.

Fridriksson EK, Shipkova PA, Sheets ED, Holowka D, Baird B, McLafferty FW. 1999. Quantitative analysis of phospholipids in functionally important membrane domains from RBL-2H3 mast cells using tandem high-resolution mass spectrometry. *Biochemistry* **38:** 8056–8063.

Garboczi DN, Ghosh P, Utz U, Fan QR, Biddison WE, Wiley DC. 1996. Structure of the complex between human T-cell receptor, viral peptide and HLA-A2. *Nature* **384:** 134–141.

Garcia KC, Adams JJ, Feng D, Ely LK. 2009. The molecular basis of TCR germline bias for MHC is surprisingly simple. *Nat Immunol* **10:** 143–147.

Garcia KC, Degano M, Stanfield RL, Brunmark A, Jackson MR, Peterson PA, Teyton L, Wilson IA. 1996. An αβ T cell receptor structure at 2.5 A and its orientation in the TCR-MHC complex. *Science* **274:** 209–219.

Garrity D, Call ME, Feng J, Wucherpfennig KW. 2005. The activating NKG2D receptor assembles in the membrane with two signaling dimers into a hexameric structure. *Proc Natl Acad Sci* **102:** 7641–7646.

Germain RN, Stefanova I. 1999. The dynamics of T cell receptor signaling: complex orchestration and the key roles of tempo and cooperation. *Annu Rev Immunol* **17:** 467–522.

Hatada MH, Lu X, Laird ER, Green J, Morgenstern JP, Lou M, Marr CS, Phillips TB, Ram MK, Theriault K, et al. 1995. Molecular basis for interaction of the protein tyrosine kinase ZAP-70 with the T-cell receptor. *Nature* **377:** 32–38.

Hayes SM, Love PE. 2006. Stoichiometry of the murine γδ T cell receptor. *J Exp Med* **203:** 47–52.

Huseby ES, Kappler JW, Marrack P. 2008. Thymic selection stifles TCR reactivity with the main chain structure of MHC and forces interactions with the peptide side chains. *Mol Immunol* **45:** 599–606.

Huseby ES, White J, Crawford F, Vass T, Becker D, Pinilla C, Marrack P, Kappler JW. 2005. How the T cell repertoire becomes peptide and MHC specific. *Cell* **122:** 247–260.

Kenworthy AK. 2001. Imaging protein-protein interactions using fluorescence resonance energy transfer microscopy. *Methods* **24:** 289–296.

Kim ST, Takeuchi K, Sun ZY, Touma M, Castro CE, Fahmy A, Lang MJ, Wagner G, Reinherz EL. 2009. The αβ T cell receptor is an anisotropic mechanosensor. *J Biol Chem* **284:** 31028–31037.

Kjer-Nielsen L, Dunstone MA, Kostenko L, Ely LK, Beddoe T, Mifsud NA, Purcell AW, Brooks AG, McCluskey J, Rossjohn J. 2004. Crystal structure of the human T cell receptor CD3 ε γ heterodimer complexed to the therapeutic mAb OKT3. *Proc Natl Acad Sci* **101:** 7675–7680.

Kuhns MS, Davis MM. 2007. Disruption of extracellular interactions impairs T cell receptor-CD3 complex stability and signaling. *Immunity* **26:** 357–369.

Kuhns MS, Davis MM, Garcia KC. 2006. Deconstructing the form and function of the TCR/CD3 complex. *Immunity* **24:** 133–139.

Lanier LL. 2009. DAP10- and DAP12-associated receptors in innate immunity. *Immunol Rev* **227:** 150–160.

Ma Z, Janmey PA, Finkel TH. 2008. The receptor deformation model of TCR triggering. *FASEB J* **22:** 1002–1008.

Manolios N, Letourneur F, Bonifacino JS, Klausner RD. 1991. Pairwise, cooperative and inhibitory interactions describe the assembly and probable structure of the T-cell antigen receptor. *Embo J* **10:** 1643–1651.

Prosser RS, Evanics F, Kitevski JL, Al-Abdul-Wahid MS. 2006. Current applications of bicelles in NMR studies of membrane-associated amphiphiles and proteins. *Biochemistry* **45:** 8453–8465.

Punt JA, Roberts JL, Kearse KP, Singer A. 1994. Stoichiometry of the T cell antigen receptor (TCR) complex: Each TCR/CD3 complex contains one TCR α, one TCR β, and two CD3 ε chains. *J Exp Med* **180:** 587–593.

Reinherz EL, Tan K, Tang L, Kern P, Liu J, Xiong Y, Hussey RE, Smolyar A, Hare B, Zhang R, et al. 1999. The crystal structure of a T cell receptor in complex with peptide and MHC class II. *Science* **286:** 1913–1921.

Reiser JB, Darnault C, Gregoire C, Mosser T, Mazza G, Kearney A, van der Merwe PA, Fontecilla-Camps JC, Housset D, Malissen B. 2003. CDR3 loop flexibility contributes to the degeneracy of TCR recognition. *Nat Immunol* **4:** 241–247.

Reiser JB, Darnault C, Guimezanes A, Gregoire C, Mosser T, Schmitt-Verhulst AM, Fontecilla-Camps JC, Malissen B, Housset D, Mazza G. 2000. Crystal structure of a T cell receptor bound to an allogeneic MHC molecule. *Nat Immunol* **1:** 291–297.

Reth M. 1989. Antigen receptor tail clue. *Nature* **338:** 383–384.

Robinson J, Waller MJ, Fail SC, McWilliam H, Lopez R, Parham P, Marsh SG. 2009. The IMGT/HLA database. *Nucleic Acids Res* **37:** D1013–1017.

Rudolph MG, Stanfield RL, Wilson IA. 2006. How TCRs bind MHCs, peptides, and coreceptors. *Annu Rev Immunol* **24:** 419–466.

Rutledge T, Cosson P, Manolios N, Bonifacino JS, Klausner RD. 1992. Transmembrane helical interactions: Zeta chain dimerization and functional association with the T cell antigen receptor. *Embo J* **11:** 3245–3254.

Samelson LE. 2002. Signal transduction mediated by the T cell antigen receptor: the role of adapter proteins. *Annu Rev Immunol* **20:** 371–394.

Scott-Browne JP, White J, Kappler JW, Gapin L, Marrack P. 2009. Germline-encoded amino acids in the αβ T-cell receptor control thymic selection. *Nature* **458:** 1043–1046.

Shin S, El-Diwany R, Schaffert S, Adams EJ, Garcia KC, Pereira P, Chien YH. 2005. Antigen recognition determinants of γδ T cell receptors. *Science* **308:** 252–255.

Siegers GM, Swamy M, Fernandez-Malave E, Minguet S, Rathmann S, Guardo AC, Perez-Flores V, Regueiro JR, Alarcon B, Fisch P, et al. 2007. Different composition of the human and the mouse γδ T cell receptor explains different phenotypes of CD3γ and CD3δ immunodeficiencies. *J Exp Med* **204:** 2537–2544.

Sigalov AB, Aivazian DA, Uversky VN, Stern LJ. 2006. Lipid-binding activity of intrinsically unstructured cytoplasmic domains of multichain immune recognition receptor signaling subunits. *Biochemistry* **45:** 15731–15739.

Stern LJ, Wiley DC. 1994. Antigenic peptide binding by class I and class II histocompatibility proteins. *Structure* **2:** 245–251.

Sun ZY, Kim ST, Kim IC, Fahmy A, Reinherz EL, Wagner G. 2004. Solution structure of the CD3εδ ectodomain and comparison with CD3εγ as a basis for modeling T cell receptor topology and signaling. *Proc Natl Acad Sci* **101:** 16867–16872.

Sun ZJ, Kim KS, Wagner G, Reinherz EL. 2001. Mechanisms contributing to T cell receptor signaling and assembly revealed by the solution structure of an ectodomain fragment of the CD3 ε γ heterodimer. *Cell* **105:** 913–923.

Wegener AM, Hou X, Dietrich J, Geisler C. 1995. Distinct domains of the CD3-γ chain are involved in surface expression and function of the T cell antigen receptor. *J Biol Chem* **270:** 4675–4680.

Weiss A, Littman DR. 1994. Signal transduction by lymphocyte antigen receptors. *Cell* **76:** 263–274.

Werlen G, Hausmann B, Palmer E. 2000. A motif in the αβ T-cell receptor controls positive selection by modulating ERK activity. *Nature* **406:** 422–426.

Xu C, Call ME, Wucherpfennig KW. 2006. A membrane-proximal tetracysteine motif contributes to assembly of CD3δε and CD3γε dimers with the T cell receptor. *J Biol Chem* **281:** 36977–36984.

Xu C, Gagnon E, Call ME, Schnell JR, Schwieters CD, Carman CV, Chou JJ, Wucherpfennig KW. 2008. Regulation of T cell receptor activation by dynamic membrane binding of the CD3ε cytoplasmic tyrosine-based motif. *Cell* **135:** 702–713.

ITAM-mediated Signaling by the T-Cell Antigen Receptor

Paul E. Love[1] and Sandra M. Hayes[2]

[1] *Eunice Kennedy Shriver* National Institute of Child Health and Human Development, National Institutes of Health, Bethesda, Maryland 20892

[2] Department of Microbiology & Immunology, SUNY Upstate Medical University, Syracuse, New York 13210

Correspondence: lovep@mail.nih.gov

Signal transduction by the T-cell antigen receptor (TCR) is initiated by phosphorylation of conserved motifs (ITAMs) contained within the cytoplasmic domains of the invariant subunits. TCR complexes contain a total of 10 ITAMs and this unusual configuration has prompted studies of the role of specific ITAMs, or of ITAM multiplicity, in regulating TCR-directed developmental and effector responses. Here, we summarize data generated during the past two decades and discuss how these findings have in some cases resolved, and in others complicated, outstanding questions relating to the function of TCR ITAMs.

Signal transduction in the immune system is regulated by a highly diverse set of cell surface receptors that are coupled by signaling intermediates to common downstream pathways. A large number of these receptors either contain within their cytoplasmic domains, or associate with subunits that contain, a conserved sequence (the ***I***mmune-receptor-***T***yrosine-based-***A***ctivation-***M***otif; ITAM) that is critical for the initiation of signaling following ligand engagement (reviewed in Underhill and Goodridge 2007). ITAMs, which were identified 20 years ago on the basis of their sequence homology (Reth 1989), consist of paired YxxL/I motifs separated by a defined interval (YxxL/I-X_{6-8}-YXXL/I). In addition, most ITAMs contain a negatively charged amino acid (D/E) in the +2 position relative to the first ITAM tyrosine (Fig. 1). ITAM containing receptors are widely expressed in hematopoietic cells including T and B lymphocytes, natural killer (NK) cells, macrophages, dendritic cells, and platelets (Fig. 2). Since their discovery, considerable progress has been made in determining how ITAMs initiate signaling cascades and thereby couple their associated receptors to distal effector responses. In this review, we focus on the TCR as a paradigm for ITAM-mediated signal initiation. In addition to being one of the most well characterized receptor complexes, the TCR exhibits several intriguing properties, such as the inclusion of an especially large number of ITAM containing subunits, which have fueled intensive study into the role of ITAMs in the TCR signaling response.

 doi: 10.1101/cshperspect.a002485
Cite this article as *Cold Spring Harb Perspect Biol* 2010;2:a002485

CD3ε	D--YEPIRKGQRDL-YSGL
CD3δ	DQLYQPLRDREDTQ-YSRL
CD3γ	EQLYQPLKDREYDQ-YSHL
Igα	ENLYEGLNLDDCSM-YEDI
Igβ	DHTYEGLDIDQTAT-YEDI
TCRζ-1	NQLYNELNLGRREE-YDVL
TCRζ-2	EGVYNALQKDKMAEAYSEI
TCRζ-3	DGLYQGLSTATKDT-YDAL
FcεRIβ	DRVYEELNIYSAT--YSEL
FcεRIγ	DGVYTGLSRNQET--YETL
DAP12	ESPYQELQGQRSDV-YSDL
Consensus	D/Ex$_{0-2}$YxxL/Ix$_{6-8}$YxxL/I

Figure 1. ITAM-containing proteins. Murine ITAM sequences were obtained from GenBank. Amino acids that align with the consensus ITAM sequence are shown in blue.

ITAMS WITHIN THE TCR SUBUNITS

TCRs are multimeric complexes composed of subunits whose function is exclusively limited to either ligand recognition or signal transduction. The predominant TCR isoform, the αβTCR, which is expressed on mature CD4^{+} and CD8^{+} T cells, contains clonotypic TCRα and TCRβ chains that form a heterodimer conferring peptide-major histocompatibility complex (MHC) binding specificity to the TCR. The TCRα/β heterodimer lacks inherent signal transducing activity but associates noncovalently with multiple signal transducing subunits: the CD3γ,-δ and -ε chains, and, in most cases, a ζ chain homodimer. The currently accepted model of αβTCR stoichiometry proposes the following subunit composition: TCRαβ, CD3γε, CD3δε, ζζ. A distinct lineage of T cells (γδ T cells) expresses a TCR complex containing a different antigen recognition heterodimer composed of chains encoded by the TCRγ and TCRδ genes. Although the invariant subunit composition of the αβTCR and the γδTCR had long been assumed to be identical, it was recently shown that the γδTCR lacks CD3δ (Hayes and Love 2002), and therefore most likely has the following stoichiometry: TCRγδ, CD3γε, CD3γε, ζζ (Hayes and Love 2006a; Siegers et al. 2007). A third complex, the pre-TCR, is expressed only during early thymocyte development at a stage that precedes rearrangement of the TCRα gene. In lieu of TCRα, TCRβ associates with an invariant chain (pre-Tα) encoded by a gene that does not undergo V-(D)-J rearrangement (Saint-Ruf et al. 1994). Although the subunit composition of the pre-TCR has not been unequivocally established, the phenotype of knockout mice lacking individual TCR subunits suggests that it likely contains at least CD3γε and ζζ dimers in addition to the TCRβ and pre-Tα chains (Hayes et al. 2003).

A striking feature of the TCR (and pre-TCR) structures is the presence of multiple distinct ITAM bearing subunits within the receptor complex. The CD3 (−γ, −δ, −ε) chains each contain a single ITAM whereas the ζ chain

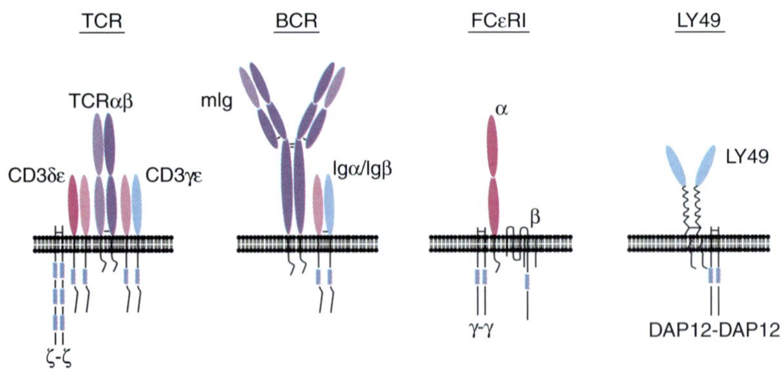

Figure 2. Comparison of the ITAM-containing T- and B-cell antigen receptors, Fc receptors and NK receptors. Subunit composition of the T-cell antigen receptor (TCR), B-cell antigen receptor (BCR), FcεR1 (an example of an activating Fc receptor) and Ly49 (an example of an activating NK receptor). Blue rectangles represent ITAMs.

contains three tandem ITAMs (Fig. 1 and Fig. 2). Consequently, the predicted octameric αβTCR and γδTCR complexes contain a total of 10 ITAMs. Moreover, although all TCR ITAMs share the conserved YxxL/I-$X_{6\text{-}8}$-YXXL/I motif, the amino acid sequence of each ITAM is distinctive (Fig. 1). The question that naturally arises is whether the individual TCR ITAMs have specific signaling responses during T-cell development and T-cell activation or instead perform equivalent but additive functions.

INITIATION OF TCR SIGNALING BY ITAMS

The essential role of ITAMs in the initiation of TCR signaling was first established by experiments demonstrating that the cytoplasmic domains of the CD3 or ζ chains, or the individual ITAM sequences from these proteins, were capable of activating T cells when fused to the extracellular domain of unrelated molecules (Irving and Weiss 1991; Letourneur and Klausner 1992; Romeo et al. 1992; Straus and Weiss 1993; Koyasu et al. 1994; Sturmhofel et al. 1995). It had been shown previously that tyrosine phosphorylation of TCR subunits, which represents one of the earliest events following TCR engagement, was mediated by the Src family tyrosine kinases Lck and Fyn (Samelson et al. 1986; June et al. 1990; Samelson et al. 1990) and that Lck associates constitutively with the CD4 and CD8 coreceptors (Veillette et al. 1988; Barber et al. 1989). An important breakthrough was the subsequent identification of ZAP-70, a Syk family tyrosine kinase that is rapidly recruited to phosphorylated ITAMs (Chan et al. 1992). Both ZAP-70 and Syk contain tandem SH2 domains that direct their specific and selective interaction with doubly phosphorylated ITAMs. The combinatorial action of Src kinases and ZAP-70 is sufficient for the full activation of downstream signaling pathways and for eliciting T-cell effector responses (Fig. 3).

A series of experiments performed over the past few years have served to clarify and refine the proximal events that take place following TCR engagement. A fundamental question has been how ligand binding promotes ITAM phosphorylation and thereby leads to the initiation

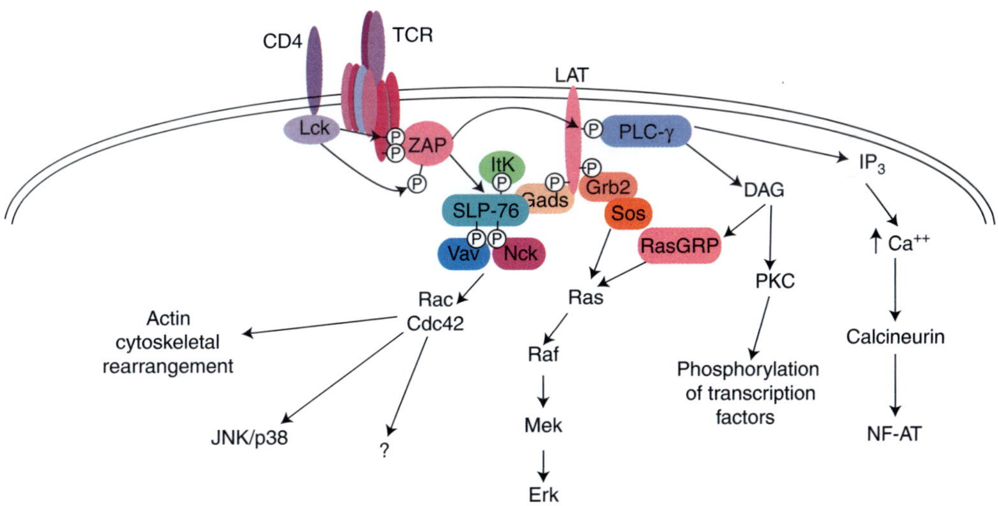

Figure 3. Proximal signaling events in the TCR-coupled signaling pathway. Interactions between the TCR and peptide-MHC results in the activation of Lck, a member of the Src family of tyrosine kinases. Lck then phosphorylates the two-tyrosine residues within the ITAMs of the CD3 and ζ chains. A second tyrosine kinase, ZAP-70, is specifically recruited to biphosphorylated ITAMs. Phosphorylation of ZAP-70 by Lck results in its activation. ZAP-70 and Lck phosphorylate and activate several downstream target proteins eventually leading to Ras activation, calcium mobilization, and actin cytoskeleton rearrangements, and ultimately to the activation of transcription factors.

of TCR signaling. One explanation, relevant to cells expressing the αβTCR, is that cobinding of the peptide-MHC complex by the TCR and by CD4 or CD8 results in the juxtaposition of Lck and the cytoplasmic tails of the invariant TCR subunits resulting in the phosphorylation of ITAMs by Lck (Chu and Littman 1994). However, this mechanism cannot explain how ITAMs within the pre-TCR and γδTCR are phosphorylated as these receptors are expressed on cells that do not express either CD4 or CD8. This paradox has led to the proposal of several new models [reviewed in (Smith-Garvin et al. 2009)] that invoke mechanisms including ligand-induced conformational changes in the TCR, receptor re-distribution on the cell surface, or subunit movement relative to the plasma membrane to explain how TCR engagement could render ITAMs more accessible to phosphorylation. One especially attractive model stems from the observation that the cytoplasmic tails of CD3ε and ζ interact with acidic phospholipids that are enriched in the inner leaflet of the plasma membrane (Aivazian and Stern 2000; Sigalov et al. 2006; Xu et al. 2008). Such interactions, which would be predicted to result in the insertion of ITAM tyrosines into the hydrophobic core of the lipid bilayer, could explain the inaccessibility of ITAMs as targets for phosphorylation in resting T cells. According to this model, ligand induced deformation of the TCR leads to dissociation of the CD3 and ζζ dimers from the plasma membrane rendering them accessible to phosphorylation by Lck and/or Fyn.

The binding of ZAP-70 to phosphorylated ITAMs and its subsequent activation has been clarified by experiments that systematically evaluated the importance of key domains and amino-acid residues within ZAP-70 (reviewed in Au-Yeung et al. 2009). Specifically, these studies have shown that ITAM binding by ZAP-70 relieves an auto-inhibitory conformation of the kinase and that phosphorylation of specific tyrosine residues within the interdomain separating the C-terminal SH2 domain and the kinase domain (interdomain B) by Lck/Fyn are important for regulating the activity of ZAP-70 (Au-Yeung et al. 2009).

FUNCTION OF INDIVIDUAL TCR CHAINS AND ITAMS

The discovery that the TCR complex contains several different ITAM-bearing subunits immediately prompted a series of studies aimed at determining whether individual ITAMs were important for particular signaling or effector responses. Experiments performed in T-cell lines indicated that the cytoplasmic domains of the CD3ε and ζ subunits were independently capable of providing activating signals similar to that of the intact TCR complex (Irving and Weiss 1991; Letourneur and Klausner 1992; Wegener et al. 1992) and suggested that TCR ITAMs are functionally redundant but additive with respect to signaling responses (Irving et al. 1993). In other studies, the binding of potential effector proteins including ZAP-70, Shc, PI3K(p85), Grb2, Fyn, and Ras-GAP to the TCR ITAMs was analyzed using synthetic phosphorylated peptides containing individual ITAM sequences (Isakov et al. 1995; Osman et al. 1996; Zenner et al. 1996). Only one of the three studies compared binding to all 6 TCR ITAM sequences (ζ1, ζ2, ζ3, CD3−γ, −δ, and −ε) (Osman et al. 1996). Data from these experiments provided evidence for preferential binding of effector molecules to individual TCR ITAMs, and in some cases (Shc, p85, Grb2, and Fyn) selective binding to a subset of TCR ITAMs (Table 1); however, additional experiments to follow up on these initial observations have not be performed. It also remains unclear if any exclusive ITAM-effector associations can be inferred from these studies because in vitro analysis of binding to isolated ITAMs may not reflect binding in the context of the intact TCR where cooperative interactions or ordered binding of effectors may take place.

The notion that the extent of TCR ITAM phosphorylation, or the order in which the individual TCR ITAMs are phosphorylated after ligand engagement, may play a role in regulating the TCR signaling response also formed the basis of intensive research over the past decade. Early experiments revealed that two prominent tyrosine phosphorylated forms of ζ-chain, p21-ζ, and p23-ζ, are detected in T cells

Table 1. Differences in the binding affinities of major signaling molecules in the TCR-coupled signaling pathway for the ITAMs within the invariant TCR subunits

Signaling Molecule	ITAM Binding Hierarchy	References
ZAP-70	$\zeta 1 \geq \zeta 2 > \varepsilon \geq \zeta 3$	Isakov et al. 1995[a]
	$\zeta 1 = \gamma = \delta > \zeta 3 > \zeta 2 = \varepsilon$	Osman et al. 1996[b]
	$\zeta 1 > \zeta 3 > \zeta 2$	Zenner et al. 1996[c]
Shc	$\gamma = \delta > \zeta 3 = \zeta 1$	Osman et al. 1996
	$\zeta 1 > \zeta 3$	Zenner et al. 1996
p85 regulatory subunit of PI3K	$\zeta 3 = \gamma = \delta > \zeta 1 = \zeta 2$	Osman et al. 1996
	$\zeta 1 \geq \zeta 2 > \zeta 3$	Zenner et al. 1996
Grb2	$\zeta 1 = \gamma = \delta$	Osman et al. 1996
	$\zeta 2$	Zenner et al. 1996
Fyn	$\zeta 2 = \gamma = \delta > \zeta 1$	Osman et al. 1996
Ras-GTPase activating protein (GAP)	$\zeta 2 \geq \zeta 1 > \zeta 3$	Zenner et al. 1996

[a]All 3 TCRζ ITAMs and the CD3ε ITAM were tested in this study.

[b]All 6 ITAMs (ζ1, ζ2, ζ3, ε, δ, and γ) were tested in this study. Only the specific ITAMs that physically interacted with the signaling molecule are listed.

[c]All 3 TCRζ ITAMs were tested in this study. Only the specific ζ ITAMs that physically interacted with the signaling molecule are listed.

(reviewed in Pitcher et al. 2003b). It was subsequently shown that p21-ζ, which is detected in unstimulated ex vivo thymocytes and T cells or in cells activated with partial agonist/antagonist ligands, contains partially phosphorylated ITAMs that are either not associated with ZAP-70 or that bind inactivated ZAP-70, whereas p23-ζ, which is generated only after TCR cross-linking or stimulation with agonist peptides, contains fully phosphorylated ITAMs that bind activated ZAP-70 (Sloan-Lancaster et al. 1994; van Oers et al. 1994; Madrenas et al. 1995). These findings led to the suggestion that partial phosphorylation of ζ-chain may be important for the induction of T-cell anergy and may in fact be inhibitory (Kersh et al. 1999). One mechanism proposed to explain ITAM-mediated inhibition is that partial phosphorylation of individual ITAMs could result in the generation of an "ITIM-like" sequence (YXXL/I/V) which then leads to the recruitment of tyrosine phosphatases including SHP-1 and SHIP to the TCR (Barrow and Trowsdale 2006; Pinheiro da Silva et al. 2008). However, other studies implied that p21-ζ may instead represent a "primed" state that facilitates rapid T-cell activation and that formation of p21-ζ through self-MHC interactions may be important for T cell survival (Witherden et al. 2000). To address these questions, van Oers and colleagues engineered transgenic mouse lines that are capable of generating p21-ζ but not p23-ζ, or that are incapable of generating either p21-ζ or p23-ζ (Pitcher et al. 2003a). Analysis of these mice demonstrated that the ability to generate p21-ζ was not essential for T-cell antagonism. In addition, they found that the ability to generate p23-ζ was not required for agonist-induced T-cell proliferation (Pitcher et al. 2003a). Similar results were obtained by other investigators using an identical but independently constructed experimental system (Ardouin et al. 1999). Whether these results indicate that partially phosphorylated TCR ITAMs (or partial phosphorylation of individual TCR ITAMs) have no inhibitory function remains unclear as the potential for redundancy with other ITAM containing subunits still exists in this model system.

To evaluate the importance of TCR signaling for T-cell development, in vivo studies were performed with gene targeted mice in which specific ITAM containing subunits were mutated. In an initial series of experiments, mice lacking expression of the CD3γ, −δ, or −ε chains (Malissen et al. 1995; Dave et al. 1997; DeJarnette et al. 1998; Haks et al. 1998) or the ζ chain (Liu et al. 1993; Love et al. 1993; Malissen et al. 1993; Ohno et al. 1993) were generated. Analysis of T-cell development in these mice revealed an important function for CD3γ, CD3ε, and ζ in regulating the transition of thymocytes from the $CD4^{-}CD8^{-}$ (Double Negative, DN) to the $CD4^{+}CD8^{+}$ (Double Positive, DP) stage, an event known to be controlled by pre-TCR signaling (von Boehmer 2005). Moreover, each of the invariant TCR subunits was found to be required for efficient transition from the DP to the mature $CD4^{+}CD8^{-}$ or $CD4^{-}CD8^{+}$ (Single Positive, SP) stage, which is dependent upon αβTCR signaling (Sebzda et al. 1999; Starr et al. 2003). Importantly, αβTCR surface expression was found to be adversely affected in each of the knockout mouse lines, results that were in agreement with in vitro studies showing that all of the TCR subunits are required for proper assembly and surface expression of the TCR (Klausner et al. 1990). Although these studies clearly demonstrated a critical role for CD3γ, −δ, −ε, and ζ in T-cell development, the effects of the gene deletions on TCR surface expression meant that knockout mice could not be used to directly evaluate the importance of ITAMs and ITAM-mediated signaling.

To specifically address the role of TCR ITAMs in thymocyte development, three groups reconstituted TCRζ knockout mice with transgenes encoding wild-type or mutant forms of ζ chain lacking one or more ITAMs, either by deleting individual ITAMs (Shores et al. 1994) or by mutation of ITAM tyrosines to phenylalanine (Ardouin et al. 1999; Pitcher et al. 2005b) (Table 2). Significantly, all three studies found that none of the TCRζ ITAMs was specifically required for thymocyte maturation. However, each study noted that the efficiency of thymocyte development was decreased in the absence of one or more ζ ITAMs, and the severity of the developmental block correlated with the number of inactivated ITAMs but not with the loss of any specific ITAM. A non-essential requirement was also demonstrated for the CD3γ ITAM (Haks et al. 2002) and CD3ε ITAM (Sommers et al. 2000) (Table 2). Together, these results strongly suggested that TCR ITAMs function collectively to amplify TCR signals (Love and Shores 2000).

POTENTIAL FUNCTIONS FOR ITAM-MEDIATED SIGNAL AMPLIFICATION DURING T-CELL DEVELOPMENT

Recent data indicate that ITAM-mediated signal amplification may have an important role in regulating αβ/γδ T lineage choice during thymocyte development. These experiments were based on two observations: 1) that αβ and γδ T cells originate from a common DN thymocyte progenitor prior to rearrangement of the TCRα locus (Petrie et al. 1992), and 2) that direct comparison of the signaling responses of the γδTCR and the αβTCR revealed that under equivalent stimulation conditions, the γδTCR delivers a stronger signal than the αβTCR (Hayes and Love 2002). Since the pre-TCR is expressed at much lower levels than the γδTCR on DN thymocytes (Hayes and Love 2006b), this raised the possibility that the relatively strong signals transmitted by the γδTCR promote commitment to the γδ lineage whereas relatively weak pre-TCR signals promote αβ lineage commitment. This differed from a model proposing that αβ/γδ lineage choice is determined before expression of the γδTCR or the pre-TCR (Kang and Raulet 1997). The importance of TCR signal strength in αβ/γδ lineage choice was tested in two studies that used an experimental system where all immature DN thymocytes express a single transgenic γδTCR (Haks et al. 2005; Hayes et al. 2005). Attenuation of γδTCR signaling potential by reducing the number of ζ ITAMs (Hayes et al. 2005) or by changing the affinity of the γδTCR-MHC interaction (Haks et al. 2005) resulted in a diversion from the γδ to the αβ lineage supporting the signal strength model of αβ/γδ lineage commitment. Consistent with these results, single cell lineage tracing experiments also demonstrated that αβ/γδ lineage choice does not take place prior to pre-TCR/γδTCR expression (Kreslavsky et al. 2008).

Table 2. Roles of ITAMs within the CD3ε, CD3γ, CD3δ, and TCRζ chains in T-cell development and function

Subunit	Experimental approach	Effects on development/function	References
CD3ε	Reconstitution of CD3ε$^{-/-}$ mice with a transgene encoding a mutant form of CD3ε containing a Y→F substitution in the distal tyrosine residue of the CD3ε ITAM	·Restores T-cell development to CD3ε$^{-/-}$ mice ·Impaired proliferative responses and cytokine responses when low doses of stimulating antibody are used ·Defects in survival of T cells expressing TCRs with low affinity to self-ligands	Sommers et al. 2000
CD3γ	Knock-in lacking the CD3γ ITAM	·Impaired efficiency of positive but not negative selection ·Defects in proximal TCR signaling events ·Proliferative responses and cytokine responses are equivalent to those of wild-type mice	Haks et al. 2001; Haks et al. 2002
TCRζ	Reconstitution of TCRζ$^{-/-}$ mice with transgene encoding truncated form of TCRζ containing no ITAMs	·Restores T-cell development to TCRζ$^{-/-}$ mice ·Defects in selection of thymocytes expressing TCRs with low affinity to self-ligands ·Impaired proliferative responses and cytokine responses when low concentrations of peptide are used	Shores et al. 1994; Love et al 2000; Ardouin et al. 1999; Pitcher et al. 2003; Pitcher et al. 2005
	Reconstitution of TCRζ$^{-/-}$ mice with transgene encoding mutated form of TCRζ containing Y→F substitutions in all ITAMs.	·Restores T-cell development to TCRζ$^{-/-}$ mice ·Defects in selection of thymocytes expressing TCRs with low affinity to self-ligands	
All	RAG1$^{-/-}$ mice reconstituted with CD3- and TCRζ-deficient bone marrow transduced with retroviral vectors encoding ITAM-mutant CD3 and TCRζ chains	·Restoration of T-cell development requires the expression of at least 7 ITAMs ·Impaired proliferative responses ·Autoimmunity is observed when two to six ITAMs are expressed	Holst et al. 2008

A large body of work derived from several independent laboratories supports the idea that ITAM-mediated signal amplification plays a critical role in selection of the mature T-cell repertoire. In DP thymocytes, productive rearrangement of the TCRα locus results in surface expression of clonally distinct αβTCR complexes. To ensure that only those cells that express TCRs with the appropriate specificities are allowed to mature, DP thymocytes are subjected to a selection process on the basis of their TCR specificity for peptide+MHC that promotes the survival and differentiation of functionally competent cells (positive selection) and triggers the deletion of overtly autoreactive cells (negative selection) (Sebzda et al. 1999; Starr et al. 2003). Positive selection is thought to be mediated by relatively weak non-agonist peptides (Hogquist et al. 1994; Santori et al. 2002) indicating that amplification of TCR signals generated from such interactions may be especially important for positive selection. Consistent with this notion, positive selection was found to be markedly impaired in ζ-deficient mice reconstituted with transgenes encoding ITAM mutant ζ chains (Shores et al. 1994; Ardouin et al. 1999; Pitcher et al. 2005b). Moreover, in each of these model systems, the extent to which positive selection was compromised was directly related to the number of ζ ITAMs that had been deleted or inactivated. Interestingly, analysis of positive selection using different αβTCR transgenes demonstrated the number of TCR ITAMs that were required for positive selection was dependent upon the presumed affinity of the TCR for its selecting self ligand. DP thymocytes expressing TCRs that are thought to bind to self peptide+MHC with relatively high affinity could be positively selected even if several ζ ITAMs were deleted, whereas positive selection of thymocytes expressing relatively low affinity TCRs required all 10 TCR ITAMs (Love et al. 2000). Thus, the requirement for ITAM-mediated signal amplification was most clearly evident under conditions where positive selection was induced by weak TCR-ligand interactions.

ITAM-MEDIATED SIGNAL AMPLIFICATION AND CENTRAL TOLERANCE

The discovery that TCR ITAM multiplicity was important for positive selection raised the question of whether negative selection is also compromised when the number of TCR ITAMs is reduced. The effect on negative selection of mutating or removing one or more of the ζ ITAMs was examined in two of the three experimental models described above. In both cases, negative selection was found to be compromised when the number of ζ ITAMs was reduced (Shores et al. 1997; Love et al. 2000; Pitcher et al. 2005a). Similar to the effects on positive selection, the impact on negative selection was directly related to the number of ITAMs that were eliminated. Importantly, although potentially auto-reactive T cells were generated in these mice, no overt signs of autoimmune disease were observed. The absence of autoimmune disease in ζ ITAM mutant mice is consistent with the hypothesis that the reactivity of T cells is "tuned" during thymocyte selection through the regulated expression of compensatory molecules such as CD5 which function to modulate the integrated TCR signaling response (Grossman and Singer 1996; Azzam et al. 2001; Wong et al. 2001; Saibil et al. 2003). However, a recent report has challenged this concept of thymocyte selection (Holst et al. 2008). In this study, retrovirus encoded wild-type or ITAM mutant CD3γ, −δ, −ε, and ζ chains (termed retrogenics) were used to reconstitute bone marrow cells from CD3ε$^{-/-}$, ζ$^{-/-}$, or CD3ε$^{-/-}$ x ζ$^{-/-}$ mice. In these experiments, mice in which T-cell development was reconstituted with constructs that yielded fewer than seven functional ITAMs per TCR developed a lethal, multi-organ autoimmune disease that was attributed to failure of central tolerance (Table 2). The reason for the striking discrepancy between these results and those of prior studies are presently unclear but may relate to the different experimental systems employed. A potential drawback of all three studies is that transgenic or retrogenic reconstitution does not replicate endogenous gene expression. Notwithstanding, these new results, which imply that genetic or environmental changes that attenuate TCR signaling in preselection thymocytes can result in failure of central tolerance, call for additional studies to resolve this important issue.

CONCLUDING REMARKS

Research over the past two decades since the discovery and initial characterization of ITAMs has served to substantially advance our understanding of the mechanisms by which ITAMs function to initiate signaling by the TCR. Currently, the preponderance of data support the idea that the multiple TCR ITAMs function mainly to amplify TCR signals and suggest that this capacity for signal amplification is especially critical for selection of the T-cell repertoire. Data generated from experiments in which the number of ITAMs per TCR are reduced have yielded conflicting results regarding the effects of these alterations on central tolerance and the risk for autoimmune disease. Because these findings impact our basic understanding of thymocyte selection, it will be both important and worthwhile to establish more physiologically appropriate models, such as "knockin" mutations to resolve this issue. Experiments revisiting the question of whether individual TCR ITAMs selectively bind to potential downsteam effectors are also overdue and are especially promising as they may yield new insights into the "raison d'être" for the multiple ITAM configuration of the TCR complex.

REFERENCES

Aivazian D, Stern LJ. 2000. Phosphorylation of T cell receptor zeta is regulated by a lipid dependent folding transition. *Nat Structural Biol* **7:** 1023–1026.

Ardouin L, Boyer C, Gillet A, Trucy J, Bernard AM, Nunes J, Delon J, Trautmann A, He HT, Malissen B, et al. 1999. Crippling of CD3-ζ ITAMs does not impair T cell receptor signaling. *Immunity* **10:** 409–420.

Au-Yeung BB, Deindl S, Hsu LY, Palacios EH, Levin SE, Kuriyan J, Weiss A. 2009. The structure, regulation, and function of ZAP-70. *Immunol Rev* **228:** 41–57.

Azzam HS, DeJarnette JB, Huang K, Emmons R, Park CS, Sommers CL, El-Khoury D, Shores EW, Love PE. 2001. Fine tuning of TCR signaling by CD5. *J Immunol* **166:** 5464–5472.

Barber EK, Dasgupta JD, Schlossman SF, Trevillyan JM, Rudd CE. 1989. The CD4 and CD8 antigens are coupled to a protein-tyrosine kinase (p56lck) that phosphorylates the CD3 complex. *Proc Natl Acad Sci* **86:** 3277–3281.

Barrow AD, Trowsdale J. 2006. You say ITAM and I say ITIM, let's call the whole thing off: the ambiguity of immunoreceptor signalling. *Eur J Immunol* **36:** 1646–1653.

Chan AC, Iwashima M, Turck CW, Weiss A. 1992. ZAP-70: A 70 kd protein-tyrosine kinase that associates with the TCR ζ chain. *Cell* **71:** 649–662.

Chu K, Littman DR. 1994. Requirement for kinase activity of CD4-associated p56lck in antibody-triggered T cell signal transduction. *J Biol Chem* **269:** 24095–24101.

Dave VP, Cao Z, Browne C, Alarcon B, Fernandez-Miguel G, Lafaille J, de la Hera A, Tonegawa S, Kappes DJ. 1997. CD3 δ deficiency arrests development of the αβ but not the γδ T cell lineage. *EMBO J* **16:** 1360–1370.

DeJarnette JB, Sommers CL, Huang K, Woodside KJ, Emmons R, Katz K, Shores EW, Love PE. 1998. Specific requirement for CD3ε in T cell development. *ProcNatl Acad Sci* **95:** 14909–14914.

Grossman Z, Singer A. 1996. Tuning of activation thresholds explains flexibility in the selection and development of T cells in the thymus. *ProcNatl Acad Sci* **93:** 14747–14752.

Haks MC, Krimpenfort P, Borst J, Kruisbeek AM. 1998. The CD3gamma chain is essential for development of both the TCRαβ and TCRγδ lineages. *EMBO J* **17:** 1871–1882.

Haks MC, Lefebvre JM, Lauritsen JP, Carleton M, Rhodes M, Miyazaki T, Kappes DJ, Wiest DL. 2005. Attenuation of γδTCR signaling efficiently diverts thymocytes to the αβ lineage. *Immunity* **22:** 595–606.

Haks MC, Pepin E, van den Brakel JH, Smeele SA, Belkowski SM, Kessels HW, Krimpenfort P, Kruisbeek AM. 2002. Contributions of the T cell receptor-associated CD3γ-ITAM to thymocyte selection. *J Exp Med* **196:** 1–13.

Hayes SM, Li L, Love PE. 2005. TCR signal strength influences αβ/γδ lineage fate. *Immunity* **22:** 583–593.

Hayes SM, Love PE. 2002. Distinct structure and signaling potential of the γδ TCR complex. *Immunity* **16:** 827–838.

Hayes SM, Love PE. 2006a. Stoichiometry of the murine γδ T cell receptor. *J Exp Med* **203:** 47–52.

Hayes SM, Love PE. 2006b. Strength of signal: a fundamental mechanism for cell fate specification. *Immunol Rev* **209:** 170–175.

Hayes SM, Shores EW, Love PE. 2003. An architectural perspective on signaling by the pre-, αβ and γδ T cell receptors. *Immunol Rev* **191:** 28–37.

Hogquist KA, Jameson SC, Heath WR, Howard JL, Bevan MJ, Carbone FR. 1994. T cell receptor antagonist peptides induce positive selection. *Cell* **76:** 17–27.

Holst J, Wang H, Eder KD, Workman CJ, Boyd KL, Baquet Z, Singh H, Forbes K, Chruscinski A, Smeyne R, et al. 2008. Scalable signaling mediated by T cell antigen receptor CD3 ITAMs ensures effective negative selection and prevents autoimmunity. *Nat Immunol* **9:** 658–666.

Irving BA, Chan AC, Weiss A. 1993. Functional characterization of a signal transducing motif present in the T cell antigen receptor ζ chain. *J Exp Med* **177:** 1093–1103.

Irving BA, Weiss A. 1991. The cytoplasmic domain of the T cell receptor ζ chain is sufficient to couple to receptor-associated signal transduction pathways. *Cell* **64:** 891–901.

Isakov N, Wange RL, Burgess WH, Watts JD, Aebersold R, Samelson LE. 1995. ZAP-70 binding specificity to T cell receptor tyrosine-based activation motifs: The tandem SH2 domains of ZAP-70 bind distinct tyrosine-based activation motifs with varying affinity. *J Exp Med* **181:** 375–380.

June CH, Fletcher MC, Ledbetter JA, Schieven GL, Siegel JN, Phillips AF, Samelson LE. 1990. Inhibition of tyrosine phosphorylation prevents T-cell receptor-mediated signal transduction. *ProcNatl Acad Sci* **87:** 7722–7726.

Kang J, Raulet DH. 1997. Events that regulate differentiation of αβ TCR+ and γδ TCR+ T cells from a common precursor. *Sem Immunol* **9:** 171–179.

Kersh EN, Kersh GJ, Allen PM. 1999. Partially phosphorylated T cell receptor ζ molecules can inhibit T cell activation. *J Exp Med* **190:** 1627–1636.

Klausner RD, Lippincott-Schwartz J, Bonifacino JS. 1990. The T cell antigen receptor: Insights into organelle biology. *Ann Rev Cell Biol* **6:** 403–431.

Koyasu S, Tse AG, Moingeon P, Hussey RE, Mildonian A, Hannisian J, Clayton LK, Reinherz EL. 1994. Delineation of a T-cell activation motif required for binding of protein tyrosine kinases containing tandem SH2 domains. *Proc Natl Acad Sci* **91:** 6693–6697.

Kreslavsky T, Garbe AI, Krueger A, von Boehmer H. 2008. T cell receptor-instructed αβ versus γδ lineage commitment revealed by single-cell analysis. *J Exp Med* **205:** 1173–1186.

Letourneur F, Klausner RD. 1992. Activation of T cells by a tyrosine kinase activation domain in the cytoplasmic tail of CD3 epsilon. *Science* **255:** 79–82.

Liu CP, Ueda R, She J, Sancho J, Wang B, Weddell G, Loring J, Kurahara C, Dudley EC, Hayday A, et al. 1993. Abnormal T cell development in CD3-ζ-/- mutant mice and identification of a novel T cell population in the intestine. *EMBO J* **12:** 4863–4875.

Love PE, Lee J, Shores EW. 2000. Critical relationship between TCR signaling potential and TCR affinity during thymocyte selection. *J Immunol* **165:** 3080–3087.

Love PE, Shores EW. 2000. ITAM multiplicity and thymocyte selection: how low can you go? *Immunity* **12:** 591–597.

Love PE, Shores EW, Johnson MD, Tremblay ML, Lee EJ, Grinberg A, Huang SP, Singer A, Westphal H. 1993. T cell development in mice that lack the ζ chain of the T cell antigen receptor complex. *Science* **261:** 918–921.

Madrenas J, Wange RL, Wang JL, Isakov N, Samelson LE, Germain RN. 1995. ζ phosphorylation without ZAP-70 activation induced by TCR antagonists or partial agonists. *Science* **267:** 515–518.

Malissen M, Gillet A, Ardouin L, Bouvier G, Trucy J, Ferrier P, Vivier E, Malissen B. 1995. Altered T cell development in mice with a targeted mutation of the CD3-έ gene. *EMBO J* **14:** 4641–4653.

Malissen M, Gillet A, Rocha B, Trucy J, Vivier E, Boyer C, Kontgen F, Brun N, Mazza G, Spanopoulou E, et al. 1993. T cell development in mice lacking the CD3-ζ/ή gene. *The EMBO J* **12:** 4347–4355.

Ohno H, Aoe T, Taki S, Kitamura D, Ishida Y, Rajewsky K, Saito T. 1993. Developmental and functional impairment of T cells in mice lacking CD3 ζ chains. *EMBO J* **12:** 4357–4366.

Osman N, Turner H, Lucas S, Reif K, Cantrell DA. 1996. The protein interactions of the immunoglobulin receptor family tyrosine-based activation motifs present in the T cell receptor ζ subunits and the CD3 γ, δ, and έ chains. *Eur J Immunol* **26:** 1063–1068.

Petrie HT, Scollay R, Shortman K. 1992. Commitment to the T cell receptor-αβ or -γδ lineages can occur just prior to the onset of CD4 and CD8 expression among immature thymocytes. *Eur J Immunol* **22:** 2185–2188.

Pinheiro da Silva F, Aloulou M, Benhamou M, Monteiro RC. 2008. Inhibitory ITAMs: A matter of life and death. *Trends Immunol* **29:** 366–373.

Pitcher LA, Mathis MA, Subramanian S, Young JA, Wakeland EK, Love PE, van Oers NS. 2005a. Selective expression of the 21-kilodalton tyrosine-phosphorylated form of TCR ζ promotes the emergence of T cells with autoreactive potential. *J Immunol* **174:** 6071–6079.

Pitcher LA, Mathis MA, Young JA, DeFord LM, Purtic B, Wulfing C, van Oers NS. 2005b. The CD3 γέ/δέ signaling module provides normal T cell functions in the absence of the TCR zeta immunoreceptor tyrosine-based activation motifs. *Eur J Immunol* **35:** 3643–3654.

Pitcher LA, Ohashi PS, van Oers NS. 2003a. T cell antagonism is functionally uncoupled from the 21- and 23-kDa tyrosine-phosphorylated TCR ζ subunits. *J Immunol* **171:** 845–852.

Pitcher LA, Young JA, Mathis MA, Wrage PC, Bartok B, van Oers NS. 2003b. The formation and functions of the 21- and 23-kDa tyrosine-phosphorylated TCR ζ subunits. *Immunol Rev* **191:** 47–61.

Reth M. 1989. Antigen receptor tail clue. *Nature* **338:** 383–384.

Romeo C, Amiot M, Seed B. 1992. Sequence requirements for induction of cytolysis by the T cell antigen/Fc receptor ζ chain. *Cell* **68:** 889–897.

Saibil SD, Ohteki T, White FM, Luscher M, Zakarian A, Elford A, Shabanowitz J, Nishina H, Hugo P, Penninger J, et al. 2003. Weak agonist self-peptides promote selection and tuning of virus-specific T cells. *Eur J Immunol* **33:** 685–696.

Saint-Ruf C, Ungewiss K, Groettrup M, Bruno L, Fehling HJ, von Boehmer H. 1994. Analysis and expression of a cloned pre-T cell receptor gene. *Science* **266:** 1208–1212.

Samelson LE, Patel MD, Weissman AM, Harford JB, Klausner RD. 1986. Antigen activation of murine T cells induces tyrosine phosphorylation of a polypeptide associated with the T cell antigen receptor. *Cell* **46:** 1083–1090.

Samelson LE, Phillips AF, Luong ET, Klausner RD. 1990. Association of the fyn protein-tyrosine kinase with the T-cell antigen receptor. *ProcNatl Acad Sci* **87:** 4358–4362.

Santori FR, Kieper WC, Brown SM, Lu Y, Neubert TA, Johnson KL, Naylor S, Vukmanovic S, Hogquist KA, Jameson SC. 2002. Rare, structurally homologous self-peptides promote thymocyte positive selection. *Immunity* **17:** 131–142.

Sebzda E, Mariathasan S, Ohteki T, Jones R, Bachmann MF, Ohashi PS. 1999. Selection of the T cell repertoire. *Ann Rev Immunol* **17:** 829–874.

Shores EW, Huang K, Tran T, Lee E, Grinberg A, Love PE. 1994. Role of TCR ζ chain in T cell development and selection. *Science* **266:** 1047–1050.

Shores EW, Tran T, Grinberg A, Sommers CL, Shen H, Love PE. 1997. Role of the multiple T cell receptor (TCR)-zeta chain signaling motifs in selection of the T cell repertoire. *J Exp Med* **185:** 893–900.

Siegers GM, Swamy M, Fernandez-Malave E, Minguet S, Rathmann S, Guardo AC, Perez-Flores V, Regueiro JR, Alarcon B, Fisch P, et al. 2007. Different composition of the human and the mouse γδ T cell receptor explains different phenotypes of CD3γ and CD3δ immunodeficiencies. *J Exp Med* **204:** 2537–2544.

Sigalov AB, Aivazian DA, Uversky VN, Stern LJ. 2006. Lipid-binding activity of intrinsically unstructured cytoplasmic domains of multichain immune recognition receptor signaling subunits. *Biochemistry* **45:** 15731–15739.

Sloan-Lancaster J, Shaw AS, Rothbard JB, Allen PM. 1994. Partial T cell signaling: Altered phospho-ζ and lack of zap70 recruitment in APL-induced T cell anergy. *Cell* **79:** 913–922.

Smith-Garvin JE, Koretzky GA, Jordan MS. 2009. T cell activation. *Ann Rev Immunol* **27:** 591–619.

Sommers CL, Dejarnette JB, Huang K, Lee J, El-Khoury D, Shores EW, Love PE. 2000. Function of CD3 ε-mediated signals in T cell development. *J Exp Med* **192:** 913–919.

Starr TK, Jameson SC, Hogquist KA. 2003. Positive and negative selection of T cells. *Ann Rev Immunol* **21:** 139–176.

Straus DB, Weiss A. 1993. The CD3 chains of the T cell antigen receptor associate with the ZAP-70 tyrosine kinase and are tyrosine phosphorylated after receptor stimulation. *J Exp Med* **178:** 1523–1530.

Sturmhofel K, Brando C, Martinon F, Shevach EM, Coligan JE. 1995. Antigen-independent, integrin-mediated T cell activation. *J Immunol* **154:** 2104–2111.

Underhill DM, Goodridge HS. 2007. The many faces of ITAMs. *Trends Immunol* **28:** 66–73.

van Oers NS, Killeen N, Weiss A. 1994. ZAP-70 is constitutively associated with tyrosine-phosphorylated TCR ζ in murine thymocytes and lymph node T cells. *Immunity* **1:** 675–685.

Veillette A, Bookman MA, Horak EM, Bolen JB. 1988. The CD4 and CD8 T cell surface antigens are associated with the internal membrane tyrosine-protein kinase p56lck. *Cell* **55:** 301–308.

von Boehmer H. 2005. Unique features of the pre-T-cell receptor α-chain: Not just a surrogate. *Nat Rev* **5:** 571–577.

Wegener AM, Letourneur F, Hoeveler A, Brocker T, Luton F, Malissen B. 1992. The T cell receptor/CD3 complex is composed of at least two autonomous transduction modules. *Cell* **68:** 83–95.

Witherden D, van Oers N, Waltzinger C, Weiss A, Benoist C, Mathis D. 2000. Tetracycline-controllable selection of CD4(+) T cells: Half-life and survival signals in the absence of major histocompatibility complex class II molecules. *J Exp Med* **191:** 355–364.

Wong P, Barton GM, Forbush KA, Rudensky AY. 2001. Dynamic tuning of T cell reactivity by self-peptide-major histocompatibility complex ligands. *J Exp Med* **193:** 1179–1187.

Xu C, Gagnon E, Call ME, Schnell JR, Schwieters CD, Carman CV, Chou JJ, Wucherpfennig KW. 2008. Regulation of T cell receptor activation by dynamic membrane binding of the CD3epsilon cytoplasmic tyrosine-based motif. *Cell* **135:** 702–713.

Zenner G, Vorherr T, Mustelin T, Burn P. 1996. Differential and multiple binding of signal transducing molecules to the ITAMs of the TCR-ζ chain. *J Cell Biochem* **63:** 94–103.

It's All about Change: The Antigen-driven Initiation of B-Cell Receptor Signaling

Wanli Liu[1], Hae Won Sohn[1], Pavel Tolar[2], and Susan K. Pierce[1]

[1]Laboratory of Immunogenetics, National Institute of Allergy and Infectious Diseases, National Institutes of Health, Rockville, Maryland 20852

[2]National Institute for Medical Research, The Ridgeway, Mill Hill, London, NW7 1AA, United Kingdom

Correspondence: spierce@niaid.nih.gov

B-cell responses are initiated by the binding of foreign antigens to the clonally distributed B-cell receptors (BCRs) resulting in the triggering of signaling cascades that activate a variety of genes associated with B-cell activation. Although we now understand the molecular nature of the signaling pathways in considerable detail what remains only poorly understood are the mechanisms by which the information that antigen has bound to the BCR ectodomain is transduced across the B-cell membrane to the BCR cytoplasmic domains to trigger signaling. To a large part this gap in knowledge is because of the paucity of techniques to temporally and spatially resolve changes in the behavior of the BCR that occur within several seconds of antigen binding. With the advent of new live-cell imaging technologies we are gaining our first clear views of the events that lead up to the triggering of BCR signaling cascades. These events may provide potential new targets for therapeutic intervention in disease involving hyper or chronic activation of B cells.

Specific, high-affinity antibody responses are the result of processes based on clonal selection (reviewed in Rajewsky 1996). In the absence of antigen, individuals generate a B-cell repertoire in which each B cell expresses a single heavy and light chain gene, the product of somatic recombination of variable and constant region gene segments. Self-reactive B cells are removed from the repertoire and when antigen enters the immune system it selects those B cells expressing BCR's with highest affinity for the antigen. Under the influence of both T cell and innate immune system regulation the antigen-selected B cells are induced to differentiate into short-lived antibody producing cells or enter germinal centers where they undergo the molecularly linked processes of somatic hypermutation and isotype switching. Antigen selection within the germinal centers results in high-affinity memory B cells expressing isotype switched BCRs. These memory B cells account, in large part, for the high titered, high affinity IgG antibody responses observed upon re-exposure to antigen. Thus, we presume that B cells are capable of initiating responses to the universe of foreign antigens to which individuals are exposed and do so through mechanisms that are sensitive to the affinity of the BCR for antigen and by which isotype switched BCRs are more effective. Until recently, the events by

 doi: 10.1101/cshperspect.a002295
Cite this article as *Cold Spring Harb Perspect Biol* 2010;2:a002295

which the binding of antigen to the BCRs triggered signaling remained largely unknown due in a large part to the paucity of experimental approaches that were able to provide the spatial and temporal resolution necessary to capture the earliest events that follow the binding of antigens to BCRs that result in triggering the B cell's signaling cascades. The conventional biochemical techniques that were used so successfully to describe the components of the BCR signaling cascades were too slow to study early events and could not provide spatial information. The application of new live-cell imaging technologies that allow resolution of single molecules over a timeframe of several seconds to the study of antigen-induced B-cell responses is providing the first views of these processes. Here we review progress in understanding the initiation of the BCR signaling using live-cell imaging technologies and how this new knowledge may explain in part the mechanisms that underlie hyper or chronic activation of B cells in autoimmunity and in B-cell cancers.

THE WHO, HOW, AND WHERE OF ANTIGEN PRESENTATION TO B CELLS (BATISTA AND HARWOOD 2009)

The responses of B cells to antigens were traditionally studied by providing B cells with multivalent soluble antigens in solution. Batista et al. (Batista et al. 2001) first made the important observation that B cells could be efficiently activated by antigen expressed by antigen presenting cells (APCs). They showed that the interaction of B cells with APCs lead to the formation of a polarized bull's eye like structure in which the BCRs were concentrated in the center, surrounded by the adherence molecule LFA-1. This structure was analogous to the immune synapse earlier described for T cells following interactions with APCs (Fooksman et al. 2010). The description of the B-cell immune synapse by Batista et al. was followed by several studies that used intravital imaging to describe the interaction of B cells with APCs in lymph nodes in vivo. These studies provided evidence that small soluble antigens are able to enter follicles and activate B cells within the follicles (Pape et al. 2007). Particulate antigens including viruses and immune complexes were observed to be captured by macrophages lining the subcapsular sinuses and transported into the cortex of the lymph node where they were presented to B cells (Carrasco and Batista 2007; Junt et al. 2007; Phan et al. 2007). In addition, B cells were also observed to engage native antigens on lymph node dendritic cells (Qi et al. 2006). These remarkable findings provided a new view of the initiation of antigen-driven BCR signaling in which BCR activation occurred at the interface of the B cell and APCs.

VIEWING B-CELL ACTIVATION ON ANTIGEN-CONTAINING FLUID LIPID BILAYERS

The observations that B cells were activated by antigen presented by APCs lead Batista and colleagues to investigate B cells engaging antigen incorporated into fluid planar lipid bilayers as surrogate APCs (Carrasco et al. 2004; Fleire et al. 2006; Weber et al. 2008). They reported B-cell activation on lipid bilayers to be a remarkably dynamic event. They observed that B cells touched the bilayer through membrane protrusions in which the BCRs engaged antigen and formed microclusters. The microclusters that were enriched in tyrosine phosphorylated proteins and were the site of Syk recruitment, actively signaled for the B cells to spread over the bilayer allowing the engagement of additional antigens. Following maximal spreading the B cell actively moved the BCR-antigen microclusters toward the center of the contact area and contracted to form an immune synapse. This dynamic process of spreading and contraction occurred within minutes of the BCR's first contact with the bilayer and was shown to be an important mechanism to allow B cells to discriminate their affinity for antigen (Fleire et al. 2006). Subsequent studies provided evidence that the requirement for B-cell activation by membrane-associated and soluble antigens were not identical. B-cell activation on antigen-containing bilayers required the B-cell coreceptor, CD19, to be present within the BCR microclusters whereas B-cell activation by soluble antigen was CD19-independent (Depoil et al. 2008). These authors suggested that CD19 preferentially partitions into signaling active antigen-BCR-Syk microclusters and amplifies signaling. As described later other fundamental differences between the requirements for B-cell activation by membrane associated versus soluble antigen have been recently described.

THE REQUIREMENT FOR PHYSICAL CROSSLINKING OF BCRS BY MULTIVALENT ANTIGENS

The BCR is composed of a membrane form of the Ig (mIg) that alone is not expressed on the cell surface and is unable to initiate signaling. Both surface expression and signaling require the noncovalent association of mIg with a disulfide-linked heterodimer of Igα and Igβ that contain ITAMs in their cytoplasmic domains (Reth 1992) in a 1 mIg:1 Igα-Igβ heterodimer stoichiometry (Schamel and Reth 2000; Tolar et al. 2005). When B cells are provided with antigen the first kinase in the signaling cascades, Lyn, is recruited to and phosphorylates the tyrosines in the Igα/Igβ cytoplasmic domains. The question is how is the information that the BCR has bound antigen to its ectodomain translated to a change in the BCR's cytoplasmic domain that can be recognized by Lyn. It is a common observation that monovalent antigens provided in solution are unable to activate B cells and that activation by antigen in solution requires the binding of BCRs to multivalent antigens, resulting in the clustering, patching, and ultimate capping of BCR-antigen complexes. Such observations lead to the conclusion that B-cell activation requires the physical crosslinking of BCRs by multivalent antigens (Metzger 1992). This crosslinking model for BCR activation fit with the interpretation of X-ray crystallographic studies of Fabs both free and bound to antigen that showed no evidence for a conformational change in the Fab on antigen binding that could be transduced down the mIg to trigger signaling (Poljak 1991). However, these influential structures of Fabs bound to antigens may be misleading in terms of signal transduction. The structure of the BCR has not yet been solved, thus, we do not understand how the mIg ectodomains and the ectodomains of the Igα–Igβ complex fit together and how antigen binding might perturb this relationship to transduce signals across the membrane. Thus, until a structure of a complete BCR complex is solved it may be too early to rule out conformational changes in BCRs on antigen binding. Indeed, the more recent crystal structures of intact IgGs provide evidence for extraordinary interdomain flexibility within Igs (Saphire et al. 2002). Additional crystal structures of the Fc of IgE (Wurzburg et al. 2000) and IgG (Radaev et al. 2001; Sondermann et al. 2000) free and bound to Fc receptors revealed a large rearrangement of the Fc upon Fc receptor binding, again indicating the conformational flexibility of the Igs. In collaboration with our National Institutes of Health (NIH) colleague, Peter Sun and members of his research group, we recently carried out structural and functional studies of the Igα-Igβ ectodomain and its assembly with the mIg. Our results provided evidence for a structural model for the BCR in which both Igα and Igβ of the heterodimer bind to the membrane proximal Cμ4 domain of mIg through an extensive contact surface involving multiple charged residues (Radaev et al. 2010). It is possible that antigen induced alteration in this interface play a role in transducing information to the cytoplasmic domains.

Nonetheless, in the absence of evidence for a mechanism by which antigen binding to mIg transduces conformational changes along the chain, the crosslinking model gained traction. Receptor crosslinking by ligands is a common mechanism for receptor activation in other biological receptor systems but for the most part these are monomeric receptors that contain kinases in their cytoplasmic domains, as in the receptor tyrosine kinases, such as the EGF receptor or stably associate with kinases such as Janus kinases in the large family of cytokine receptors (Schlessinger 2000). In these cases, receptor crosslinking brings kinases in the cytoplasmic domains into close proximity allowing cross phosphorylation and receptor activation. In contrast, the BCR has no inherent kinase activity but rather relies on recruitment of Lyn following antigen binding. Receptors that rely on ligand-induced crosslinking also differ from the BCR in a second important way, they are activated by one or a small set of structurally related ligands. These receptors have evolved to recognize their ligands and the ligands can reproducibly bring the receptors into a geometry that allows transphosphorylation. In contrast, the BCRs must be able to respond to a universe of foreign antigens of all sizes and shapes with no preconception of what these antigens will be. It is difficult to imagine how antigen binding to BCRs could in all cases result in BCRs crosslinked into a specified geometry. In addition, antibody responses can be generated to small monovalent toxins and to highly mobile cell membrane components such as phospholipids that have no obvious mechanism to crosslink BCRs.

Reth and colleagues (Reth et al. 2000) appreciated this difficulty with the BCR crosslinking model and proposed that in resting cells BCRs self assembled into clusters that antigens disrupt to initiate signaling. Their hypothesis was based in part on biochemical evidence for high molecular weight BCR oligomers in detergent lysates of resting B cells in the absence of antigen (Schamel and Reth 2000). However, recent studies using fluorescent resonance energy transfer (FRET) failed to provide evidence that BCR oligomers exist in the membranes of live resting B cells (Tolar et al. 2005). Nonetheless, the observation of BCR oligomers in detergent solution is of interest, suggesting that BCRs may have an intrinsic propensity to oligomerize.

Our recent studies characterizing the early events in the initiation of B-cell signaling led to the surprising finding that B cells are activated by monovalent antigens but only when the B cells engaged antigen in fluid lipid bilayers mimicking an APC (Tolar et al. 2009). We observed that the responses of splenic B cells specific for the hapten NIP placed on a fluid lipid bilayer containing NIP_1-conjugated peptides attached to fluid lipid bilayers were indistinguishable from B cells responding to lipid bilayers containing the multivalent antigen NIP_{14}-BSA. In solution the NIP_1-peptide, as predicted, had no effect on B cells indicating a qualitative difference between monovalent antigen in solution and monovalent antigen incorporated into fluid lipid bilayers. Neither NIP_1-peptide in solution nor in a fluid lipid bilayer has the ability to physically crosslink the BCR, and yet monovalent haptenated peptides incorporated into fluid lipid bilayers but not in solution activate B cells. A fundamental difference between monovalent antigens in solution and in fluid bilayers is the potential of antigen in membranes in to exert a force or torque on the antigen-bound BCR. Such a strain on the BCR could conceivably be transduced into changes that would affect either the mIg ectodomains or the transmembrane domains directly or their relationship with the associated Igα/Igβ heterodimer, ultimately resulting in changes in the cytoplasmic domains of the BCR. Of course conformational changes are easier to propose than to prove and direct evidence for a change in the BCR structure upon engagement of antigen on a membrane will likely await the solution of a complete BCR structure.

THE ROLE OF THE BCR'S mIg ECTODOMAIN IN BCR SIGNALING

Our discovery that B cells responded to monovalent membrane-associated antigens lead us to carry out detailed analyses of the behavior of the BCRs as they first encountered antigen on a fluid lipid bilayer using live-cell imaging at the single molecule level to delineate the mechanism underlying the observed B-cell activation. What we found was consistent with a model we have termed "the conformation-induced oligomerization model" for BCR microclustering (Tolar and Pierce, 2010) (Fig. 1). The key observation from single-molecule imaging was that the BCR when engaging monovalent antigen-containing fluid bilayer stopped. Thus, the binding of monovalent antigen dramatically altered the behavior of the freely diffusing BCR before the formation of microclusters. Two additional results provided evidence that stopping was not the direct result of antigen binding to the BCR but rather that antigen-bound BCRs were mobile in the membrane and only stopped when they encountered other antigen-bound

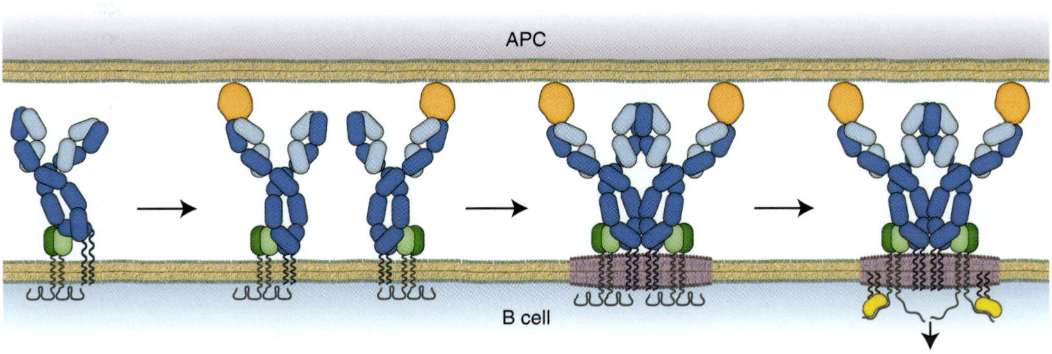

Figure 1. The "conformation-induced oligomerization" model for the initiation of BCR signaling. From *left* to *right*: In the absence of antigen the BCR (blue H chain, cyan L chains, green IgαIgβ) is freely diffusing in the B-cell membrane in a conformation that is not permissive for oligomerization. Binding to monovalent antigen on an APC surface reveals an oligomerization face in the Cμ4 ectodomain of the BCR. The antigen bound BCR is still freely diffusing but stops when it encounters another antigen bound BCR and oligomerizes. The BCR oligomer perturbs the local lipid environment resulting in the coalescence of saturated lipids (in purple) around the oligomer resulting in the simultaneous recruitment of the membrane tethered Lyn (yellow color) and the "opening" of the BCR cytoplasmic domains. Lyn's phosphorylation of the BCR ITAMs triggers the signaling pathways.

BCRs and oligomerized. First, BCR stopping was highly dependent on the concentration of the monovalent antigen in the bilayer. Monovalent antigens diffuse more rapidly in the lipid bilayer than do BCRs in the B-cell membrane so that by imaging the monovalent antigens in bilayers we determined that BCRs bound to and slowed the movement of the antigens (Tolar et al. 2009). However, the antigens stopped only when the concentrations of the antigen were sufficiently high to ensure a high probability that antigen-bound BCRs would encounter each other. Second, imaging the movement of individual BCR in areas where immobile BCR microclusters had already formed, we observed that BCRs were far more likely to stop when they entered immobile BCR clusters (Tolar et al. 2009). We concluded from these results that monovalent antigen binding altered the BCR in some way that allowed the antigen bound BCR to oligomerize with and form microclusters with other antigen bound BCRs.

Through an extensive mutational analyses we determined that BCR clustering in response to monovalent membrane antigens is a BCR intrinsic event that does not require a signaling competent BCR and was mediated by the membrane proximal ectodomains of the mIg, Cμ4 (Tolar et al. 2009). Thus, BCRs expressing mIgM that contained in their transmembrane domain a YS→VV mutation allowing the mIgM to be expressed on the plasma membrane alone, in the absence of Igα/Igβ, stopped and oligomerized in response to monovalent antigen in the lipid bilayer. However, mIgM (YS→VV) that lacked the Cμ4 ectodomain failed to stop and oligomerize in response to monovalent membrane antigen. We hypothesized that Cμ4 contained a homo-oligomerization domain that was in some way hidden or blocked and not accessible in the unligated mIgM but became accessible when the mIgM bound antigen on the opposing fluid lipid bilayer (Fig. 1). We reasoned that if this were indeed the case then the Cμ4 domain expressed alone may spontaneously oligomerize. We found that in contrast to Cμ2-4 (YS→VV) and Cμ3-4 (YS→VV) that when expressed on cell surfaces freely diffused in the membrane, Cμ4 (YS→VV) expressed alone on the cell surface stopped and formed microscopic clusters in the membrane. When expressed with Igα-Igβ, Cμ4 was signaling active as measured by the recruitment of Syk and activation of the cells to express CD69. Similar results were obtained for mIgG in which case the membrane proximal Cγ3 domain was necessary and sufficient for oligomerization. These results changed our thinking considerably about the role of the mIg in BCR signaling. The mIg was not simply an inert, passive antigen binding domain that provided a means of physical crosslinking of the BCRs but rather a dynamic active component of the BCR.

THE AFFINITY AND ISOTYPE OF THE BCR INFLUENCE EARLY BCR INTRINSIC EVENTS IN THE INITIATION OF SIGNALING

Immunological memory or the ability to respond more rapidly and robustly to a secondary challenge with antigen is a hallmark of adaptive immunity. For B cells memory responses are characterized by the rapid production of high-affinity, isotype-switched antibodies of which IgGs are the predominant isotypes (Gray 1993). Current evidence indicates that B cells expressing high affinity, isotype switched BCRs are generated in germinal center reactions through the coupled processes of somatic hyper-mutation and switch recombination (McHeyzer-Williams and McHeyzer-Williams 2005). Presumably high affinity, IgG expressing B cells have an advantage and ultimately dominate the memory B-cell compartment. However, the molecular mechanisms by which high-affinity and isotype-switched BCRs confer such an advantage are incompletely understood.

High-affinity B cells have been shown in adoptive transfer experiments in vivo to out compete low affinity B cells for survival, presumably reflecting a competition at the clonal level for limiting quantities of antigen, appropriate niches or T cell help (Dal Porto et al. 2002; Paus et al. 2006; Phan et al. 2006; Shih et al. 2002a,b; Takahashi et al. 1998). At the level of individual B cells evidence has been provided that signaling through high affinity BCRs is qualitatively different from that through low affinity BCRs (Kouskoff et al. 1998) and more recently that the affinity of B cells for antigen influence the degree to which B cells spread over antigen containing bilayers allowing increased accumulation of antigen and enhanced responses (Fleire et al. 2006). Concerning the benefit of the IgG- versus IgM-BCRs it appears clear that the cytoplasmic tail of IgG leads to enhanced B-cell responses (Horikawa et al. 2007; Kaisho et al. 1997; Martin and Goodnow 2002; Waisman et al. 2007; Wakabayashi et al. 2002) because of qualitatively different signaling (Horikawa et al. 2007) or through mechanisms that involve the phosphorylation of the mIgG tail promoting the recruitment of Grb2 (Engels et al. 2009).

We asked are the BCR intrinsic events that result in BCR clustering in response to monovalent antigen influenced by the affinity or the isotype of the BCR? To do so cell lines were generated that expressed chimeric BCRs that contained mIgG or mIgM engineered to express variable regions that were either high or low affinity for the hapten NIP. We discovered that high affinity BCRs formed immobile, signaling active clusters more efficiently as compared with low affinity BCRs (Fig. 2). Imaging these clusters for 120 s we observed that as compared to low affinity BCRs clusters, the high affinity clusters grew more rapidly in both fluorescence intensity and in size. Ultimately, the rapid growth in the BCR clusters resulted in enhanced downstream signaling to Syk (Liu et al. 2010).

In addition to the sensitivity of BCR clustering to BCR antigen affinity we observed that independent of affinity IgG-BCRs were more efficient than IgM-BCRs in forming immobile signaling active BCRs (Fig. 3). The IgG-BCRs microclusters also grew in fluorescence intensity and size more rapidly as compared to IgM-BCRs resulting in enhanced downstream signaling. Moreover, the enhanced oligomerization and clustering of the IgG-BCRs in response to membrane associated antigens resulted in significant differences in downstream signaling events measured 10 min later including the recruitment of the Syk (Liu unpublished observation). These observations immediately raised the question: What feature of the IgG-BCR accounted for its ability to oligomerize more rapidly? Earlier studies established an essential role for the cytoplasmic tail of mIgG in enhancing B-cell responses

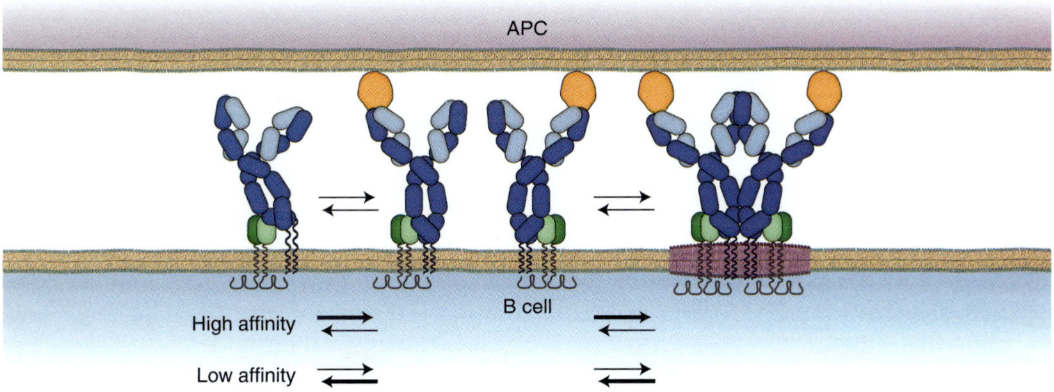

Figure 2. The early intrinsic events in the initiation of signaling are affinity dependent. The BCR transitions reversibly between free, antigen bound and oligomerized states (from *left* to *right*). These states are highly sensitive to the affinity of the BCR for antigen. High affinity BCRs are readily driven to oligomers whereas low affinity receptors only inefficiently oligomerize.

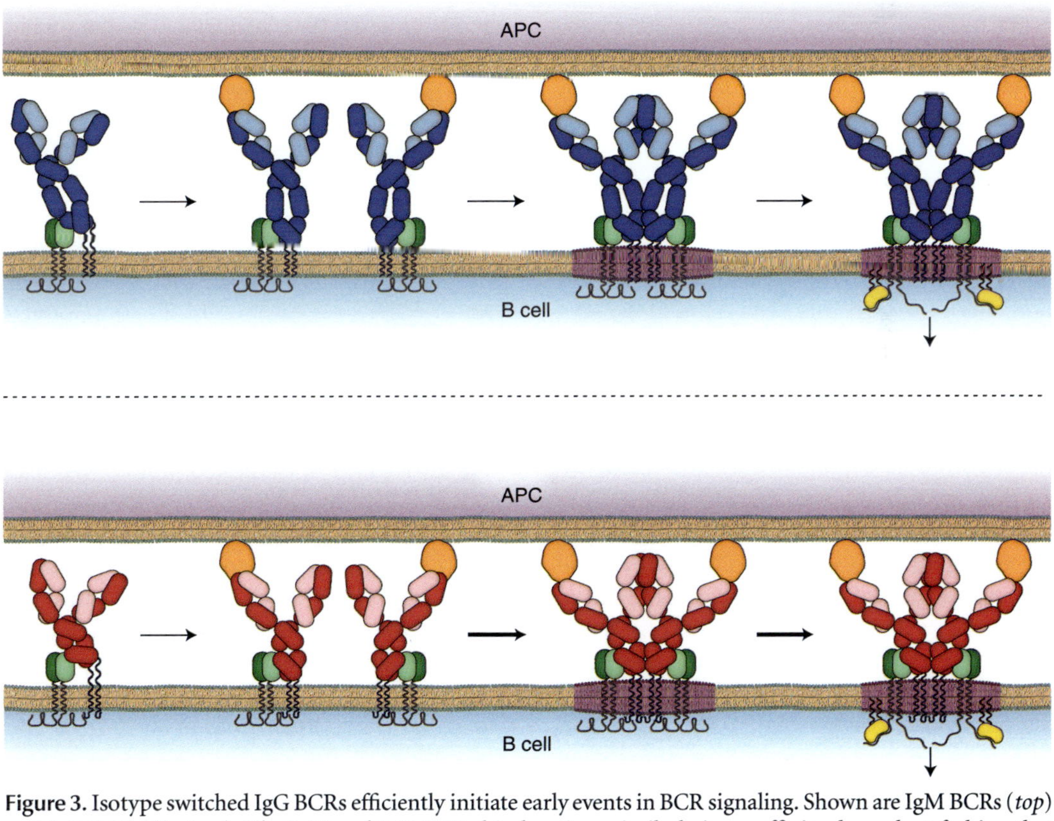

Figure 3. Isotype switched IgG BCRs efficiently initiate early events in BCR signaling. Shown are IgM BCRs (*top*) and IgG BCRs (*bottom*). The IgM and IgG BCRs bind antigen similarly in an affinity dependent fashion that brings the BCRs into an oligomerization-permissive conformation. However, because of the properties of its cytoplasmic tail the IgG BCRs oligomerize and initiate signaling more efficiently.

(Horikawa et al. 2007; Kaisho et al. 1997; Martin and Goodnow 2002; Waisman et al. 2007; Wakabayashi et al. 2002). Our recent studies indicated that the ability of the μ and γ ectodomains to oligomerize was comparable (Tolar et al. 2009) suggesting that the BCR-intrinsic advantage of IgG-BCRs may also be dependent on the mIgG cytoplasmic tail. Indeed, we determined that the enhanced clustering and signaling of IgG BCRs was dependent on the IgG tail. Because these BCR-intrinsic events precede the triggering of signaling cascades, the mIgG tail is presumably functioning in BCR oligomerization by a mechanism distinct from that described by Engels et al. (Engels et al. 2009) involving phosphorylation of the IgG tail.

TRANSLATING BCR OLIGOMERIZATION INTO BCR SIGNALING

We began this review by asking what changes occur in the BCRs, as the BCRs engage antigen that allows the Lyn to discriminate antigen-free from the antigen-bound receptors? It is possible that the structures of the oligomeric BCR cytoplasmic domains themselves are distinct from that of the monomer. Our earlier published results using quantitative FRET provided evidence that antigen binding resulted in the clustering of the BCR ectodomains and cytoplasmic domains into close molecular proximity and that although the ectodomains remained in close proximity the intracellular domains moved apart, losing FRET (Tolar et al. 2005). We referred to this phenomenon as a transition of the cytoplasmic domains from a "closed" to an "open" form. The temporal relationship between "opening" of the cytoplasmic domains, the phosphorylation of the BCRs by Lyn and the recruitment of Syk could not be resolved leaving the cause-and-effect relationships between these events an open question. Thus, although it is possible that BCR monomers and oligomers are discriminated by Lyn, at present there is little direct experimental support of such a model.

Another possibility for a mechanism by which BCR oligomers and monomers might be discriminated comes from studies of the lipid environment of the BCR monomer and antigen-bound clusters. The unligated BCRs were determined to be more readily solubilized by detergents as compared with antigen-bound BCRs suggesting that the local lipid environment of BCRs changed after antigen binding altering the solubility characteristics (Dykstra et al. 2003). Based on a variety of studies in other receptor systems the differential solubility suggested that the BCRs in resting cells were in relatively disordered membrane microdomains whereas antigen-clustered BCRs were in more ordered membrane microdomains composed of saturated lipids and cholesterol, coined membrane "lipid rafts" (Simons and Toomre 2000). The significance of this observation was that the first kinases in the BCR signal cascade, namely, Src-family kinases were tethered to the membrane by saturated lipids and were concentrated in lipid raft microdomains. Thus, the change in local lipid environment would have the indirect effect of bringing the first kinase into close proximity to the clustered BCR. However, the interpretation of detergent solubility studies are fraught with potential artifacts and if the proposed changes in lipid environment were to be considered seriously as a mechanism for signal transduction, the possibility would need to be tested in intact living cells. To do so we used quantitative FRET analyses of cells that expressed BCRs containing a FRET donor fluorescent protein in their cytoplasmic domains and a series of FRET acceptor fluorescent proteins that were tethered to the cytoplasmic face of the B-cell plasma membrane by saturated or unsaturated lipids. We also studied the association of Lyn itself engineered to contain a FRET acceptor fluorescent protein with the BCR. We observed that on BCR clustering the saturated lipid probe selectively and transiently associated with the BCR microclusters, an interaction that was followed by a prolonged association of the BCR clusters with Lyn (Sohn et al. 2006; Sohn et al. 2008b). The association of the clustered BCRs with Lyn correlated both spatially and temporally with the BCR's cytoplasmic domains coming into a signaling active "open" form. We concluded from these observations that the BCR oligomers transiently perturbed the local lipid environment that had the repercussion of bringing the clustered BCRs into close molecular proximity with the Src family kinases promoting protein-protein interactions between Lyn and the cytoplasmic domains of Igα/Igβ (Fig. 1) (Tolar et al. 2008). We also suggested that the change in the local lipid environment could have the effect of directly, physically altering the configuration of the receptor's cytoplasmic domains (Fig. 1). Saturated lipids cause thickening and curvature of membranes that could cause the BCR cytoplasmic domains to move to adjust to this change. It is also possible that the cytoplasmic domains of the Igα/Igβ heterodimer associate directly with the membrane lipids and that changes in lipids alter this association. Recent studies by Xu et al. (Xu et al. 2008) showed that the cytoplasmic domain of the T-cell receptor CD3ε chain associated with the inner leaflet of plasma membrane resulting in the deep insertion of two key tyrosine residues sequestering these from Src kinases in the unligated receptor that were presumably available in the antigen-bound receptor. These results provide an elegant mechanism by which receptor protein-membrane lipid interactions can control the initiation of signaling. The ability of BCR clusters to perturb the local lipid environment resulting in Lyn's association with the BCR may provide a common mechanism by which BCRs physically crosslinked by antigen or engaged by monovalent membrane associated antigens initiate signaling. The BCR clusters that form in response to monovalent antigens on membranes that cannot physically crosslink the BCRs versus multivalent antigen may not be the same in their molecular conformations but may share the common feature of perturbing the membrane, facilitating the interactions of Lyn with the BCR.

REGULATING ANTIGEN-DRIVEN BCR-INTRINSIC EVENTS

The outcome of signaling through the BCR is regulated by both the developmental stage of the B cell and by a variety of B-cell coreceptors that both promote and attenuate BCR signaling. This raises the question are the antigen-driven BCR intrinsic events described here regulated either developmentally or by B-cell coreceptors and if so, how? We have begun to address this question by analyzing the effect of a potent B-cell inhibitory receptor, the FcγRIIB, on early antigen-driven BCR intrinsic events. The FcγRIIB is a well characterized coreceptor that when coligated to the BCR through the binding of immune complexes, recruits the lipid phosphatase SHIP that serves to dampen BCR signaling by reversing the BCR-induced activity of PI3 kinase (Ravetch and Bolland 2001). We determined that the FcγRIIB had the additional effect of destabilizing BCR oligomers, not by physically disrupting BCR oligomerization but by perturbing the local lipid environment of the BCR oligomer, blocking the BCR's association with saturated lipids and Lyn (Liu et al. 2009; Sohn et al. 2008a) (Fig. 4). Interestingly, an FcγRIIB loss of function mutation in the transmembrane domain of the human protein that is associated with systemic lupus erythematosus, a systemic autoimmune disease (Floto et al. 2005), failed to block the early BCR oligomerization. The transmembrane mutation was previously shown to block the association of FcγRIIB with saturated lipids (Floto et al. 2005). We believe that further studies to elucidate the effect of various coreceptors and the developmental state of the B cell on early BCR intrinsic events may provide new strategies to block hyper-BCR activation in autoimmune diseases.

THE POTENTIAL REPERCUSSIONS OF SPONTANEOUS, CHRONIC BCR OLIGOMERIZATION

Here we reviewed our current understanding of the molecular mechanisms that underlie the BCR intrinsic events that lead to BCR oligomerization and ultimately to BCR signaling. We showed that mutant BCRs that were composed of mIgs that contained only the membrane proximal domain of mIgM, Cμ4, spontaneously oligomerized and signaled (Tolar et al. 2009). This observation raises the question, are there naturally occurring mutations that result

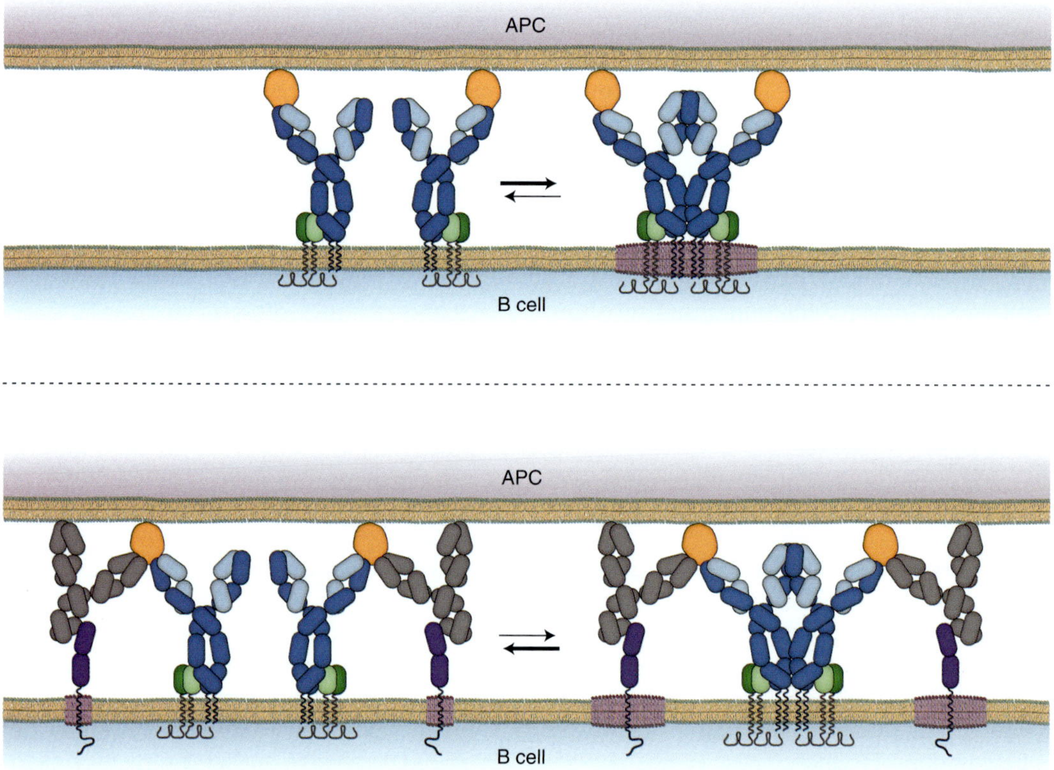

Figure 4. The early BCR intrinsic events in B-cell activation are a novel target for FcγRIIB inhibition. When coengaged with the FcγRIIB (*bottom*) through binding of immune complexes (grey colored IgG and orange colored antigen) the FcγRIIB (bright blue color) blocks the BCR's ability to form stable oligomers (Liu et al. 2009) and associate with raft lipids (Sohn et al. 2008a). Blocking these early events does not require molecular proximity of the cytoplasmic domains of the BCR and FcγRIIB but relies on the rapid and sustained association of the FcγRIIB with raft lipids.

in spontaneous clustering or activation and if so what are the repercussions of chronic activation? Recently, in collaboration with our NIH colleague, Dr. Louis Staudt and members of his laboratory, we analyzed the behavior of the BCRs on several activated B cell (ABC)-like subtype of diffuse large B-cell lymphomas (DLBCLs) that were dependent on the BCR for their survival. We determined that the BCRs on these tumors formed immobile BCR oligomers within prominent clusters similar to those observed on antigen stimulated normal B cells (Davis et al. 2009). Somatic mutations affecting the cytoplasmic domains of Igα and Igβ were frequent in these tumors but rarely in other DLBCLs and never in Burkitts or MALT lymphomas. The full sequences of the mIg of ABC DLBCLs and the correlation with the spontaneous clustering phenotype ought to provide new insights into the mechanisms underlying oligomerization and provide strategies for therapeutic intervention.

SUMMARY

New live-cell imaging technologies are providing the first views into the extraordinarily dynamic events that accompany the initiation of B cell signaling. Understanding how these events are regulated during development and by coreceptors and how natural mutations affect these events may provide new avenues for therapeutic intervention in disease that involve hyper or chronic B-cell activation such as systemic autoimmune diseases and B-cell malignancies.

REFERENCES

Batista FD, Harwood NE. 2009. The who, how and where of antigen presentation to B cells. *Nat Rev Immunol* **9:** 15–27.

Batista FD, Iber D, Neuberger MS. 2001. B cells acquire antigen from target cells after synapse formation. *Nature* **411:** 489–494.

Carrasco YR, Batista FD. 2007. B cells acquire particulate antigen in a macrophage-rich area at the boundary between the follicle and the subcapsular sinus of the lymph node. *Immunity* **27:** 160–171.

Carrasco YR, Fleire SJ, Cameron T, Dustin ML, Batista FD. 2004. LFA-1/ICAM-1 interaction lowers the threshold of B cell activation by facilitating B cell adhesion and synapse formation. *Immunity* **20:** 589–599.

Dal Porto JMHaberman AMKelsoe G, Shlomchik MJ. 2002. Very low affinity B cells form germinal centers, become memory B cells, and participate in secondary immune responses when higher affinity competition is reduced. *J Exp Med* **195:** 1215–1221.

Davis RE, Ngo VN, Lenz G, Tolar P, Young R, Romesser PB, Kohlhammer H, Lamy L, Zhao H, Yang Y. 2009. Chronic active B cell receptor signaling in diffuse large B cell lymphoma. *Nature*: in press.

Depoil D, Fleire S, Treanor BL, Weber M, Harwood NE, Marchbank KL, Tybulewicz VL, Batista FD. 2008. CD19 is essential for B cell activation by promoting B cell receptor-antigen microcluster formation in response to membrane-bound ligand. *Nat Immunol* **9:** 63–72.

Dykstra M, Cherukuri A, Sohn HW, Tzeng SJ, Pierce SK. 2003. Location is everything: Lipid rafts and immune cell signaling. *Annu Rev Immunol* **21:** 457–481.

Engels N, Konig LM, Heemann C, Lutz J, Tsubata T, Griep S, Schrader V, Wienands J. 2009. Recruitment of the cytoplasmic adaptor Grb2 to surface IgG and IgE provides antigen receptor-intrinsic costimulation to class-switched B cells. *Nat Immunol* **10:** 1018–1025.

Fleire SJ, Goldman JP, Carrasco YR, Weber M, Bray D, Batista FD. 2006. B cell ligand discrimination through a spreading and contraction response. *Science* **312:** 738–741.

Floto RA, Clatworthy MR, Heilbronn KR, Rosner DR, MacAry PA, Rankin A, Lehner PJ, Ouwehand WH, Allen JM, Watkins NA. 2005. Loss of function of a lupus-associated FcγRIIb polymorphism through exclusion from lipid rafts. *Nat Med* **11:** 1056–1058.

Fooksman DR, Vardhana S, Vasiliver-Shamis G, Liese J, Blair D, Waite J, Sacristan C, Victora G, Zanin-Zhorov A, Dustin ML. 2010. Functional Anatomy of T Cell Activation and Synapse Formation. *Annu Rev Immunol* **28:** 79–105.

Gray D. 1993. Immunological memory. *Annu Rev Immunol* **11:** 49–77.

Horikawa K, Martin SW, Pogue SL, Silver K, Peng K, Takatsu K, Goodnow CC. 2007. Enhancement and suppression of signaling by the conserved tail of IgG memory-type B cell antigen receptors. *J Exp Med* **204:** 759–769.

Junt T, Moseman EA, Iannacone M, Massberg S, Lang PA, Boes M, Fink K, Henrickson SE, Shayakhmetov DM, Di Paolo NC. 2007. Subcapsular sinus macrophages in lymph nodes clear lymph-borne viruses and present them to antiviral B cells. *Nature* **450:** 110–114.

Kaisho T, Schwenk F, Rajewsky K. 1997. The roles of gamma 1 heavy chain membrane expression and cytoplasmic tail in IgG1 responses. *Science* **276:** 412–415.

Kouskoff V, Famiglietti S, Lacaud G, Lang P, Rider JE, Kay BK, Cambier JC, Nemazee D. 1998. Antigens varying in affinity for the B cell receptor induce differential B lymphocyte responses. *J Exp Med* **188:** 1453–1464.

Liu W, Meckel T, Tolar P, Sohn HW, Pierce SK. 2009. Antigen-induced oligomerization of the B cell receptor is an early target of FcγRIIB inhibition. *J Immunol* **184:** 1977–1989.

Liu W, Meckel T, Tolar P, Sohn HW, Pierce SK. 2010. Antigen affinity discrimination is an intrinsic function of the B cell receptor. *J Exp Med* DOI: 10.1084/jem.20092123.

Martin SW, Goodnow CC. 2002. Burst-enhancing role of the IgG membrane tail as a molecular determinant of memory. *Nat Immunol* **3:** 182–188.

McHeyzer-Williams LJ, McHeyzer-Williams MG. 2005. Antigen-specific memory B cell development. *Annu Rev Immunol* **23:** 487–513.

Metzger H. 1992. Transmembrane signaling: The joy of aggregation. *J Immunol* **149:** 1477–1487.

Pape KA, Catron DM, Itano AA, Jenkins MK. 2007. The humoral immune response is initiated in lymph nodes by B cells that acquire soluble antigen directly in the follicles. *Immunity* **26:** 491–502.

Paus D, Phan TG, Chan TD, Gardam S, Basten A, Brink R. 2006. Antigen recognition strength regulates the choice between extrafollicular plasma cell and germinal center B cell differentiation. *J Exp Med* **203:** 1081–1091.

Phan TG, Grigorova I, Okada T, Cyster JG. 2007. Subcapsular encounter and complement-dependent transport of immune complexes by lymph node B cells. *Nat Immunol* **8:** 992–1000.

Phan TG, Paus D, Chan TD, Turner ML, Nutt SL, Basten A, Brink R. 2006. High affinity germinal center B cells are actively selected into the plasma cell compartment. *J Exp Med* **203:** 2419–2424.

Poljak RJ. 1991. Structure of antibodies and their complexes with antigens. *Molecular Immunology* **28:** 1341–1345.

Qi H, Egen JG, Huang AY, Germain RN. 2006. Extrafollicular activation of lymph node B cells by antigen-bearing dendritic cells. *Science* **312:** 1672–1676.

Radaev S, Motyka S, Fridman WH, Sautes-Fridman C, Sun PD. 2001. The structure of a human type III Fcgamma receptor in complex with Fc. *J Biol Chem* **276:** 16469–16477.

Radaev S, Zou Z, Tolar P, Nguyen K, Nguyen AT, Krueger PD, Pierce S, Sun PD. 2010. Structural and functional studies of Ig alpha/beta and its assembly with the B cell antigen receptor. *Structure* (in press).

Rajewsky K. 1996. Clonal selection and learning in the antibody system. *Nature* **381:** 751–758.

Ravetch JV, Bolland S. 2001. IgG Fc receptors. *Annu Rev Immunol* **19:** 275–290.

Reth M. 1992. Antigen receptors on B lymphocytes. *Annu Rev Immunol* **10:** 97–121.

Reth M, Wienands J, Schamel WW. 2000. An unsolved problem of the clonal selection theory and the model of an oligomeric B-cell antigen receptor. *Immunol Rev* **176:** 10–18.

Saphire EO, Stanfield RL, Crispin MD, Parren PW, Rudd PM, Dwek RA, Burton DR, Wilson IA. 2002. Contrasting IgG structures reveal extreme asymmetry and flexibility. *J Mol Biol* **319:** 9–18.

Schamel WW, Reth M. 2000. Monomeric and oligomeric complexes of the B cell antigen receptor. *Immunity* **13:** 5–14.

Schlessinger J. 2000. Cell Signaling by Receptor Tyrosine Kinases. *Cell* **103:** 211–225.

Shih TA, Roederer M, Nussenzweig MC. 2002b. Role of antigen receptor affinity in T cell-independent antibody responses in vivo. *Nat Immunol* **3:** 399–406.

Shih TA, Meffre E, Roederer M, Nussenzweig MC. 2002a. Role of BCR affinity in T cell dependent antibody responses in vivo. *Nat Immunol* **3:** 570–575.

Simons K, Toomre D. 2000. Lipid rafts and signal transduction. *Nat Rev Mol Cell Biol* **1:** 31–39.

Sohn HW, Pierce SK, Tzeng SJ. 2008a. Live cell imaging reveals that the inhibitory FcgammaRIIB destabilizes B cell receptor membrane-lipid interactions and blocks immune synapse formation. *J Immunol* **180:** 793–799.

Sohn HW, Tolar P, Pierce SK. 2008b. Membrane heterogeneities in the formation of B cell receptor-Lyn kinase microclusters and the immune synapse. *J Cell Biol* **182:** 367–379.

Sohn HW, Tolar P, Jin T, Pierce SK. 2006. Fluorescence resonance energy transfer in living cells reveals dynamic membrane changes in the initiation of B cell signaling. *Proc Natl Acad Sci* **103:** 8143–8148.

Sondermann P, Huber R, Oosthuizen V, Jacob U. 2000. The 3.2-A crystal structure of the human IgG1 Fc fragment-Fc gammaRIII complex. *Nature* **406:** 267–273.

Takahashi Y, Dutta PR, Cerasoli DM, Kelsoe G. 1998. In situ studies of the primary immune response to (4-hydroxy-3-nitrophenyl)acetyl. V. Affinity maturation develops in two stages of clonal selection. *J Exp Med* **187:** 885–895.

Tolar P, Pierce SK. 2010. A conformation-induced oligomerization model for B cell receptor microclustering and signaling. *Curr Top Microbiol Immunol* **340:** 155–169.

Tolar P, Sohn HW, Pierce SK. 2005. The initiation of antigen-induced B cell antigen receptor signaling viewed in living cells by fluorescence resonance energy transfer. *Nat Immunol* **6:** 1168–1176.

Tolar P, Sohn HW, Pierce SK. 2008. Viewing the antigen-induced initiation of B-cell activation in living cells. *Immunol Rev* **221:** 64–76.

Tolar P, Hanna J, Krueger PD, Pierce SK. 2009. The constant region of the membrane immunoglobulin mediates B cell-receptor clustering and signaling in response to membrane antigens. *Immunity* **30:** 44–55.

Waisman A, Kraus M, Seagal J, Ghosh S, Melamed D, Song J, Sasaki Y, Classen S, Lutz C, Brombacher F. 2007. IgG1 B cell receptor signaling is inhibited by CD22 and promotes the development of B cells whose survival is less dependent on Ig alpha/beta. *J Exp Med* **204:** 747–758.

Wakabayashi C, Adachi T, Wienands J, Tsubata T. 2002. A distinct signaling pathway used by the IgG-containing B cell antigen receptor. *Science* **298:** 2392–2395.

Weber M, Treanor B, Depoil D, Shinohara H, Harwood NE, Hikida M, Kurosaki T, Batista FD. 2008. Phospholipase C-γ2 and Vav cooperate within signaling microclusters to propagate B cell spreading in response to membrane-bound antigen. *J Exp Med* **205:** 853–868.

Wurzburg BA, Garman SC, Jardetzky TS. 2000. Structure of the Human IgE-Fc Cε3-Cε4 Reveals Conformational Flexibility in the Antibody Effector Domains. *Immunity* **13:** 375–385.

Xu C, Gagnon E, Call ME, Schnell JR, Schwieters CD, Carman CV, Chou JJ, Wucherpfennig KW. 2008. Regulation of T cell receptor activation by dynamic membrane binding of the CD3epsilon cytoplasmic tyrosine-based motif. *Cell* **135:** 702–713.

The Cytoskeleton Coordinates the Early Events of B-Cell Activation

Naomi E. Harwood and Facundo D. Batista

Lymphocyte Interaction Laboratory, Cancer Research UK London Research Institute, Lincoln's Inn Fields Laboratories, London WC2A 3LY, United Kingdom

Correspondence: facundo.batista@cancer.org.uk

B cells contribute to protective adaptive immune responses through generation of antibodies and long-lived memory cells, following engagement of the B-cell receptor (BCR) with specific antigen. Recent imaging investigations have offered novel insights into the ensuing molecular and cellular events underlying B-cell activation. Following engagement with antigen, BCR microclusters form and act as sites of active signaling through the recruitment of intracellular signaling molecules and adaptors. Signaling through these "microsignalosomes" is propagated and enhanced through B-cell spreading in a CD19-dependent manner. Subsequently, the mature immunological synapse is formed, and functions as a platform for antigen internalization, enabling the antigen presentation to helper T cells required for maximal B-cell activation. In this review, we discuss the emerging and critical role for the cytoskeleton in the coordination and regulation of these molecular events during B-cell activation

An individual is continually bombarded by potentially disease-causing agents. However, under normal circumstances, the action of the immune system ensures that these encounters relatively infrequently result in the development of symptomatic illness. The immune system can be broadly divided into two component parts, the innate and the adaptive branches. The innate branch mediates rapid inflammatory responses following the recognition of motifs typically associated with pathogens through a collection of pattern recognition receptors. In contrast, adaptive immune responses tend to emerge a few days after initial infection and show four central characteristics: memory, specificity, diversity, and self-nonself discrimination. In spite of these two divisions, the effective elimination of pathogens most often requires intricate collaboration between the innate and adaptive immune responses.

Small white blood cells, known as lymphocytes, are the fundamental participants mediating adaptive immune responses. Lymphocytes originate from hematopoietic stem cells in the bone marrow and although a subset migrate to the thymus to form T cells, others remain in the bone marrow to complete their development into B cells (Halin et al. 2005). Mature lymphocytes circulate throughout the body and are often localized in secondary lymphoid organs such as the lymph nodes and the spleen, which are specialized sites for lymphocyte activation (Junt et al. 2008). T cells are responsible for cell-mediated immunity and are commonly

 doi: 10.1101/cshperspect.a002360
Cite this article as *Cold Spring Harb Perspect Biol* 2011;3:a002360

classified into $CD4^+$ helper and $CD8^+$ cytotoxic cells according to their expression of surface coreceptors. On the other hand, B cells mediate humoral immunity through the secretion of antibodies that recognize and neutralize invading pathogens.

To become activated to produce antibodies, B cells must first recognize specific antigen through the B-cell receptor (BCR). This specific binding event initiates intracellular signaling leading to altered gene expression, reorganization of the B-cell cytoskeleton, and antigen internalization. Importantly, BCR-mediated internalization targets antigen to endosomes containing newly synthesized major histocompatibility complex (MHC) (Aluvihare et al. 1997; Amigorena et al. 1994), such that processed antigen can be presented to $CD4^+$ T cells, thereby recruiting help to facilitate maximal B-cell activation (Lanzavecchia 1985; Rock et al. 1984). Activated B cells can either rapidly mediate the secretion of low affinity antibodies (MacLennan et al. 2003), or can enter into a specialized structure known as a germinal center (GC) to undergo affinity maturation, producing plasma cells capable of high-affinity antibody production and long-lasting memory cells (MacLennan 1994; Rajewsky 1996).

The molecular events underlying BCR-mediated signaling have historically been characterized using standard biochemical analysis methods in vitro. As such, the BCR has been identified as comprising of a membrane immunoglobulin (mIg) responsible for binding extracellular antigen, in complex with an Igα/β sheath containing immunoreceptor tyrosine activation motifs (ITAMs) in the intracellular domains (Reth 1989). Cross-linking of the BCR by multivalent antigen triggers phosphorylation of the ITAMs through Src family kinases such as Lyn and Syk. This early phosphorylation leads to the recruitment of intracellular effectors including PLCγ2, Vav, Btk, and PI3K, and adaptors including Blnk and Grb2, to form a multicomponent assembly known as the signalosome (Dal Porto et al. 2004; DeFranco 1997; Kurosaki 2002; Scharenberg et al. 2007). Cellular readouts of the coordinated activity of the signalosome include calcium signaling and activation of transcription factors such as NF-κB. Although these classical strategies have provided an essential foundation of intracellular mediators involved in signaling downstream of the BCR, they contribute little insight into the spatiotemporal dynamics and organization of molecular events within the cell (Treanor and Batista 2007).

To address the questions of how and where molecular events occur within the cell, imaging-based strategies have been developed and applied to visualize directly molecular events underlying B-cell activation. Importantly, these approaches not only allow verification of molecular pathways identified previously, but also shed light on the spatiotemporal dynamics of these pathways within the cell. In addition, imaging approaches have led to the identification of unexpected and novel roles for molecules such as CD19 in mounting B-cell responses to antigen. An essential tenet that has been revealed by these imaging studies is the importance of molecular segregation as a means of coordinating and regulating interactions within the cell. Indeed, a pivotal role for the cytoskeleton as a means of mediating segregation is emerging, offering a versatile mechanism for the regulation of numerous molecular interactions. The importance of the cytoskeleton is evidenced by the recent observation that mutations in various cytoskeleton regulators are associated with the development of antibody deficiency syndromes in humans (Conley et al. 2009). In this review, we discuss the contribution that recent imaging investigations have made to our understanding of the process and regulation of B-cell activation. In particular, we highlight the central role for the cytoskeleton in the organization and control of molecular events during B-cell activation.

BCR DISTRIBUTION IN THE RESTING B CELL IS SHAPED BY THE CYTOSKELETON

As activation is initiated following cross-linking of the BCR by antigen, it is important to characterize the distribution and dynamics of the BCR in the resting state to fully appreciate the molecular events regulating the initiation of B-cell activation. In view of the fact that multivalent, but not monovalent, soluble antigens trigger activation (S. Minguet, M. Reth, W. Schamel, pers. comm.; Tolar et al. 2009), it was widely assumed that the BCR exists as a monomer in resting B cells. However, as native gel electrophoresis showed that BCRs are able to interact with one another, it was suggested that BCRs form oligomers on the surface of the resting B cell (Schamel and Reth 2000). On the contrary, a fluorescence resonance energy transfer approach was unable to detect an interaction between individual BCRs in the membrane, indicating that BCRs are monomeric in the absence of antigen (Tolar et al. 2005). Thus, the precise state of the BCR in the membrane before antigen stimulation remains controversial.

Recently, we have sought to visualize the dynamics of BCR on the surface of the resting B cells using Dual-View total internal reflection microscopy (TIRFM) (Treanor et al. 2010) (Fig. 1A). Using this method, we were able to track single particles of BCR in the resting B-cell membrane. We observed that BCRs were not able to diffuse freely in the plane of the membrane and instead show a range of diffusion behaviors, with a proportion of receptors much less mobile (diffusion coefficient less than $<0.01\ \mathrm{um^2s^{-1}}$). This restriction in diffusion is contrary to the predictions of the original fluid-mosaic model proposed by Singer and Nicholson (1972); however, it is in line with more recent investigations tracking single molecules of transmembrane proteins in fibroblasts (Kusumi et al. 2005; Sako and Kusumi 1994; Simson et al. 1995). Indeed, these observations led the authors to suggest that the cell membrane contains discrete confinement zones around 30–700 nm in diameter that restrict long-range diffusion between adjacent zones. This is known as the "picket-fence" model, in which selected transmembrane proteins act as pickets defining confinement zones and are attached to the underlying cytoskeleton fence (Fujiwara et al. 2002). Furthermore, it has

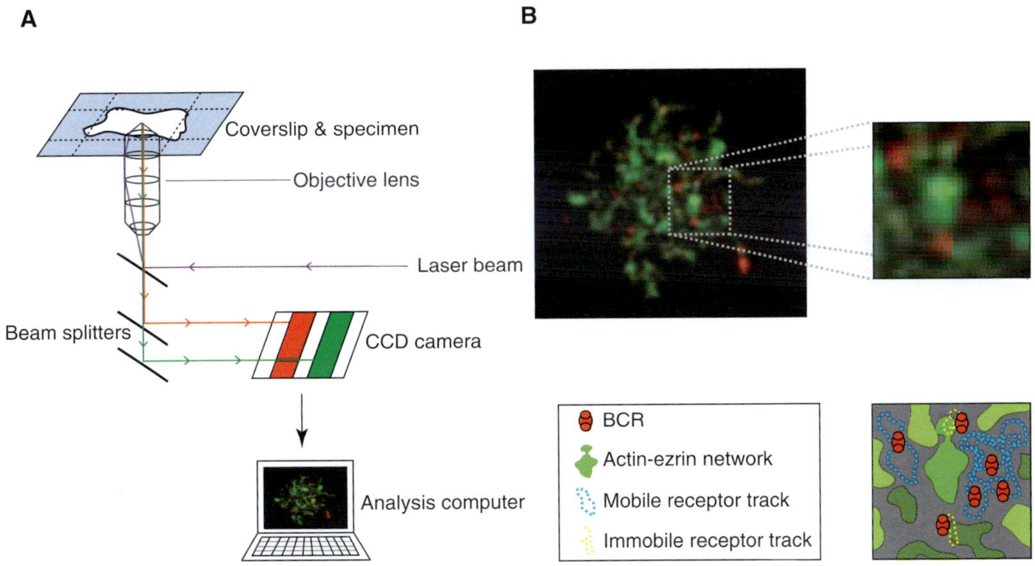

Figure 1. Dual-View TIRFM for simultaneous tracking of BCR alongside the actin cytoskeleton. (*A*) Schematic representation of the setup of the Dual-View TIRFM. The total internal reflection of incoming laser beam (purple) generates evanescent waves capable of exciting flurophores in close proximity to the sample interface. The two colored emission beams (green and red) are passed through beam splitters, detected simultaneously by a CCD camera and overlaid in an analysis computer. (*B*) Snapshot of a movie collected tracking single molecules of BCR (red, anti-BCR Fab) and actin (green, LifeAct a marker of filamentous actin). BCRs located in actin-rich areas are more restricted in their diffusion (yellow tracks in schematic zoom), whereas BCRs located in actin-poor areas tend to be less restricted in their diffusion (blue tracks in schematic zoom).

been suggested that lipid microdomains and/or protein islands may contribute to the observed restriction in diffusion of transmembrane proteins (Lillemeier et al. 2006; Simons and Ikonen 1997). Interestingly, we have used our methodology to simultaneously visualize the BCR alongside components of the cytoskeleton, and showed that slow-moving or immobile BCRs were often co-incident with regions with high density of an ezrin-defined actin network (Treanor et al. 2010) (Fig. 1B). Indeed, disruption of the cytoskeleton network through pharmacological agents results in intracellular signaling similar to that observed following antigen stimulation, likely as a result of BCR cluster formation. Thus, it appears that the cytoskeleton plays an important role in regulating the distribution and dynamics of the BCR in the resting B cell.

RECOGNITION OF ANTIGEN TRIGGERS B-CELL ACTIVATION BY AN UNKNOWN MECHANISM

The BCR can recognize and respond to native antigens both in soluble or membrane-bound forms, though the latter have a lower threshold and thus represent a more effective means of triggering B-cell activation (Batista et al. 2001; Batista and Neuberger 1998, 2000). Although it has been known for some time that dendritic cells (Huang et al. 2005; Wykes et al. 1998), follicular dendritic cells (Chen et al. 1978; Mandel et al. 1980; Tew et al. 1980), and macrophages (Koppel et al. 2005) can present intact antigen to B cells in vitro, very recent multiphoton microscopy investigations have verified the physiological significance of antigen on the surface of presenting cells in mediating B-cell activation in vivo (Carrasco and Batista 2007; Junt et al. 2007; Phan et al. 2007; Phan et al. 2009; Qi et al. 2006; Suzuki et al. 2009). In view of the predominance of membrane-bound antigen in initiating activation, the original hypotheses developed to describe BCR triggering in response to soluble antigen stimulation must be re-examined. Indeed, it has been observed that monovalent antigen attached to the surface of a presenting cell, unlike monovalent soluble antigen, can trigger B-cell activation (Tolar et al. 2009). Furthermore, the additional constraints that are imposed on the B cell to recognize antigen immobilized on the surface of a presenting cell must be considered. As T cells absolutely require the recognition of antigen in complex with MHC on the surface of a presenting cell, models developed to describe the initiation of T-cell activation may prove useful in the derivation of a general description of activation of B lymphocytes.

As the modules responsible for ligand binding and intracellular signaling in immunoreceptors such as the BCR and TCR are separated, the molecular mechanism by which the external binding of antigen is communicated across the membrane is not immediately obvious. Indeed, this has been the subject of a number of investigations and still remains controversial. Two most likely models have emerged to explain the triggering of activation in T cells (Fig. 2). The first of these is perhaps the more simple conceptually and proposes that the binding of antigen to the extracellular region is communicated to the intracellular signaling region through conformational change(s) in the intervening domains (Kuhns et al. 2006; Schamel et al. 2006) (Fig. 2A). Although high-resolution static crystallographic data have not provided evidence in support of this view (reviewed in Garcia et al. 1999; Hennecke and Wiley 2001; Rudolph et al. 2006; van der Merwe 2001), a number of recent studies have identified conformational changes in the CD3ε domain of the TCR following ligation such that CD3ε ITAM motifs are exposed to activating Src-family kinases and binding of intracellular effectors (Gil et al. 2002; Xu et al. 2008). The second model is known as kinetic-segregation and explains the initiation of T-cell triggering through the alterations in the membrane distribution of inhibitory phosphatases, such as CD45 (Choudhuri et al. 2005; Davis and van der Merwe 2006) (Fig. 2B). This model assumes that, before antigen stimulation, the TCR is at equilibrium such that it is equally likely to be phosphorylated or dephosphorylated. However, on contact with an antigen-presenting cell, small adhesion molecules such as CD2 on the T-cell surface

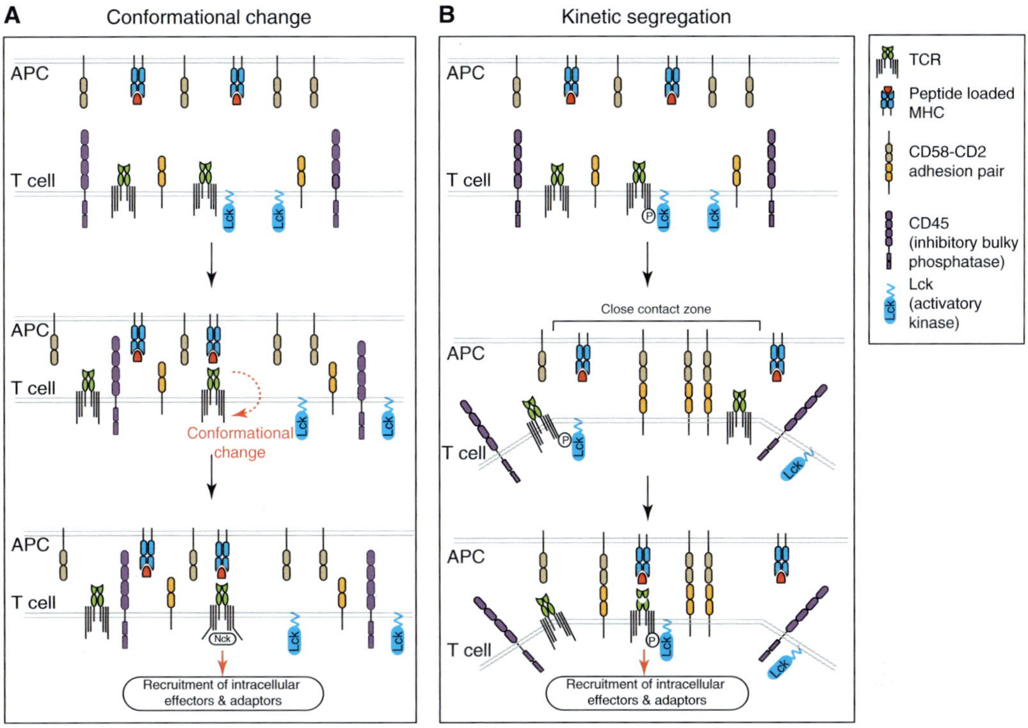

Figure 2. Two models of T-cell triggering. (*A*) The conformational change model states that extracellular pMHC binding to the TCR triggers intracellular conformational changes in the associated CD3 complex, opening up binding sites for the intracellular adaptor Nck. Subsequently, the "opened" TCR recruits activatory kinases (such as Lck) to phosphorylate the TCR, leading to assembly of the signalosome. (*B*) The kinetic segregation model postulates that, in the resting state, TCR is equally likely to be phosphorylated by activatory kinases or dephosphorylated inhibitory phosphatases. On proximity of an antigen-presenting cell, small adhesion molecules such as CD2 and CD58 establish a "close contact zone" that excludes bulky surface molecules such as CD45, but permits diffusion of TCR and Lck. TCR in the "close contact zone" is more likely to be found in a phosphorylated state, and the half-life of phosphorylated TCR in this region is increased following recognition of specific pMHC, triggering the assembly of the signalosome and T-cell activation.

establish close-contact zones that exclude more bulky cell-surface molecules, including CD45. Given that the extracellular domain of the TCR is relatively small, its diffusion into and out of the close contact zone is not restricted. Accordingly, TCR within the close contact zone is more likely to be phosphorylated because of the size-dependent exclusion of CD45. When the TCR residing in the close contact zone specifically recognizes antigen on the surface of the presenting cell, it will be held in this zone, increasing the lifetime of phosphorylated TCR, such that triggering occurs. There are a number of lines of experimental evidence supporting this viewpoint, including the importance of the balance of phosphorylation (Secrist et al. 1993) and the size-dependence of various extracellular domains for triggering T-cell activation (Choudhuri et al. 2005; Irles et al. 2003). Taken together, it seems likely that some features of both the conformational change and the kinetic-segregation models participate during the initiation of TCR triggering.

Can these two models prove useful in the description of analogous processes during B-cell activation? At this stage, there is no high-resolution structural data that supports a rapid conformational change to initiate triggering of the

BCR following antigen binding; however, this does not necessarily eliminate the possibility that conformational changes may play a role in the initiation of BCR triggering. Indeed, quantitative FRET imaging has revealed conformational changes in the cytoplasmic domains of the BCR complex that are dependent on Lyn recruitment and activity, and thus are assumed to occur later on during BCR signaling (Tolar et al. 2005). In terms of the kinetic-segregation model, it is clear that the balance of phosphorylation is important for the initiation of B-cell activation (Rolli et al. 2002). Interestingly, although CD45 is the predominant cell-surface phosphatase in T cells (Mustelin et al. 1989), B cells express significant amounts of another phosphatase CD148 (Zhu et al. 2008). Thus, though CD45-deficient B cells are not significantly impaired in their B-cell activation (Byth et al. 1996; Depoil et al. 2008; Kishihara et al. 1993), it is possible that the bulkier CD148 may play a redundant role in size-dependent exclusion during B-cell triggering. As such, it will be necessary to investigate B cells deficient in CD45 and CD148 to assess the validity of the kinetic-segregation model to describe the initiation of BCR triggering. Future imaging investigations will provide insight into the dynamic distribution of various molecules in the B-cell membrane, and thus shed light on the precise mechanism underlying BCR triggering.

BCR MICROCLUSTERS FORM AND MEDIATE INTRACELLULAR SIGNALING AFTER ANTIGEN STIMULATION

Although imaging methods have not yet provided a definitive molecular mechanism underlying the initiation of BCR triggering, they have proved incredibly informative in the characterization of early events of B-cell activation. Currently, the earliest observable event associated with successful antigen stimulation is the formation of microclusters of BCR in the membrane (Fig. 3). BCR microclusters have been visualized using TIRFM and time-lapse wide-field fluorescence microscopy following settling of B cells on planar lipid bilayers containing antigen (Depoil et al. 2008; Fleire et al. 2006). These microclusters are formed of approximately 50 to 500 BCR molecules and can be comprised of either or both of the BCR istoypes expressed by primary B cells, IgM, and IgD (Depoil et al. 2008).

Interestingly, the formation of BCR microclusters was not impaired in Lyn-deficient B cells (Weber et al. 2008), indicating that this process is not dependent on signaling through the BCR. In contrast, the formation of BCR microclusters is absolutely dependent on alterations in the underlying cytoskeleton (Depoil et al. 2008). In line with this, we have very recently observed that the ezrin-actin network underlying the B-cell membrane is reorganized following antigen stimulation, such that defined "corrals" form around BCR microclusters (B. Treanor and F.D. Batista, unpubl. data). Furthermore, as these corrals restrict BCR diffusion, it seems likely that the cytoskeleton network may contribute to the stability and integrity of BCR microclusters. Thus, it could be that BCR microclusters form as a result of diffusion trapping following restriction of diffusion by the B-cell cytoskeleton. In addition, we expect that other factors play a role in formation of BCR microclusters. In line with this, Pierce and colleagues observed that the membrane-proximal Cμ4 domains of IgM have the propensity to oligomerize, suggesting that these domains play an important role in microcluster formation (Tolar et al. 2009). The authors do, however, note that in order to mediate this function, the IgM molecules must first be confined within the membrane, potentially through the action of the B-cell cytoskeleton.

The assembly of BCR microclusters was observed to occur on a timescale coincident with the initiation of calcium signaling, suggesting that they may play a role in mediating intracellular signaling. In line with this, early observations indicated that microclusters of antigen often were coincident with staining for phosphotyrosine (Fleire et al. 2006) and excluded inhibitory phosphatases such as CD45 (Depoil et al. 2008). We used an approach involving the expression of fluorescently labeled intracellular effectors in B cells and observed the impact of settling on antigen-containing lipid bilayers by

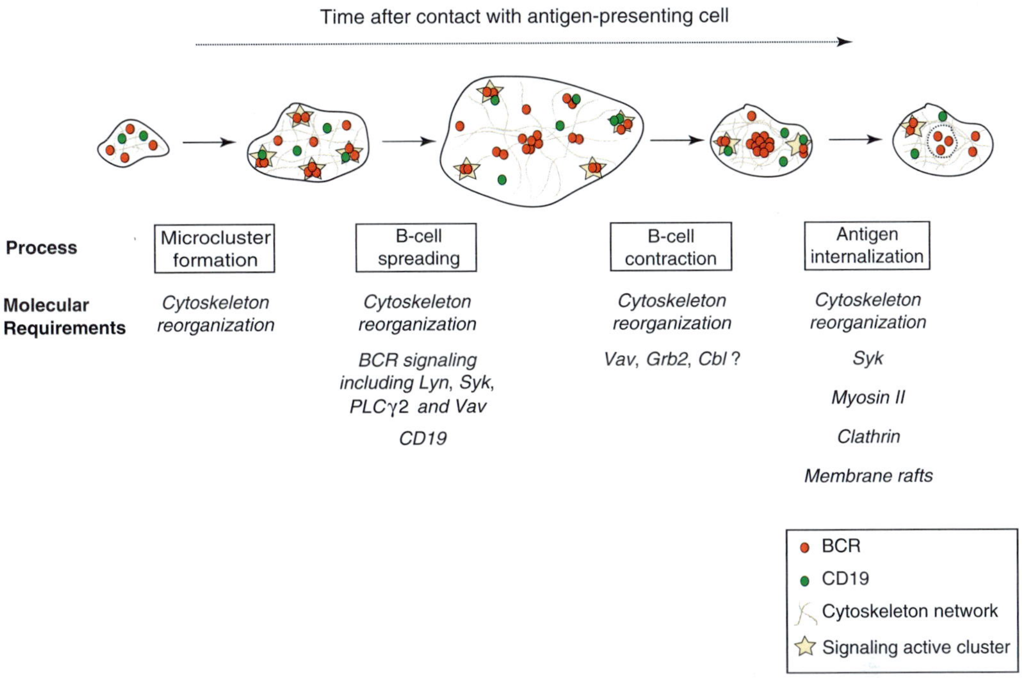

Figure 3. B-cell activation by antigen on the surface of a presenting cell. Schematic representation of the processes occurring at the B-cell surface following recognition of antigen on a presenting surface and what is currently known about the molecular requirements of these processes. On initial contact with antigen, BCR microclusters form, driving B-cell spreading. During spreading, signaling active BCR microclusters are moved (possibly via reterograde actin flow) toward the central cluster. Following achievement of maximum spread, the more prolonged contraction phase occurs, resulting in the formation of the mature immunological synapse. The central cluster of the immunological synapse acts as a platform for internalization of antigen. It appears that at every stage, cytoskeleton reorganizations play an important role in orchestrating and coordinating the molecular processes.

three-color TIRFM. In this way, we visualized the sequential and highly coordinated recruitment of the initiating kinases Lyn and Syk, followed by the downstream mediators Vav and PLCγ2 (Weber et al. 2008). A subsequent FRET-based investigation confirmed the recruitment of Lyn to BCR microclusters and suggested that membrane lipids may play a role in the regulation of this association (Sohn et al. 2008). In addition, we also showed that not only the recruitment but also the enzymatic activity of particular mediators was required to initiate calcium signaling (Weber et al. 2008). These observations indicate that BCR microclusters function to mediate intracellular signaling. As this is in agreement with previous observations concerning the role of TCR microclusters (Bunnell et al. 2002; Campi et al. 2005; Yokosuka et al. 2005), we have redefined microclusters as microsignalosomes and suggested that they are a common unit of signaling in lymphocytes (Depoil et al. 2008; Harwood and Batista 2008). Importantly, microsignalosomes allow intricate regulation of immunoreceptor signaling through the potential for dynamic recruitment of positive and negative regulators. In line with this, we have observed that components of microsignalosomes can cooperate to facilitate the recruitment and retention of the other (Weber et al. 2008). This type of elegant mechanism for regulating BCR signaling is not possible from "cap" structures previously observed following stimulation with soluble antigen (Schreiner et al. 1976; Stackpole et al.

1974; Unanue et al. 1972). Thus, given that the predominant form of antigen-mediating activation in vivo is membrane-bound, B cells can very precisely regulate the extent of activation according to the context in which the antigen is presented.

SPREADING PROPAGATES MICROSIGNALOSOMES NECESSARY FOR B-CELL ACTIVATION

Although microsignalosomes are required for mediating BCR signaling their formation alone, is not sufficient to elicit B-cell activation. In addition, it is necessary for the B-cell to amplify the number of microsignalosomes such that the threshold for antigen stimulation is exceeded (Weber et al. 2008). The B cell achieves this amplification by spreading out its membrane across the presenting cell, allowing contact with greater amounts of antigen on the surface (Fleire et al. 2006) (Fig. 3). As BCR engagement of antigen at the leading edge drives the spreading response, the extent of spreading depends on the affinity and density of antigen on the presenting surface and determines how much antigen the B cell accumulates for presentation to helper T cells (Fleire et al. 2006). As spreading is abrogated in the presence of inhibitors of actin polymerization and Src-family kinases, this cellular response absolutely requires intracellular signaling and cytoskeleton reorganization.

The signaling events downstream of the BCR important for mediating spreading have recently been investigated (Weber et al. 2008) through a screen of an existing panel of mutants in a chicken B-cell line (Shinohara and Kurosaki 2006). The DT40 B-cell line shows a high rate of homologous recombination, enabling the rapid generation of B cells deficient in a particular mediator (Kurosaki 1999). In addition, this strategy has the advantage of allowing investigation of molecules whose deletion is embryonically lethal in a knockout animal system. It was observed that Lyn and Syk were absolutely required for the initiation of B-cell spreading in response to antigen-containing membranes (Weber et al. 2008). Furthermore, although B cells deficient in Vav and PLCγ2 were severely impaired in mediating spreading, it is clear that these two mediators play a critical role in the propagation of the cellular response. B cells deficient in Btk and Blnk were markedly impaired in spreading, suggesting that the activation and retention of PLCγ2 is important during spreading. In contrast, B cells lacking all three IP_3 receptors were fully competent in mediating spreading, a somewhat surprising observation given the calcium-dependence of the analogous process in T cells (Bunnell et al. 2001).

Although the key intracellular mediators involved in B-cell spreading have been established, the molecular mechanisms underlying the necessary cytoskeleton rearrangements currently remain unclear. However, as Vav is known to be an important regulator of the cytoskeleton and has been established as critical for propagation of B-cell spreading (Weber et al. 2008), this offers insight into potential mechanisms for reorganization of the cytoskeleton. Alongside its role as a molecular adaptor, Vav functions as a guanine nucleotide exchange factor (GEF) to activate RhoGTPases such as Cdc42 and Rac that are involved in the regulation of cytoskeleton reorganizations (Jaffe and Hall 2005). In line with this, the GEF activity of Vav has been shown to be essential during B-cell spreading (Weber et al. 2008). In addition, in T cells, Vav has been implicated as a regulator of ezrin dephosphorylation in response to TCR stimulation (Faure et al. 2004). This is of particular interest as dephosphorylation of ezrin leads to release of the association between the plasma membrane and the actin cytoskeleton, facilitating the formation of cell conjugates. Interestingly, a rapid and global dephosphorylation of ezrin has been observed following stimulation of B cells with soluble antigen (Gupta et al. 2006). In conjunction with Vav, a number of other intracellular B-cell effectors have been implicated in the regulation of cytoskeleton reorganizations following stimulation of the BCR. These include the leukocyte-specific homolog of cortactin HS1 (Gomez et al. 2006; Hao et al. 2004; Yamanashi et al. 1997), cofilin

(DesMarais et al. 2005; Yonezawa et al. 1991), DOCK8 (Randall et al. 2009), and B-lymphocyte adaptor molecule of 32kDa (Bam32) (Fournier et al. 2003; Han et al. 2003; Marshall et al. 2000). Bam32 represents a particularly interesting candidate as it is recruited to the plasma membrane and phosphorylated following BCR ligation (Marshall et al. 2000), and B cells deficient in Bam32 are impaired in actin polymerization and antigen internalization (Niiro et al. 2004). Given the importance of the spreading response in shaping the outcome of B-cell activation, detailed investigations of these and other mediators will be invaluable in deriving a complete description of the molecular mechanism communicating BCR ligation with cytoskeleton reorganizations.

CD19 IS ESSENTIAL FOR B-CELL SPREADING AND ACTIVATION

During a dissection of the requirements for B-cell spreading, we uncovered a novel and unexpected role for the B-cell coreceptor CD19 (Depoil et al. 2008). Until this discovery, CD19 was most commonly known as a component of the CD21-CD19-CD81-leu13 complex responsible for enhanced BCR signaling in response to antigen coated with complement fragments (Fearon and Carroll 2000). In these circumstances, the ligation of CD21 (also known as complement receptor 2) triggers CD19 to recruit intracellular effectors that facilitate signaling through the BCR. However, the observation that CD19-deficient mice show more severely impaired B-cell responses to antigen than mice lacking CD21 (Ahearn et al. 1996; Engel et al. 1995; Rickert et al. 1995) suggests that CD19 performs an alternative role independent of CD21 during B-cell activation. In line with this, we observed that B cells lacking CD19 were unable to initiate calcium signaling and thus could not mediate spreading and activation in response to antigen on the surface of presenting cells (Depoil et al. 2008). Interestingly, and in agreement with previous investigations, we observed that CD19-deficient B cells respond to soluble antigen in a similar manner to wild-type B cells (Depoil et al. 2008; Fujimoto et al. 1999; Sato et al. 1997). These observations nicely complement recent intravital imaging investigations, demonstrating the central role of antigen bound to presenting cells as the predominant form triggering B-cell activation in vivo (Carrasco and Batista 2007; Junt et al. 2007; Phan et al. 2007; Phan et al. 2009; Qi et al. 2006; Suzuki et al. 2009).

Though the molecular mechanism by which CD19 mediates this essential role in B-cell spreading has not yet been fully elucidated, imaging strategies have allowed us to gain some insight into this process. The cytoplasmic domain of CD19 contains several motifs that can mediate the recruitment of numerous intracellular effectors, including PI3K and Vav (Li et al. 1997; Tuveson et al. 1993). Furthermore, it has been suggested that CD19 forms oligomers in the membrane (Brooks et al. 2004), potentially allowing cooperation and prolonged retention of recruited molecules. Thus, it seems plausible to suggest that CD19 mediates the recruitment and stabilization of additional intracellular effectors in the microsignalosome to facilitate signaling through the BCR. In support of this mechanism, it was observed that CD19 becomes transiently associated with BCR microclusters following stimulation with membrane-bound antigen (Depoil et al. 2008). More recently, we have used Dual-View TIRFM to visualize single particles of CD19 moving between signaling BCR microclusters following antigen stimulation (D. Depoil and F.D. Batista, unpubl. data). In addition, we observed that mediators recruited through CD19, such as PI3K, are laterally segregated from those recruited solely through the BCR in the B-cell membrane. Thus, at this stage, it appears that CD19 plays a role in B cells similar to that of LAT in T cells (Zhang and Samelson 2000), providing the boost to BCR signaling necessary to stimulate cytoskeleton reorganization. Interestingly, mutations in CD19 and other potential mediators of cytoskeleton reorganization have been associated with antibody deficiency syndromes in humans (Conley et al. 2009).

B-CELL CONTRACTION RESULTS IN FORMATION OF THE IMMUNOLOGICAL SYNAPSE

B cells attain maximum spread a few minutes after initial contact with an antigen-presenting cell and subsequently the B-cell membrane undergoes a more prolonged contraction phase (Fleire et al. 2006) (Fig. 3). The mechanism underlying the initiation of B-cell contraction remains unclear; however, a genetic screen to identify mediators involved in the spreading response revealed that B cells deficient in Vav, Grb2, or Cbl showed impaired contraction (Weber et al. 2008; T. Schnyder and F.D. Batista, unpubl. data). B-cell contraction results in the dramatic reorganization of the membrane forming a structure known as the immunological synapse (IS) (Batista et al. 2001; Carrasco et al. 2004). The IS was originally identified in $CD4^+$ T cells (Grakoui et al. 1999; Krummel et al. 2000; Monks et al. 1998) and has since become recognized as a common feature associated with lymphocyte activation (Davis et al. 1999; Potter et al. 2001; Stinchcombe et al. 2001). The IS comprises a cluster of immunoreceptors in a central supramolecular activation cluster (cSMAC) surrounded by a ring of adhesion molecules, such as the integrins LFA-1 and VLA-4, in the peripheral SMAC (pSMAC) (Monks et al. 1998). It is thought that during spreading, clusters of signaling immunoreceptors associated with the underlying actin cytoskeleton are passively translocated to the center of contact via retrograde actin flow to assemble the cSMAC (Kaizuka et al. 2007). Interestingly, this would provide an elegant mechanism to explain the segregation of molecules within the IS according to their particular association with components of the cytoskeleton. In addition, recent data suggested that in T cells, this transportation is also mediated through an active process involving the actin-based molecular motor myosin IIA (Ilani et al. 2009). However, the contribution of the two mechanisms or alternatives during contraction and formation of the IS in B cells remains to be established.

Though imaging was used to visualize the structure more than a decade ago, the function of the IS has remained controversial. It was originally put forward that given the concentration of immunoreceptors in the cSMAC, this would likely represent the site for active signaling. However, this assumption must be revisited given that intracellular calcium signaling is observed more rapidly than the formation of the mature IS, and in view of the more recent observation that microclusters of immunoreceptors actively signal the cell periphery during spreading (Bunnell et al. 2002; Campi et al. 2005; Depoil et al. 2008; Weber et al. 2008; Yokosuka et al. 2005). In line with this, a recent investigation tracking microcluster movement in T cells showed that slowing the movement of microclusters toward the cSMAC prolongs signaling through the TCR (Nguyen et al. 2008). Thus, these observations support the proposal that the cSMAC forms a platform for internalization of immunoreceptors and attenuation of signaling (Lee et al. 2002). However, it seems that the division of labor between components of the IS may not be absolute since under certain conditions, such as stimulation with weak agonists, signaling in the cSMAC can be observed (Čemerski et al. 2008).

The role of the IS in antigen internalization is of particular significance in B cells, as this is known to be critical for the recruitment of $CD4^+$ T-helper cells required for maximal B-cell activation (Batista et al. 2001). It is clear that the BCR functions not only to mediate intracellular signaling, but also to target antigen to the correct compartment for intracellular processing (Aluvihare et al. 1997; Amigorena et al. 1994). Indeed, it has been shown that uptake through the BCR represents a more efficient means to accumulate antigen for presentation to T cells compared with fluid-phase endocytosis (Aluvihare et al. 1997; Cheng et al. 1999). BCR engagement initiates the reorganization of intracellular compartments to form a central multivesicular compartment enriched in MHC-II, antigen, and accessory molecules necessary for processing (Boes et al. 2004; Drake et al. 1999; Lankar et al. 2002; Siemasko and Clark 2001; Siemasko et al. 1998; Vascotto et al. 2007). Though the precise molecular requirements for the BCR-mediated internalization of antigen are not yet clear, actin-depolymerizing agents disrupt antigen internalization, establishing a critical role for the actin cytoskeleton during this step of B-cell activation (Barois et al. 1998; Brown and Song 2001). As B cells deficient in Syk show altered intracellular trafficking and impaired presentation of antigen to $CD4^+$ T helper cells, it has been suggested that Syk plays a key role in regulating interactions between intracellular vesicles and actin filaments (Le Roux et al. 2007). In addition, the actin-based motor protein myosin II has been shown to be required for antigen processing and presentation (Vascotto et al. 2007). It has been put forward that most efficient BCR-mediated antigen internalization occurs when the actin cytoskeleton cooperates with clathrin- and membrane-raft-mediated pathways (Stoddart et al. 2005). Although it is clear that the IS effectively coordinates the processes enabling antigen internalization, future imaging investigations will be vital to establish the molecular mechanisms underlying antigen internalization, presentation, and thus B-cell activation.

CONCLUDING REMARKS

It is evident from our discussion of the molecular events of B-cell activation that the actin cytoskeleton plays a key role at each stage, including: regulating the distribution of BCR in the resting cell membrane; mediating the assembly and stability of BCR microclusters; underlying the cellular spreading and contraction responses; and finally coordinating antigen internalization and processing. Thus, the B cell uses the actin cytoskeleton as a framework to successfully manage numerous molecular interactions. Interestingly, the cytoskeleton structure provides a scaffold to establish polarization of intracellular effectors and compartments, such as has been observed before cell division in T cells (Chang et al. 2007). This is a particularly intriguing issue in lymphocyte biology, as this type of polarization can lead to asymmetric segregation of components after cell division, potentially allowing generation of two daughter cells with different effector functions.

ACKNOWLEDGMENTS

Cancer Research UK supports research in the Lymphocyte Interaction Laboratory. We would like to thank members of the Lymphocyte Interaction Laboratory for critical reading of the manuscript particularly B. Treanor for providing the imaging data shown in Figure 1B.

REFERENCES

Ahearn J, Fischer M, Croix D, Goerg S, Ma M, Xia J, Zhou X, Howard R, Rothstein T, Carroll M. 1996. Disruption of the Cr2 locus results in a reduction in B-1a cells and in an impaired B cell response to T-dependent antigen. *Immunity* **4:** 251–262.

Aluvihare VR, Khamlichi AA, Williams GT, Adorini L, Neuberger MS. 1997. Acceleration of intracellular targeting of antigen by the B-cell antigen receptor: Importance depends on the nature of the antigen-antibody interaction. *EMBO J* **16:** 3553–3562.

Amigorena S, Drake JR, Webster P, Mellman I. 1994. Transient accumulation of new class II MHC molecules in a novel endocytic compartment in B lymphocytes. *Nature* **369:** 113–120.

Barois N, Forquet F, Davoust J. 1998. Actin microfilaments control the MIIC class II antigen presentation pathway in B cells. *J Cell Sci* **111:** 1791–1800.

Batista F, Neuberger M. 1998. Affinity dependence of the B cell response to antigen: A threshold, a ceiling, and the importance of off-rate. *Immunity* **8:** 751–759.

Batista F, Neuberger M. 2000. B cells extract and present immobilized antigen: Implications for affinity discrimination. *EMBO J* **19:** 513–520.

Batista F, Iber D, Neuberger M. 2001. B cells acquire antigen from target cells after synapse formation. *Nature* **411:** 489–494.

Boes M, Cuvillier A, Ploegh H. 2004. Membrane specializations and endosome maturation in dendritic cells and B cells. *Trends Cell Biol* **14:** 175–183.

Brooks S, Kirkham P, Freeberg L, Carter R. 2004. Binding of cytoplasmic proteins to the CD19 intracellular domain is high affinity, competitive, and multimeric. *J Immunol* **172:** 7556–7564.

Brown BK, Song W. 2001. The actin cytoskeleton is required for the trafficking of the B cell antigen receptor to the late endosomes. *Traffic* **2:** 414–427.

Bunnell S, Hong D, Kardon J, Yamazaki T, McGlade C, Barr V, Samelson L. 2002. T cell receptor ligation induces the formation of dynamically regulated signaling assemblies. *J Cell Biol* **158:** 1263–1275.

Bunnell S, Kapoor V, Trible R, Zhang W, Samelson L. 2001. Dynamic actin polymerization drives T cell receptor-induced spreading: A role for the signal transduction adaptor LAT. *Immunity* **14:** 315–329.

Byth K, Conroy L, Howlett S, Smith A, May J, Alexander D, Holmes N. 1996. CD45-null transgenic mice reveal a positive regulatory role for CD45 in early thymocyte development, in the selection of CD4+CD8+ thymocytes, and B cell maturation. *J Exp Med* **183:** 1707–1718.

Čemerski S, Das J, Giurisato E, Markiewicz MA, Allen PM, Chakraborty AK, Shaw AS. 2008. The balance between T cell receptor signaling and degradation at the center of the immunological synapse is determined by antigen quality. *Immunity* **29:** 414–422.

Campi G, Varma R, Dustin M. 2005. Actin and agonist MHC-peptide complex-dependent T cell receptor microclusters as scaffolds for signaling. *J Exp Med* **202:** 1031–1036.

Carrasco Y, Batista F. 2007. B cells acquire particulate antigen in a macrophage-rich area at the boundary between the follicle and the subcapsular sinus of the lymph node. *Immunity* **27:** 160–171.

Carrasco Y, Fleire S, Cameron T, Dustin M, Batista F. 2004. LFA-1/ICAM-1 interaction lowers the threshold of B cell activation by facilitating B cell adhesion and synapse formation. *Immunity* **20:** 589–599.

Chang JT, Palanivel VR, Kinjyo I, Schambach F, Intlekofer AM, Banerjee A, Longworth SA, Vinup KE, Mrass P, Oliaro J, et al. 2007. Asymmetric T lymphocyte division in the initiation of adaptive immune responses. *Science* **315:** 1687–1691.

Chen L, Frank A, Adams J, Steinman R. 1978. Distribution of horseradish peroxidase (HRP)-anti-HRP immune complexes in mouse spleen with special reference to follicular dendritic cells. *J Cell Biol* **79:** 184–199.

Cheng PC, Steele CR, Gu L, Song W, Pierce SK. 1999. MHC class II antigen processing in B cells: accelerated intracellular targeting of antigens. *J Immunol* **162:** 7171–7180.

Choudhuri K, Wiseman D, Brown M, Gould K, van der Merwe P. 2005. T-cell receptor triggering is critically dependent on the dimensions of its peptide-MHC ligand. *Nature* **436:** 578–582.

Conley ME, Dobbs AK, Farmer DM, Kilic S, Paris K, Grigoriadou S, Coustan-Smith E, Howard V, Campana D. 2009. Primary B cell immunodeficiencies: comparisons and contrasts. *Annu Rev Immunol* **27:** 199–227.

Dal Porto J, Gauld S, Merrell K, Mills D, Pugh-Bernard A, Cambier J. 2004. B cell antigen receptor signaling 101. *Mol Immunol* **41:** 599–613.

Davis S, van der Merwe P. 2006. The kinetic-segregation model: TCR triggering and beyond. *Nat Immunol* **7:** 803–809.

Davis D, Chiu I, Fassett M, Cohen G, Mandelboim O, Strominger J. 1999. The human natural killer cell immune synapse. *Proc Natl Acad Sci* **96:** 15062–15067.

DeFranco A. 1997. The complexity of signaling pathways activated by the BCR. *Curr Opin Immunol* **9:** 296–308.

Depoil D, Fleire S, Treanor BL, Weber M, Harwood NE, Marchbank KL, Tybulewicz VL, Batista FD. 2008. CD19 is essential for B cell activation by promoting B cell receptor-antigen microcluster formation in response to membrane-bound ligand. *Nat Immunol* **9:** 63–72.

DesMarais V, Ghosh M, Eddy R, Condeelis J. 2005. Cofilin takes the lead. *J Cell Sci* **118:** 19–26.

Drake JR, Lewis TA, Condon KB, Mitchell RN, Webster P. 1999. Involvement of MIIC-like late endosomes in B cell receptor-mediated antigen processing in murine B cells. *J Immunol* **162:** 1150–1155.

Engel P, Zhou L, Ord D, Sato S, Koller B, Tedder T. 1995. Abnormal B lymphocyte development, activation, and differentiation in mice that lack or overexpress the CD19 signal transduction molecule. *Immunity* **3:** 39–50.

Faure S, Salazar-Fontana L, Semichon M, Tybulewicz V, Bismuth G, Trautmann A, Germain R, Delon J. 2004. ERM proteins regulate cytoskeleton relaxation promoting T cell-APC conjugation. *Nat Immunol* **5:** 272–279.

Fearon D, Carroll M. 2000. Regulation of B lymphocyte responses to foreign and self-antigens by the CD19/CD21 complex. *Annu Rev Immunol* **18:** 393–422.

Fleire S, Goldman J, Carrasco Y, Weber M, Bray D, Batista F. 2006. B cell ligand discrimination through a spreading and contraction response. *Science* **312:** 738–741.

Fournier E, Isakoff S, Ko K, Cardinale C, Inghirami G, Li Z, Curotto de Lafaille M, Skolnik E. 2003. The B cell SH2/PH domain-containing adaptor Bam32/DAPP1 is required for T cell-independent II antigen responses. *Curr Biol* **13:** 1858–1866.

Fujimoto M, Bradney A, Poe J, Steeber D, Tedder T. 1999. Modulation of B lymphocyte antigen receptor signal transduction by a CD19/CD22 regulatory loop. *Immunity* **11:** 191–200.

Fujiwara T, Ritchie K, Murakoshi H, Jacobson K, Kusumi A. 2002. Phospholipids undergo hop diffusion in compartmentalized cell membrane. *J Cell Biol* **157:** 1071–1081.

Garcia KC, Teyton L, Wilson IA. 1999. Structural basis of T cell recognition. *Annu Rev Immunol* **17:** 369–397.

Gil D, Schamel WW, Montoya M, Sánchez-Madrid F, Alarcón B. 2002. Recruitment of Nck by CD3 epsilon reveals a ligand-induced conformational change essential for T cell receptor signaling and synapse formation. *Cell* **109:** 901–912.

Gomez T, McCarney S, Carrizosa E, Labno C, Comiskey E, Nolz J, Zhu P, Freedman B, Clark M, Rawlings D, et al. 2006. HS1 functions as an essential actin-regulatory adaptor protein at the immune synapse. *Immunity* **24:** 741–752.

Grakoui A, Bromley S, Sumen C, Davis M, Shaw A, Allen P, Dustin M. 1999. The immunological synapse: A molecular machine controlling T cell activation. *Science* **285:** 221–227.

Gupta N, Wollscheid B, Watts J, Scheer B, Aebersold R, DeFranco A. 2006. Quantitative proteomic analysis of B cell lipid rafts reveals that ezrin regulates antigen receptor-mediated lipid raft dynamics. *Nat Immunol* **7:** 625–633.

Halin C, Rodrigo Mora J, Sumen C, von Andrian U. 2005. In vivo imaging of lymphocyte trafficking. *Annu Rev Cell Dev Biol* **21:** 581–603.

Han A, Saijo K, Mecklenbräuker I, Tarakhovsky A, Nussenzweig M. 2003. Bam32 links the B cell receptor to ERK and JNK and mediates B cell proliferation but not survival. *Immunity* **19:** 621–632.

Hao J, Carey G, Zhan X. 2004. Syk-mediated tyrosine phosphorylation is required for the association of hematopoietic lineage cell-specific protein 1 with lipid rafts and B cell antigen receptor signalosome complex. *J Biol Chem* **279:** 33413–33420.

Harwood NE, Batista FD. 2008. New insights into the early molecular events underlying B cell activation. *Immunity* **28:** 609–619.

Hennecke J, Wiley DC. 2001. T cell receptor-MHC interactions up close. *Cell* **104:** 1–4.

Huang N, Han S, Hwang I, Kehrl J. 2005. B cells productively engage soluble antigen-pulsed dendritic cells: Visualization of live-cell dynamics of B cell-dendritic cell interactions. *J Immunol* **175:** 7125–7134.

Ilani T, Vasiliver-Shamis G, Vardhana S, Bretscher A, Dustin ML. 2009. T cell antigen receptor signaling and immunological synapse stability require myosin IIA. *Nat Immunol* **10:** 531–539.

Irles C, Symons A, Michel F, Bakker T, van der Merwe P, Acuto O. 2003. CD45 ectodomain controls interaction with GEMs and Lck activity for optimal TCR signaling. *Nat Immunol* **4:** 189–197.

Jaffe A, Hall A. 2005. Rho GTPases: Biochemistry and biology. *Annu Rev Cell Dev Biol* **21:** 247–269.

Junt T, Scandella E, Ludewig B. 2008. Form follows function: Lymphoid tissue microarchitecture in antimicrobial immune defence. *Nat Rev Immunol* **8:** 764–775.

Junt T, Moseman E, Iannacone M, Massberg S, Lang P, Boes M, Fink K, Henrickson S, Shayakhmetov D, Di Paolo N, et al. 2007. Subcapsular sinus macrophages in lymph nodes clear lymph-borne viruses and present them to antiviral B cells. *Nature* **450:** 110–114.

Kaizuka Y, Douglass AD, Varma R, Dustin ML, Vale RD. 2007. Mechanisms for segregating T cell receptor and adhesion molecules during immunological synapse formation in Jurkat T cells. *Proc Natl Acad Sci* **104:** 20296–20301.

Kishihara K, Penninger J, Wallace V, Kündig T, Kawai K, Wakeham A, Timms E, Pfeffer K, Ohashi P, Thomas M. 1993. Normal B lymphocyte development but impaired T cell maturation in CD45-exon6 protein tyrosine phosphatase-deficient mice. *Cell* **74:** 143–156.

Koppel E, Wieland C, van den Berg V, Litjens M, Florquin S, van Kooyk Y, van der Poll T, Geijtenbeek T. 2005. Specific ICAM-3 grabbing nonintegrin-related 1 (SIGNR1) expressed by marginal zone macrophages is essential for defense against pulmonary *Streptococcus pneumoniae* infection. *Eur J Immunol* **35:** 2962–2969.

Krummel M, Sjaastad M, Wülfing C, Davis M. 2000. Differential clustering of CD4 and CD3ζ during T cell recognition. *Science* **289:** 1349–1352.

Kuhns MS, Davis MM, Garcia KC. 2006. Deconstructing the form and function of the TCR/CD3 complex. *Immunity* **24:** 133–139.

Kurosaki T. 1999. Genetic analysis of B cell antigen receptor signaling. *Annu Rev Immunol* **17:** 555–592.

Kurosaki T. 2002. Regulation of B-cell signal transduction by adaptor proteins. *Nat Rev Immunol* **2:** 354–363.

Kusumi A, Ike H, Nakada C, Murase K, Fujiwara T. 2005. Single-molecule tracking of membrane molecules: Plasma membrane compartmentalization and dynamic assembly of raft-philic signaling molecules. *Semin Immunol* **17:** 3–21.

Lankar D, Vincent-Schneider H, Briken V, Yokozeki T, Raposo G, Bonnerot C. 2002. Dynamics of major histocompatibility complex class II compartments during B cell receptor-mediated cell activation. *J Exp Med* **195:** 461–472.

Lanzavecchia A. 1985. Antigen-specific interaction between T and B cells. *Nature* **314:** 537–539.

Le Roux D, Lankar D, Yuseff M-I, Vascotto F, Yokozeki T, Faure-André G, Mougneau E, Glaichenhaus N, Manoury B, Bonnerot C, et al. 2007. Syk-dependent actin dynamics regulate endocytic trafficking and processing of antigens internalized through the B-cell receptor. *Mol Biol Cell* **18:** 3451–3462.

Lee K, Holdorf A, Dustin M, Chan A, Allen P, Shaw A. 2002. T cell receptor signaling precedes immunological synapse formation. *Science* **295:** 1539–1542.

Li X, Sandoval D, Freeberg L, Carter R. 1997. Role of CD19 tyrosine 391 in synergistic activation of B lymphocytes by coligation of CD19 and membrane Ig. *J Immunol* **158:** 5649–5657.

Lillemeier BF, Pfeiffer JR, Surviladze Z, Wilson BS, Davis MM. 2006. Plasma membrane-associated proteins are clustered into islands attached to the cytoskeleton. *Proc Natl Acad Sci* **103:** 18992–18997.

MacLennan I. 1994. Germinal centers. *Annu Rev Immunol* **12:** 117–139.

MacLennan IC, Toellner KM, Cunningham AF, Serre K, Sze DM, Zúñiga E, Cook MC, Vinuesa CG. 2003. Extrafollicular antibody responses. *Immunol Rev* **194:** 8–18.

Mandel T, Phipps R, Abbot A, Tew J. 1980. The follicular dendritic cell: long term antigen retention during immunity. *Immunol Rev* **53:** 29–59.

Marshall A, Niiro H, Lerner C, Yun T, Thomas S, Disteche C, Clark E. 2000. A novel B lymphocyte-associated adaptor protein, Bam32, regulates antigen receptor signaling downstream of phosphatidylinositol 3-kinase. *J Exp Med* **191:** 1319–1332.

Monks C, Freiberg B, Kupfer H, Sciaky N, Kupfer A. 1998. Three-dimensional segregation of supramolecular activation clusters in T cells. *Nature* **395:** 82–86.

Mustelin T, Coggeshall K, Altman A. 1989. Rapid activation of the T-cell tyrosine protein kinase pp56lck by the CD45 phosphotyrosine phosphatase. *Proc Natl Acad Sci* **86:** 6302–6306.

Nguyen K, Sylvain NR, Bunnell SC. 2008. T cell costimulation via the integrin VLA-4 inhibits the actin-dependent centralization of signaling microclusters containing the adaptor SLP-76. *Immunity* **28:** 810–821.

Niiro H, Allam A, Stoddart A, Brodsky F, Marshall A, Clark E. 2004. The B lymphocyte adaptor molecule of 32 kilodaltons (Bam32) regulates B cell antigen receptor internalization. *J Immunol* **173:** 5601–5609.

Phan T, Grigorova I, Okada T, Cyster J. 2007. Subcapsular encounter and complement-dependent transport of immune complexes by lymph node B cells. *Nat Immunol* **8:** 992–1000.

Phan TG, Green JA, Gray EE, Xu Y, Cyster JG. 2009. Immune complex relay by subcapsular sinus macrophages and noncognate B cells drives antibody affinity maturation. *Nat Immunol* **10:** 786–793.

Potter T, Grebe K, Freiberg B, Kupfer A. 2001. Formation of supramolecular activation clusters on fresh ex vivo CD8+ T cells after engagement of the T cell antigen receptor and CD8 by antigen-presenting cells. *Proc Natl Acad Sci* **98:** 12624–12629.

Qi H, Egen JG, Huang AY, Germain RN. 2006. Extrafollicular activation of lymph node B cells by antigen-bearing dendritic cells. *Science* **312:** 1672–1676.

Rajewsky K. 1996. Clonal selection and learning in the antibody system. *Nature* **381:** 751–758.

Randall K, Lambe T, Johnson A, Kucharska E, Domaschenz H, Whittle B, Tze L, Enders A, Crockford T, Bouriez-Jones T, et al. 2009. Dock8 mutations cripple B cell immunological synapses, germinal centers and long-lived antibody production. *Nat Immunol* **10:** 1283–1291.

Reth M. 1989. Antigen receptor tail clue. *Nature* **338:** 383–384.

Rickert R, Rajewsky K, Roes J. 1995. Impairment of T-cell-dependent B-cell responses and B-1 cell development in CD19-deficient mice. *Nature* **376:** 352–355.

Rock K, Benacerraf B, Abbas A. 1984. Antigen presentation by hapten-specific B lymphocytes. I. Role of surface immunoglobulin receptors. *J Exp Med* **160:** 1102–1113.

Rolli V, Gallwitz M, Wossning T, Flemming A, Schamel W, Zürn C, Reth M. 2002. Amplification of B cell antigen receptor signaling by a Syk/ITAM positive feedback loop. *Mol Cell* **10:** 1057–1069.

Rudolph MG, Stanfield RL, Wilson IA. 2006. How TCRs bind MHCs, peptides, and coreceptors. *Annu Rev Immunol* **24:** 419–466.

Sako Y, Kusumi A. 1994. Compartmentalized structure of the plasma membrane for receptor movements as revealed by a nanometer-level motion analysis. *J Cell Biol* **125:** 1251–1264.

Sato S, Miller A, Howard M, Tedder T. 1997. Regulation of B lymphocyte development and activation by the CD19/CD21/CD81/Leu 13 complex requires the cytoplasmic domain of CD19. *J Immunol* **159:** 3278–3287.

Schamel W, Reth M. 2000. Monomeric and oligomeric complexes of the B cell antigen receptor. *Immunity* **13:** 5–14.

Schamel WW, Risueño RM, Minguet S, Ortíz AR, Alarcón B. 2006. A conformation- and avidity-based proofreading mechanism for the TCR-CD3 complex. *Trends Immunol* **27:** 176–182.

Scharenberg A, Humphries L, Rawlings D. 2007. Calcium signalling and cell-fate choice in B cells. *Nat Rev Immunol* **7:** 778–789.

Schreiner GF, Braun J, Unanue ER. 1976. Spontaneous redistribution of surface immunoglobulin in the motile B lymphocyte. *J Exp Med* **144:** 1683–1688.

Secrist JP, Burns LA, Karnitz L, Koretzky GA, Abraham RT. 1993. Stimulatory effects of the protein tyrosine phosphatase inhibitor, pervanadate, on T-cell activation events. *J Biol Chem* **268:** 5886–5893.

Shinohara H, Kurosaki T. 2006. Genetic analysis of B cell signaling. *Subcell Biochem* **40:** 145–187.

Siemasko K, Clark MR. 2001. The control and facilitation of MHC class II antigen processing by the BCR. *Curr Opin Immunol* **13:** 32–36.

Siemasko K, Eisfelder BJ, Williamson E, Kabak S, Clark MR. 1998. Cutting edge: Signals from the B lymphocyte antigen receptor regulate MHC class II containing late endosomes. *J Immunol* **160:** 5203–5208.

Simons K, Ikonen E. 1997. Functional rafts in cell membranes. *Nature* **387:** 569–572.

Simson R, Sheets ED, Jacobson K. 1995. Detection of temporary lateral confinement of membrane proteins using single-particle tracking analysis. *Biophys J* **69:** 989–993.

Singer SJ, Nicolson GL. 1972. The fluid mosaic model of the structure of cell membranes. *Science* **175:** 720–731.

Sohn H, Pierce S, Tzeng S. 2008. Live cell imaging reveals that the inhibitory Fc{gamma}RIIB destabilizes B cell receptor membrane-lipid interactions and blocks immune synapse formation. *J Immunol* **180:** 793–799.

Stackpole CW, Jacobson JB, Lardis MP. 1974. Two distinct types of capping of surface receptors on mouse lymphoid cells. *Nature* **248:** 232–234.

Stinchcombe J, Bossi G, Booth S, Griffiths G. 2001. The immunological synapse of CTL contains a secretory domain and membrane bridges. *Immunity* **15:** 751–761.

Stoddart A, Jackson AP, Brodsky FM. 2005. Plasticity of B cell receptor internalization upon conditional depletion of clathrin. *Mol Biol Cell* **16:** 2339–2348.

Suzuki K, Grigorova I, Phan TG, Kelly LM, Cyster JG. 2009. Visualizing B cell capture of cognate antigen from follicular dendritic cells. *J Exp Med* **206:** 1485–1493.

Tew JG, Phipps RP, Mandel TE. 1980. The maintenance and regulation of the humoral immune response: Persisting antigen and the role of follicular antigen-binding dendritic cells as accessory cells. *Immunol Rev* **53:** 175–201.

Tolar P, Sohn H, Pierce S. 2005. The initiation of antigen-induced B cell antigen receptor signaling viewed in living cells by fluorescence resonance energy transfer. *Nat Immunol* **6:** 1168–1176.

Tolar P, Hanna J, Krueger PD, Pierce SK. 2009. The constant region of the membrane immunoglobulin mediates B cell-receptor clustering and signaling in response to membrane antigens. *Immunity* **30:** 44–55.

Treanor B, Batista F. 2007. Mechanistic insight into lymphocyte activation through quantitative imaging and theoretical modelling. *Curr Opin Immunol* **19:** 476–483.

Treanor B, Depoil D, Gonzalez-Granja A, Barral P, Weber M, Dushek O, Bruckbauer A, Batista FD. 2010. The membrane skeleton controls diffusion dynamics and signaling through the B cell receptor. *Immunity* **32:** 187–199.

Tuveson D, Carter R, Soltoff S, Fearon D. 1993. CD19 of B cells as a surrogate kinase insert region to bind phosphatidylinositol 3-kinase. *Science* **260:** 986–989.

Unanue ER, Perkins WD, Karnovsky MJ. 1972. Ligand-induced movement of lymphocyte membrane macromolecules. I. Analysis by immunofluorescence and ultrastructural radioautography. *J Exp Med* **136:** 885–906.

van der Merwe PA. 2001. The TCR triggering puzzle. *Immunity* **14:** 665–668.

Vascotto F, Lankar D, Faure-André G, Vargas P, Diaz J, Le Roux D, Yuseff M-I, Sibarita J-B, Boes M, Raposo G, et al. 2007. The actin-based motor protein myosin II regulates MHC class II trafficking and BCR-driven antigen presentation. *J Cell Biol* **176:** 1007–1019.

Weber M, Treanor B, Depoil D, Shinohara H, Harwood NE, Hikida M, Kurosaki T, Batista FD. 2008. Phospholipase C-γ2 and Vav cooperate within signaling microclusters to propagate B cell spreading in response to membrane-bound antigen. *J Exp Med* **205:** 853–868.

Wykes M, Pombo A, Jenkins C, MacPherson G. 1998. Dendritic cells interact directly with naive B lymphocytes to transfer antigen and initiate class switching in a primary T-dependent response. *J Immunol* **161:** 1313–1319.

Xu C, Gagnon E, Call ME, Schnell JR, Schwieters CD, Carman CV, Chou JJ, Wucherpfennig KW. 2008. Regulation of T cell receptor activation by dynamic membrane binding of the CD3epsilon cytoplasmic tyrosine-based motif. *Cell* **135:** 702–713.

Yamanashi Y, Fukuda T, Nishizumi H, Inazu T, Higashi K, Kitamura D, Ishida T, Yamamura H, Watanabe T, Yamamoto T. 1997. Role of tyrosine phosphorylation of HS1 in B cell antigen receptor-mediated apoptosis. *J Exp Med* **185:** 1387–1392.

Yokosuka T, Sakata-Sogawa K, Kobayashi W, Hiroshima M, Hashimoto-Tane A, Tokunaga M, Dustin M, Saito T. 2005. Newly generated T cell receptor microclusters initiate and sustain T cell activation by recruitment of Zap70 and SLP-76. *Nat Immunol* **6:** 1253–1262.

Yonezawa N, Homma Y, Yahara I, Sakai H, Nishida E. 1991. A short sequence responsible for both phosphoinositide binding and actin binding activities of cofilin. *J Biol Chem* **266:** 17218–17221.

Zhang W, Samelson L. 2000. The role of membrane-associated adaptors in T cell receptor signalling. *Semin Immunol* **12:** 35–41.

Zhu J, Brdicka T, Katsumoto T, Lin J, Weiss A. 2008. Structurally distinct phosphatases CD45 and CD148 both regulate B cell and macrophage immunoreceptor signaling. *Immunity* **28:** 183–196.

Src-Family and Syk Kinases in Activating and Inhibitory Pathways in Innate Immune Cells: Signaling Cross Talk

Clifford A. Lowell

Department of Laboratory Medicine, University of California, San Francisco, California 94143

Correspondence: clifford.lowell@ucsf.edu

The response of innate immune cells to growth factors, immune complexes, extracellular matrix proteins, cytokines, pathogens, cellular damage, and many other stimuli is regulated by a complex net of intracellular signal transduction pathways. The majority of these pathways are either initiated or modulated by Src-family or Syk tyrosine kinases present in innate cells. The Src-family kinases modulate the broadest range of signaling responses, including regulating immunoreceptors, C-type lectins, integrins, G-protein-coupled receptors, and many others. Src-family kinases also modulate the activity of other kinases, including the Tec-family members as well as FAK and Pyk2. Syk kinase is required for initiation of signaling involving receptors that utilize immunoreceptor tyrosine activation (ITAM) domains. This article reviews the major activating and inhibitory signaling pathways regulated by these cytoplasmic tyrosine kinases, illuminating the many examples of signaling cross talk between pathways.

Innate immune cells, including macrophages, dendritic cells, granulocytes, and mast cells, function as the first line of defense against pathogens. These cells use a dizzying array of cell-surface receptors, which are connected to an equally complicated intracellular signal transduction network, to sense pathogen molecules and then orchestrate the appropriate immune response. Among the intracellular signaling molecules that are most crucial for innate immune cells are the cytoplasmic tyrosine kinases. Two major kinase families that operate in the proximal intracellular signaling pathways in innate cells are the Src-family kinases and the Syk-ZAP70 family. A third family of kinases, the Tek family, also have important roles in innate cells. They are not discussed in detail in this article, but are reviewed elsewhere in articles on the subject.

There are eight members of the Src family; innate immune cells primarily express Hck, Fgr, Lyn, and to a lesser extent, Src (Lowell 2004). The Syk-ZAP70 family has only two members and only Syk is found in innate cells. Most innate cell types express the same spectrum of kinases with some specific cellular differences. For example, mast cells express a broader range of Src-family kinases than macrophages or dendritic cells (Colgan and Hankel 2010). In general, Src-family and Syk kinases tend to operate together in signaling pathways,

 doi: 10.1101/cshperspect.a002352
Cite this article as *Cold Spring Harb Perspect Biol* 2011;3:a002352

with the Src-family being "upstream" or activated first in response to pathogen detection. These enzymes then communicate downstream to Tec-family members. The Tec-family kinases expressed in innate cells include Btk, Bmx, and Tec (Koprulu and Ellmeier 2009; Tohyama and Yamamura 2009). Additionally, Src-family kinases activate yet another family of PTKs, the FAK/Pyk2 tyrosine kinases, which play a major role in integrin signaling (Hauck et al. 2000).

Though primarily studied in activating pathways, Src-family and Syk kinases also activate inhibitory signaling pathways (Nimmerjahn and Ravetch 2008). In many situations, inhibitory signaling often overrides the activating signal. Pathways can also be initiated at different times or rates. Finally, to add even more complexity, activating and inhibitory pathways often interact indirectly, for example, through the production of cytokines and growth factors and not through direct intracellular biochemical interactions; Hence the term signaling "cross talk," which now appears commonly in the literature (O'Neill 2008; Ivashkiv 2009; Page et al. 2009).

OVERVIEW OF ACTIVATING PATHWAYS

Classical Immunoreceptor Pathways

In the prototypical immunoreceptor pathway, engagement of the receptor leads to activation of Src-family kinases, which in turn phosphorylate immunoreceptor tyrosine-based activation motifs (ITAMs) present on either the receptor or associated subunits (Fig. 1). This leads to recruitment of Syk, by binding of the Syk SH2 protein domain to the phospho-ITAM residues, and activation of Syk allowing it to phosphorylate downstream substrates. One of the enzymes activated downstream is phosphoinositide 3-kinase (PI3-kinase), which generates membrane-associated phosphatidylinositol (3,4,5)-triphosphate (PIP3). The FAK/Pyk2 kinases are activated directly via the Src-family kinases, where they contribute to downstream responses involving cell adhesion and migration. Though usually depicted as a linear signaling pathway with Src-kinases at the top and FAK/Pyk2 at the bottom, there are many points of interaction and cross regulation. Together, these pathways impinge on downstream factors, such as MAPK kinases, which have broad effects on gene transcription; the Rac/Rho pathway to modulate cytoskeletal function; the inositol trisphosphate (IP3), and diacylglycerol pathway (DAG), which regulates Ca^{2+} entry into cells and activation of various isoforms of protein kinase C (PKC). Overall, the outline of the prototypical immunoreceptor pathway as described here is similar in both innate and adaptive immune cells (Smith-Garvin et al. 2009; Kurosaki et al. 2010).

New Concepts in Immunoreceptor Signaling: CARD9 and Receptor Diversity

Recent and exciting developments in the immunoreceptor paradigm include recent progress in delineating how this pathway is connected to NF-κB and the demonstration that many innate immune receptors utilize the "immunoreceptor pathway" even though they lack ITAMs and therefore are not technically immunoreceptors.

The adapter protein CARD9, which contains a caspase-recruitment domain (CARD), has now been shown to be the link between a variety of ITAM-containing receptors involved in recognition of fungal and probably other pathogen structures and NF-κB (Fig. 2) (Gross et al. 2006; Gross et al. 2009). CARD9 is closely related to the lymphocyte protein CARMA-1, which forms a complex with the adapter proteins Bcl-10 and MALT1, and thus links the T- and B-cell receptors to the NF-κB pathway (Rawlings et al. 2006). CARD9 forms the same complex in innate cells. In lymphoid cells, CARMA-1 is activated by PKC isoforms (PKCθ in T-cells and PKCβ in B-cells), which phosphorylate CARMA-1, resulting in a conformational change that allows it to interact with IKK, leading to IκB turnover and NF-κB activation. In myeloid cells, it remains unclear if CARD9 activation is directly downstream of PKC activation (Hara and Saito 2009). Nevertheless, it is clear that in the absence of CARD9, receptors involved in fungal pathogen recognition (Dectin-1, Dectin-2) are unable to activate NF-κB, and more importantly, the

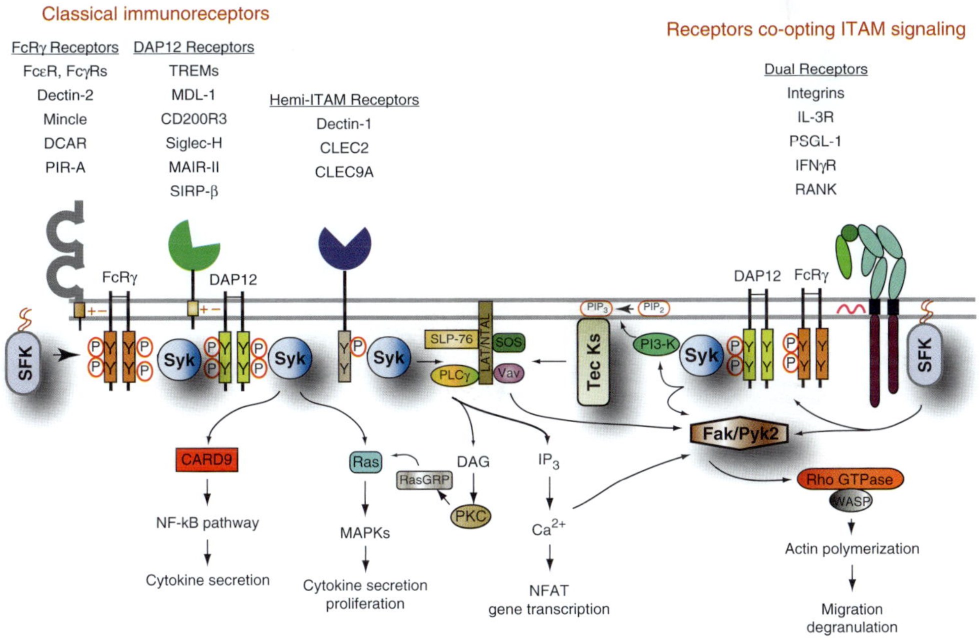

Figure 1. Cytoplasmic tyrosine kinases in the activating signaling pathways utilizing ITAM-containing adapters. Examples of immunoreceptors, hemi-ITAM C-type lectin receptors, and nonimmunoreceptors that utilize ITAM-signaling adapters and the cytoplasmic tyrosine kinases (indicated as shaded molecules) discussed in this article are shown. "Classical immunoreceptors" refers to those signaling molecules that are directly coupled to ITAM adapters FcRγ (group shown on the left) or DAP12 (group shown on the right) through transmembrane charged residues, shown as "+" and "−" in the figure. These immunoreceptors consist of immunoglobulin superfamily-containing proteins (such as the Fc receptors, PIR-A, or the TREMs) or the C-type lectin receptors (Dectin-2, Mincle, or MDL-1). In some cases, receptors may utilize either signaling adapter (see Hamerman et al. 2009). Examples of the C-type lectin receptors that have ITAM-like sequences as imbedded domains within their cytoplasmic tails are Dectin-1, CLEC2, and CLEC9A. The sequences with the ITAM domain of these receptors differs in that the membrane distal tyrosine resides in a YxxxL sequence (or YxxxI for mouse), as opposed to the traditional YxxL ITAM sequence found around the proximal tyrosine. As a result, the membrane distal tyrosine is dispensable for signaling, leading to the designation of these receptors as "hemi-ITAM" molecules (Kerrigan and Brown 2010). For a more complete list of FcRγ- and DAP12-associated receptors, see Lanier (2009), Graham and Brown (2009), and Kanazawa (2007). Not shown is the human FcγRIIA receptor, which is unique among the Fc receptors for having an ITAM sequence directly imbedded in its cytoplasmic tail (see Nimmerjahn and Ravetch 2008). To the right are shown examples of receptors that link to or co-opt the ITAM pathway, with the best example being the leukocyte integrins (see Abram and Lowell 2009). For all of these receptors, it remains unclear how they are coupled to the ITAM adapters; hence this association is indicated as a "∼" within the membrane region. Despite the difference in the coupling of these receptors to the initial activating pathways, the overall signaling events that follow receptor engagement by their respective ligands are quite similar (as an examples, see Gilfillan and Rivera 2009; Mocsai et al. 2010). The first step involves activation of Src-family kinases (shown as "SFK"), which are anchored in the membrane by their N-terminal acetylation sites. The SFK phosphorylate the ITAM adapters, leading to docking sites for Syk kinase, which then phosphorylates a number of substrates, including signaling scaffolding proteins such as SLP76, LAT, or NTAL, which in turn recruit other molecules (Vav family members) to initiate downstream responses. Syk function is also critical for PI3-kinase activation, which in turn leads to generation of membrane-bound PIP3 lipids that serve as membrane binding sites for the Tec-family kinases. Tec-family members contribute to scaffold protein phosphorylation. Together, these three cytoplasmic kinases contribute to the main downstream pathways of Ca^{2+} signaling, MAPK activation, and NF-κB activation. Syk and Src-family kinases also contribute to activation of the FAK/Pyk2 kinases that feed primarily into the Rho/WASP pathway of actin polymerization, leading to cytoskeletal changes in innate immune cells required for adhesion, migration, and degranulation responses.

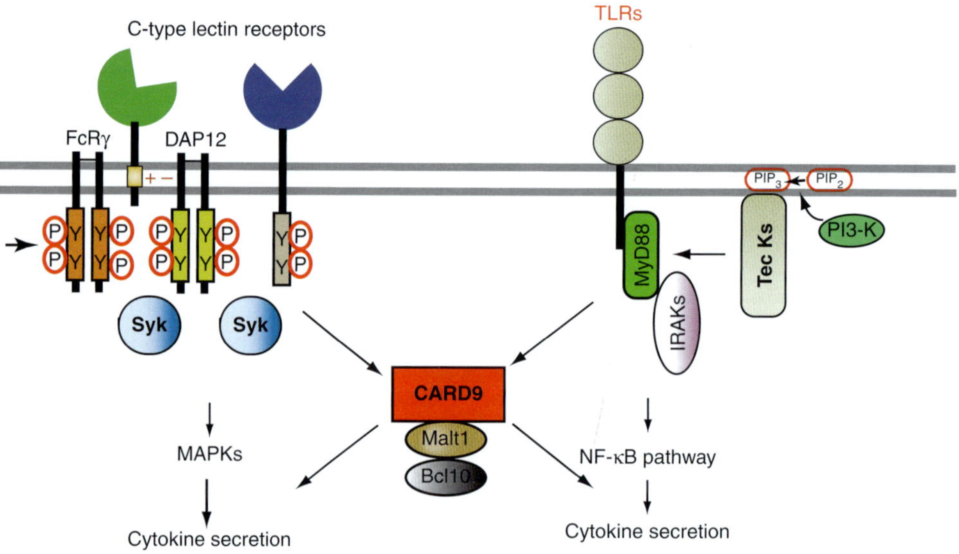

Figure 2. Tyrosine kinase pathways leading to CARD9 and TLR signaling responses. The central role of the CARD9/Malt1/Bcl10 complex in activating primarily the NF-κB pathway and to a lesser extent the MAPK pathway is shown (see Colonna 2007). Engagement of C-type lectin receptors, leading to Syk activation, is coupled to the CARD9 complex through an unclear pathway, but may involve PKCs. This in turn directly leads to activation of IKK, which in turn activates downstream NF-κB pathways (see Hara and Saito 2009). The importance of this pathway in innate immune cell recognition has been revealed by the similar defect in host defense in mice lacking either the C-type lectin receptors, Syk, or CARD9 (Gross et al. 2006). The CARD9 complex may also facilitate the ability of the TLRs to engage downstream MAPK pathway activation. Upstream of the TLRs, the major cytoplasmic tyrosine kinase involved in this innate immune receptor function is Tec-family members, as revealed by poor TLR signaling function in cells derived from either mice or humans lacking various Tec kinases. It remains unclear how the Tec kinases operate in the TLR pathway, but could involve phosphorylation of signaling adapters MyD88 or Mal.

entire repertoire of ITAM-containing receptors in myeloid cells are uncoupled from NF-κB (Robinson et al. 2006; Yamasaki et al. 2008). This results in profound defects in cytokine responses, which in vivo translates to poor responses to pathogens, specifically fungi such as *Candida albicans*, bacteria such as *Listeria monocytogenes*, and *M. tuberculosis* (Ruland 2008; Dorhoi et al. 2010).

Immunoreceptor signaling has also recently been shown to play a role downstream of innate receptors that lack ITAM sequences. This concept evolved from studies of innate cells lacking the ITAM-signaling adapters DAP12 and the FcεRIγ chain (referred to as FcRγ) (Nimmerjahn and Ravetch 2008; Lanier 2009). Typically, DAP12 and FcRγ are coupled to immunoreceptors through charged amino acid interactions within the transmembrane regions of each protein (Fig. 1). Surprisingly, a number of nonimmunoreceptor pathways, such as neutrophil integrin signaling or IL-3 responses in basophils, are lost in innate cells derived from mice lacking DAP12 or FcRγ (Mocsai et al. 2006; Hida et al. 2009). Examples of receptors that coopt the classical immunoreceptor pathway are shown in Figure 1. The mechanisms by which these nonimmunoreceptors couple to DAP12 and/or FcRγ to initiate signaling remain unclear.

OVERVIEW OF INHIBITORY PATHWAYS

Classical Inhibitory Pathways

Most inhibitory signals are mediated in a fashion very similar to the immunoreceptor pathway, except that inhibitory pathways are initiated through phosphorylation of immunoreceptor tyrosine-based inhibitory motifs (ITIMs) defined by the sequence amino acids I/V/L/SxYxxL/V (Fig. 3) (Munitz 2010). The Src-family kinase Lyn is primarily responsible for ITIM phosphorylation. Phospho-ITIM domains serve as docking sites for several types of phosphatases, such as SHP-1 (Src homology 2 domain containing protein tyrosine phosphatase-1/2) or SHIP-1 (SH-2-containing inositol phosphatase), which then down-modulate signaling responses by dephosphorylating downstream substrates. Like the activating pathways,

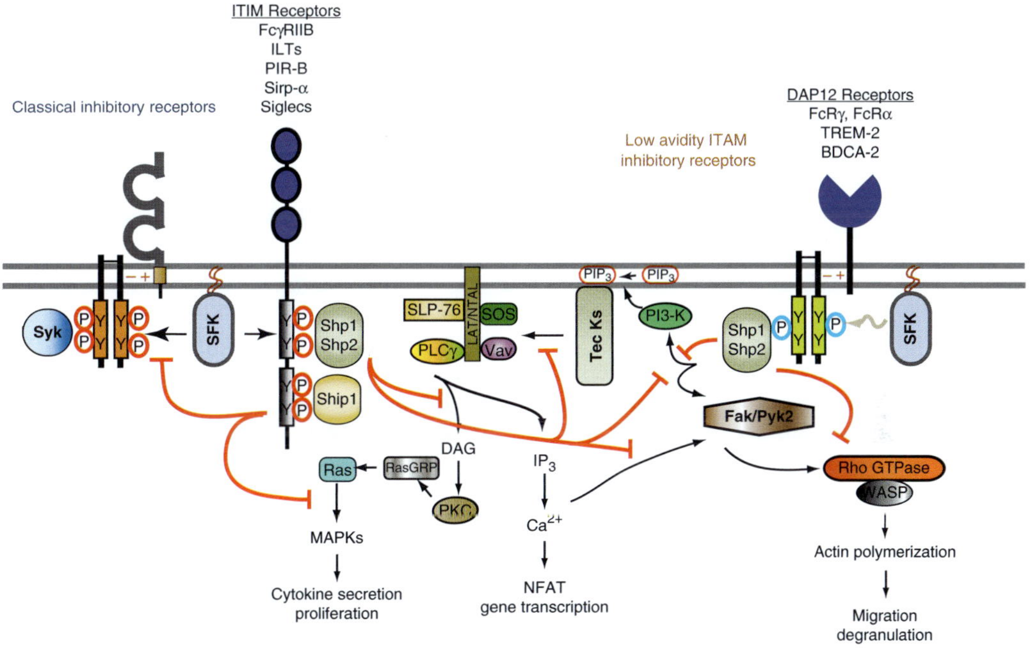

Figure 3. Inhibitory pathways in innate cells utilizing ITIM or "inhibitory" ITAM signaling pathways. Shown to the left are classical inhibitory receptors that contain ITIM binding sequences within their cytoplasmic tails. Like activating receptors, the inhibitory receptors can be both immunoglobulin superfamily proteins or C-type lectin receptors. For a more exhaustive list of inhibitory receptors in innate cells, see Munitz (2010). Inhibitory receptors are often engaged simultaneously with activating receptors, though they recognize distinct sets of ligands, during innate immune recognition of complex pathogens such as bacteria and yeast. The ITIM domains on these proteins are then phosphorylated by SFKs, which results in the recruitment of tyrosine or lipid phosphatase (Shp1/2 or Ship, respectively), which in turn dampen signaling by dephosphorylation of a number of substrates (the ITAMs themselves or downstream substrates in the MAPK, Ca^{2+}, or Rho/WASP pathways). Ship works directly to convert PIP3 back to PIP2 and hence opposes PI3-kinase function (Sly et al. 2007). Not shown are inhibitory DOK proteins, which are cytoplasmic adapter proteins that also recruit RasGAP and Ship to down-regulate signaling responses (see Mashima et al. 2009). This inhibitory receptor pathway also functions to down-modulate other signaling responses besides just ITAM/immunoreceptor pathways, in particular G-protein-coupled pathways and TLR responses (Zhang et al. 2005; O'Neill 2008). To the right are shown examples of inhibitory receptors that function through "inhibitory" ITAM signaling. This pathway is believed to be engaged following low avidity interactions of these receptors with various ligands, which results in only weak phosphorylation of associated ITAM adapters (shown as the ∼P on FcRγ and DAP12). The partial phosphorylation of the ITAMs leads to Shp-1/2 recruitment instead of Syk recruitment, hence engaging an inhibitory response analogous to an ITIM pathway (see Kanamaru et al. 2008; Hamerman et al. 2009). This more novel mechanism of signal inhibition has mainly been revealed by studies of innate cells lacking DAP12 or FcRγ. Like the classical ITIM pathway, the low avidity inhibitory ITAM pathway also regulates TLR responses (Hamerman et al. 2005). However, when these receptors are engaged by more high avidity ligands, they can induce activating signaling responses.

this inhibitory pathway was first worked out in lymphocytes and NK cells, mainly through the study of the inhibitory Fc receptor, FcγRIIb (Ravetch and Lanier 2000). The number of newly recognized ITIM-containing inhibitory receptors in innate immune cells has grown to include a number of Ig superfamily molecules (PIR-B, Sirp-1α), sialic binding proteins, referred to as Siglecs, and a few C-type lectin receptors (Robinson et al. 2006; Crocker et al. 2007). In many cases, the ligands for these inhibitory receptors are unknown. In a broad sense, the inhibitory receptors function by limiting innate immune cell functions, including proliferation, cytokine responses, and cell adhesion. Since inhibitory pathways are often engaged at the same time as the activating pathways, the resulting cellular response depends on a balance between the activating signal and the inhibitory signal. Mutagenesis studies suggest that inhibitory signaling is usually dominant; for example, innate immune cells from Lyn kinase-deficient mice display hyperproliferative responses to cytokines and a hyperadhesive phenotype that correlates with the lack of phosphatase activation in these cells (reviewed by Scapini et al. 2009).

Inhibitory ITAMs

There is increasing evidence that ITAM-coupled activating receptors also contribute to inhibitory signaling (Fig. 3). This was first demonstrated in mast cell experiments that compared engagement of IgE-FcεR receptors by low versus high affinity haptens (Torigoe et al. 1998). The low affinity haptens suppressed the signaling responses. A similar observation was made by Pasquier et al., who found that a Fab fragment of an anti-FcαRI mAb inhibited IgG-mediated phagocytosis in a transfected macrophage cell line, suggesting that low affinity or partial activation of one ITAM-containing receptor can inhibit the function of another ITAM-containing immunoreceptor (Pasquier et al. 2005). Subsequently, this same group demonstrated that monovalent engagement of FcαRI inhibits chemotaxis to MCP-1 and TNF, as well as downstream signal transduction, in an SHP-1-dependent fashion, resulting in the attenuation of tissue injury in an Ab-induced glomerulonephritis model (Kanamaru et al. 2008). This has led to the model whereby low affinity or partial (i.e., monomeric) engagement of activating receptors results in only modest ITAM phosphorylation events, leading to the recruitment of phosphatases (SHP-1 in particular) instead of Syk kinase (Pinheiro da Silva et al. 2008), resulting in an overall diminution of cellular response (Fig. 3).

The activation of ITAM signaling pathways also functions to inhibit other signaling responses, such as TLR signaling (Hamerman et al. 2009). In particular, both macrophages and dendritic cells derived from DAP12 and FcRγ-deficient mice display increased responses to various TLR ligands, such as CpG, endotoxin (LPS), or yeast zymosan (Hamerman et al. 2005; Chu et al. 2008). In macrophages, this inhibitory signaling may be mediated through TREM-2, which is a DAP12 coupled receptor thought to be involved in recognition of various bacterial structures and potentially apoptotic cells (Hamerman et al. 2006; Takahashi et al. 2007; Hsieh et al. 2009; N'Diaye et al. 2009). Additional FcRγ and DAP12-associated receptors on both mouse and human plasmacytoid DCs have also been found to inhibit interferon production following TLR stimulation (Blasius et al. 2004; Fuchs et al. 2005; Swiecki and Colonna 2007).

The molecular mechanisms by which ITAM-based pathways cross-inhibit TLR responses remain unclear (reviewed by Ivashkiv 2008). In addition to phosphatase recruitment, activation of PI3-kinase-mediated pathways that both suppress NF-κB activation and alter recruitment of MyD88 and Mal adaptors to TLRs have been proposed (Kagan and Medzhitov 2006). Activation of calcineurin by ITAM signaling, leading to dephosphorylation of TLR signaling pathway molecules, may also be involved (Kang et al. 2007). Recently, Han et al. have provided an elegant model for ITAM-mediated down regulation of TLR responses (Han et al. 2010). Building on the observation that Mac-1-deficient mice (lacking the integrin αm protein that dimerizes with the β2 chain to form Mac-1) demonstrate increased inflammatory responses to dextran sodium sulfate-induced colitis (Abdelbaqi et al. 2006), this group and others (Wang et al. 2010) found that strong ligation of β2 integrins in macrophages and dendritic cells induces resistance to subsequent stimulation by various TLR ligands. Since signaling through β2 integrins proceeds in a DAP12/FcRγ to Syk pathway, Han et al. (2010) investigated potential intracellular Syk substrates. They found that Syk phosphorylates the TLR adapter proteins MyD88 and TRIF on specific tyrosine residues (Y277 and Y375, respectively), and that phosphorylation of these residues results in recruitment of the E3 ubiquitin ligase Cbl-b, which targets both the adapter proteins and Syk kinase itself for ubiquitin-mediated proteolysis, thus attenuating TLR responses (reviewed by Means and Luster 2010). This unique pathway provides a direct intracellular mechanism for signaling cross talk between ITAM and TLR pathways. It is likely that under physiologic situations of high affinity integrin ligation, such as during diapedisis through vascular endothelium, signaling through this pathway limits innate immune cell activation and reduces inadvertent immunopathology.

Indirect Down-modulation of Signaling: Pathway Cross Talk

While most studies have focused on intracellular mechanisms for inhibitory signaling, there is growing recognition that immunoreceptor activation can also be limited by post-transcriptional events. Following engagement of ITAM pathways in macrophages or dendritic cells, production of inhibitory cytokines such IL-10 and TGF-β can lead to subsequent down-modulation of signaling, both for the ITAM pathway itself and for other pathways, such as the TLRs. For example, engagement of the C-type lectin receptor DC-SIGN by ManLAM, a cell-wall component present in *Mycobacterium tuberculosis*, down-modulates subsequent stimulation of DCs by LPS (Geijtenbeek et al. 2003). The mechanism involves acetylation of the p65 subunit of NF-κB, leading to prolonged and increased production of IL-10, which in turn inhibits LPS and other TLR responses (Gringhuis et al. 2007). More recently, IL-10 production by neutrophils, following activation through C-type lectin receptors or other ITAM-Syk pathways, has been found to have an immunomodulatory effect on subsequent inflammatory responses (Zhang et al. 2009). Hence, in states of chronic *mycobacterial* infection, engagement of the C-type lectin/Syk pathway leads to enough neutrophil-derived IL-10 to limit responses of macrophages and DCs. Thus, neutrophil depletion can sometimes increase the immunopathology associated with infection. Similarly, high affinity engagement of β2 integrins on human macrophages can lead to sufficient production of IL-10 to down-modulate subsequent TLR stimulation (Wang et al. 2010). In all of these cases, the activating stimulation (either through the C-type lectins or the β2 integrins) needs to be present for hours to days to allow for sufficient production of IL-10 and other inhibitors. For integrin signaling, this indirect pathway is probably additive with the direct biochemical mechanism of inhibitory cross talk via MyD88/TRIF phosphorylation, but would dominate during prolonged periods of integrin activation potentially to limit chronic inflammation.

SRC-FAMILY AND SYK TYROSINE KINASES AND INNATE SIGNALING

The history of tyrosine kinase involvement in innate immune responses to pathogen molecules (LPS, bacterial cell wall components, and foreign DNA) is long and confusing. Many of these studies pre-date the identification of the Toll-like receptors (TLRs) as the major receptors involved in recognition of pathogens. These studies were based on the findings that LPS treatment of innate immune cells increases overall protein tyrosine phosphorylation and various tyrosine kinase inhibitors, particularly those that broadly target Src-family kinases, and could block cellular responses to LPS (Orlicek et al. 1999; Smolinska et al. 2008). The hypothesis that TLRs could activate Src-family kinases were refuted by the finding that macrophages lacking Hck, Fgr, and Lyn displayed normal (or even enhanced) responses to LPS treatment (Meng and Lowell 1997). Indeed, recent studies have suggested that Lyn kinase functions as a negative regulator of TLR responses, based on the finding that Lyn-deficient macrophages show increased cytokine responses to TLR4 and TLR2 stimulation (Keck et al. 2010). The mechanism by which Lyn functions as an inhibitor of TLR responses is still unclear (Kagan and Medzhitov 2006).

FUNCTIONS OF SPECIFIC KINASE FAMILIES

Src-Family Kinases

The Src-family kinases make up the largest family of cytoplasmic tyrosine kinases expressed in innate cells. Of the nine Src-family kinases, most are present in one type of innate cell or another, with Hck, Fgr, and Lyn kinase being most heavily expressed in monocytes, macrophages, granulocytes, and DCs. These kinases have been implicated as primary signaling molecules downstream of a host of immune cell receptors, including immunoreceptors, cytokine receptors, integrins, and various pathogen receptors (TREMs and Dectins, as examples). Inhibitor studies and use of knockout mice have demonstrated a critical role for Src-family kinases in a variety of host defense and inflammatory conditions (Okutani et al. 2006; Abram and Lowell 2008; Ingley 2008). In some cases, these kinases play clear and direct roles in innate immune signaling, in both activating and inhibitory pathways (Figs. 1 and 3), but in other cases their role may be more indirect (for example, by modulating cytokine responses that impact TLR pathways). The Src-family kinases are also responsible for phosphorylation and direct activation of other cytoplasmic tyrosine kinases, in particular Tec-family and FAK/Pyk2 and less directly Syk; hence Src-family members are often at the top of most signaling pathways in innate cells.

Src-Family Kinases in Immunoreceptor Pathways

The most well-understood function of these kinases is in the classical immunoreceptor activating (ITAM) pathway (Fig. 1) utilized by many immunoreceptors (Fc receptors, NK activating receptors, and pathogen recognition molecules such as TREM family members), where various Src-family members are known to phosphorylate residues on the receptor associated ITAM adapter to initiate downstream signaling responses. As stated above, the major new finding in this area has been the demonstration that many innate immune receptors that are not classical immunoreceptors in fact utilize this same ITAM-mediated mechanism for intracellular signaling, and hence loss of Src-family activity directly affects these pathways as well. The best example is the integrin pathway—loss of Src-family kinase activity (or removal of the ITAM adapters FcRγ and DAP12, the substrates of the kinases) results in a complete deficiency of β1, β2, and β3 integrin function in innate cells (Abram and Lowell 2009). In all of these pathways, it is uncertain how the Src-family kinases are activated following ligand binding by the receptor. Dephosphorylation of the regulatory C-terminal tyrosine in Src-family kinases, which activates these enzymes, occurs through the action of either receptor tyrosine phosphatases such as CD45 or CD148, or potentially other cytoplasmic tyrosine phosphatases (LMW-PTP and Lyp/PEP, as examples) (Hermiston et al. 2009; Saunders and Johnson 2010; Zambuzzi et al. 2010); however, whether these phosphatases are directly recruited to the signaling receptors remains unclear.

Src-Family Kinases in Other Innate Pathways

Src-family kinases also play important roles in pathways where ITAM/immunoreceptor molecules are not involved. A number of studies suggest that Src-family kinases work with Jak kinases in supporting cytokine responses, either by phosphorylation of receptor subunits or potentially phosphorylating Stat molecules (Reddy et al. 2000; Hayakawa and Naoe 2006). For the classical growth factor receptors, such as the G-CSF or the GM-CSF receptor, Src-kinases have been found to be physically associated with the receptor through SH3/

receptor interactions involving membrane proximal regions (Sampson et al. 2007; Perugini et al. 2010). In other cytokine responses, Src-family kinases are thought to act more downstream of the receptor, often through interactions with TRAF signaling molecules. Such is the case in the RANK/RANKL signaling response, where interaction of Src and TRAF6 lead to enhancement of downstream responses (particularly activation of PI3-kinase activity and downstream Akt function) (Leibbrandt and Penninger 2008). Formation of TRAF6/Src signaling complexes has also been reported downstream of IL-1 and TNFα signaling pathways, as well as in CD40/CD40L interactions, again with the implication that this complex is involved in stimulating downstream PI3-kinase signaling (Mukundan et al. 2005; Wang et al. 2006).

Src-family kinases have also been shown to be involved in IL-6 signaling pathways, through a direct interaction with the IL-6 receptor signaling protein gp130 (Hallek et al. 1997; Haushserr et al. 2007). This signaling function of Src-family kinases may be particularly important in myeloma cells, where IL-6 serves as a growth factor for these cells. A number of studies have shown involvement of Src-family kinases downstream of the M-CSF receptor in macrophages and myeloid progenitors, with their function being to couple receptor activation to downstream PI3-kinase pathways (Lee and States 2000; Bourgin-Hierle et al. 2008). As in the case of the IL-3 receptor, the M-CSF receptor pathway may also co-opt ITAM-containing adapters to activate downstream functions, since DAP12 has been recently found to be required for optimal M-CSF receptor signaling (Zou et al. 2008). Finally, as in the immunoreceptor pathways, not all the signaling functions of Src kinases in cytokine pathways are activating, since the deficiency of Lyn kinase actually leads to enhanced responses to some cytokines, such as G-CSF, potentially through reduced recruitment of phosphatases (Mermel et al. 2006). The mechanism through which Lyn kinase functions as an inhibitor, while other Src-family members function as activators of cytokine signaling, remains unclear.

Src-Family Kinases and GPCRs

Several Src-family kinases have been implicated in regulating chemokine or chemoattractant receptor signaling, which are mediated by G-protein-coupled receptors (GPCRs) (Luttrell and Luttrell 2004). Lyn kinase is activated in macrophages downstream of both CXCR4 and CCR5, and is thought to couple these GPCRs to the MAPK and PI3-kinase pathways (Ptasznik et al. 2002; Tomkowicz et al. 2006; Cheung et al. 2009). Inhibitor studies have also placed Src-family kinases downstream of monocyte-chemoattractant protein (MCP-1) signaling via CCR2 (Arefieva et al. 2005) and IL-8 signaling via CXCR1 (Sai et al. 2008). Similarly, neutrophils derived from $hck^{-/-}fgr^{-/-}$ double mutant mice show significant functional defects following formyl peptide (fMLF) stimulation, which involves signaling through GPCR coupled formyl peptide receptors (Fumagalli et al. 2007). But even in the GPCR signaling pathway, the same paradigm of Src-kinases acting both in a positive and inhibitory fashion seems to be established. While Hck/Fgr-deficient neutrophils have reduced responses to foryml peptide stimulation, they manifest hyper-responsive signaling to chemokines that signal through CCR1 and CXCR2 (Zhang et al. 2005). This inhibitory function in the CCR1/CXCR2 pathways involves phosphorylation of the ITIM-containing inhibitory receptor PIR-B, which in turn is involved in recruiting tyrosine phosphatases that modulate downstream responses from the receptors.

Src-Family Kinases and Selectin Signaling

Src-family kinases have been implicated in the innate immune cell response to selectin engagement, primarily recognition of E-selectin by leukocyte PSGL-1 and CD44 counter-receptors. Engagement of these receptors leads to activation of tyrosine phosphorylation that is inhibited by PP2 and other relative Src-family-kinase-specific inhibitors (Hidari et al. 1997; Kumar et al. 2001). The Fgr kinase seems dominant in this signaling response (Zarbock et al. 2008), although there is clearly some redundancy with other Src-family members (Yago et al. 2010). However, like the integrin pathway, it appears that selectin recognition also depends on ITAM adapters, since this signaling is reduced in DAP12-deficient neutrophils (Zarbock et al. 2008).

Src-Family Kinases and Membrane Bound Receptors

A number of GPI-linked proteins in innate cells are known to signal through Src-family kinases, including TLR4-associated CD14 and the urokinase plasminogen activator receptor (uPA-R) (Stefanova et al. 1993; D'Alessio and Blasi 2009). In innate immune cells (mainly macrophages), engagement of these receptors leads to increased tyrosine phosphorylation and subsequent signaling events that affect adhesion and migration. It is likely that clustering of these receptors leads to aggregation of lipid raft membrane structures, through the GPI linkages in the receptors, which in turn brings the kinases together, since they too are located in the raft structures.

Src-Family Kinase Connections to Tec Kinases

Besides their function in TLR pathways, Tec-family kinases also play important roles in downstream immunoreceptor signaling, where their activation is mediated both through PIP3 generation by PI3-kinase and by direct phosphorylation by Src-family kinases. Monocytes from patients with Bruton's agammaglobulinemia show reduced uptake of both Ig and complement opsonized particles (Amoras et al. 2003; Jongstra-Bilen et al. 2008). Studies in macrophage cell lines suggest that Btk and Tec communicate to the actin polymerization pathway, since these kinases localize to phagocytic cups near sites of actin polymerization. Btk and Tec also play a significant role in macrophage/osteoclast RANK/RANKL signaling. Remarkably, like Src or Src/Hck-deficient mice, the double mutant $Tec^{-/-}Btk^{-/-}$ mice have severe osteopetrosis due to impaired osteoclast maturation (Shinohara et al. 2008). Since it is known that RANK signaling is mediated, in part, through DAP12 (Koga et al. 2004), the effect of Tec/Btk deficiency may reflect impaired activation of downstream actin polymerization responses.

Src-Family Kinase Connections to FAK/Pyk2 in Integrin Pathways

Based on studies in nonimmune cells, Src-family kinases can directly modulate the activity of FAK/Pyk2 kinases (Fig. 1). FAK is widely expressed in most cells of the body, while Pyk2 is expressed mainly in the nervous system, T-cells, and various innate cells. FAK kinase has been very extensively studied in fibroblasts and various tumor cell types, where it plays an important role in cell adhesion signaling downstream of integrin activation (Tomar and Schlaepfer 2009). More limited studies of Pyk2 suggest a similar function. Both of these kinases are substrates for Src-family kinases following integrin ligation. Phosphorylation of FAK or Pyk2 by Src-family members leads to protein unfolding and activation of their enzymatic activity. In innate cells, FAK and Pyk2 are found in podosomes, which are the main contact sites in leukocytes (Calle et al. 2006). Macrophages lacking FAK exhibit elevated protrusive activity, altered adhesion dynamics, impaired chemotaxis, and elevated basal Rac1 activity, leading to a marked inability to form stable lamellipodia necessary for directional locomotion (Okigaki et al. 2003; Owen et al. 2007a). These defects point to an alteration in integrin outside-in signaling, consistent with the role of this kinase in fibroblasts and tumor cells. As a result of this defect, recruitment of FAK-deficient monocytes to sites of inflammation is impaired. A similar phenotype is observed in macrophages derived from Pyk2-deficient animals (Okigaki et al. 2003). The involvement of FAK and Pyk2 in macrophage integrin signaling may also affect bacterial phagocytosis by these receptors, since siRNA-mediated knockdown of FAK and/or Pyk2 can reduce uptake of various Yersina strains of bacteria (Owen et al. 2007b). FAK deficiency also reduces neutrophil adhesion, migration, and antibacterial uptake, again potentially through defects in integrin-mediated signaling events (Kasorn et al. 2009). Inhibition of Pyk2 function in neutrophils, through use of inhibitory peptides transduced into the cells, reduces cell adhesion and spreading following integrin activation, again suggesting a similar role for this kinase in neutrophil integrin signaling (Han et al. 2003). Pyk2 has also been implicated in the integrin signaling responses leading to cytokine (in particular IL-10) production (Wang et al. 2010). Despite these initial studies, the specific individual functions of FAK versus Pyk2 in innate immune integrin signaling remains to be defined.

Syk Kinase

The repertoire of signaling pathways that Syk has been implicated in (reviewed by Mocsai et al. 2010) is more restricted than Src-family kinases. Syk kinase is activated by engagement of its two SH2 domains by phospho-ITAM domains; hence it functions only in ITAM-like pathways. This includes, of course, the novel "inhibitory ITAM" pathway described above. It is likely that Syk is only involved in the subset of signaling pathways that also depend on Src-family kinases. However, this doesn't mean that blockade of Syk kinases always produces the same phenotype in a given immunoreceptor signaling response that blockade of Src-family members does. For example, the defect in macrophage phagocytosis in Src-family-kinase-deficient cells is different than in Syk-deficient cells. Lack of Src-family kinases produces a moderate to severe defect in FcR-mediated particle uptake due to a reduction in initial actin polymerization at the phagocytic cup (Fitzer-Attas et al. 2000). However, Syk-deficient cells show a complete block in FcR-mediated particle uptake due to a block in fusion of the arms of the phagocytic cup, a step subsequent to actin polymerization events (Crowley et al. 1997). Hence, it is likely that the more proximal Src-family kinases signal to many other pathways besides just ITAM-mediated Syk activation. Finally, it should be noted that Syk has been implicated in some pathways in which the role of Src-family kinases are unknown. This is particularly true for the C-type lectin receptors in innate cells.

Syk and ITAM Pathways

Syk kinase plays a critical role in all innate immune cell signaling pathways involving ITAM or ITAM-like signaling adapters. This includes signaling from classical immunoreceptors (FcRs, TREMs, Dectin-2, and others), nonimmunoreceptors that co-opt the ITAM adapters for signaling (integrins, selectins, IL-3 receptor), and those C-type lectin receptors that have hemi-ITAM sequences embedded in their C-terminal tails (Dectin-1) (Fig. 1). Since many of the C-type lectin and other immunoreceptor-like pathways converge on Syk, loss of Syk activity produces a more profound block than loss of any specific receptor (LeibundGut-Landmann et al. 2007). These same C-type lectin receptors also are involved in innate immune recognition of mycobacterial and viral pathogen molecules, all of which are affected by Syk blockade in either macrophages or dendritic cells (Chen et al. 2008; Werninghaus et al. 2009).

Syk and the Inflammasome

In addition to signaling downstream through CARD9 to NF-κB, Syk-dependent signaling events have been linked to activation of intracellular pattern recognition receptors, which in turn activate the inflammasome complex leading to IL-1β production. For example, activation of the Nod-like receptor protein NLRP3 requires upstream Syk activity during innate immune cell responses to fungal molecules (Gross et al. 2009). Similarly, Syk-deficient dendritic cells fail to respond to monosodium urate crystals (present in the joints of gout patients), a pathway also known to require NLRP3 (Ng et al. 2008; Martinon 2010). It remains unclear how Syk activation is coupled to the NLRP3/inflammasome complex.

In contrast to Src-family kinases, it is likely that Syk does not signal in cytokine or GPCR-linked pathways. Lack of Syk has no impact on neutrophil, macrophage, or mast cell recognition of various cytokine growth factors or GPCR agonists, such as formyl peptides, ATP, or other agents (Mocsai et al. 2003).

CONCLUSIONS

The roles of cytoplasmic tyrosine kinases in innate immune responses are complex, ranging from direct signaling involvement in very defined pathways (such as Syk in C-type lectin receptor signaling) to more diffuse interactions (Src-family kinases somehow regulating GPCR responses) to indirect secondary effects (Src-family and Syk function in immunoreceptor pathways leading to IL-10 production that feeds back on TLR-mediated responses). It is clear that Src-family members have the broadest effects on overall signaling, while Syk has more defined roles. This diffuse, versus very defined, functional role is mirrored in the fact that deficiencies in single Src-family kinases tend to produce limited signaling defects in innate cells, while loss of Syk has very defined broad functional effects. Because of redundancy, studying the role of Src-kinases in any pathway often requires blocking multiple family members, after which one often finds that many signaling responses are affected. This overall broad function for Src-family members is analogous to how a rheostat controls lighting: It dials up and dials down responses in a graded fashion. In contrast, Syk acts more like a signaling switch: It is critically required in an absolute way in a limited number of pathways.

When viewed in this fashion, it becomes obvious that if we are to design therapeutics that target these kinases for use in inflammatory or autoimmune disease, we are better off focusing on the switch kinases (Syk) rather than the rheostat kinases (Src-family), since targeting the former will produce defined effects. Indeed, a number of companies are close to producing/releasing highly active Syk kinase inhibitors that have strong potential for treatment of immune-mediated disease (Cohen and Fleischmann 2010; Colonna et al. 2010). In contrast, most of the Src-family inhibitors produced and used clinically so far are rather broadly acting, and hit enzymes besides just Src-family members. This may be useful for when treating cancer, which is where most of the anti-Src kinase drugs have been used (Kim et al. 2009), but is likely problematic for chronic treatment of immune-mediated disease. Moreover, distinguishing between individual Src-family members may be chemically impossible to achieve. Nevertheless, given our ever-expanding understanding of kinase-mediated signaling pathways in innate cells, it is quite likely that in the very near future we will see highly active drugs that block these pathways, resulting potentially in clinical benefit.

ACKNOWLEDGMENTS

CAL is supported by NIH grants AI065495 and AI068150. We thank Clare Abram, Patrizia Scapini, and Chrystelle Lamagna for critical reading of the manuscript.

REFERENCES

Abdelbaqi M, Chidlow JH, Matthews KM, Pavlick KP, Barlow SC, Linscott AJ, Grisham MB, Fowler MR, Kevil CG. 2006. Regulation of dextran sodium sulfate induced colitis by leukocyte beta 2 integrins. *Lab Invest* **86:** 380–390.

Abram CL, Lowell CA. 2008. The diverse functions of Src family kinases in macrophages. *Front Biosci* **13:** 4426–4450.

Abram CL, Lowell CA. 2009. The ins and outs of leukocyte integrin signaling. *Annu Rev Immunol* **27:** 339–362.

Amoras AL, Kanegane H, Miyawaki T, Vilela MM. 2003. Defective Fc-, CR1- and CR3-mediated monocyte phagocytosis and chemotaxis in common variable immunodeficiency and X-linked agammaglobulinemia patients. *J Investig Allergol Clin Immunol* **13:** 181–188.

Arefieva TI, Kukhtina NB, Antonova OA, Krasnikova TL. 2005. MCP-1-stimulated chemotaxis of monocytic and endothelial cells is dependent on activation of different signaling cascades. *Cytokine* **31:** 439–446.

Blasius A, Vermi W, Krug A, Facchetti F, Cella M, Colonna M. 2004. A cell-surface molecule selectively expressed on murine natural interferon-producing cells that blocks secretion of interferon-alpha. *Blood* **103:** 4201–4206.

Bourgin-Hierle C, Gobert-Gosse S, Therier J, Grasset MF, Mouchiroud G. 2008. Src-family kinases play an essential role in differentiation signaling downstream of macrophage colony-stimulating factor receptors mediating persistent phosphorylation of phospholipase C-gamma2 and MAP kinases ERK1 and ERK2. *Leukemia* **22:** 161–169.

Calle Y, Burns S, Thrasher AJ, Jones GE. 2006. The leukocyte podosome. *Eur J Cell Biol* **85:** 151–157.

Chen ST, Lin YL, Huang MT, Wu MF, Cheng SC, Lei HY, Lee CK, Chiou TW, Wong CH, Hsieh SL. 2008. CLEC5A is critical for dengue-virus-induced lethal disease. *Nature* **453:** 672–676.

Cheung R, Malik M, Ravyn V, Tomkowicz B, Ptasznik A, Collman RG. 2009. An arrestin-dependent multi-kinase signaling complex mediates MIP-1beta/CCL4 signaling and chemotaxis of primary human macrophages. *J Leukoc Biol* **86:** 833–845.

Chu CL, Yu YL, Shen KY, Lowell CA, Lanier LL, Hamerman JA. 2008. Increased TLR responses in dendritic cells lacking the ITAM-containing adapters DAP12 and FcRγ. *Eur J Immunol* **38:** 166–173.

Cohen S, Fleischmann R. 2010. Kinase inhibitors: a new approach to rheumatoid arthritis treatment. *Curr Opin Rheumatol* **22:** 330–335.

Colgan JD, Hankel IL. 2010. Signaling pathways critical for allergic airway inflammation. *Curr Opin Allergy Clin Immunol* **10:** 42–47.

Colonna L, Catalano G, Chew C, D'Agati V, Thomas JW, Wong FS, Schmitz J, Masuda ES, Reizis B, Tarakhovsky A, et al. 2010. Therapeutic Targeting of Syk in Autoimmune Diabetes. *J Immunol* **185:** 1532–1543.

Colonna M. 2007. All roads lead to CARD9. *Nat Immunol* **8:** 554–555.

Crocker PR, Paulson JC, Varki A. 2007. Siglecs and their roles in the immune system. *Nat Rev Immunol* **7:** 255–266.

Crowley MT, Costello PS, Fitzer-Attas CJ, Turner M, Meng F, Lowell C, Tybulewicz VL, DeFranco AL. 1997. A critical role for Syk in signal transduction and phagocytosis mediated by Fcgamma receptors on macrophages. *J Exp Med* **186:** 1027–1039.

D'Alessio S, Blasi F. 2009. The urokinase receptor as an entertainer of signal transduction. *Front Biosci* **14:** 4575–4587.

Dorhoi A, Desel C, Yeremeev V, Pradl L, Brinkmann V, Mollenkopf HJ, Hanke K, Gross O, Ruland J, Kaufmann SH. 2010. The adaptor molecule CARD9 is essential for tuberculosis control. *J Exp Med* **207:** 777–792.

Fitzer-Attas CJ, Lowry M, Crowley MT, Finn AJ, Meng F, DeFranco AL, Lowell CA. 2000. Fcγ receptor-mediated phagocytosis in macrophages lacking the Src family tyrosine kinases Hck, Fgr, and Lyn. *J Exp Med* **191:** 669–682.

Fuchs A, Cella M, Kondo T, Colonna M. 2005. Paradoxic inhibition of human natural interferon-producing cells by the activating receptor NKp44. *Blood* **106:** 2076–2082.

Fumagalli L, Zhang H, Baruzzi A, Lowell CA, Berton G. 2007. The Src family kinases Hck and Fgr regulate neutrophil responses to N-formyl-methionyl-leucyl-phenylalanine. *J Immunol* **178:** 3874–3885.

Geijtenbeek TB, Van Vliet SJ, Koppel EA, Sanchez-Hernandez M, Vandenbroucke-Grauls CM, Appelmelk B, Van Kooyk Y. 2003. Mycobacteria target DC-SIGN to suppress dendritic cell function. *J Exp Med* **197:** 7–17.

Gilfillan AM, Rivera J. 2009. The tyrosine kinase network regulating mast cell activation. *Immunol Rev* **228:** 149–169.

Graham LM, Brown GD. 2009. The Dectin-2 family of C-type lectins in immunity and homeostasis. *Cytokine* **48:** 148–155.

Gringhuis SI, den Dunnen J, Litjens M, van Het Hof B, van Kooyk Y, Geijtenbeek TB. 2007. C-type lectin DC-SIGN modulates Toll-like receptor signaling via Raf-1 kinase-dependent acetylation of transcription factor NF-kappaB. *Immunity* **26:** 605–616.

Gross O, Gewies A, Finger K, Schafer M, Sparwasser T, Peschel C, Forster I, Ruland J. 2006. Card9 controls a non-TLR signalling pathway for innate anti-fungal immunity. *Nature* **442:** 651–656.

Gross O, Poeck H, Bscheider M, Dostert C, Hannesschlager N, Endres S, Hartmann G, Tardivel A, Schweighoffer E, Tybulewicz V, et al. 2009. Syk kinase signalling couples to the Nlrp3 inflammasome for anti-fungal host defence. *Nature* **459:** 433–436.

Hallek M, Neumann C, Schaffer M, Danhauser-Riedl S, von Bubnoff N, de Vos G, Druker BJ, Yasukawa K, Griffin JD, Emmerich B. 1997. Signal transduction of interleukin-6 involves tyrosine phosphorylation of multiple cytosolic proteins and activation of Src-family kinases Fyn, Hck, and Lyn in multiple myeloma cell lines. *Exp Hematol* **25:** 1367–1377.

Hamerman JA, Jarjoura JR, Humphrey MB, Nakamura MC, Seaman WE, Lanier LL. 2006. Cutting edge: inhibition of TLR and FcR responses in macrophages by triggering receptor expressed on myeloid cells (TREM)-2 and DAP12. *J Immunol* **177:** 2051–2055.

Hamerman JA, Ni M, Killebrew JR, Chu CL, Lowell CA. 2009. The expanding roles of ITAM adapters FcRgamma and DAP12 in myeloid cells. *Immunol Rev* **232:** 42–58.

Hamerman JA, Tchao NK, Lowell CA, Lanier LL. 2005. Enhanced Toll-like receptor responses in the absence of signaling adaptor DAP12. *Nat Immunol* **6:** 579–586.

Han C, Jin J, Xu S, Liu H, Li N, Cao X. 2010. Integrin CD11b negatively regulates TLR-triggered inflammatory responses by activating Syk and promoting degradation of MyD88 and TRIF via Cbl-b. *Nat Immunol* **11:** 734–742.

Han H, Fuortes M, Nathan C. 2003. Critical role of the carboxyl terminus of proline-rich tyrosine kinase (Pyk2) in the activation of human neutrophils by tumor necrosis factor: separation of signals for the respiratory burst and degranulation. *J Exp Med* **197:** 63–75.

Hara H, Saito T. 2009. CARD9 versus CARMA1 in innate and adaptive immunity. *Trends Immunol* **30:** 234–242.

Hauck CR, Klingbeil CK, Schlaepfer DD. 2000. Focal adhesion kinase functions as a receptor-proximal signaling component required for directed cell migration. *Immunol Res* **21:** 293–303.

Hausherr A, Tavares R, Schaffer M, Obermeier A, Miksch C, Mitina O, Ellwart J, Hallek M, Krause G. 2007. Inhibition of IL-6-dependent growth of myeloma cells by an acidic peptide repressing the gp130-mediated activation of Src family kinases. *Oncogene* **26:** 4987–4998.

Hayakawa F, Naoe T. 2006. SFK-STAT pathway: an alternative and important way to malignancies. *Ann N Y Acad Sci* **1086:** 213–222.

Hermiston ML, Zikherman J, Zhu JW. 2009. CD45, CD148, and Lyp/Pep: critical phosphatases regulating Src family kinase signaling networks in immune cells. *Immunol Rev* **228:** 288–311.

Hida S, Yamasaki S, Sakamoto Y, Takamoto M, Obata K, Takai T, Karasuyama H, Sugane K, Saito T, Taki S. 2009.

Fc receptor γ-chain, a constitutive component of the IL-3 receptor, is required for IL-3-induced IL-4 production in basophils. *Nat Immunol* **10:** 214–222.

Hidari KI, Weyrich AS, Zimmerman GA, McEver RP. 1997. Engagement of P-selectin glycoprotein ligand-1 enhances tyrosine phosphorylation and activates mitogen-activated protein kinases in human neutrophils. *J Biol Chem* **272:** 28750–28756.

Hsieh CL, Koike M, Spusta SC, Niemi EC, Yenari M, Nakamura MC, Seaman WE. 2009. A role for TREM2 ligands in the phagocytosis of apoptotic neuronal cells by microglia. *J Neurochem* **109:** 1144–1156.

Ingley E. 2008. Src family kinases: regulation of their activities, levels and identification of new pathways. *Biochim Biophys Acta* **1784:** 56–65.

Ivashkiv LB. 2008. A signal-switch hypothesis for cross-regulation of cytokine and TLR signalling pathways. *Nat Rev Immunol* **8:** 816–822.

Ivashkiv LB. 2009. Cross-regulation of signaling by ITAM-associated receptors. *Nat Immunol* **10:** 340–347.

Jongstra-Bilen J, Puig Cano A, Hasija M, Xiao H, Smith CI, Cybulsky MI. 2008. Dual functions of Bruton's tyrosine kinase and Tec kinase during Fcgamma receptor-induced signaling and phagocytosis. *J Immunol* **181:** 288–298.

Kagan JC, Medzhitov R. 2006. Phosphoinositide-mediated adaptor recruitment controls Toll-like receptor signaling. *Cell* **125:** 943–955.

Kanamaru Y, Pfirsch S, Aloulou M, Vrtovsnik F, Essig M, Loirat C, Deschenes G, Guerin-Marchand C, Blank U, Monteiro RC. 2008. Inhibitory ITAM signaling by FcαRI-FcRγ chain controls multiple activating responses and prevents renal inflammation. *J Immunol* **180:** 2669–2678.

Kanazawa N. 2007. Dendritic cell immunoreceptors: C-type lectin receptors for pattern-recognition and signaling on antigen-presenting cells. *J Dermatol Sci* **45:** 77–86.

Kang YJ, Kusler B, Otsuka M, Hughes M, Suzuki N, Suzuki S, Yeh WC, Akira S, Han J, Jones PP. 2007. Calcineurin negatively regulates TLR-mediated activation pathways. *J Immunol* **179:** 4598–4607.

Kasorn A, Alcaide P, Jia Y, Subramanian KK, Sarraj B, Li Y, Loison F, Hattori H, Silberstein LE, Luscinskas WF, et al. 2009. Focal adhesion kinase regulates pathogen-killing capability and life span of neutrophils via mediating both adhesion-dependent and -independent cellular signals. *J Immunol* **183:** 1032–1043.

Keck S, Freudenberg M, Huber M. 2010. Activation of murine macrophages via TLR2 and TLR4 is negatively regulated by a Lyn/PI3K module and promoted by SHIP1. *J Immunol* **184:** 5809–5818.

Kerrigan AM, Brown GD. 2010. Syk-coupled C-type lectin receptors that mediate cellular activation via single tyrosine based activation motifs. *Immunol Rev* **234:** 335–352.

Kim LC, Song L, Haura EB. 2009. Src kinases as therapeutic targets for cancer. *Nat Rev Clin Oncol* **6:** 587–595.

Koga T, Inui M, Inoue K, Kim S, Suematsu A, Kobayashi E, Iwata T, Ohnishi H, Matozaki T, Kodama T, et al. 2004. Costimulatory signals mediated by the ITAM motif cooperate with RANKL for bone homeostasis. *Nature* **428:** 758–763.

Koprulu AD, Ellmeier W. 2009. The role of Tec family kinases in mononuclear phagocytes. *Crit Rev Immunol* **29:** 317–333.

Kumar P, Hosaka S, Koch AE. 2001. Soluble E-selectin induces monocyte chemotaxis through Src family tyrosine kinases. *J Biol Chem* **276:** 21039–21045.

Kurosaki T, Shinohara H, Baba Y. 2010. B cell signaling and fate decision. *Annu Rev Immunol* **28:** 21–55.

Lanier LL. 2009. DAP10- and DAP12-associated receptors in innate immunity. *Immunol Rev* **227:** 150–160.

Lee AW, States DJ. 2000. Both src-dependent and -independent mechanisms mediate phosphatidylinositol 3-kinase regulation of colony-stimulating factor 1-activated mitogen-activated protein kinases in myeloid progenitors. *Mol Cell Biol* **20:** 6779–6798.

Leibbrandt A, Penninger JM. 2008. RANK/RANKL: regulators of immune responses and bone physiology. *Ann N Y Acad Sci* **1143:** 123–150.

LeibundGut-Landmann S, Gross O, Robinson MJ, Osorio F, Slack EC, Tsoni SV, Schweighoffer E, Tybulewicz V, Brown GD, Ruland J, et al. 2007. Syk- and CARD9-dependent coupling of innate immunity to the induction of T helper cells that produce interleukin 17. *Nat Immunol* **8:** 630–638.

Lowell CA. 2004. Src-family kinases: rheostats of immune cell signaling. *Mol Immunol* **41:** 631–643.

Luttrell DK, Luttrell LM. 2004. Not so strange bedfellows: G-protein-coupled receptors and Src family kinases. *Oncogene* **23:** 7969–7978.

Martinon F. 2010. Mechanisms of uric acid crystal-mediated autoinflammation. *Immunol Rev* **233:** 218–232.

Mashima R, Hishida Y, Tezuka T, Yamanashi Y. 2009. The roles of Dok family adapters in immunoreceptor signaling. *Immunol Rev* **232:** 273–285.

Means TK, Luster AD. 2010. Integrins limit the Toll. *Nat Immunol* **11:** 691–693.

Meng F, Lowell CA. 1997. Lipopolysaccharide (LPS)-induced macrophage activation and signal transduction in the absence of Src-family kinases Hck, Fgr, and Lyn. *J Exp Med* **185:** 1661–1670.

Mermel CH, McLemore ML, Liu F, Pereira S, Woloszynek J, Lowell CA, Link DC. 2006. Src family kinases are important negative regulators of G-CSF-dependent granulopoiesis. *Blood* **108:** 2562–2568.

Mocsai A, Abram CL, Jakus Z, Hu Y, Lanier LL, Lowell CA. 2006. Integrin signaling in neutrophils and macrophages uses adaptors containing immunoreceptor tyrosine-based activation motifs. *Nat Immunol* **7:** 1326–1333.

Mocsai A, Ruland J, Tybulewicz VL. 2010. The SYK tyrosine kinase: a crucial player in diverse biological functions. *Nat Rev Immunol* **10:** 387–402.

Mocsai A, Zhang H, Jakus Z, Kitaura J, Kawakami T, Lowell CA. 2003. G-protein-coupled receptor signaling in Syk-deficient neutrophils and mast cells. *Blood* **101:** 4155–4163.

Mukundan L, Bishop GA, Head KZ, Zhang L, Wahl LM, Suttles J. 2005. TNF receptor-associated factor 6 is an essential mediator of CD40-activated proinflammatory pathways in monocytes and macrophages. *J Immunol* **174:** 1081–1090.

Munitz A. 2010. Inhibitory receptors on myeloid cells: new targets for therapy? *Pharmacol Ther* **125:** 128–137.

N'Diaye EN, Branda CS, Branda SS, Nevarez L, Colonna M, Lowell C, Hamerman JA, Seaman WE. 2009. TREM-2 (triggering receptor expressed on myeloid cells 2) is a phagocytic receptor for bacteria. *J Cell Biol* **184:** 215–223.

Ng G, Sharma K, Ward SM, Desrosiers MD, Stephens LA, Schoel WM, Li T, Lowell CA, Ling CC, Amrein MW, et al. 2008. Receptor-independent, direct membrane binding leads to cell-surface lipid sorting and Syk kinase activation in dendritic cells. *Immunity* **29:** 807–818.

Nimmerjahn F, Ravetch JV. 2008. Fcγ receptors as regulators of immune responses. *Nat Rev Immunol* **8:** 34–47.

O'Neill LA. 2008. When signaling pathways collide: positive and negative regulation of toll-like receptor signal transduction. *Immunity* **29:** 12–20.

Okigaki M, Davis C, Falasca M, Harroch S, Felsenfeld DP, Sheetz MP, Schlessinger J. 2003. Pyk2 regulates multiple signaling events crucial for macrophage morphology and migration. *Proc Natl Acad Sci U S A* **100:** 10740–10745.

Okutani D, Lodyga M, Han B, Liu M. 2006. Src protein tyrosine kinase family and acute inflammatory responses. *Am J Physiol Lung Cell Mol Physiol* **291:** 129–141.

Orlicek SL, Hanke JH, English BK. 1999. The src family-selective tyrosine kinase inhibitor PP1 blocks LPS and IFN-gamma-mediated TNF and iNOS production in murine macrophages. *Shock* **12:** 350–354.

Owen KA, Pixley FJ, Thomas KS, Vicente-Manzanares M, Ray BJ, Horwitz AF, Parsons JT, Beggs HE, Stanley ER, Bouton AH. 2007a. Regulation of lamellipodial persistence, adhesion turnover, and motility in macrophages by focal adhesion kinase. *J Cell Biol* **179:** 1275–1287.

Owen KA, Thomas KS, Bouton AH. 2007b. The differential expression of Yersinia pseudotuberculosis adhesins determines the requirement for FAK and/or Pyk2 during bacterial phagocytosis by macrophages. *Cell Microbiol* **9:** 596–609.

Page TH, Smolinska M, Gillespie J, Urbaniak AM, Foxwell BM. 2009. Tyrosine kinases and inflammatory signalling. *Curr Mol Med* **9:** 69–85.

Pasquier B, Launay P, Kanamaru Y, Moura IC, Pfirsch S, Ruffie C, Henin D, Benhamou M, Pretolani M, Blank U, et al. 2005. Identification of FcαRI as an inhibitory receptor that controls inflammation: dual role of FcRγ ITAM. *Immunity* **22:** 31–42.

Perugini M, Brown AL, Salerno DG, Booker GW, Stojkoski C, Hercus TR, Lopez AF, Hibbs ML, Gonda TJ, D'Andrea RJ. 2010. Alternative modes of GM-CSF receptor activation revealed using activated mutants of the common beta-subunit. *Blood* **115:** 3346–3353.

Pinheiro da Silva F, Aloulou M, Benhamou M, Monteiro RC. 2008. Inhibitory ITAMs: a matter of life and death. *Trends Immunol* **29:** 366–373.

Ptasznik A, Urbanowska E, Chinta S, Costa MA, Katz BA, Stanislaus MA, Demir G, Linnekin D, Pan ZK, Gewirtz AM. 2002. Crosstalk between BCR/ABL oncoprotein and CXCR4 signaling through a Src family kinase in human leukemia cells. *J Exp Med* **196:** 667–678.

Ravetch JV, Lanier LL. 2000. Immune inhibitory receptors. *Science* **290:** 84–89.

Rawlings DJ, Sommer K, Moreno-Garcia ME. 2006. The CARMA1 signalosome links the signalling machinery of adaptive and innate immunity in lymphocytes. *Nat Rev Immunol* **6:** 799–812.

Reddy EP, Korapati A, Chaturvedi P, Rane S. 2000. IL-3 signaling and the role of Src kinases, JAKs and STATs: a covert liaison unveiled. *Oncogene* **19:** 2532–2547.

Robinson MJ, Sancho D, Slack EC, LeibundGut-Landmann S, Reis e Sousa C. 2006. Myeloid C-type lectins in innate immunity. *Nat Immunol* **7:** 1258–1265.

Ruland J. 2008. CARD9 signaling in the innate immune response. *Ann N Y Acad Sci* **1143:** 35–44.

Sai J, Raman D, Liu Y, Wikswo J, Richmond A. 2008. Parallel phosphatidylinositol 3-kinase (PI3K)-dependent and Src-dependent pathways lead to CXCL8-mediated Rac2 activation and chemotaxis. *J Biol Chem* **283:** 26538–26547.

Sampson M, Zhu QS, Corey SJ. 2007. Src kinases in G-CSF receptor signaling. *Front Biosci* **12:** 1463–1474.

Saunders AE, Johnson P. 2010. Modulation of immune cell signalling by the leukocyte common tyrosine phosphatase, CD45. *Cell Signal* **22:** 339–348.

Scapini P, Pereira S, Zhang H, Lowell CA. 2009. Multiple roles of Lyn kinase in myeloid cell signaling and function. *Immunol Rev* **228:** 23–40.

Shinohara M, Koga T, Okamoto K, Sakaguchi S, Arai K, Yasuda H, Takai T, Kodama T, Morio T, Geha RS, et al. 2008. Tyrosine kinases Btk and Tec regulate osteoclast differentiation by linking RANK and ITAM signals. *Cell* **132:** 794–806.

Sly LM, Ho V, Antignano F, Ruschmann J, Hamilton M, Lam V, Rauh MJ, Krystal G. 2007. The role of SHIP in macrophages. *Front Biosci* **12:** 2836–2848.

Smith-Garvin JE, Koretzky GA, Jordan MS. 2009. T cell activation. *Annu Rev Immunol* **27:** 591–619.

Smolinska MJ, Horwood NJ, Page TH, Smallie T, Foxwell BM. 2008. Chemical inhibition of Src family kinases affects major LPS-activated pathways in primary human macrophages. *Mol Immunol* **45:** 990–1000.

Stefanova I, Corcoran ML, Horak EM, Wahl LM, Bolen JB, Horak ID. 1993. Lipopolysaccharide induces activation of CD14-associated protein tyrosine kinase p53/56lyn. *J Biol Chem* **268:** 20725–20728.

Swiecki MK, Colonna M. 2007. Running to stand still: BCR-like signaling suppresses type I IFN responses in pDC. *Eur J Immunol* **37:** 3327–3329.

Takahashi K, Prinz M, Stagi M, Chechneva O, Neumann H. 2007. TREM2-transduced myeloid precursors mediate nervous tissue debris clearance and facilitate recovery in an animal model of multiple sclerosis. *PLoS Med* **4:** e124.

Tohyama Y, Yamamura H. 2009. Protein tyrosine kinase, syk: a key player in phagocytic cells. *J Biochem* **145:** 267–273.

Tomar A, Schlaepfer DD. 2009. Focal adhesion kinase: switching between GAPs and GEFs in the regulation of cell motility. *Curr Opin Cell Biol* **21:** 676–683.

Tomkowicz B, Lee C, Ravyn V, Cheung R, Ptasznik A, Collman RG. 2006. The Src kinase Lyn is required for CCR5 signaling in response to MIP-1beta and R5 HIV-1 gp120 in human macrophages. *Blood* **108:** 1145–1150.

Torigoe C, Inman JK, Metzger H. 1998. An unusual mechanism for ligand antagonism. *Science* **281:** 568–572.

Wang KZ, Wara-Aswapati N, Boch JA, Yoshida Y, Hu CD, Galson DL, Auron PE. 2006. TRAF6 activation of PI 3--kinase-dependent cytoskeletal changes is cooperative with Ras and is mediated by an interaction with cytoplasmic Src. *J Cell Sci* **119(Pt 8):** 1579–1591.

Wang L, Gordon RA, Huynh L, Su X, Park Min KH, Han J, Arthur JS, Kalliolias GD, Ivashkiv LB. 2010. Indirect inhibition of Toll-like receptor and type I interferon responses by ITAM-coupled receptors and integrins. *Immunity* **32:** 518–530.

Werninghaus K, Babiak A, Gross O, Holscher C, Dietrich H, Agger EM, Mages J, Mocsai A, Schoenen H, Finger K, et al. 2009. Adjuvanticity of a synthetic cord factor analogue for subunit Mycobacterium tuberculosis vaccination requires FcRγ-Syk-Card9-dependent innate immune activation. *J Exp Med* **206:** 89–97.

Yago T, Shao B, Miner JJ, Yao L, Klopocki AG, Maeda K, Coggeshall KM, McEver RP. 2010. E-selectin engages PSGL-1 and CD44 through a common signaling pathway to induce integrin αLβ2-mediated slow leukocyte rolling. *Blood* **116:** 485–494.

Yamasaki S, Ishikawa E, Sakuma M, Hara H, Ogata K, Saito T. 2008. Mincle is an ITAM-coupled activating receptor that senses damaged cells. *Nat Immunol* **9:** 1179–1188.

Zambuzzi WF, Milani R, Teti A. 2010. Expanding the role of Src and protein-tyrosine phosphatases balance in modulating osteoblast metabolism: lessons from mice. *Biochimie* **92:** 327–332.

Zarbock A, Abram CL, Hundt M, Altman A, Lowell CA, Ley K. 2008. PSGL-1 engagement by E-selectin signals through Src kinase Fgr and ITAM adapters DAP12 and FcRγ to induce slow leukocyte rolling. *J Exp Med* **205:** 2339–2347.

Zhang H, Meng F, Chu CL, Takai T, Lowell CA. 2005. The Src family kinases Hck and Fgr negatively regulate neutrophil and dendritic cell chemokine signaling via PIR-B. *Immunity* **22:** 235–246.

Zhang X, Majlessi L, Deriaud E, Leclerc C, Lo-Man R. 2009. Coactivation of Syk kinase and MyD88 adaptor protein pathways by bacteria promotes regulatory properties of neutrophils. *Immunity* **31:** 761–771.

Zou W, Reeve JL, Liu Y, Teitelbaum SL, Ross FP. 2008. DAP12 couples c-Fms activation to the osteoclast cytoskeleton by recruitment of Syk. *Mol Cell* **31:** 422–431.

ZAP-70: An Essential Kinase in T-Cell Signaling

Haopeng Wang[1], Theresa A. Kadlecek[1], Byron B. Au-Yeung[1], Hanna E. Sjölin Goodfellow[1], Lih-Yun Hsu[1], Tanya S. Freedman[1], and Arthur Weiss[1]

[1]Howard Hughes Medical Institute, Rosalind Russell Medical Research Center for Arthritis, Department of Medicine, University of California, San Francisco, San Francisco, California 94143

Correspondence: aweiss@medicine.ucsf.edu

ZAP-70 is a cytoplasmic protein tyrosine kinase that plays a critical role in the events involved in initiating T-cell responses by the antigen receptor. Here we review the structure of ZAP-70, its regulation, its role in development and in disease. We also describe a model experimental system in which ZAP-70 function can be interrupted by a small chemical inhibitor.

Approximately 18 years ago, a T-cell receptor (TCR)–ζ chain-associated 70 kDa tyrosine phosphoprotein (therefore termed ZAP-70) was identified in TCR-stimulated Jurkat cells (Chan et al. 1991). The importance of ZAP-70, a cytoplasmic tyrosine kinase mainly expressed in T cells, was rapidly revealed by the following observations: ZAP-70-deficient patients have no functional T cells in their peripheral blood and suffer from severe combined immunodeficiency (SCID) (Arpaia et al. 1994; Chan et al. 1994; Elder et al. 1994); T-cell development in ZAP-70 deficient mice is arrested at the transition from the CD4 + CD8 + double positive (DP) stage, where positive selection occurs (Negishi et al. 1995b; Kadlecek et al. 1998); and a Jurkat-derived cell line (p116) deficient in ZAP-70 expression fails to activate downstream signaling events after TCR stimulation (Williams et al. 1998). Because ZAP-70 plays a critical role in T-cell development and activation, much effort has been focused on studying the regulation, structure, and function of ZAP-70.

How does ZAP-70 mediate TCR signaling? Our lab proposed a model by which the TCR initiates signaling via sequential interactions with cytoplasmic tyrosine kinases (Weiss 1993). When the TCR interacts with peptide antigen bound to a MHC complex molecule on antigen presenting cells, coreceptor-associated Lck is brought into proximity of the CD3 complex and phosphorylates tyrosines in the immunoreceptor tyrosine-based activation motifs (ITAMs). When doubly phosphorylated, ITAMs recruit ZAP-70 via a relatively high affinity interaction by binding to the tandem SH2 domains of ZAP-70. Recent studies suggest that this binding event leads to the release of ZAP-70 from its autoinhibited conformation, which exposes the regulatory phosphorylation sites for Lck-mediated

 doi: 10.1101/cshperspect.a002279
Cite this article as *Cold Spring Harb Perspect Biol* 2010;2:a002279

phosphorylation. In addition, tyrosines in the activation loop of the ZAP-70 kinase domain are then phosphorylated by Lck or by ZAP-70 itself in *trans* to further promote its catalytic activity. A number of signaling proteins, including the linker for the activation of T cells (LAT) and the SH2-domain-containing leukocyte protein of 76 kDa (SLP-76), are subsequently phosphorylated by active ZAP-70. The phosphorylated LAT and SLP-76 proteins function as scaffolds to recruit many other signaling molecules (Fig. 1). The consequences of these early signaling events eventually lead to T-cell activation, proliferation, and differentiation.

Here we discuss the structure and regulation of ZAP-70, as well as the role of ZAP-70 in T-cell development and disease. We also highlight a recently developed chemical-genetic approach of inhibiting the ZAP-70 kinase activity and its potential usage in studying the role of ZAP-70 in T-cell responses, both in vitro and in vivo, which will facilitate and justify further effort in developing an inhibitor of wild-type ZAP-70 protein.

STRUCTURE OF ZAP-70

ZAP-70 is composed of two SH2 domains and a carboxy-terminal kinase domain (Fig. 2A). Unlike Src-family kinases with an amino-terminal SH3-domain and a SH2-domain, the amino terminus of the spleen tyrosine kinase (Syk) family of kinases, of which ZAP-70 is a member, consists of two tandem SH2 domains that are separated by a linker region, termed interdomain A. In ZAP-70, these SH2 domains bind to doubly phosphorylated ITAMs of TCR ζ chain, and are responsible for bridging ZAP-70 to the activated TCR. Another linker region, interdomain B, connects the SH2 domains to the kinase domain (Chan et al. 1992).

The crystal structure of a truncated version of ZAP-70 containing the tandem SH2 domains complexed with a doubly tyrosine phosphorylated ITAM peptide reveals interesting features about its structure and clues to its regulation. The most unusual feature of the ZAP-70 SH2 ITAM complex is that the phosphotyrosine binding (PTB) pocket of the amino-terminal

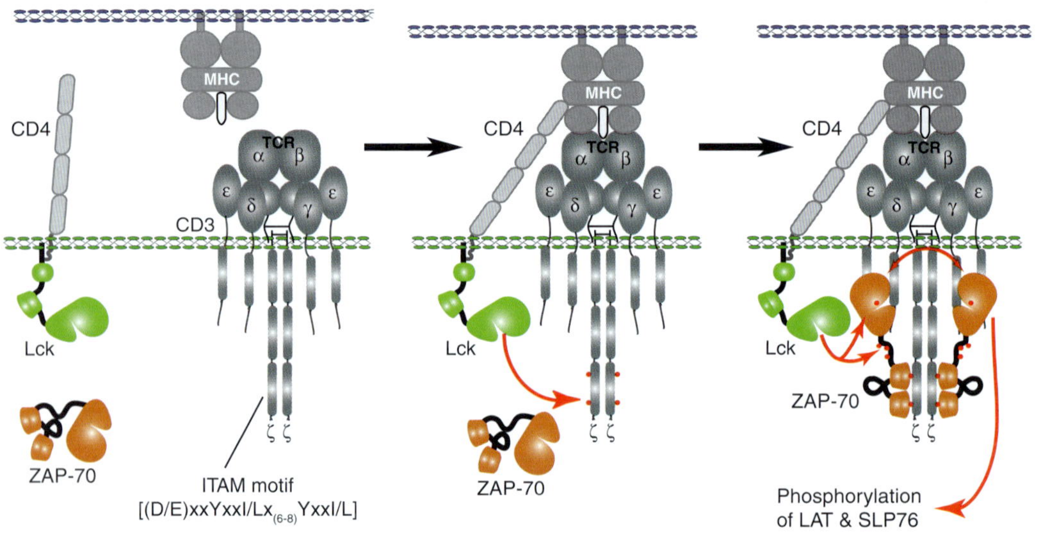

Figure 1. A sequential model for T-cell activation. Following TCR engagement, CD4-associated Lck is brought into proximity of the CD3 complex and phosphorylates the ITAMs (phoshorylation depicted as red dots). Doubly phosphorylated ITAMs then interact with the tandem SH2 domains of ZAP-70. After ITAM binding, ZAP-70 can be phosphorylated by Lck, which results in activation of ZAP-70 catalytic activity and its autophosphorylation. Active ZAP-70 subsequently phosphorylates LAT and SLP-76, which function as scaffolds to recruit many other signaling molecules and lead to T-cell activation, proliferation, and differentiation (not shown).

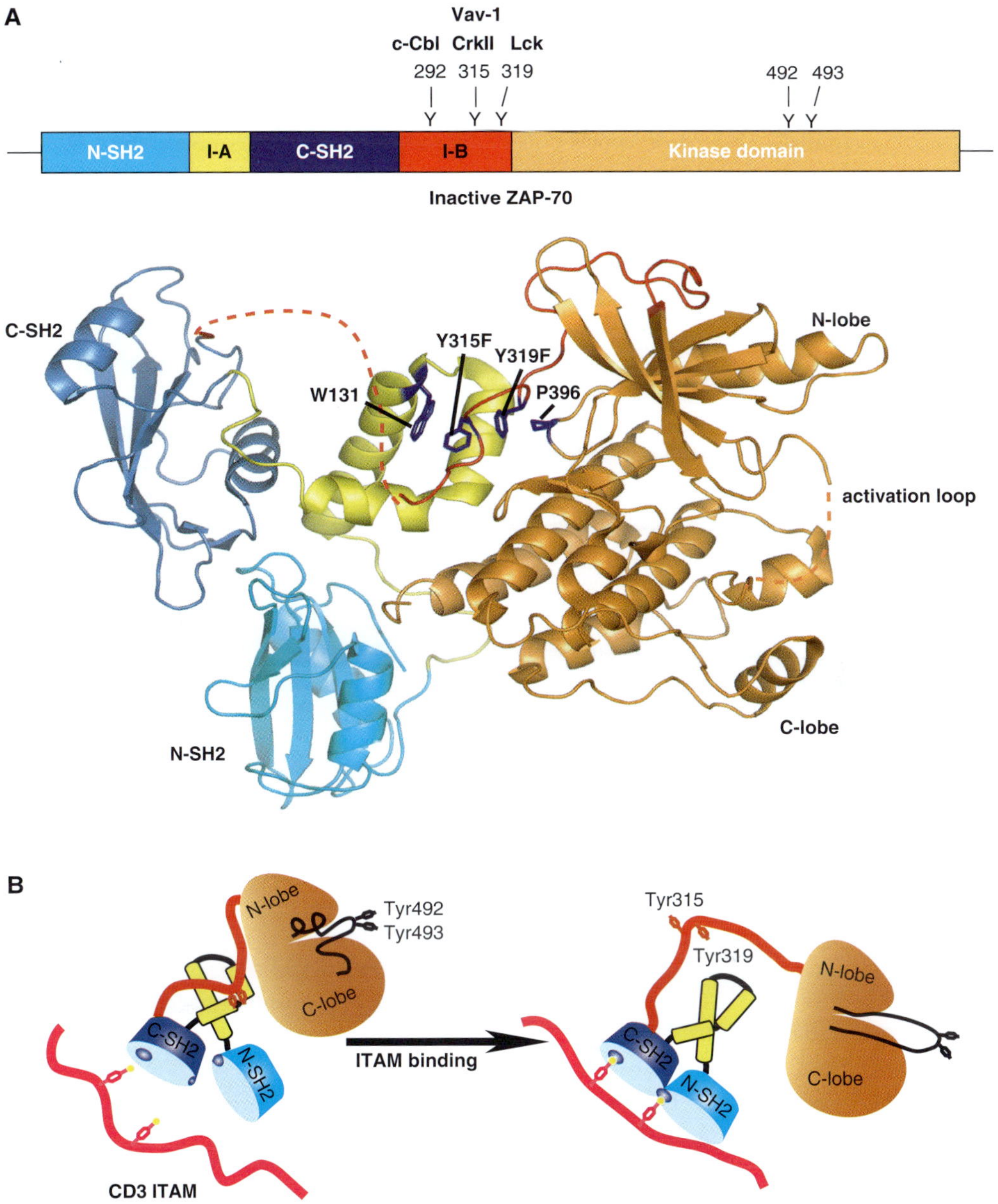

Figure 2. Structural organization of autoinhibited ZAP-70, and model of autoinhibition. (*A*) Schematic of ZAP-70 domains and phosphorylated tyrosine residues with binding molecules (*top*) and a crystal structure of autoinhibited ZAP-70 (PDB code 2OZO, *bottom*). The amino-terminal SH2 domain, interdomain A, carboxy-terminal SH2 domain, interdomain B, and the kinase domain are shown in cyan, yellow, blue, red, and orange, respectively. The side chains of residues involved in the aromatic–aromatic interactions between interdomain A, interdomain B, and the kinase domain are labeled and colored in dark blue. Dotted lines represent disordered regions. (*B*) Model for the activation of ZAP-70 following ITAM binding. The SH2 domains, interdomain A, interdomain B, and kinase domains are colored as in (*A*). The doubly phosphorylated ITAM is depicted in pink. When ZAP-70 shows an autoinhibited conformation it is incompatible with binding to phosphorylated ITAMs. Following ITAM binding, conformational changes in ZAP-70 promote disassembly of the interface mediating the autoinhibited conformation, and exposure of tyrosines in interdomain B, leading to their phosphorylation that further destabilizes the interface.

SH2 domain is formed by residues contributed from both the amino- and carboxy-terminal SH2 domains that are brought into juxtaposition by ITAM binding. As in most SH2 domains, the carboxy-terminal SH2 PTB pocket is completely composed of residues within the carboxy-terminal SH2 domain. In addition, interdomain A appears to form a coiled-coil domain that facilitates extensive contact between the two SH2 domains and stabilizes their interaction (Hatada et al. 1995). This structure suggests that the SH2 domains do not function independently, and that interactions between the domains are important for specific recognition of the phospho-ITAMs of the activated TCR. Mutational studies addressing minimal spatial requirements between tyrosines in ITAM sequences for signaling also support this model (Ottinger et al. 1998).

Comparison of the crystal structures and NMR analysis of the bound and unbound SH2 domains reveals two very different conformations. Binding to phospho-ITAM peptide induces large movements between the SH2 domains, and the formation of the complete amino-terminal PTB pocket. Compared to a more rigid structure in the bound form, interdomain A shows a flexible structure in the unbound form. This more rigid structure is likely to contribute to the specificity of binding to the appropriately spaced and phosphorylated ITAM (Folmer et al. 2002). Additionally, studies using antibodies against regions within interdomain A show binding to ZAP-70 only when it is not associated with the ζ chain, further enforcing the idea of a conformational change upon docking to the ITAM (Grazioli et al. 1998).

Following TCR engagement, two tyrosine residues located in the activation loop of the kinase domain (Y492 and Y493) are phosphorylated. It is thought that this phosphorylation displaces the activation loop from the catalytic site of the kinase, leading to complete activation of the kinase (Watts et al. 1994). The crystal structure of a protein consisting of the kinase domain (aa327-606) in complex with the ATP-competitive inhibitor Staurosporin, shows an active-like kinase conformation, despite the fact that Y492 and Y493 are unphosphorylated. Its bilobed structure is well conserved compared to the catalytic domains of many protein kinases. Common to these kinases, ZAP-70 is composed of an amino-terminal lobe containing a five-stranded antiparallel β-sheet and a single helix, and a carboxy-terminal lobe that is predominantly helical with short strands of β-sheets. An extended linker connects these lobes, and the ATP-binding site is located at the junction between the two lobes (Jin et al. 2004). The ability of the catalytic domain to adopt an active conformation without activation loop phosphorylation suggests that additional mechanisms are used in the full-length protein to regulate its activation.

ZAP-70 contains several tyrosine residues that are phosphorylated following TCR stimulation and are known to be important in its regulation, activity, and association with other signaling molecules. Y292, Y315, and Y319 are located within interdomain B and when phosphorylated are thought to serve as docking sites for downstream signaling molecules (Fig. 2A). Regulation of ZAP-70 kinase activity is also dependent on tyrosine residues in interdomain B. Mutation of both Y315 and Y319 to phenylananine renders the kinase inactive. In contrast, mutation of these same residues to alanine results in increased basal kinase activity (Brdicka et al. 2005). Additionally, some function is maintained in a mutant lacking most of interdomain B, suggesting that it may be involved in an autoinhibitory mechanism (Zhao and Weiss 1996). It was hypothesized that mutation of Y315 and Y319 to phenylalanine stabilizes an autoinhibited conformation and tyrosine phosphorylation may impede the autoinhibited conformation (Brdicka et al. 2005).

Indeed, the crystal structure of autoinhibited ZAP-70 has also been solved at 2.6 Angstrom resolution using a recombinant protein with Y315 and Y319 in interdomain B mutated to phenylalanine (Fig. 2A). In the crystal structure the catalytic domain in autoinhibited ZAP-70 adopts a conformation that is characterized by many of the features of the inactive conformation of Src. Helix αC in the N-lobe of the kinase domain is rotated away from the active site, the salt bridge between Glu386 and Lys369 that is required for catalytic activity is disrupted, and a short 3_{10}-helix at the base of the activation loop packs against the displaced helix αC. Although this structure contains many common features of autoinhibited Src family kinases, the mechanism for its autoinhibitory regulation is novel. A hydrogen bonding network involving interdomains A and B and the back of the kinase domain serves to stabilize the hinge region of the catalytic domain, thereby stabilizing the inactive conformation. This notion is supported by comparison of energy displacement data from the active and inactive kinase domains. Two interactions serve to stabilize the autoinhibited conformation. First, two α-helices from interdomain A (viewed as a single helix in the tandem SH2-ITAM bound structure) are docked onto the carboxy-terminal lobe of the kinase domain. A proline residue at the junction of these two is wedged between two tyrosines on the back of the C-lobe of the catalytic domain. Interdomain B is positioned between the kinase domain and interdomain A. Additionally, critical residues stabilizing this conformation include a set of perpendicular aromatic-aromatic interactions between W131 and P296 with F315 and F319 that create a rather stable hydrophobic environment (Deindl et al. 2007). The hydroxyls of Y315 and Y319 in the wild-type kinase would presumably decrease the stability of the hydrophobic interactions in this region and allow for more conformational flexibility in wild-type ZAP-70. Of considerable interest, the alignment of the tandem SH2 domains in the autoinhibited structure is incompatible with binding to phosphorylated ITAMs, because the PTB pocket of the amino-terminal SH2 domain is not appropriately apposed to the carboxy-terminal SH2 domain to complete the PTB site. Studies of mutant kinases, predicted to interfere with the interface interactions, that were overexpressed in 293T epithelial cells and in the Jurkat cell line deficient in ZAP-70 validate this structure. In vitro kinase activity, TCR-induced tyrosine phosphorylation, intracellular Ca^{2+} response, as well as ITAM peptide binding were all increased compared to WT ZAP-70 in mutants when the interface is disrupted (Deindl et al. 2009).

A multistep model for activation of ZAP-70 has been proposed that involves spatial and temporal coordination of several key events (Fig. 2B). In resting cells ZAP-70 is in a dynamic equilibrium, and transitions in and out of the autoinhibited conformation. Following TCR stimulation and Lck-mediated phosphorylation of tyrosine residues of the ITAM, the tandem SH2 domains of ZAP-70 can then bind the doubly phosphorylated ITAM. The binding of the two SH2 domains requires their realignment from the autoinhibited conformation. Thus, a conformation of ZAP-70 that disrupts the interactions stabilizing interdomain B interaction with interdomain A as well as with the catalytic domain is favored, thereby destabilizing the interactions key to the maintenance of the autoinhibited state of ZAP-70. This allows for increased accessibility and Lck-mediated phosphorylation of Y315 and Y319 in interdomain B, phosphorylation of Y493 of the activation loop of the kinase, and subsequent full activation of the kinase.

REGULATION OF ZAP-70

Recruitment of ZAP-70 to the TCR

ZAP-70 activation can be regulated both by binding to phosphorylated ITAMs of the TCR and by phosphorylation of multiple tyrosine residues on ZAP-70 (Fig. 2A). Imaging studies have shown that in resting T cells, ZAP-70 is distributed throughout the cytoplasm, but is rapidly recruited to the plasma membrane following TCR stimulation and ITAM phosphorylation (Sloan-Lancaster et al. 1997; Sloan-Lancaster et al. 1998). ZAP-70 mobility in the cytoplasm is rapid, and decreases as ZAP-70 associates with the plasma membrane (Sloan-Lancaster et al. 1998). Whereas the recruitment to the plasma membrane is enhanced by Lck activity and phosphorylation of the ITAMs, involvement of other proteins such as the activated forms of ezrin and RhoH, a member of the Rho GTPase family, has also been reported

(Sloan-Lancaster et al. 1997; Gu et al. 2006; Ilani et al. 2007).

Following TCR ligation, TCR microclusters are formed in which signaling is initiated and sustained. The microclusters continue to form in the periphery of the immune synapse (IS) and later translocate to the center of the synapse. ZAP-70 accumulates at these microclusters. Although the initial clustering of TCR is independent of tyrosine phosphorylation events, the activity of Lck and ITAM phosphorylation are required for ZAP-70 accumulation (Sloan-Lancaster et al. 1998; Campi et al. 2005; Yokosuka et al. 2005; Varma et al. 2006).

Positive Regulation of ZAP-70 Function

Although critical, binding to TCR-ζ and perturbation of the autoinhibitory conformation alone is insufficient for complete ZAP-70 activation (van Oers et al. 1994; Chan et al. 1995; Zhao et al. 1999; Jin et al. 2004). Phosphorylation of tyrosine residues, by the action of Src family kinases such as Lck, and/or by ZAP-70 itself, is also important. Requirements for full catalytic activity include phosphorylation of tyrosine residue Y493 in the activation loop of the kinase domain of ZAP-70. Y493 is a preferred phosphorylation site for Lck but can also be a substrate for ZAP-70 autophosphorylation (Watts et al. 1994; Chan et al. 1995; Kong et al. 1996). If phosphorylation is prevented by mutation of this residue to phenylalanine, Lck is unable to activate ZAP-70 (Chan et al. 1995; Wange et al. 1995). Reconstitution studies of the Syk deficient DT40 B-cell line showed that Lck mediates phosphorylation of Y493 and that catalytic activity of ZAP-70 is required for downstream signaling (Chan et al. 1995; Kong et al. 1996). When overexpressed in the Jurkat T-cell line, ZAP-70 mutated both on Y492 and Y493 to phenylalanines (Y492F, Y493F) impaired TCR-mediated signaling. However, the mutant was able to bind Lck, suggesting that Lck association with ZAP-70 is not dependent on activation loop phosphorylation (Mege et al. 1996).

Following TCR engagement, the tyrosine residues Y315 and Y319 in interdomain B are phosphorylated. In addition to stabilization of the active conformation, these are docking sites for other molecules controlling the function of ZAP-70 and downstream signaling. Phosphorylated Y319 is a positive regulator of ZAP-70 function and is well documented to bind Lck. Overexpression of Y319F ZAP-70 mutant in Jurkat cells resulted in a dominant-negative effect, with reduced TCR-triggered activation of NFAT and IL-2 induction. Further, it conferred impaired up-regulation of ZAP-70 catalytic activity and reduced phosphorylation of its substrates LAT and SLP-76 (Di Bartolo et al. 1999). Additionally, Y319F ZAP-70 expressed in ZAP-70 deficient Jurkat cells failed to bind Lck and to reconstitute TCR triggered signaling (Williams et al. 1999). Further, mutations of the Lck SH2 domain abrogated binding of Lck to ZAP-70 and T-cell activation (Pelosi et al. 1999; Williams et al. 1999). The binding of Lck to phosphorylated Y319 may thus help prevent reversion to the autoinhibitory conformation, promote Lck mediated phosphorylation of Y493, and facilitate activation of downstream signaling. The importance of Y319 phosphorylation in positive regulation of ZAP-70 was confirmed in a mouse model expressing Y319F ZAP-70, which resulted in severe defects in calcium mobilization and thymocyte selection (Gong et al. 2001).

Vav-1 has been shown to interact with ZAP-70 via its SH2 domain, and mutation of Y315 to phenylalanine abrogates binding of Vav to ZAP-70 in DT40 cells and causes a decrease in antigen receptor mediated tyrosine phosphorylation of Vav, SLP-76, Shc and ZAP-70 itself (Katzav et al. 1994; Wu et al. 1997). Further, although not having large effects on intrinsic kinase activity, the Y315F mutation was reported to alter ZAP-70 binding affinity for phosphorylated ITAMs, reducing its capacity to trigger Erk activation in murine T cells (Di Bartolo et al. 1999). This suggested that the Vav/ZAP-70 interaction via phospho-Y315 promotes ZAP-70 function. However, because Vav phosphorylation was not affected by the ZAP-70 mutation in vivo, the importance of Vav SH2 binding to ZAP-70 at Y315 for its phosphorylation is not clear (Magnan et al. 2001).

Actin polymerization may be influenced by ZAP-70 interaction with members of the CT10 regulator of kinase (Crk) family. Crk adaptor protein can bind to phospho-Y315 on ZAP-70. A model has been proposed in which phosphorylated ZAP-70 associates with Crk-like (CrkL)-WIP-WASP complexes allowing redistribution of the complex to lipid rafts and the IS following TCR triggering. PKCθ, whose activity is indirectly dependent on Lck and ZAP-70, can phosphorylate WIP thereby releasing WASP from WIP-mediated inhibition. WASP can then be activated by membrane bound Cdc42, leading to actin polymerization, whereas WIP binds newly formed actin and thus helps to stabilize the IS (Sasahara et al. 2002). Studies of mouse models have further confirmed a detectable, though modest, positive regulatory role of Y315 in T-cell activation and thymocyte selection (Gong et al. 2001; Magnan et al. 2001). Whether this is mediated by specific interactions of ZAP-70 with Vav, CrkL or other molecules, in addition to stabilization of an active conformation, has yet to be clarified.

Negative Regulation of ZAP-70 Function

Phosphorylation of specific tyrosine residues has also been implicated in negative regulation of ZAP-70 function. Y492 is phosphorylated following TCR triggering. Mutation of this tyrosine to phenylalanine elevates ZAP-70 kinase activity, showing a negative regulatory role for this site. Furthermore, Y492F mutant ZAP-70 was able to restore antigen receptor signaling in Syk deficient DT40 B cell line, indicating that phosphorylation of this tyrosine residue is not required for ZAP-70 function (Watts et al. 1994; Chan et al. 1995; Wange et al. 1995). Y492, like Y493, is positioned in the activation loop of the kinase domain, and could thus potentially influence accessibility of the catalytic site. However, the exact mechanism for Y492 influence on kinase activity has yet to be revealed (Chan et al. 1995; Wange et al. 1995; Jin et al. 2004).

Although able to recruit positive effectors of TCR signaling, Y315 was also shown to be involved in ZAP-70 association with the Crk family member, CrkII, which has been implicated in negative regulation of TCR signaling. Lck-dependent CrkII interaction with ZAP-70 was detected after pervanadate treatment of Jurkat cells, and mutation of Y315 to phenylalanine eliminated this association (Gelkop and Isakov 1999; Gelkop et al. 2005). One possibility is that negative regulation by CrkII is exerted by complex formation between Crk, p85 PI3K, and the E3 ubiquitin ligase Cbl (Gelkop et al. 2001).

Several studies have also described a negative regulatory role for Y292 in interdomain B. Mutation of Y292 to phenylalanine enhanced NFAT induction in DT40 cells in response to BCR stimulation, without directly affecting kinase activity or TCR-ζ binding. Phosphorylation of Y292 may therefore promote association of ZAP-70 with a negative regulatory protein (Zhao and Weiss 1996). One strong candidate is c-Cbl, known to diminish TCR-mediated activation (Murphy et al. 1998; Naramura et al. 1998). ZAP-70 has been reported to associate with Cbl following TCR stimulation and phosphorylation of Y292 was required for this interaction, as well as for the negative effect of Cbl on signaling (Lupher et al. 1996; Lupher et al. 1997; Rao et al. 2000; Magnan et al. 2001). Cbl may promote ubiquitination and degradation of TCR-ζ using ZAP-70 as an adaptor protein, possibly also affecting ZAP-70 function (Wang et al. 2001; Naramura et al. 2002; Davanture et al. 2005). However, because of discrepancies in phenotypes of c-Cbl knockout mice, mice expressing the ZAP70 Y292F mutation, and mice expressing a mutation in the tyrosine kinase binding domain of c-Cbl, the mechanism for the regulatory effect of Y292 remains undefined (Murphy et al. 1998; Naramura et al. 1998; Magnan et al. 2001; Thien et al. 2003).

SYK FAMILY KINASES AND THEIR ROLE IN T-CELL DEVELOPMENT

ZAP-70 and Syk, the only two members of the Syk kinase family, provide both overlapping and distinct functions in immunoreceptor signaling. Both ZAP-70 and Syk consist of two tandem N-terminal SH2 domains and a carboxy-terminal kinase domain. Although the tandem SH2 domains of the two kinases share 57% sequence identity, crystal structures of the tandem SH2 domains complexed with a phosphorylated ITAM peptide reveal important differences between ZAP-70 and Syk. As discussed earlier, in the ITAM-bound state, both SH2 domains of ZAP-70 are tightly apposed, making extensive contacts with each other and the phosphorylated ITAM. The amino-terminal SH2 domain of ZAP-70, unlike its carboxy-terminal SH2 domain, has an incomplete PTB pocket. The phosphotyrosine of the ITAM peptide in the amino-terminal PTB pocket is closely associated with the side chain of Y238 and K242 from the C-terminal SH2 domain (Fig. 3A), revealing that the amino-terminal SH2 domain requires residues of the carboxy-terminal SH2 domain to complete this PTB pocket. These features of the ZAP-70 structure confer high selectivity for interaction with the activated TCR:CD3 complex (Hatada et al. 1995). Surprisingly, the PTB site of Syk is self-contained within the amino-terminal SH2 domain. The interaction between phosphotyrosine in the amino-terminal PTB site and

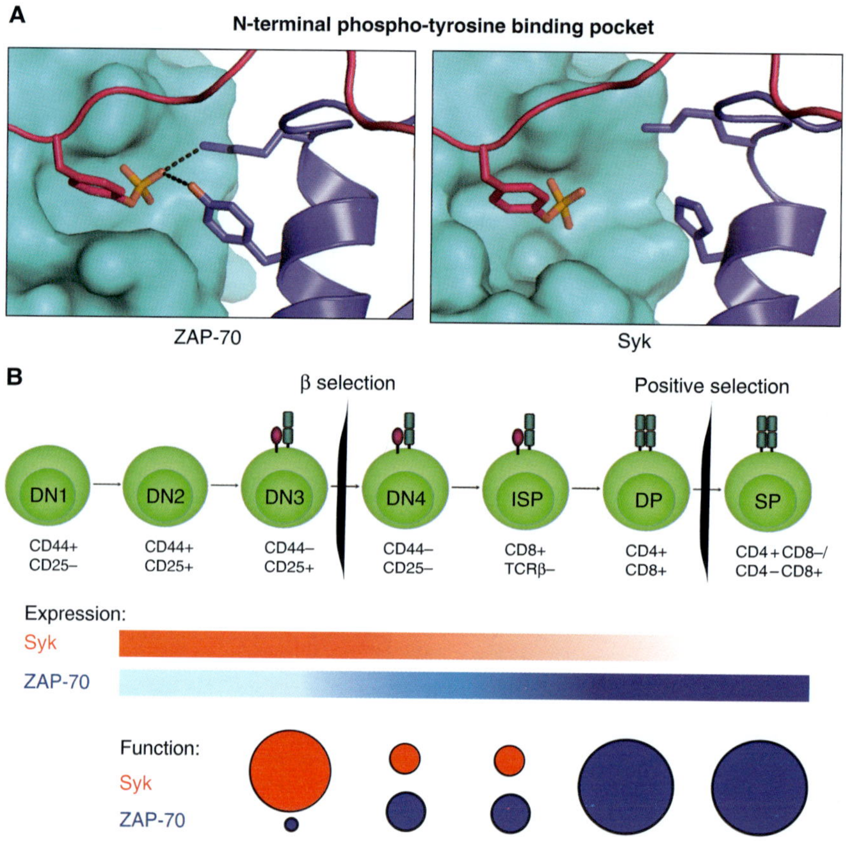

Figure 3. Comparison of ZAP-70 with Syk. (*A*) Cartoon structures of the amino-terminal phosphotyrosine binding pocket in ZAP-70 (PDB code 2OQ1, *left*) and in Syk (PDB code 1A81, *right*). The amino-terminal SH2 domain, carboxy-terminal SH2 domain, and phosphorylated ITAM peptide are shown in cyan, blue, and pink, respectively. There are six different conformations in the crystal asymmetric unit of Syk. The Syk structure shown here is representative of three of six conformations. In the other three conformations, only one residue (Lysine) in the carboxy-terminal SH2 domain seems to interact with the phosphotyrosine of the ITAM. However, the electron density for the side-chain of this Lysine is absent in all three conformations (not shown), suggesting that this interaction may not contribute significantly to the binding. (*B*) Expression and function of Syk family kinases throughout T-cell development.

residues of carboxy-terminal SH2 domain is significantly reduced compared to ZAP-70 (Fig. 3A), suggesting that the amino- and carboxy-terminal SH2 domains of Syk can function independently (Futterer et al. 1998; Kumaran et al. 2003). In addition, the asymmetric unit of Syk crystals contains six different conformations, reflecting a remarkable flexibility in the relative orientation of the two tandem SH2 domains. These features may provide the molecular basis for Syk interaction with a greater variety of ITAMs and other binding partners that vary considerably in the length of the spacer region between the two phosphotyrosines (Futterer et al. 1998; Kumaran et al. 2003).

Although both Syk and ZAP-70 are expressed during T-cell development, they have distinct expression patterns and play different roles. The early thymic progenitors transit through four stages as CD4/CD8 double negative (DN) cells (DN1–4) before up-regulating CD8 and CD4 to become double positive (DP) thymocytes, and finally become single positive (SP) thymocytes. Thymocyte maturation requires two critical checkpoints: pre-TCR mediated β-selection at the DN3 stage and TCR-mediated positive/negative selection at the DP stage. Syk and ZAP-70 proteins are inversely expressed across T-cell development. Whereas Syk is highly expressed in early development (DN1-DN3), ZAP-70 is highly expressed in later development (after DN4). However, there is overlapping expression of both kinases during some transitory stages, including the DN3, DN4, and "immature single-positive" (ISP) stages (Palacios and Weiss 2007). The positive/negative selection at the DP stage is completely arrested in ZAP-70 knockout mice but is not affected by Syk deficiency (Negishi et al. 1995b). Neither ZAP-70- nor Syk-deficient thymocytes have any obvious defect at the β-selection stage; however, the double knockout of both kinases results in an absolute block at this stage, indicating that Syk and ZAP-70 kinases can be functionally redundant during pre-TCR signaling at the DN3 stage (Cheng et al. 1997).

To determine whether Syk or ZAP-70 is preferentially used to mediate pre-TCR signaling, competitive repopulation assays using ZAP-70-deficient or Syk-deficient versus WT hematopoietic stem cells were performed in irradiated host mice. The Syk-deficient, but not ZAP-70-deficient, thymocytes are specifically impaired in initial pre-TCR signaling and cell-cycle entry at the DN3 stage. Despite overlapping expression of both kinases, only ZAP-70 is responsible for sustained pre-TCR/TCR signaling and cellular proliferation during the DN4, ISP and DP stages (Palacios and Weiss 2007). These data suggest a model where Syk-dependent pre-TCR signaling is replaced with ZAP-70-dependent pre-TCR signaling in early T-cell development (Fig. 3B). To further test this model, the generation of a knockin mouse in which Syk is targeted into the ZAP-70 locus could provide further insights. ZAP-70-like expression of Syk within DN4, ISP and DP stages would allow testing of whether ZAP-70 has a unique function in generating DP thymocytes before positive selection.

ZAP-70 AND DISEASE

Chronic Lymphocytic Leukemia (CLL)

B-chronic leukemia (B-CLL) is a B-cell malignancy caused by accumulation of monoclonal CD5+ B lymphocytes in the blood, bone marrow, lymph nodes, and spleen (Chiorazzi et al. 2005). Early studies showed that the mutational status of the immunoglobulin heavy chains of B-CLL was associated with a distinctly different prognosis, with ~50 percent of cases in each group. Subsequent gene expression profiling showed that ZAP-70 expression was the most discriminating feature between the two subtypes (Rosenwald et al. 2001). In fact, ZAP-70 protein expression in B-CLL patients appears to have more predictive value than IgVh mutations, and its expression level appears to be constant during the course of disease (Rassenti et al. 2004).

Despite the prognostic value of ZAP-70 expression in B-CLL, the presence of what might be considered a T-lineage specific gene in B-cell lineage derived leukemic cells is unexpected. Whether ZAP-70 expression plays a functional role in B-CLL progression is unclear. ZAP-70 expression in B-CLL is associated with enhanced BCR signaling (Chen et al. 2002). ZAP-70 can couple to the BCR and up-regulate Syk phosphorylation and activation as well as downstream signaling events. Transduction of ZAP-70-deficient B-CLL cells with adenoviral vectors expressing ZAP-70 increased BCR-induced tyrosine phosphorylation of several BCR signaling molecules as well as mobilization of intracellular calcium (Chen et al. 2005). Together, these results suggest that the function of ZAP-70 in CLL cells may be to enhance BCR signaling.

The finding that ZAP-70 can enhance BCR signaling in CLL B cells seems counterintuitive, given that there are similar levels of Syk expression in both ZAP-70+ or ZAP-70− CLL B cells and Syk has approximately 100-fold greater in vitro kinase activity than does ZAP-70 (Latour et al. 1996). A recent study showed that ZAP-70 promotes BCR signaling independently of its kinase activity, as both wild-type ZAP-70 and a catalytically inactive ZAP-70 mutant induced similar increases in calcium and protein tyrosine phosphorylation (Chen et al. 2008). Consistent with these observations, the interaction between ZAP-70 and downstream effector molecules appeared unchanged in the CLL B cells (Gobessi et al. 2007). Together, these results suggest that ZAP-70 may function as an adaptor protein to facilitate BCR signaling or compete for a negative regulator of Syk.

How might kinase-inactive ZAP-70 enhance Syk activation? Previous studies of DP thymocytes showed that ZAP-70, rather than Syk, promotes more efficient TCR-ζ chain phosphorylation by recruiting Lck into the TCR-ζ complex (Ashe et al. 1999). The role of ZAP-70 in augmenting TCR ITAM phosphorylation occurs when expression of Lck and the TCR are limiting in DP thymocytes, and is independent of its kinase activity. Similarly, B-CLL cells express lower levels of IgM and CD79b. Therefore, it is conceivable that, when limiting numbers of CD79b ITAMs are available in B-CLL cells, ZAP-70 positively affects Syk activity by facilitating BCR recruitment to Lyn. In support of this notion, ZAP-70 expression in B-CLL is associated with efficient translocation of BCR to lipid rafts in response to BCR engagement. As a result, phosphorylation of the BCR by lipid-raft-associated Lyn kinase is significantly stronger in ZAP-70+ CLL cells (Allsup et al. 2005). Signaling enhancement of Syk by ZAP-70 in B-CLL may represent an amplification mechanism that compensates for the low expression of IgM and CD79 in B-CLL.

In summary, ZAP-70 can enhance the intracellular signaling capacity of the Ig expressed in CLL. Whether such increased intracellular signaling influences the survival or proliferation of CLL cells, leading to disease progression remains unknown. However, ZAP-70 is an attractive candidate for treatment, owing to its participation in BCR signaling in CLL.

SCID

Loss of function or expression of ZAP-70 in humans leads to SCID. This form of SCID is characterized by normal numbers of nonfunctional CD4+ T cells and an absence of CD8+ T cells in the peripheral blood of these patients (Elder et al. 1994; Hivroz and Fischer 1994). As shown in Figure 4, the ZAP-70 mutations occurring in human SCID, which abolish ZAP-70 expression, are mostly located in the kinase domain (Arpaia et al. 1994; Chan et al. 1994; Elder et al. 1994; Gelfand et al. 1995; Matsuda et al. 1999; Noraz et al. 2000; Toyabe et al. 2001). Mutations causing transcriptional loss or destabilized protein have also been reported (Fig. 4) (Gelfand et al. 1995; Matsuda et al. 1999; Noraz et al. 2000). A patient with a missense mutation within the highly conserved DLAARN motif of kinase domain has been reported, resulting in a moderate decrease in ZAP-70 stability and complete absence of kinase activity (Elder et al. 2001). Interestingly, an identical spontaneous mutation in DLAARN motif has also been reported in mice (Wiest et al. 1997), but still leads to a discordant phenotype of the peripheral T-cell compartment of humans and mice.

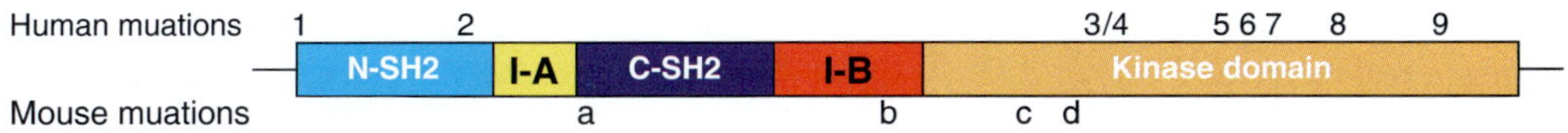

Human SCID mutations

Mutant allele	DNA mutation	Consequence to protein
1	transcript defect	No protein
2	C to A transition at nucl. 448	Missense (P to Q at residue 80, temperature-sensitive allele)
3	C to T transition at nucl. 1602	Missense (R to C at residue 465)
4	G to A transition at nucl. 1603	Missense (R to H at residue 465)
5	13 bp deletion (nucl. 1719-1731)	Deletion (premature termination at residue 538)
6	C to T transition at nucl. 1729	Missense (A to V at residue 507)
7	C to A transition at nucl. 1763	Missense (S to R at residue 518)
8	G to A transition at nucl. 1833	Splicing error (9 bp in-frame addition; L-E-Q 3 addition at residue 541)
9	A to T transition at nucl. 1923	Missense (M to L at residue 572, temperature-sensitive allele)

Hypomorphic ZAP-70 mutants

Mutant mouse	Mutant Name	Mutation	Autoimmune phenotype
a	SKG	W to C at residue 163	autoimmune arthritis, hyperglobulinemia, RF autoantibodies
b	YYAA	Y to A at residue 315 and 319	no arthritis, RF autoantibodies
c	ZAP-70$^{mrd/mrd}$	I to F at residue 367	no anti-DNA autoantibody, no hyper-IgE
d	ZAP-70$^{mrt/mrt}$	W to R at residue 504	no anti-DNA autoantibody, no hyper-IgE
c&d	ZAP-70$^{mrd/mrt}$	both I367F and W504R	anti-DNA autoantibody, hyper-IgE

Figure 4. Mutations of ZAP-70 in Human SCID and hypomorphic mutant mice.

Regardless of the mutation, the lack of ZAP-70 or complete loss of its function affects T-cell maturation differently in humans and in mice. In ZAP-70-deficient mice, T-cell development is blocked at the DP stage of thymocyte differentiation, resulting in no mature T cells in the periphery (Negishi et al. 1995a; Kadlecek et al. 1998). The selective deficiency of peripheral CD8+ cells in humans is caused by a defect in thymic T-cell development. In a thymic section from a ZAP-70 deficient SCID patient, DP T cells are present in the thymic cortex, but only CD8+SP cells are absent from the medulla (Arpaia et al. 1994). This observation suggests that positive selection of CD8+ cells is specifically blocked. It is possible that ZAP-70 plays different roles in selection of human CD4+ and CD8+ T cells and is indispensable for maturation into CD8+ T cells.

Although positive selection of CD4 SP cells occurs in these patients, the peripheral CD4+ T cells do not respond normally to mitogens or to stimulation by anti-CD3 antibodies, as shown by a failure to proliferate and produce IL-2. Interestingly, unlike peripheral CD4+ T cells, CD4+ thymocytes from the patients are able to transduce signals via their TCRs as measured by calcium flux and tyrosine phosphorylation (Gelfand et al. 1995). The differential signaling capacity between CD4+ thymocytes and peripheral T cells may be explained by the observations that Syk is present at higher levels and later in development in human thymocytes compared to mouse thymocytes (Chu et al. 1998; Chu et al. 1999). These findings suggest that higher levels of Syk in human thymocytes may be able to partially compensate for the loss of ZAP-70 in SCID patients' CD4+ thymocytes but not in peripheral T cells. Further down-regulation of Syk expression in the human peripheral T cells and/or the absolute requirement for ZAP-70 may explain the signaling defect seen in peripheral CD4+ T cells lacking ZAP-70. The mechanism for the preferential selection of CD4+ cells remains unclear. It is possible that greater association of Lck with CD4 rather than CD8 may allow more efficient activation of ZAP-70, sufficient to overcome a threshold leading to CD4 but not CD8 lineage selection (Wiest et al. 1993). Taken together, studies on human ZAP-70 immunodeficiency reveal the critical role of ZAP-70 signaling both in mature T cells and in thymic development.

Hypomorphic ZAP-70 Mutant Mice

Although ZAP-70 deficiency typically leads to the complete absence of T cells in the periphery of mice, several ZAP-70 mutant mice with partial defects in TCR signaling have been reported. The SKG mouse has a spontaneous missense mutation (W163C) in the C-terminal SH2 domain of ZAP-70 and develops autoimmune arthritis in response to innate immune activation (Sakaguchi et al. 2003). This mutation greatly impairs the binding of ZAP-70 to the TCR-ζ chain, thereby altering TCR signaling and affecting thymic development, particularly positive and negative selection, and the differentiation of T cells. As a consequence of impaired negative selection, it is thought that the thymocytes showing the strongest affinity for self-peptide/MHC ligands are positively selected instead of being deleted. This presumably leads to a shift of T-cell repertoire that would otherwise have been eliminated during negative selection in WT mice. Adoptive transfer studies have suggested that expansion of these autoreactive CD4+ T cells that escaped negative selection is a key cause of autoimmune arthritis in SKG mice. SKG arthritis resembles human rheumatoid arthritis (RA) in many ways, such as presentation of autoimmune arthritis, hyper-γ-globulinemia, vasculitis, nodule formation, and rheumatoid factor production (Sakaguchi et al. 2006). Therefore, SKG mice may be a suitable model for understanding the cellular mechanism for autoimmune arthritis and might also provide useful insights into the development of autoimmunity in general.

Another hypomorphic ZAP-70 mutant mouse has recently been reported by our lab and has revealed mechanistic insights into how signaling defects may result in autoimmunity or autoimmune disease (Hsu et al. 2009). This ZAP-70 YYAA strain has knockin alanine mutations of Y315 and Y319 in the interdomain B of ZAP-70. YYAA mice show defects similar to SKG mice. For instance, YYAA mice show diminished TCR signaling, impaired T-cell development, and defective positive and negative selection. However, in contrast to the SKG mice, which develop both arthritis and rheumatoid factor autoantibodies on the Balb/c background, YYAA mice only develop rheumatoid factor autoantibody. Quite surprisingly, in addition to impaired negative selection seen in the YYAA mice, the suppressive activity of T regulatory cells in these mice is also profoundly defective. This raises the question of why very limited autoimmunity is observed in the YYAA mice. Based on the extent of deletion of Vβ TCR genes to endogenous superantigens, our findings suggest that TCR-mediated negative selection is more impaired in the SKG than in the YYAA mice. Thus, it appears that different TCR repertoires result from different sensitivity to negative selection in these mice. The difference in TCR repertoires may be responsible for their different susceptibility to arthritis.

Recently, a hypomorphic allelic series of ZAP-70 have been generated through *N*-ethyl-*N*-nitrosourea (ENU)-induced mutations, and these mutations are located within the kinase domain of ZAP-70 (Siggs et al. 2007). A graded change in TCR signaling potential created by combining two ZAP-70 hypomorphic mutations (*mrd* and *mrt*) has revealed that the balance between thymic selection and function of effector T cells is differentially regulated. In these ZAP-70 allelic mutants, this results in the coincident occurrence of both autoimmunity (spontaneous production of anti-DNA autoantibodies) and immune dysregulation (hyper-IgE syndrome). Interestingly, although both SKG mice and $ZAP^{mrt/mrd}$ mice show defective negative selection, they show different autoimmune syndromes. SKG mice develop arthritis but not anti-DNA autoantibody, whereas $ZAP^{mrt/mrd}$ mice develop anti-DNA antibody but not arthritis. There are two possibilities that might explain the difference in autoimmune syndromes in SKG, YYAA and $ZAP^{mrt/mrd}$ mice. First, the genetic background could play a role in the resultant phenotypes.

Alternatively, quantitative and qualitative differences in TCR signaling resulting from ZAP-70 mutations may skew the TCR repertoire in ways that manifest as different autoimmune syndromes in these mice. Collectively, ZAP-70 hypomorphic alleles extend our understanding of how a quantitative difference in T-cell signaling may influence the outcome of immune responses as well as identify a threshold between susceptibility toward autoimmunity versus autoimmune disease.

ZAP-70 AS A THERAPEUTIC TARGET

Because of its critical role in TCR signaling, ZAP-70 is an attractive target for therapy. A specific inhibitor for ZAP-70 could potentially block T-cell activation at a level of signal transduction very proximal to the TCR. Additionally, ZAP-70 expression is relatively restricted to T cells, NK cells, and basophils (Au-Yeung et al. 2009). Thus, effects of specific ZAP-70 inhibition would be limited to these cell types. However, NK cells and basophils also express Syk, which may be able to compensate for the lack of ZAP-70. This appears to be the case for ZAP-70 deficient NK cells, which are still able to exert cytolytic activity (Negishi et al. 1995b). This suggests that inhibition of ZAP-70 activity may specifically target T-cell responses, while leaving innate immunity intact. Such an inhibitor could be particularly useful in settings of allograft rejection or T-cell-mediated autoimmunity.

To date, the development of a highly specific, cell permeable small molecule inhibitor of ZAP-70 has been difficult. Although some potential ZAP-70 inhibitors have been described (Hirabayashi et al. 2009), their specificity and potency in T cells or in vivo have not been shown.

In the absence of an available inhibitor, our lab has engineered a ZAP-70 mutant that retains catalytic activity, yet can be inhibited by an analog of the kinase inhibitor PP1. This chemical-genetic approach takes advantage of the presence of a conserved, bulky "gatekeeper" residue within the ATP-binding region of tyrosine kinases. Mutation of the bulky gatekeeper residue to glycine or alanine allows the mutant kinase to accommodate a larger analog of PP1, one that would be too large to inhibit the WT kinase (Bishop et al. 1998). This technique has allowed our lab to study the requirements for ZAP-70 kinase activity by T cells for productive TCR signaling leading to T-cell activation.

An analog-sensitive ZAP-70 (ZAP-70AS) mutant was generated by mutating the gatekeeper residue M414, to alanine, along with a second mutation, C405V, to stabilize the ATP-binding domain (Levin et al. 2008) (Fig. 5). Expression of ZAP-70AS in ZAP-70 deficient P116 Jurkat T cells showed that, indeed, expression of ZAP-70AS could reconstitute TCR signaling. Additionally, stimulation of ZAP-70AS-expressing Jurkat cells in the presence of the inhibitor 3-MB-PP1 resulted in a rapid, dose-dependent inhibition of downstream signals, including phosphorylation of LAT and SLP-76, as well as impaired calcium mobilization and ERK phosphorylation. Furthermore, the ZAP-70AS system showed the dynamic role of ZAP-70 catalytic activity in the regulation of the calcium flux, as inhibitor addition at the plateau of the calcium response resulted in a rapid decrease in intracellular free Ca^{2+} down to basal levels within seconds.

These studies further highlight the critical importance of ZAP-70 kinase activity in the transduction of signals from the TCR. In the future, it will be of great interest to test the ZAP-70AS system in mature, primary T cells. Such experiments will enable investigation of whether ZAP-70 activity is required for naïve, effector, and memory T-cell responses. Additionally, using the ZAP-70AS system may eventually allow us to model the effects of a ZAP-70 inhibitor in pathological T-cell responses in vivo. Until then, however, the analog-sensitive approach could be applied to the study of other signaling kinases to analyze the temporal requirements for kinase activity in TCR signal transduction and T-cell activation.

ACKNOWLEDGMENTS

H.W., T.A.K., B.B.A-Y., H.E.S.G., and L-Y.H. contributed equally to this work.

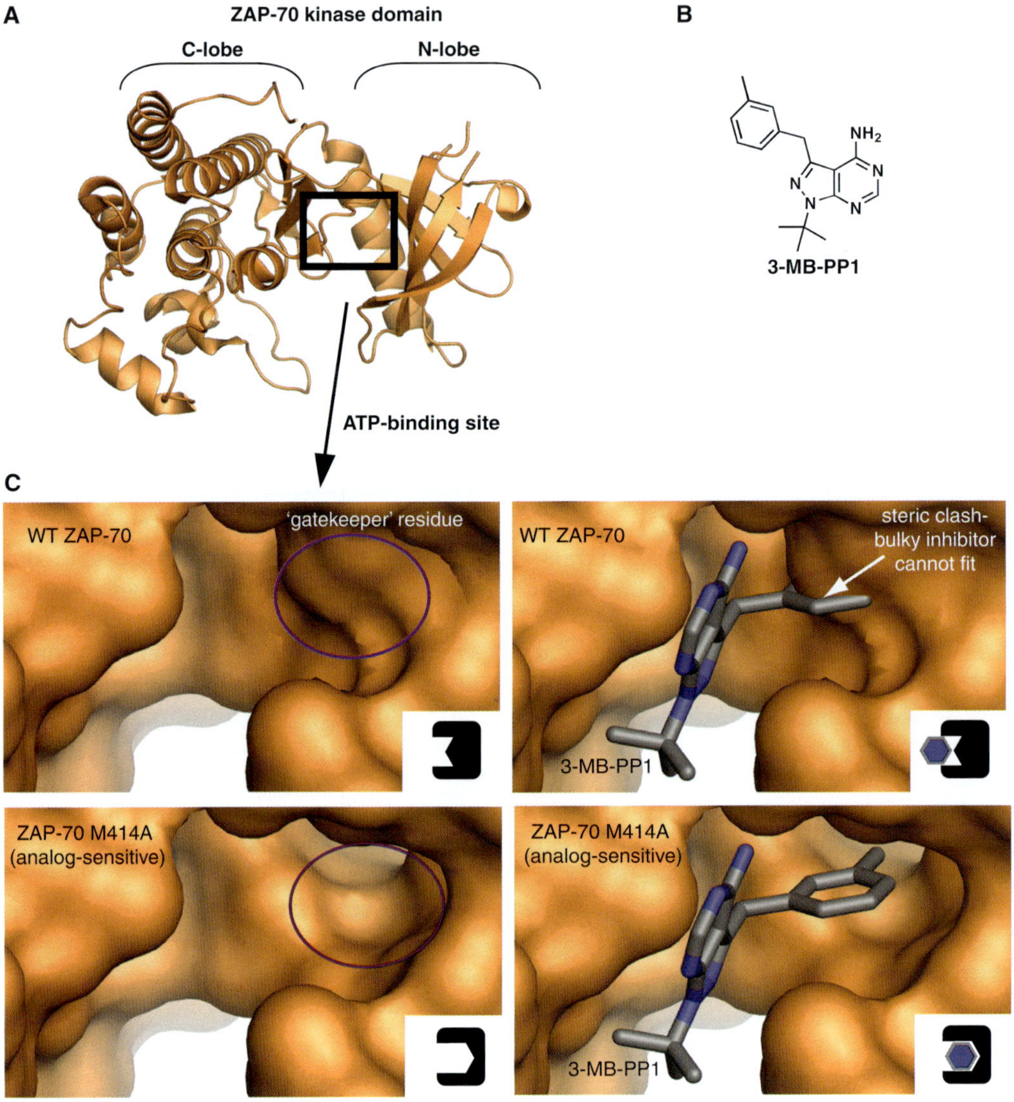

Figure 5. Chemical-genetic strategy to generate an inhibitor-sensitive ZAP-70. (*A*) Structure of the ZAP-70 kinase domain (PDB code 1U59). C- and N-lobes are indicated. The ATP-binding site is boxed. (*B*) Chemical structure of the inhibitor 3-MB-PP1. (*C*) Schematic of the analog-sensitive strategy. *Left*, Diagrams show the WT (*top*) and analog-sensitive mutant (*bottom*) ZAP-70 ATP-binding regions, modeled using PyMol. Circles identify the region of the gatekeeper residue. *Right*, Diagrams show the model of 3-MB-PP1 binding within the WT (*top*) and analog-sensitive mutant (*bottom*) ZAP-70. The presence of the "gatekeeper" residue in the ATP-binding site of WT ZAP-70 prevents the binding of the bulky 3-MB-PP1 inhibitor. In the analog-sensitive ZAP-70 mutant (bottom), the enlarged ATP-binding pocket allows binding of 3-MB-PP1. The 3-MB-PP1 model is derived from the crystal structure of the kinase domain of cryptosporidium parvum calcium dependent protein kinase in complex with 3-MB-PP1 (PDB code 2WEI). Top, white arrow points to the steric clash between 3-MB-PP1 and the gatekeeper residue, indicating that 3-MB-PP1 is not compatible with binding to WT ZAP-70.

REFERENCES

Allsup DJ, Kamiguti AS, Lin K, Sherrington PD, Matrai Z, Slupsky JR, Cawley JC, Zuzel M. 2005. B-cell receptor translocation to lipid rafts and associated signaling differ between prognostically important subgroups of chronic lymphocytic leukemia. *Cancer Res* **65:** 7328–7337.

Arpaia E, Shahar M, Dadi H, Cohen A, Roifman CM. 1994. Defective T cell receptor signaling and CD8+ thymic selection in humans lacking zap-70 kinase. *Cell* **76:** 947–958.

Ashe JM, Wiest DL, Abe R, Singer A. 1999. ZAP-70 protein promotes tyrosine phosphorylation of T cell receptor signaling motifs (ITAMs) in immature CD4(+)8(+) thymocytes with limiting p56(lck). *J Exp Med* **189:** 1163–1168.

Au-Yeung BB, Deindl S, Hsu LY, Palacios EH, Levin SE, Kuriyan J, Weiss A. 2009. The structure, regulation, and function of ZAP-70. *Immunol Rev* **228:** 41–57.

Bishop AC, Shah K, Liu Y, Witucki L, Kung C, Shokat KM. 1998. Design of allele-specific inhibitors to probe protein kinase signaling. *Curr Biol* **8:** 257–266.

Brdicka T, Kadlecek TA, Roose JP, Pastuszak AW, Weiss A. 2005. Intramolecular regulatory switch in ZAP-70: Analogy with receptor tyrosine kinases. *Mol Cell Biol* **25:** 4924–4933.

Campi G, Varma R, Dustin ML. 2005. Actin and agonist MHC-peptide complex-dependent T cell receptor microclusters as scaffolds for signaling. *J Exp Med* **202:** 1031–1036.

Chan AC, Dalton M, Johnson R, Kong GH, Wang T, Thoma R, Kurosaki T. 1995. Activation of ZAP-70 kinase activity by phosphorylation of tyrosine 493 is required for lymphocyte antigen receptor function. *Embo J* **14:** 2499–2508.

Chan AC, Irving BA, Fraser JD, Weiss A. 1991. The zeta chain is associated with a tyrosine kinase and upon T-cell antigen receptor stimulation associates with ZAP-70, a 70-kDa tyrosine phosphoprotein. *Proc Natl Acad Sci* **88:** 9166–9170.

Chan AC, Iwashima M, Turck CW, Weiss A. 1992. ZAP-70: A 70 kd protein-tyrosine kinase that associates with the TCR ζ chain. *Cell* **71:** 649–662.

Chan AC, Kadlecek TA, Elder ME, Filipovich AH, Kuo WL, Iwashima M, Parslow TG, Weiss A. 1994. ZAP-70 deficiency in an autosomal recessive form of severe combined immunodeficiency. *Science* **264:** 1599–1601.

Chen L, Apgar J, Huynh L, Dicker F, Giago-McGahan T, Rassenti L, Weiss A, Kipps TJ. 2005. ZAP-70 directly enhances IgM signaling in chronic lymphocytic leukemia. *Blood* **105:** 2036–2041.

Chen L, Huynh L, Apgar J, Tang L, Rassenti L, Weiss A, Kipps TJ. 2008. ZAP-70 enhances IgM signaling independent of its kinase activity in chronic lymphocytic leukemia. *Blood* **111:** 2685–2692.

Chen L, Widhopf G, Huynh L, Rassenti L, Rai KR, Weiss A, Kipps TJ. 2002. Expression of ZAP-70 is associated with increased B-cell receptor signaling in chronic lymphocytic leukemia. *Blood* **100:** 4609–4614.

Cheng AM, Negishi I, Anderson SJ, Chan AC, Bolen J, Loh DY, Pawson T. 1997. The Syk and ZAP-70 SH2-containing tyrosine kinases are implicated in pre-T cell receptor signaling. *Proc Natl Acad Sci* **94:** 9797–9801.

Chiorazzi N, Rai KR, Ferrarini M. 2005. Chronic lymphocytic leukemia. *N Engl J Med* **352:** 804–815.

Chu DH, Morita CT, Weiss A. 1998. The Syk family of protein tyrosine kinases in T-cell activation and development. *Immunol Rev* **165:** 167–180.

Chu DH, van Oers NS, Malissen M, Harris J, Elder M, Weiss A. 1999. Pre-T cell receptor signals are responsible for the down-regulation of Syk protein tyrosine kinase expression. *J Immunol* **163:** 2610–2620.

Davanture S, Leignadier J, Milani P, Soubeyran P, Malissen B, Malissen M, Schmitt-Verhulst AM, Boyer C. 2005. Selective defect in antigen-induced TCR internalization at the immune synapse of CD8 T cells bearing the ZAP-70(Y292F) mutation. *J Immunol* **175:** 3140–3149.

Deindl S, Kadlecek TA, Brdicka T, Cao X, Weiss A, Kuriyan J. 2007. Structural basis for the inhibition of tyrosine kinase activity of ZAP-70. *Cell* **129:** 735–746.

Deindl S, Kadlecek T, Cao X, Kuriyan J, Weiss A. 2009. Stability of an autoinhibitory interface in the structure of the tyrosine kinase ZAP-70 impacts T cell receptor response. *Proc Natl Acad Sci* **106:** 20699–20704.

Di Bartolo V, Mege D, Germain V, Pelosi M, Dufour E, Michel F, Magistrelli G, Isacchi A, Acuto O. 1999. Tyrosine 319, a newly identified phosphorylation site of ZAP-70, plays a critical role in T cell antigen receptor signaling. *J Biol Chem* **274:** 6285–6294.

Elder ME, Lin D, Clever J, Chan AC, Hope TJ, Weiss A, Parslow TG. 1994. Human severe combined immunodeficiency due to a defect in ZAP-70, a T cell tyrosine kinase. *Science* **264:** 1596–1599.

Elder ME, Skoda-Smith S, Kadlecek TA, Wang F, Wu J, Weiss A. 2001. Distinct T cell developmental consequences in humans and mice expressing identical mutations in the DLAARN motif of ZAP-70. *J Immunol* **166:** 656–661.

Folmer RH, Geschwindner S, Xue Y. 2002. Crystal structure and NMR studies of the apo SH2 domains of ZAP-70: Two bikes rather than a tandem. *Biochemistry* **41:** 14176–14184.

Futterer K, Wong J, Grucza RA, Chan AC, Waksman G. 1998. Structural basis for Syk tyrosine kinase ubiquity in signal transduction pathways revealed by the crystal structure of its regulatory SH2 domains bound to a dually phosphorylated ITAM peptide. *J Mol Biol* **281:** 523–537.

Gelfand EW, Weinberg K, Mazer BD, Kadlecek TA, Weiss A. 1995. Absence of ZAP-70 prevents signaling through the antigen receptor on peripheral blood T cells but not on thymocytes. *J Exp Med* **182:** 1057–1065.

Gelkop S, Isakov N. 1999. T cell activation stimulates the association of enzymatically active tyrosine-phosphorylated ZAP-70 with the Crk adapter proteins. *J Biol Chem* **274:** 21519–21527.

Gelkop S, Babichev Y, Isakov N. 2001. T cell activation induces direct binding of the Crk adapter protein to the regulatory subunit of phosphatidylinositol 3-kinase (p85) via a complex mechanism involving the Cbl protein. *J Biol Chem* **276:** 36174–36182.

Gelkop S, Gish GD, Babichev Y, Pawson T, Isakov N. 2005. T cell activation-induced CrkII binding to the Zap70 protein tyrosine kinase is mediated by Lck-dependent phosphorylation of Zap70 tyrosine 315. *J Immunol* **175:** 8123–8132.

Gobessi S, Laurenti L, Longo PG, Sica S, Leone G, Efremov DG. 2007. ZAP-70 enhances B-cell-receptor signaling despite absent or inefficient tyrosine kinase activation in chronic lymphocytic leukemia and lymphoma B cells. *Blood* **109:** 2032–2039.

Gong Q, Jin X, Akk AM, Foger N, White M, Gong G, Bubeck Wardenburg J, Chan AC. 2001. Requirement for tyrosine residues 315 and 319 within ζchain-associated protein 70 for T cell development. *J Exp Med* **194:** 507–518.

Grazioli L, Germain V, Weiss A, Acuto O. 1998. Anti-peptide antibodies detect conformational changes of the inter-SH2 domain of ZAP-70 due to binding to the ζ chain and to intramolecular interactions. *J Biol Chem* **273:** 8916–8921.

Gu Y, Chae HD, Siefring JE, Jasti AC, Hildeman DA, Williams DA. 2006. RhoH GTPase recruits and activates Zap70 required for T cell receptor signaling and thymocyte development. *Nat Immunol* **7:** 1182–1190.

Hatada MH, Lu X, Laird ER, Green J, Morgenstern JP, Lou M, Marr CS, Phillips TB, Ram MK, Theriault K et al. 1995. Molecular basis for interaction of the protein tyrosine kinase ZAP-70 with the T-cell receptor. *Nature* **377:** 32–38.

Hirabayashi A, Mukaiyama H, Kobayashi H, Shiohara H, Nakayama S, Ozawa M, Miyazawa K, Misawa K, Ohnota H, Isaji M. 2009. Structure-activity relationship studies of 5-benzylaminoimidazo[1,2-c]pyrimidine-8-carboxamide derivatives as potent, highly selective ZAP-70 kinase inhibitors. *Bioorg Med Chem* **17:** 284–294.

Hivroz C, Fischer A. 1994. Immunodeficiency diseases. Multiple roles for ZAP-70. *Curr Biol* **4:** 731–733.

Hsu LY, Tan YX, Xiao Z, Malissen M, Weiss A. 2009. A hypomorphic allele of ZAP-70 reveals a distinct thymic threshold for autoimmune disease versus autoimmune reactivity. *J Exp Med* **206:** 2527–2541.

Ilani T, Khanna C, Zhou M, Veenstra TD, Bretscher A. 2007. Immune synapse formation requires ZAP-70 recruitment by ezrin and CD43 removal by moesin. *J Cell Biol* **179:** 733–746.

Jin L, Pluskey S, Petrella EC, Cantin SM, Gorga JC, Rynkiewicz MJ, Pandey P, Strickler JE, Babine RE, Weaver DT et al. 2004. The three-dimensional structure of the ZAP-70 kinase domain in complex with staurosporine: Implications for the design of selective inhibitors. *J Biol Chem* **279:** 42818–42825.

Kadlecek TA, van Oers NS, Lefrancois L, Olson S, Finlay D, Chu DH, Connolly K, Killeen N, Weiss A. 1998. Differential requirements for ZAP-70 in TCR signaling and T cell development. *J Immunol* **161:** 4688–4694.

Katzav S, Sutherland M, Packham G, Yi T, Weiss A. 1994. The protein tyrosine kinase ZAP-70 can associate with the SH2 domain of proto-Vav. *J Biol Chem* **269:** 32579–32585.

Kong G, Dalton M, Bubeck Wardenburg J, Straus D, Kurosaki T, Chan AC. 1996. Distinct tyrosine phosphorylation sites in ZAP-70 mediate activation and negative regulation of antigen receptor function. *Mol Cell Biol* **16:** 5026–5035.

Kumaran S, Grucza RA, Waksman G. 2003. The tandem Src homology 2 domain of the Syk kinase: A molecular device that adapts to interphosphotyrosine distances. *Proc Natl Acad Sci* **100:** 14828–14833.

Latour S, Chow LM, Veillette A. 1996. Differential intrinsic enzymatic activity of Syk and Zap-70 protein-tyrosine kinases. *J Biol Chem* **271:** 22782–22790.

Levin SE, Zhang C, Kadlecek TA, Shokat KM, Weiss A. 2008. Inhibition of ZAP-70 kinase activity via an analog-sensitive allele blocks T cell receptor and CD28 superagonist signaling. *J Biol Chem* **283:** 15419–15430.

Lupher ML Jr, Reedquist KA, Miyake S, Langdon WY, Band H. 1996. A novel phosphotyrosine-binding domain in the N-terminal transforming region of Cbl interacts directly and selectively with ZAP-70 in T cells. *J Biol Chem* **271:** 24063–24068.

Lupher ML Jr, Songyang Z, Shoelson SE, Cantley LC, Band H. 1997. The Cbl phosphotyrosine-binding domain selects a D(N/D)XpY motif and binds to the Tyr292 negative regulatory phosphorylation site of ZAP-70. *J Biol Chem* **272:** 33140–33144.

Magnan A, Di Bartolo V, Mura AM, Boyer C, Richelme M, Lin YL, Roure A, Gillet A, Arrieumerlou C, Acuto O et al. 2001. T cell development and T cell responses in mice with mutations affecting tyrosines 292 or 315 of the ZAP-70 protein tyrosine kinase. *J Exp Med* **194:** 491–505.

Matsuda S, Suzuki-Fujimoto T, Minowa A, Ueno H, Katamura K, Koyasu S. 1999. Temperature-sensitive ZAP70 mutants degrading through a proteasome-independent pathway. Restoration of a kinase domain mutant by Cdc37. *J Biol Chem* **274:** 34515–34518.

Mege D, Di Bartolo V, Germain V, Tuosto L, Michel F, Acuto O. 1996. Mutation of tyrosines 492/493 in the kinase domain of ZAP-70 affects multiple T-cell receptor signaling pathways. *J Biol Chem* **271:** 32644–32652.

Murphy MA, Schnall RG, Venter DJ, Barnett L, Bertoncello I, Thien CB, Langdon WY, Bowtell DD. 1998. Tissue hyperplasia and enhanced T-cell signalling via ZAP-70 in c-Cbl-deficient mice. *Mol Cell Biol* **18:** 4872–4882.

Naramura M, Jang IK, Kole H, Huang F, Haines D, Gu H. 2002. c-Cbl and Cbl-b regulate T cell responsiveness by promoting ligand-induced TCR down-modulation. *Nat Immunol* **3:** 1192–1199.

Naramura M, Kole HK, Hu RJ, Gu H. 1998. Altered thymic positive selection and intracellular signals in Cbl-deficient mice. *Proc Natl Acad Sci* **95:** 15547–15552.

Negishi I, Motoyama N, Nakayama K, Nakayama K, Senju S, Hatakeyama S, Zhang Q, Chan AC, Loh DY. 1995a. Essential role for ZAP-70 in both positive and negative selection of thymocytes. *Nature* **376:** 435–438.

Negishi I, Motoyama N, Nakayama K, Senju S, Hatakeyama S, Zhang Q, Chan AC, Loh DY. 1995b. Essential role for ZAP-70 in both positive and negative selection of thymocytes. *Nature* **376:** 435–438.

Noraz N, Schwarz K, Steinberg M, Dardalhon V, Rebouissou C, Hipskind R, Friedrich W, Yssel H, Bacon K, Taylor N. 2000. Alternative antigen receptor (TCR) signaling in T cells derived from ZAP-70-deficient patients expressing high levels of Syk. *J Biol Chem* **275:** 15832–15838.

Ottinger EA, Botfield MC, Shoelson SE. 1998. Tandem SH2 domains confer high specificity in tyrosine kinase signaling. *J Biol Chem* **273:** 729–735.

Palacios EH, Weiss A. 2007. Distinct roles for Syk and ZAP-70 during early thymocyte development. *J Exp Med* **204:** 1703–1715.

Pelosi M, Di Bartolo V, Mounier V, Mege D, Pascussi JM, Dufour E, Blondel A, Acuto O. 1999. Tyrosine 319 in the interdomain B of ZAP-70 is a binding site for the Src homology 2 domain of Lck. *J Biol Chem* **274:** 14229–14237.

Rao N, Lupher ML Jr, Ota S, Reedquist KA, Druker BJ, Band H. 2000. The linker phosphorylation site Tyr292 mediates the negative regulatory effect of Cbl on ZAP-70 in T cells. *J Immunol* **164:** 4616–4626.

Rassenti LZ, Huynh L, Toy TL, Chen L, Keating MJ, Gribben JG, Neuberg DS, Flinn IW, Rai KR, Byrd JC et al. 2004. ZAP-70 compared with immunoglobulin heavy-chain gene mutation status as a predictor of disease progression in chronic lymphocytic leukemia. *N Engl J Med* **351:** 893–901.

Rosenwald A, Alizadeh AA, Widhopf G, Simon R, Davis RE, Yu X, Yang L, Pickeral OK, Rassenti LZ, Powell J et al. 2001. Relation of gene expression phenotype to immunoglobulin mutation genotype in B cell chronic lymphocytic leukemia. *J Exp Med* **194:** 1639–1647.

Sakaguchi S, Sakaguchi N, Yoshitomi H, Hata H, Takahashi T, Nomura T. 2006. Spontaneous development of autoimmune arthritis due to genetic anomaly of T cell signal transduction: Part 1. *Semin Immunol* **18:** 199–206.

Sakaguchi N, Takahashi T, Hata H, Nomura T, Tagami T, Yamazaki S, Sakihama T, Matsutani T, Negishi I, Nakatsuru S et al. 2003. Altered thymic T-cell selection due to a mutation of the ZAP-70 gene causes autoimmune arthritis in mice. *Nature* **426:** 454–460.

Sasahara Y, Rachid R, Byrne MJ, de la Fuente MA, Abraham RT, Ramesh N, Geha RS. 2002. Mechanism of recruitment of WASP to the immunological synapse and of its activation following TCR ligation. *Mol Cell* **10:** 1269–1281.

Siggs OM, Miosge LA, Yates AL, Kucharska EM, Sheahan D, Brdicka T, Weiss A, Liston A, Goodnow CC. 2007. Opposing functions of the T cell receptor kinase ZAP-70 in immunity and tolerance differentially titrate in response to nucleotide substitutions. *Immunity* **27:** 912–926.

Sloan-Lancaster J, Presley J, Ellenberg J, Yamazaki T, Lippincott-Schwartz J, Samelson LE. 1998. ZAP-70 association with T cell receptor ζ(TCRζ): Fluorescence imaging of dynamic changes upon cellular stimulation. *J Cell Biol* **143:** 613–624.

Sloan-Lancaster J, Zhang W, Presley J, Williams BL, Abraham RT, Lippincott-Schwartz J, Samelson LE. 1997. Regulation of ZAP-70 intracellular localization: Visualization with the green fluorescent protein. *J Exp Med* **186:** 1713–1724.

Thien CB, Scaife RM, Papadimitriou JM, Murphy MA, Bowtell DD, Langdon WY. 2003. A mouse with a loss-of-function mutation in the c-Cbl TKB domain shows perturbed thymocyte signaling without enhancing the activity of the ZAP-70 tyrosine kinase. *J Exp Med* **197:** 503–513.

Toyabe S, Watanabe A, Harada W, Karasawa T, Uchiyama M. 2001. Specific immunoglobulin E responses in ZAP-70-deficient patients are mediated by Syk-dependent T-cell receptor signalling. *Immunology* **103:** 164–171.

van Oers NS, Killeen N, Weiss A. 1994. ZAP-70 is constitutively associated with tyrosine-phosphorylated TCR ζ in murine thymocytes and lymph node T cells. *Immunity* **1:** 675–685.

Varma R, Campi G, Yokosuka T, Saito T, Dustin ML. 2006. T cell receptor-proximal signals are sustained in peripheral microclusters and terminated in the central supramolecular activation cluster. *Immunity* **25:** 117–127.

Wang HY, Altman Y, Fang D, Elly C, Dai Y, Shao Y, Liu YC. 2001. Cbl promotes ubiquitination of the T cell receptor ζ through an adaptor function of Zap-70. *J Biol Chem* **276:** 26004–26011.

Wange RL, Guitian R, Isakov N, Watts JD, Aebersold R, Samelson LE. 1995. Activating and inhibitory mutations in adjacent tyrosines in the kinase domain of ZAP-70. *J Biol Chem* **270:** 18730–18733.

Watts JD, Affolter M, Krebs DL, Wange RL, Samelson LE, Aebersold R. 1994. Identification by electrospray ionization mass spectrometry of the sites of tyrosine phosphorylation induced in activated Jurkat T cells on the protein tyrosine kinase ZAP-70. *J Biol Chem* **269:** 29520–29529.

Weiss A. 1993. T cell antigen receptor signal transduction: A tale of tails and cytoplasmic protein-tyrosine kinases. *Cell* **73:** 209–212.

Wiest DL, Ashe JM, Howcroft TK, Lee HM, Kemper DM, Negishi I, Singer DS, Singer A, Abe R. 1997. A spontaneously arising mutation in the DLAARN motif of murine ZAP-70 abrogates kinase activity and arrests thymocyte development. *Immunity* **6:** 663–671.

Wiest DL, Yuan L, Jefferson J, Benveniste P, Tsokos M, Klausner RD, Glimcher LH, Samelson LE, Singer A. 1993. Regulation of T cell receptor expression in immature CD4+CD8+ thymocytes by p56lck tyrosine kinase: Basis for differential signaling by CD4 and CD8 in immature thymocytes expressing both coreceptor molecules. *The Journal of experimental medicine* **178:** 1701–1712.

Williams BL, Irvin BJ, Sutor SL, Chini CC, Yacyshyn E, Bubeck Wardenburg J, Dalton M, Chan AC, Abraham RT. 1999. Phosphorylation of Tyr319 in ZAP-70 is required for T-cell antigen receptor-dependent phospholipase C-γ1 and Ras activation. *Embo J* **18:** 1832–1844.

Williams BL, Schreiber KL, Zhang W, Wange RL, Samelson LE, Leibson PJ, Abraham RT. 1998. Genetic evidence for differential coupling of Syk family kinases to the T-cell receptor: Reconstitution studies in a ZAP-70-deficient Jurkat T-cell line. *Mol Cell Biol* **18:** 1388–1399.

Wu J, Zhao Q, Kurosaki T, Weiss A. 1997. The Vav binding site (Y315) in ZAP-70 is critical for antigen receptor-mediated signal transduction. *J Exp Med* **185:** 1877–1882.

Yokosuka T, Sakata-Sogawa K, Kobayashi W, Hiroshima M, Hashimoto-Tane A, Tokunaga M, Dustin ML, Saito T. 2005. Newly generated T cell receptor microclusters initiate and sustain T cell activation by recruitment of Zap70 and SLP-76. *Nat Immunol* **6:** 1253–1262.

Zhao Q, Weiss A. 1996. Enhancement of lymphocyte responsiveness by a gain-of-function mutation of ZAP-70. *Mol Cell Biol* **16:** 6765–6774.

Zhao Q, Williams BL, Abraham RT, Weiss A. 1999. Interdomain B in ZAP-70 regulates but is not required for ZAP-70 signaling function in lymphocytes. *Mol Cell Biol* **19:** 948–956.

The LAT Story: A Tale of Cooperativity, Coordination, and Choreography

Lakshmi Balagopalan, Nathan P. Coussens, Eilon Sherman, Lawrence E. Samelson, and Connie L. Sommers

LCMB/CCR/NCI/NIH, Bethesda, Maryland 20892-4256

Correspondence: samelson@helix.nih.gov

The adapter molecule LAT is a nucleating site for multiprotein signaling complexes that are vital for the function and differentiation of T cells. Extensive investigation of LAT in multiple experimental systems has led to an integrated understanding of the formation, composition, regulation, dynamic movement, and function of LAT-nucleated signaling complexes. This review discusses interactions of signaling molecules that bind directly or indirectly to LAT and the role of cooperativity in stabilizing LAT-nucleated signaling complexes. In addition, it focuses on how imaging studies visualize signaling assemblies as signaling clusters and demonstrate their dynamic nature and cellular fate. Finally, this review explores the function of LAT based on the interpretation of mouse models using various LAT mutants.

INTRODUCTION

Linker for activation of T cells (LAT) was cloned a little more than a decade ago. Since then, a multitude of studies have revealed that LAT-based complexes catalyze critical TCR-mediated signaling reactions and enable activation of multiple downstream pathways that control almost all TCR-initiated cellular responses. This article spotlights diverse experimental systems in which LAT function has been studied. Information gained from these studies has led to an integrated understanding of the cellular function of LAT.

Cloning and Structural Features of LAT

The study of the tyrosine phosphorylation of proteins induced by immunoreceptor and growth factor receptor stimulation has led to critical insights into mechanisms of signal transduction (Hunter 2009). Early studies showed that a number of proteins became phosphorylated on tyrosine residues following TCR stimulation in Jurkat T-cell leukemia cells and in normal T cells (June et al. 1990). Many of these proteins, such as ZAP-70, SLP-76, and PLC-γ1, have been shown to be critical elements for TCR signal transduction (Kane et al. 2000). A protein with an apparent molecular weight of 36 and 38 kDa was also prominently phosphorylated on tyrosine in response to TCR stimulation. Several preliminary studies showed that this protein, known then as pp36/38, was membrane-associated and capable of binding SH2 domains of Grb2, Grap, PLC-γ1, and the p85 subunit of phosphatidylinositol 3-kinase (PI3K) (June et al. 1990; Gilliland et al. 1992; Buday et al. 1994; Sieh et al. 1994; Fukazawa

Cite this article as *Cold Spring Harb Perspect Biol* 2010;2:a005512

et al. 1995a; Trub et al. 1997). Although pp36/38 was first observed in 1990, it proved challenging to isolate. It was not until 1998 that the Samelson laboratory cloned it by large-scale membrane purification of activated Jurkat cells (Zhang et al. 1998a). Shortly after, Weber et al. reported the cloning of the rat and human proteins from thymocytes (Weber et al. 1998). The Samelson lab named the protein product LAT, for Linker for Activation of T cells based on several of its characteristics. LAT is expressed in T cells and in a limited number of other immune cell types (mast cells, natural killer cells, megakaryocytes, platelets, and immature B cells) (Facchetti et al. 1999; Oya et al. 2003). Furthermore, as detailed below, LAT facilitates the recruitment of many signaling proteins to the plasma membrane where it links receptors, tyrosine kinases and their substrates and other effector molecules together, functioning as a critical activator of T cells.

Sequencing of human LAT cDNA identified an open reading frame encoding a protein predicted to contain 233 amino acids. The mouse and rat homologs of LAT encode 242 and 241 amino acid proteins, respectively, and have 65%–70% sequence identity with human LAT. The predicted molecular mass of LAT is 25 kDa. However, LAT is strikingly acidic and its charge may account for slower migration on SDS-PAGE leading to its apparent molecular weight of 36/38 kDa. Structurally, LAT is a type III transmembrane protein. It has a cytosolic carboxyl terminus (like type I proteins), but lacks a signal sequence (Brown 2006). LAT contains only a four-amino-acid extracellular region, a single transmembrane spanning region and a long intracellular region with no apparent intrinsic enzymatic activity or protein–protein interaction domains. However, consistent with the strong tyrosine phosphorylation of pp36/38 observed upon TCR stimulation, the intracellular domain of LAT contains nine tyrosines conserved between humans, mice, and rats. Examination of LAT amino-acid sequence also revealed two conserved cysteine residues (C26 and C29 in human LAT), which are located adjacent to the predicted transmembrane domain of LAT and are subject to posttranslational palmitoylation (Zhang et al. 1998b). Palmitoylation of LAT is necessary for LAT function (Lin et al. 1999; Zhang et al. 1999a), but the role of palmitoylation in specific localization of LAT within the plasma membrane has been controversial and is discussed below. More recently, studies have shown that LAT is also subject to ubiquitylation (Brignatz et al. 2005; Balagopalan et al. 2007), a modification that might be involved in activation-induced internalization of LAT complexes and regulation of LAT protein levels. Inspection of LAT amino acid sequence reveals two lysines (K52 and K204 in human LAT), which might serve as potential sites for ubiquitylation. A schematic of LAT structural features is shown in Figure 1A.

Palmitoylation and Membrane Localization of LAT

Palmitoylation is thought to enhance the association of transmembrane proteins, including LAT, with regions of membrane heterogeneity sometimes called lipid rafts (Brown 2006). In T cells, several signaling molecules in addition to LAT, including the TCR, Lck, Vav, Grb2, PLC-γ1, and Ras are associated with lipid rafts (Brdicka et al. 1998; Montixi et al. 1998; Zhang et al. 1998b). Thus, rafts have been postulated to function as important platforms to initiate signaling cascades (Brown and London 1998). However, the physiological role of this membrane heterogeneity has been intensely debated (Munro 2003; He and Marguet 2008; Kenworthy 2008). Reflecting this debate, the importance of raft localization for LAT function has been controversial.

Early studies reported that a cysteine-mutated form of LAT was not recruited to lipid rafts and could not reconstitute signaling in a LAT-deficient Jurkat cell line, leading to the conclusion that localization of LAT to lipid rafts was required for its function (Lin et al. 1999; Zhang et al. 1999a). However, more recent studies indicated that palmitoylation of LAT was required for the protein to be transported efficiently to the plasma membrane and that, in the absence of palmitoylation, LAT was

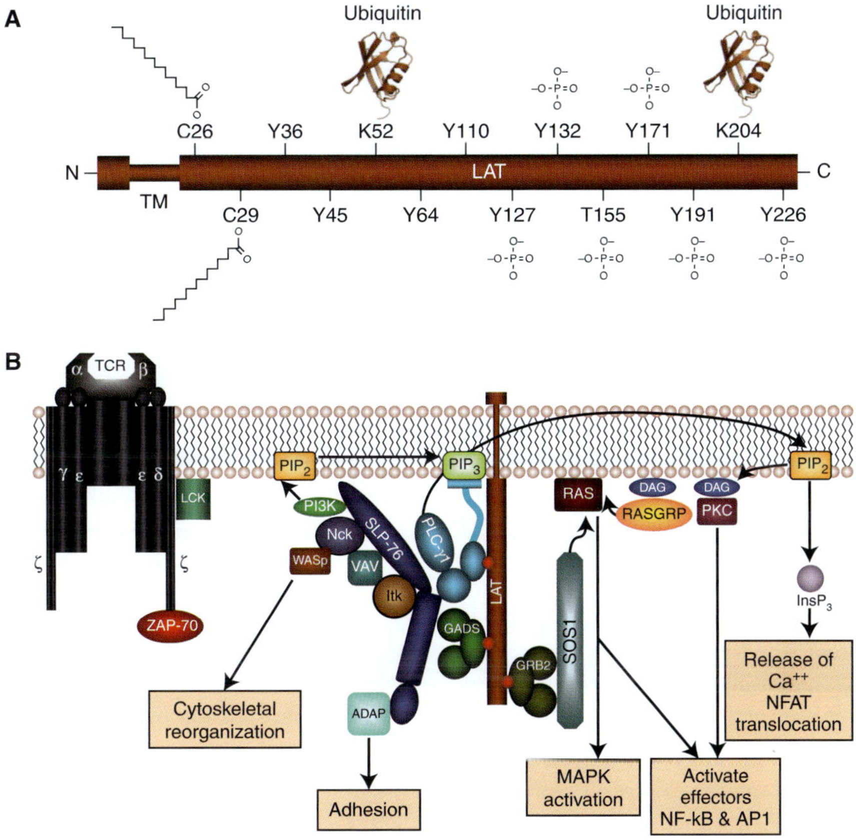

Figure 1. LAT in TCR signal transduction. (*A*) Human LAT is a 233 amino-acid type III transmembrane protein with four extracellular amino acids, a single transmembrane region, and a cytosolic region that undergoes multiple posttranslational modifications. Modifications include palmitoylation at cysteine residues C26 and C29 and possible ubiquitylation at lysine residues K52 and K204 (Ubiquitin [Protein Data Bank ID 1UBQ] is represented as a ribbon diagram generated with the program PyMOL). Nine tyrosine residues are conserved between human and mouse LAT sequences, five of which have been shown to undergo phosphorylation (Y127, Y132, Y171, Y191, and Y226). Threonine 155 is also phosphorylated by Erk. (*B*) Ligation of the TCR induces tyrosine phosphorylation of numerous adapter and effector proteins leading to the activation of multiple signaling pathways important for gene transcription, cytoskeletal reorganization, and cell adhesion. LAT is central to this process by nucleating multiprotein signaling complexes that are important for enzyme activation and signal propagation.

susceptible to degradation (Gringhuis et al. 2000; Hundt et al. 2006; Tanimura et al. 2006; Hundt et al. 2009). Furthermore, LAT fusion proteins targeted to nonraft domains reconstituted LAT function in LAT-deficient Jurkat cells or LAT-deficient mice (Zhu et al. 2005; Hundt et al. 2009). These data raise the possibility that the signaling defects initially observed for a LAT palmitoylation mutant might result from defects in plasma membrane transport, rather than displacement from lipid domains. Collectively, these data suggest that targeting of LAT to the plasma membrane of the cell is sufficient for its function, regardless of specific localization within the membrane.

LAT Is Central to T-Cell Signaling

The essential role of LAT in T-cell signal transduction has been demonstrated in a variety of experimental settings. Initial work showed that Jurkat T-cell lines lacking expression of LAT were defective in several TCR-mediated signaling events including Ca^{2+} mobilization, Ras activation, CD69 expression, Erk activation, and AP-1/NFAT-directed gene transcription (Finco et al. 1998; Zhang et al. 1999a; Samelson 2002). Reintroduction of LAT rescued all these defects indicating that LAT is indispensable for T-cell activation via the TCR. Soon after, targeted disruption of the LAT gene in mice revealed a requirement for LAT in T-cell development. Animals that lack LAT exhibited an early arrest of thymocyte development and no mature αβ T cells were found in their peripheral lymphoid organs (Zhang et al. 1999b). Additional reports highlighted the adapter function of LAT by showing that multiple signaling proteins bind phosphorylated LAT once the TCR is engaged. These LAT binding proteins subsequently attract multiple cytosolic protein partners and further waves of tyrosine phosphorylation and protein interactions occur. Thus LAT functions as a crucial nucleating site at the plasma membrane for multiprotein signaling complexes.

LAT AS A NUCLEATION CENTER FOR T-CELL SIGNAL TRANSDUCTION

LAT functions as a classic adapter protein by facilitating the formation of multiprotein complexes (Fig. 1B). Numerous studies have investigated the composition of LAT-nucleated complexes. This section will describe the direct and indirect associations of enzymes, adapters and effectors to LAT and provide insights into how these interactions lead to activation of specific intracellular pathways.

Phosphorylation of LAT

LAT is rapidly phosphorylated on tyrosine residues upon TCR engagement. Though a full understanding of how the kinases phosphorylate the individual tyrosines on LAT in vivo remains unresolved, overexpression and in vitro studies have implicated primarily ZAP-70, but also Lck and Itk in LAT phosphorylation (Zhang et al. 1998a; Paz et al. 2001; Perez-Villar et al. 2002; Jiang and Cheng 2007). Even less is known about the phosphatases that dephosphorylate LAT, however CD148 has been implicated (Baker et al. 2001).

PLC-γ1 Binds Y132 of LAT and Mediates Transcriptional Activation in T Cells

The phospholipase PLC-γ1 is an important mediator of TCR signal transduction. PLC-γ1 hydrolyzes phosphatidylinositol 4,5-bisphosphate (PIP_2) to produce inositol 1,4,5-trisphosphate (IP_3) and diacylglycerol (DAG) (Rhee and Bae 1997). DAG stimulates the Ras activator RasGrp (Ras guanyl-releasing protein) and the serine/threonine kinase PKC (protein kinase C). The production of IP_3 by PLC-γ1 promotes the release of stored intracellular calcium, causing an extracellular Ca^{2+} influx and a sustained increase in intracellular Ca^{2+} concentrations important for NFAT activation (Oh-hora 2009).

PLC-γ1 is a multidomain enzyme that includes a pleckstrin homology (PH) domain that can bind membrane phosphoinositides, two phosphotyrosine-binding SH2 (src homology 2) domains and an SH3 (src homology 3) domain, which binds specific proline-rich sequences. PLC-γ1 binding to LAT was observed even before LAT was cloned and was shown to be dependent on the amino-terminal SH2 domain of PLC-γ1 (Gilliland et al. 1992; Weber et al. 1992; Stoica et al. 1998; Irvin et al. 2000). Interaction with LAT was shown to be required for both PLC-γ1 activation and localization near its substrate PIP_2 at the plasma membrane (Zhang et al. 1999a; Zhang et al. 2000; Lin and Weiss 2001). Subsequently, the amino-terminal SH2 domain of PLC-γ1 was shown to bind phosphorylated LAT Y132 with high affinity (Paz et al. 2001; Houtman et al. 2004). Furthermore, a single Y132F mutation abolished the association between LAT and PLC-γ1 in Jurkat T cells (Zhang et al. 2000). However, further experimentation revealed that full PLC-γ1 activation required additional binding interactions to LAT and LAT-associated molecules.

Several studies showed that LAT Y171 and Y191 also contributed to PLC-γ1 binding and that all four distal LAT tyrosines (132, 171, 191, and 226) were required for optimal PLC-γ1 phosphorylation and downstream calcium influx (Zhang et al. 2000; Lin and Weiss 2001; Zhu et al. 2003). The adapter protein SLP-76 (that binds to Gads, which binds to LAT Y171 and Y191) was also shown to be required for PLC-γ1 activation (Yablonski et al. 1998). Braiman and colleagues demonstrated that multiple domains of PLC-γ1 were required for association with LAT and all three SH domains were essential for enzyme activation. Furthermore, PLC-γ1 activity depended on interactions with other LAT-associated proteins including SLP-76, the guanine nucleotide exchange factor Vav and the E3-ubiquitin ligase c-Cbl (Braiman et al. 2006).

Grb2 Family Members Bind Y171, Y191, and Y226 of Phosphorylated LAT and Recruit a Number of Signaling Proteins to the LAT Complex

The Grb2 family members Grb2, Gads, and Grap can potentially bind phosphorylated LAT at the distal three tyrosines: Y171, Y191, and Y226 (Wange 2000). Grb2 (Growth-factor-receptor-bound protein 2) is a ubiquitously expressed adapter protein that contains an SH2 domain flanked by amino- and carboxy-terminal SH3 domains. The Grb2 adapter is involved in Ras and MAP kinase pathway activation in many cell surface receptor systems, primarily by association with the guanine nucleotide exchange factor (GEF) Sos1. Interestingly, Grb2 association with LAT requires the SH2 domain of Grb2 and pairs of LAT phosphotyrosines: Y171/Y191, Y171/Y226, or Y191/Y226 (Zhang et al. 2000; Zhu et al. 2003). Gads (Grb2-related adapter protein downstream of Shc) contains a structure similar to Grb2, but also has a unique glutamine- and proline-rich domain between the SH2 domain and the carboxy-terminal SH3 domain. In addition to binding LAT, Gads binds SLP-76, which recruits a number of other signaling proteins including PLC-γ1 (see Fig. 1B) (Liu et al. 1999; Zhang et al. 2000; Yablonski et al. 2001). Grap (Grb2-related adapter protein) is similar in structure to Grb2, but its role in T-cell signal transduction is not as well defined. It also associates with LAT through its SH2 domain (Trub et al. 1997; Zhang et al. 1998a). Studies of T cells from Grap-deficient mice indicate that Grap may be a negative regulator of TCR signal transduction (Shen et al. 2002).

Sos1 (Son of Sevenless homolog) is a multidomain GEF that promotes activation of the small G protein Ras, which in turn activates the MAP kinase Erk (Quilliam 2007). Sos1 binds Grb2 constitutively (Wittekind et al. 1994; Houtman et al. 2006). Following TCR ligation, Grb2 recruits Sos1 to the site of Ras localization at the membrane. Early experiments suggested that a trimolecular complex formed between Grb2, Sos1, and LAT following TCR engagement (Buday et al. 1994; Sieh et al. 1994). Consistent with this, Zhang and colleagues reported that Sos1 did not associate with the LAT mutant Y171/191F, which also did not associate with Grb2 (Zhang et al. 1998a). Interestingly, the significance of a Grb2/Sos1 complex likely extends beyond the activation of Ras. Houtman and colleagues have described a mechanism whereby a 2:1 Grb2:Sos1 complex can facilitate clustering of LAT-based signaling complexes (to be discussed in more detail later) (Houtman et al. 2006).

The proto-oncogene *c-Cbl* encodes a 120 kDa multidomain E3-ubiquitin ligase, which is phosphorylated and associates with LAT following TCR activation. c-Cbl also binds Grb2 constitutively (Donovan et al. 1994; Fukazawa et al. 1995b). Grb2 seems to be required for the interaction between c-Cbl and LAT, because c-Cbl does not associate with the LAT mutant Y171/191F, which fails to bind Grb2 (Zhang et al. 1998a). However, there is evidence that c-Cbl is also stabilized by interactions with other LAT-associated molecules, such as Nck (Rivero-Lezcano et al. 1994; Wunderlich et al. 1999), the p85 subunit of PI3K (Fukazawa et al. 1995b), and PLC-γ1 (Rellahan et al. 2003; Chiang et al. 2004; Braiman et al. 2006). By stabilizing the associations of molecules that contribute to signal activation, c-Cbl could promote positive TCR-induced signals; however, c-Cbl also appears to play a negative regulatory role in signaling.

In LAT-deficient Jurkat cells, c-Cbl is hyperphosphorylated upon TCR stimulation (Finco et al. 1998; Mingueneau et al. 2009) and phosphorylated c-Cbl has been implicated in negative regulation of TCR signal transduction (Boussiotis et al. 1997; Murphy et al. 1998). In addition, consistent with its role as an E3-ubiquitin ligase in several signaling systems, c-Cbl has been implicated in ubiquitylation of the TCRζ and CD3δ chains of the TCR as well as the signaling proteins ZAP-70, LAT, PLC-γ1, PI3K, Vav, and PKC-θ in T cells (Weissman 2001; Duan et al. 2004; Balagopalan et al. 2007). The effects of LAT ubiquitylation on controlling internalization and cellular LAT levels are discussed in the imaging section later.

Molecules Recruited to LAT via SLP-76 Promote T-Cell Activation

SLP-76 is an adapter protein containing three tyrosine motifs, a central proline-rich domain and a carboxy-terminal SH2 domain (Koretzky et al. 2006). The proline-rich domain of SLP-76 binds SH3 domains of Grb2 and Gads; however SLP-76 binds Gads with a 40–50-fold stronger affinity than Grb2 (Berry et al. 2002; Harkiolaki et al. 2003). Gads recruits SLP-76 to LAT by way of the interaction of its SH2 domain with phospho-LAT (Liu et al. 1999). SLP-76 recruits multiple effector molecules to the LAT complex. The three phosphotyrosines of SLP-76 bind PI3K, Vav, Itk, and Nck. The central proline-rich region binds to Gads with high affinity and to PLC-γ1 with low affinity (Yablonski et al. 2001; Houtman et al. 2004). However, SLP-76 only associates weakly with LAT in the absence of the LAT residue Y132, suggesting that SLP-76 binding to LAT is stabilized by the presence of PLC-γ1 (Zhang et al. 2000). The carboxy-terminal SH2 domain of SLP-76 interacts with the proteins HPK1, ADAP, and Shb. These interactions result in positive and negative effects on TCR signaling. Interactions that lead to negative effects on TCR signaling will be discussed in the following section.

PI3K is comprised of a 110 kDa catalytic subunit and an 85 kDa regulatory subunit (p85) and catalyzes the phosphorylation of PIP_2 to generate PIP_3 (Fruman and Bismuth 2009). The accumulation of PIP_3 facilitates membrane recruitment of proteins that contain PH domains, which recognize PIP_3 (Fruman et al. 1999). In T cells, PI3K regulates the recruitment of proteins that facilitate TCR signal transduction such as PLC-γ1, Sos1, RasGAP, Itk, Tec, and Vav. Early studies showed that PI3K bound directly to LAT (Fukazawa et al. 1995a; Lahesmaa et al. 1995; Paz et al. 2001); however more recent studies showed binding of p85 to tyrosine-phosphorylated SLP-76 (Shim et al. 2004). Inhibition of PI3K in T cells resulted in reduced TCR-induced calcium influx, Rac1 activation, and Erk1/2 phosphorylation. However, PI3K inhibition did not reduce the phosphorylation of PH domain-containing proteins such as PLC-γ1 and Vav (Cruz-Orcutt and Houtman 2009). Therefore PI3K plays a role in TCR activation, but its exact mechanism is still unclear.

Other SLP-76 interactors have clearer roles on T-cell activation. Vav is a guanine nucleotide exchange factor that activates the Rho family G-proteins Rac1, RhoA, and Cdc42. A number of studies have demonstrated the importance of the association of the Vav SH2 domain with SLP-76 phosphotyrosines (Koretzky et al. 2006). In addition to its catalytic role, Vav acts as a scaffold to stabilize PLC-γ1 and Itk interactions with SLP-76. In Vav-deficient T cells, the interaction between SLP-76 and PLC-γ1 was substantially reduced and Itk was not phosphorylated (Reynolds et al. 2002; Braiman et al. 2006). The Tec family kinase Itk phosphorylates PLC-γ1 (Readinger et al. 2009) and LAT (Perez-Villar et al. 2002). Itk is recruited to the LAT complex by binding of its SH2 domain to tyrosine phosphorylated SLP-76 (Bunnell et al. 2000). Itk plays a major role in T-cell activation by virtue of its phosphorylation of PLC-γ1, which is required for the activation of PLC-γ1.

Vav and the adapters Nck and WASp cooperate to effect cytoskeletal changes in activated T cells. Nck binds phosphorylated SLP-76 (Wunderlich et al. 1999) and WASp (Wiskott-Aldrich syndrome protein) (Rohatgi et al. 2001; Barda-Saad et al. 2005), bringing WASp in proximity to Vav. Vav recruits and activates Cdc42, which is necessary for WASp activation. Activated

WASp interacts with the Arp2/3 complex to initiate actin polymerization and cytoskeletal changes (Zeng et al. 2003).

Two other adapter molecules that bind SLP-76 and mediate positive signal transduction in T cells are ADAP and Shb. Phosphorylated ADAP binds the SH2 domain of SLP-76 (da Silva et al. 1997; Musci et al. 1997). This interaction is crucial for adhesion and integrin function in T cells (Baker et al. 2009; Horn et al. 2009; Wang et al. 2009). Tyrosine phosphorylated Shb also binds the SH2 domain of SLP-76 and promotes LAT, SLP-76, Vav, and PLC-γ1 phosphorylation (Lindholm et al. 1999; Lindholm et al. 2002). HPK1 is another molecule that binds the SH2 domain of SLP-76 but mediates negative signal transduction.

LAT Interactions Also Inhibit TCR Signal Transduction

Several molecules that bind to the LAT complex can inhibit TCR signal transduction. HPK1 (hematopoietic progenitor kinase 1) is a MAP4 kinase (Hu et al. 1996; Kiefer et al. 1996) that down-regulates activation of the MAP kinase Erk2 and the transcription factor AP-1 in a kinase-dependent manner (Liou et al. 2000). HPK-1 has a complex relationship with SLP-76. It can phosphorylate SLP-76; however, this ultimately leads to decreased SLP-76 and PLC-γ1 phosphorylation and IL-2 production (Di Bartolo et al. 2007; Shui et al. 2007).

SHIP-1 is a lipid phosphatase that mediates Fc receptor-dependent negative signaling (Daeron and Lesourne 2006). In T cells, SHIP-1 also acts as an adapter to recruit Dok-1 and Dok-2 to LAT complexes, which results in attenuated TCR signaling. SHIP-1-associated Dok-1 and Dok-2 negatively regulate ZAP-70 activity (Dong et al. 2006; Yasuda et al. 2007) and recruit RasGAP, which opposes Ras-mediated signaling (Tamir et al. 2000; Ott et al. 2002). The adapter protein Gab2 also has several potential mechanisms for inhibiting signaling. It can bind SHP-2 (Yamasaki et al. 2001) and it may compete with SLP-76 for binding to Grb2 and Gads (Yamasaki et al. 2003). Other molecules that may mediate inhibitory signaling are the adapter SLAP (Src-like adapter protein) (Myers et al. 2006), the tyrosine phosphatase SHP-1 (Lorenz 2009) and 4.1R (Diakowski et al. 2006). 4.1R binds LAT directly and is thought to prevent LAT phosphorylation by ZAP-70 (Kang et al. 2009).

COOPERATIVITY AMONG LAT-ASSOCIATED PROTEINS STABILIZES SIGNALING COMPLEXES

It is clear that LAT plays a central role in T-cell activation downstream of the TCR by directly or indirectly recruiting kinases, effectors, and adapters to facilitate the coordination of highly regulated signal transduction pathways. Biophysical approaches have begun to reveal important insights into some of the molecular mechanisms that govern the stabilization of LAT-associated complexes. These studies indicate that numerous weak interactions cooperatively coordinate the formation of spatially and temporally specific signaling complexes.

Qualitative Results Indicate Cooperativity Among LAT-associated Proteins

Cooperativity is an altered affinity between proteins because of interaction with other molecules and/or posttranslational modifications. A number of results suggest that LAT complexes are stabilized by cooperative interactions. SH2 domain-containing proteins including PLC-γ1, Grb2, and Gads bind directly to LAT phosphotyrosine motifs with a preference for PLC-γ1 at pY132, Gads at pY191 and pY171, and Grb2 at pY171, pY191, and pY226 (Samelson 2002). Nevertheless, mutations of LAT Gads/Grb2-binding residues (Y171, Y191, and Y226) reduced PLC-γ1 binding (Zhang et al. 2000; Hartgroves et al. 2003). At a minimum, Y132 and Y191 were required for an association between LAT and PLC-γ1, indicating that PLC-γ1 association with LAT was stabilized by Gads binding to LAT (Zhu et al. 2003). Likewise, there was evidence that PLC-γ1 stabilized the binding of Grb2 to LAT. LAT Y132F mutation resulted in substantial losses of both Grb2 and PLC-γ1 binding (Hartgroves et al. 2003).

Experiments with LAT-nucleated signaling complexes on liposomes by Sangani and colleagues also indicated cooperativity. In these experiments Grb2 alone did not stably bind liposome-associated LAT; however, after incubation with Jurkat lysate, Grb2 associated along with Gads, PLC-γ1, SLP-76, Itk, and Sos1 (Sangani et al. 2009). Interestingly, other studies showed that the LAT phosphorylation sites were required within a single molecule and could not be added *in trans* among multiple LAT proteins (Lin and Weiss 2001). This suggests that signaling complexes are stabilized around individual LAT molecules.

LAT Binding Preferences Are Not Solely Driven by Affinities

Binding studies were performed to better understand the basis for the specificity of the interactions of PLC-γ1, Grb2, and Gads with individual LAT motifs. Interestingly, Grb2 and Gads did not have substantially different affinities for the pY171, pY191, or pY226 motifs, results that did not explain the preference of Gads for pY171 and pY191. However, the proteins had a 50–100-fold weaker affinity for pY132, which explained the lack of binding to that site. The PLC-γ1 SH2 domain bound the pY132 motif with stronger affinity than the pY171, pY191, or pY226 motif; however, it was thought that the affinity difference alone would not account for its selectivity (Houtman et al. 2004). Overall, these studies indicated that the binding affinities alone were not sufficient to drive the apparent specificity of LAT motifs for PLC-γ1 and Gads.

In addition to affinity, other factors likely contribute to the specificity of PLC-γ1 for the LAT residue Y132. For example, the PLC-γ1 PH domain would position PLC-γ1 near the membrane, likely favoring interactions with pY132, as it is the most membrane proximal of the four motifs (Houtman et al. 2004). However, SLP-76 may also contribute to the stabilization of PLC-γ1 binding to LAT. Binding studies had revealed a weak interaction between the SLP-76 proline-rich region and PLC-γ1; however, Houtman and colleagues showed that the interaction required structural changes within SLP-76. They showed that the high affinity interaction between Gads and SLP-76 stabilized the PLC-γ1-binding conformation of SLP-76. Therefore, positive cooperativity mediated by the Gads/SLP-76 pair allowed PLC-γ1 to bind SLP-76 (Houtman et al. 2004). Interestingly, LAT residue Y191 was phosphorylated substantially earlier than Y132 in T cells. Therefore, it is possible that the Gads/SLP-76 complex associates with LAT before PLC-γ1. Moreover, SLP-76 is among the earliest proteins to be phosphorylated (Houtman et al. 2005). Phosphorylated SLP-76 binds Itk, which has been shown to phosphorylate the PLC-γ1 residue Y783 (Bunnell et al. 2000; Serrano et al. 2005; Beach et al. 2007). Y783 of PLC-γ1 is phosphorylated before LAT Y132, suggesting that PLC-γ1 associates with the Gads/SLP-76/Itk complex before binding to pY132 (Houtman et al. 2005). It was suggested that these cooperative binding events preceding PLC-γ1 binding to LAT could allow for tight control of PLC-γ1 activity.

Cooperative Interactions Promote LAT Oligomerization

Because LAT can potentially bind two or more Grb2 molecules and because the Grb2 SH3 ligands Sos1 and c-Cbl also contain multiple Grb2 binding sites, Houtman and colleagues investigated whether Grb2 complexes could link multiple LAT molecules into higher-order structures (Fig. 2). Biophysical studies were carried out to investigate the binding, composition and stiochiometry of complexes formed from mixtures of LAT with Grb2 and Sos1 or c-Cbl (Houtman et al. 2006; Houtman et al. 2007). The studies provided direct evidence that LAT was able to bind multiple Grb2 molecules and that the proline-rich regions of Sos1 and c-Cbl each bound two Grb2 molecules. Interestingly, Grb2 bound the two regions of Sos1 with either significantly different affinities or a twofold negative cooperativity in binding to the second site, whereby binding of the first Grb2 weakened the affinity of Sos1 for a second Grb2 molecule. These results indicated that the 2:1 complex was

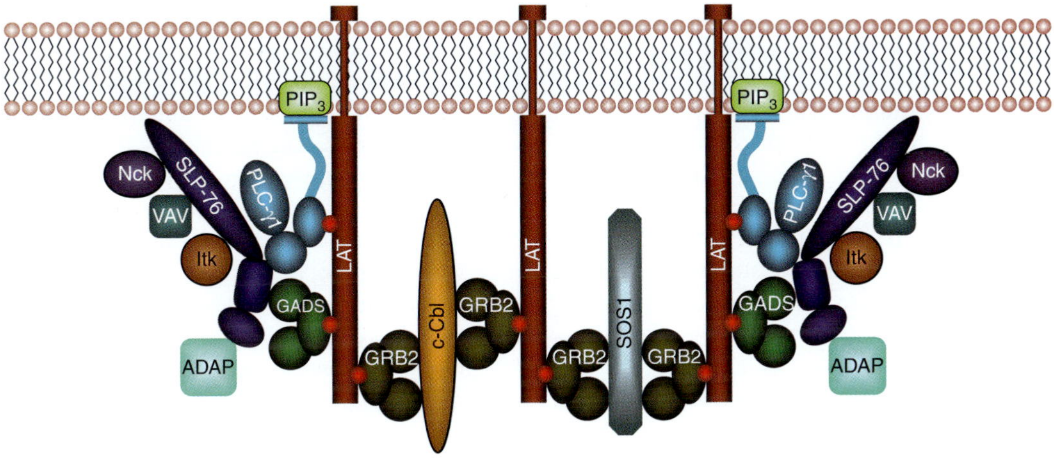

Figure 2. LAT complexes. Multiple LAT molecules are oligomerized by complexes of 2:1 Grb2:Sos1 or Grb2:Cbl in vitro and in cells. The clustering of LAT complexes is thought to be crucial for signal transduction and possibly for the control of TCR sensitivity to strong or weak stimuli.

only formed when Grb2 was in excess of Sos1. It was further demonstrated that Grb2 is in ~1000-fold excess of Sos1 in T cells, suggesting that the 2:1 complex of Grb2 and Sos1, which forms in vitro could also be physiologically relevant in T cells (Houtman et al. 2006; Houtman et al. 2007). Next, it was shown that Grb2/Sos complexes could oligomerize LAT into large molecular complexes. Detailed studies of the energetics contributing to the complex formation showed that intermediate complexes, such as a 2:1 complex of Grb2:LAT, were slightly destabilized, consistent with negative cooperativity in their formation. In contrast, positive cooperative interactions resulted in a large stabilized molecular complex of LAT, Grb2, and Sos1. It was suggested that the presence of both negative and positive cooperativity in the formation of these complexes promoted the assembly of physiologically relevant complexes, while suppressing intermediates (Houtman et al. 2007). The relevance of Grb2-mediated LAT clustering was verified in vivo and was shown to have functional consequences on signaling and T-cell activation (Houtman et al. 2006).

The biophysical analyses of LAT and LAT-associated proteins have revealed a role for intermolecular cooperativity in stabilizing LAT-nucleated complexes. Cooperativity allows for a high degree of organization of signaling molecules into macromolecular complexes visualized as "microclusters" using imaging techniques as described later.

IMAGING OF LAT SIGNALING COMPLEXES AND CLUSTERS

In the past decade, advanced microscopic approaches have provided striking images of the early events in T-cell activation. This section will review the contribution of imaging studies in developing a clearer understanding of the initiation, composition, dynamics, and spatiotemporal regulation of LAT-nucleated signaling complexes.

LAT Microclusters: Sites of Nucleation of Signaling Complexes

A little over a decade ago the immunological synapse (IS) was identified as a specialized junction between T cells and antigen presenting surfaces, which is characterized by a central cluster enriched for TCR-CD3 (cSMAC) and an integrin-rich peripheral cluster (pSMAC) (Monks et al. 1998). Subsequent studies identified the distal SMAC (dSMAC), an outer region that contains large molecules such as CD43 and CD45 (Delon et al. 2001; Freiberg et al. 2002).

Real time visualization of the early events of T-cell activation has given greater temporal and spatial accuracy. Studies in which images were collected at the onset of T-cell activation showed that within the first 30 s of contact of a T cell with a stimulatory surface, small TCR-rich structures form at the periphery of the T-cell contact zone (Grakoui et al. 1999; Krummel et al. 2000). These discrete structures enriched for TCR have subsequently been called "microclusters."

The first high resolution images of LAT microclusters came from a method in which T cells were dropped onto stimulatory coverslips coated with CD3 antibody and visualized using confocal microscopy (Bunnell et al. 2002). Using this technique, LAT colocalized with TCR-rich microclusters within seconds of activation. In addition to LAT, a large number of components of the TCR signaling pathway including protein tyrosine kinases ZAP-70 and Lck, adapters SLP-76, Grb2, Gads, Nck and WASp, and enzymes including PLC-γ1, Vav, c-Cbl, and Sos1 were rapidly recruited to microclusters. Larger glycoproteins such as CD43 and the protein tyrosine phosphatase CD45 were excluded from these structures (Bunnell et al. 2002; Barda-Saad et al. 2005; Braiman et al. 2006; Houtman et al. 2006). These rapidly assembled clusters of signaling complexes were the predominant sites of TCR-induced tyrosine phosphorylation and their appearance was coincident with cytosolic Ca^{2+} elevations, establishing TCR microclusters as sites of signal initiation (Krummel et al. 2000; Bunnell et al. 2002; Seminario and Bunnell 2008). Thus microclusters are the smallest structures visible by confocal microscopy capable of driving effective TCR signaling and are made up of signaling complexes that contain adapters and effectors required for T-cell activation (Fig. 3).

Subsequent studies using peptide-MHC-containing lipid bilayers and TIRF microscopy verified and extended these results (Campi et al. 2005; Yokosuka et al. 2005; Varma et al. 2006; Yokosuka et al. 2008; Kaizuka et al. 2009). Under these experimental conditions, microclusters were continuously generated at the periphery of the T-cell peptide-MHC interface and were transported and consolidated at the center of the contact to form a cSMAC (Varma et al. 2006). Microclusters play a predominant role in the generation of sustained signals. Current models also indicate that cluster centralization plays an important role in signal termination and that the cSMAC is a site both for TCR down-regulation yet also for amplification of weak signals (Lee et al. 2003; Mossman et al. 2005; Varma et al. 2006; Balagopalan et al. 2007; Cemerski et al. 2008; Nguyen et al. 2008). A recent study using a photoactivatable agonist in a lipid bilayer system has provided even greater precision in the temporal resolution of these events and demonstrated that adapters are recruited to microclusters within four seconds (Huse et al. 2007).

Fine spatial imaging studies indicated that signaling protein clusters may actually represent discrete interdigitating domains (Douglass and Vale 2005; Nguyen et al. 2008). These structures resembled the distinct protein and lipid domains observed by transmission electron microscopy (TEM) (Wilson et al. 2001; Lillemeier et al. 2006). TEM studies in mast cells and T cells have shown formation of "primary signaling domains" that include the receptor, and "secondary signaling domains" that include LAT and PLC-γ1, but not the receptor. Apparently, these domains, submicron in size, preexist before receptor activation. Upon receptor activation, these domains were shown to concatenate to form larger patches that are likely equivalent to the microclusters observed by light microscopy (Wilson et al. 2001; Lillemeier et al. 2006; Lillemeier et al. 2009).

Because the resolution limit of the microscope (≅200 nm) is much larger than the size of individual proteins or protein complexes (on the order of a few nanometers), colocalization observed by fluorescence microscopy does not prove that proteins interact. Instead, Forster resonance energy transfer (FRET) can be employed to study colocalization on the scale of a few nanometers and therefore provides information about molecular interactions (Gascoigne et al. 2009). Using this technique, FRET was detected between PLC-γ1 and either LAT, c-Cbl, Vav or SLP-76, as well as SLP-76

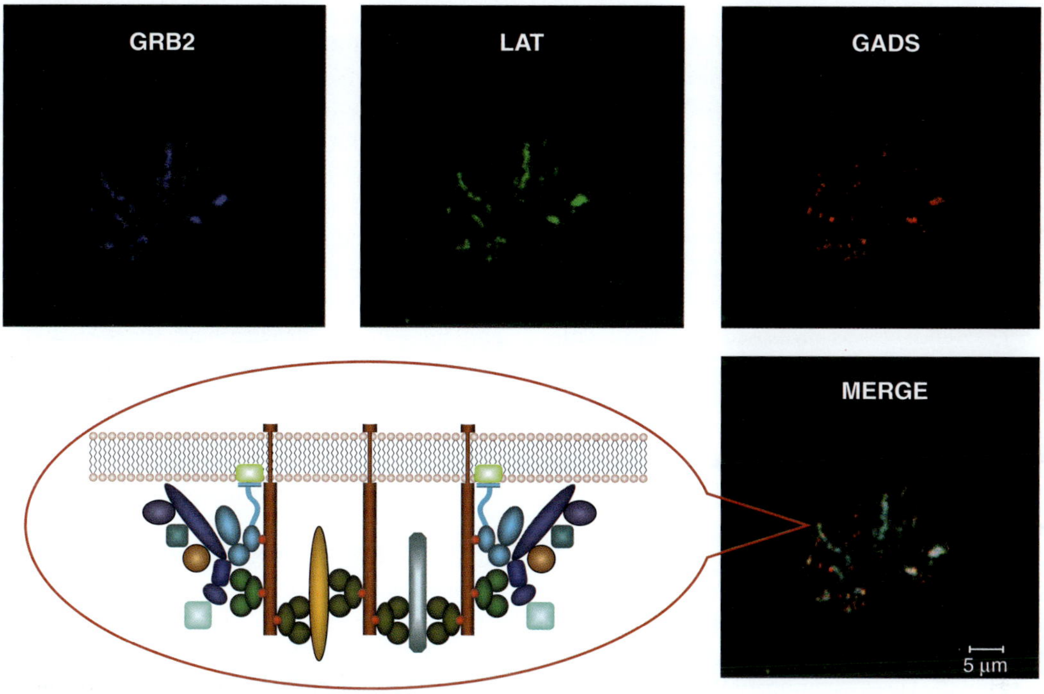

Figure 3. LAT signaling complexes and microclusters. A confocal image of the surface of a Jurkat E6.1 T cell dropped onto a TCR stimulatory coverslip shows LAT clusters, which colocalize with the LAT-binding proteins Grb2 and Gads. Grb2 (upper *left*) is in blue, LAT (upper *middle*) is in green and Gads (upper *right*) is in red. LAT signaling complexes (represented in the cartoon) form a pronounced network of heterogeneous and dynamic microclusters on the plasma membrane (see merge).

and Nck within microclusters, suggesting direct interactions of these proteins within microclusters (Barda-Saad et al. 2005; Braiman et al. 2006). In comparison, measurable but low FRET was observed between LAT and either SLP-76 or Nck, consistent with previous data showing that these molecules interact indirectly (Barda-Saad et al. 2005).

Another microscopic technique, fluorescence recovery after photobleaching (FRAP), has been used to evaluate LAT dynamics at the population level. This method allows evaluation of mobility and population-level diffusion dynamics of molecules (Tanimura et al. 2003a). Studies on LAT dynamics at the membrane revealed the fast exchange of molecules between LAT clusters or patches localized at sites of stimulation (Tanimura et al. 2003b; Douglass and Vale 2005). Single-molecule imaging studies have revealed that LAT molecules at the plasma membrane display abrupt changes between immobility and rapid diffusion (Douglass and Vale 2005). Transient immobilization of LAT correlated strongly with encounter with clusters. These data indicate that LAT molecules diffuse between clusters with occasional trapping within clusters.

Mechanisms of Assembly of LAT Clusters and LAT-Nucleated Complexes

Data from various experimental systems have shown that microclusters play a crucial role in TCR-mediated signaling pathways because inhibition of cluster formation resulted in reduced levels of TCR signaling (Singer et al. 2004; Bunnell et al. 2006; Houtman et al. 2006). The importance of clustering may be to concentrate activators and exclude inhibitors creating microdomains that shift the equilibrium to favor downstream signaling in T cells. However, the mechanisms that govern the assembly of these structures have remained elusive. Because LAT is both a lipid raft resident and an adapter with multiple tyrosines that serve as docking sites for signaling proteins, the contribution of both lipid rafts as well as protein scaffolds in LAT clustering has been investigated.

Early studies indicated that lipid raft patches form at the sites of TCR activation and copatched with proteins essential for T-cell activation (Harder and Simons 1999; Janes et al. 1999). Hence it was proposed that lipid rafts provide a platform for assembly of signaling domains during T-cell activation. However, data from multiple studies indicated that membrane recruitment, not raft localization was required for LAT function (Zhu et al. 2005; Hundt et al. 2009). Furthermore, Harder and Kuhn showed that LAT, but not other raft-associated molecules such as Lck and GM1, was selectively enriched within plasma membrane fragments containing activated TCR, calling into question the view that upon TCR activation coalescence of raft-associated membrane proteins in the vicinity of activated TCR leads to T-cell triggering (Harder and Kuhn 2000). Imaging studies also showed that LAT did not colocalize with a general raft marker GFP-GPI (Bunnell et al. 2002). Together, these studies indicate that lipid rafts are not a primary factor in the assembly of LAT-containing signaling complexes.

Instead, several studies have indicated that phosphotyrosine-dependent protein interactions play the predominant role in organization of TCR-dependent signaling complexes. First, LAT was not recruited to TCR-enriched immunoisolates in cells treated with the potent Src kinase inhibitor PP2 (Harder and Kuhn 2000). Second, imaging approaches using planar substrates demonstrated that LAT variants with mutated tyrosines in the cytoplasmic tail were not recruited to signaling microclusters (Douglass and Vale 2005; Bunnell et al. 2006; Houtman et al. 2006). Finally, single particle tracking techniques as well as FRAP methods that measured diffusion rates of LAT revealed that LAT cytoplasmic tyrosines are required for confinement of LAT within transient subdomains (Tanimura et al. 2003b; Douglass and Vale 2005). In contrast, interfering with lipid raft association of LAT through mutations did not alter its diffusion behavior (Douglass and Vale 2005).

The earlier-described results are most consistent with LAT clusters being formed and maintained by a network of protein–protein interactions. The combination of biochemical and biophysical studies has shown that cooperative and multivalent interactions drive the assembly of productive signaling complexes and imaging approaches have supported this scenario. Houtman et al. used confocal microscopy to verify the physiological relevance of Grb2-Sos1-mediated oligomerization of LAT, observed by biophysical techniques (Houtman et al. 2006). A LAT mutant unable to bind Grb2 did not localize to signaling clusters. In addition, expression of a Sos1 fragment unable to oligomerize LAT inhibited LAT clustering and downstream signaling. Thus, multivalent interactions between these three molecules appeared to promote the assembly of multiprotein complexes important for TCR activation. In another study using confocal approaches, Bunnell and colleagues demonstrated that cooperative interactions between LAT, Gads and SLP-76 stabilized SLP-76 microclusters (Bunnell et al. 2006). Another report demonstrated that although the SH3 and C-terminal SH2 domains of PLC-γ1 do not bind directly to LAT, they participate in the stabilization of PLC-γ1-LAT association via other proteins in the LAT-nucleated signaling complex (Braiman et al. 2006).

Finally, several studies have argued that actin is required for segregating molecules on the T-cell surface (Wulfing and Davis 1998; Delon et al. 2001). F actin was present at microclusters and the contact sites of T cells and stimulatory surfaces at early time points (Bunnell et al. 2001; Barda-Saad et al. 2005). TEM studies of Wilson et al. in mast cells also support the interaction of the actin cytoskeleton with receptor clusters (Wilson et al. 2001). Furthermore, inhibitor studies showed that actin is essential for the formation of microclusters (Campi et al. 2005; Douglass and Vale 2005; Varma et al. 2006; Nguyen et al. 2008). An intact cytoskeleton is also required for dynamic translocation of clusters once they are formed and it has been proposed that retrograde actin flow drives centripetal cluster movement (Barr et al. 2006; Varma et al. 2006; Kaizuka et al. 2007; Nguyen et al. 2008; Ilani et al. 2009). In this regard, microclusters must be connected to actin filaments via linker proteins, and LAT-mediated recruitment of actin-regulating adapters Nck and WASp to these structures may play a central role in this process (Barda-Saad et al. 2005).

Dynamics of LAT Clusters, LAT Internalization, and LAT Ubiquitylation

The dynamic nature and changing composition of TCR microclusters after they are generated has been extensively demonstrated. Across all modeling systems, following initial recruitment in close proximity of the TCR, components of the complexes rapidly exit microclusters with distinct dynamics (Bunnell et al. 2002; Barda-Saad et al. 2005; Braiman et al. 2006). The adapters LAT and SLP-76 departed the initial complexes in what appeared to be small vesicular intermediates and dissipated soon thereafter (Bunnell et al. 2002). Recent studies have shown that microclusters containing LAT and SLP-76 undergo endocytosis upon TCR stimulation (Barr et al. 2006; Balagopalan et al. 2007). SLP-76 is endocytosed in a lipid-raft dependent pathway that requires the association of the endocytic machinery with ubiquitylated proteins. LAT endocytosis is a complex process and internalized LAT is found in several intracellular compartments (Bonello et al. 2004; Brignatz et al. 2005; Balagopalan et al. 2007). Insights into the molecular mechanisms that regulate endocytosis of microcluster-associated proteins came from studies in which internalization of the microclusters was inhibited. First, expression of versions of c-Cbl that are defective in the RING domain that mediates ubiquitin ligase activity severely inhibited LAT and SLP-76 movement and endocytosis. Furthermore, the ubiquitin-interacting motif (UIM) of eps15, which is known to block clathrin-independent internalization of the EGFR (Sigismund et al. 2005), severely inhibited the internalization of SLP-76 clusters (Barr et al. 2006). Thus, the E3 ligase c-Cbl and ubiquitin appear to be intimately involved in the sorting of LAT and SLP-76 into mobile endocytic structures.

In addition to containing tyrosines, the LAT cytoplasmic domain contains two lysines that could serve as sites of ubiquitylation. Indeed LAT ubiquitylation has been observed both in Jurkat T cells and in COS-7 cells (Brignatz et al. 2005; Balagopalan et al. 2007). Consistent with c-Cbl being an E3 ligase for LAT ubiquitylation, expression of c-Cbl caused a modest increase in LAT ubiquitylation. In contrast, the RING finger mutant caused a significant decrease in ubiquitylated LAT species. These data are consistent with a model in which c-Cbl-mediated ubiquitylation is required for rapid internalization of LAT-nucleated signaling clusters. Given the essential scaffolding role of the adapter protein LAT in T cell activation, the regulated internalization of activated LAT signaling complexes may be one efficient strategy by which to control the duration and localization of signaling from microclusters and, thus, regulate the kinetics, intensity, and specificity of T-cell signaling. For a detailed discussion on endocytic regulation of T-cell microclusters, see Balagopalan et al. 2009.

ANIMAL STUDIES REVEAL THE IN VIVO FUNCTIONS OF LAT

Although T-cell lines have been enormously useful for discerning the role of LAT in T-cell function within single T cells, animal model systems provide the opportunity to study the function of LAT during T-cell development as well as its function in T cells in the context of the entire immune system.

Studies of LAT Null Mice

Studies of LAT null mice show the crucial importance of LAT in pre-TCR and TCR signaling. $\alpha\beta$ T-cell development proceeds through ordered stages that can be characterized by expression of the cell surface markers CD4 and CD8: DN (double negative for CD4 and CD8)$\rightarrow$DP (double positive for CD4 and CD8)$\rightarrow$SP (single positive for either CD4 or CD8). The DN stage can further be divided into four stages based on the expression of the cell surface markers CD44 and CD25: DN1 ($CD44^{+}CD25^{-}$)→DN2 ($CD44^{+}CD25^{+}$)→ DN3 ($CD44^{-}CD25^{+}$)→DN4 ($CD44^{-}$ $CD25^{-}$). Signaling through the immature, nonclonotypic pre-TCR drives development from the DN3 to the DN4 stage. Mature TCR signaling drives T-cell development from the DP to SP stage. During the DP→SP transition, thymocytes are tested based on their TCR specificities and are selected for survival (by positive selection, based on MHC specificity and nonself reactivity) or death (by negative selection, based on potential self-reactivity) (Sommers et al. 2004).

In LAT knockout mice, the endogenous wild-type LAT gene was replaced with a mutant form of the gene that could not be expressed by homologous recombination (Zhang et al. 1999b). LAT knockout mice showed a complete block in αβ T-cell development at the immature, DN3 stage indicating an essential role for LAT in pre-TCR signaling. No peripheral αβ T cells were observed. To investigate the role of LAT in mature TCR signaling, Shen et al. used a conditional knockout approach (Shen et al. 2009). Their approach used Cre-Lox technology (Wang 2009), whereby conditional expression of the Cre recombinase allowed deletion of a LoxP-flanked sequence that resulted in T cell-specific disruption of expression of LAT at the DP stage. The authors observed a severe block in the DP→SP stage of thymocyte development indicating a strong role for LAT in TCR-mediated thymocyte selection. However, some CD4 and CD8 SP T cells developed in these mice and it was unclear whether these T cells developed because of residual LAT in some developing DPs or because LAT is not completely essential for the DP→SP transition (Shen et al. 2009).

In Vivo Functions of LAT Phosphotyrosines and Cysteines

Previous in vitro studies with C26/29A LAT mutants had shown that the mutant LAT molecules could not be tyrosine phosphorylated, did not localize to plasma membrane rafts and could not mediate TCR signaling (Zhang et al. 1998b). In the animal experiments, LAT knockout bone marrow cells were infected with retroviruses containing mutant LAT C26/29A. The infected bone marrow cells were transferred to irradiated wild type B6 mice and the mutant LAT was unable to mediate T-cell development (Hundt et al. 2009). To assess whether raft localization (in addition to plasma membrane localization) was necessary for the in vivo function of LAT, several approaches have been taken using LAT fusion proteins. LAT-LAX and Src-LAT fusion proteins localized to the plasma membrane but not to rafts. Both of these fusion proteins mediated T-cell development in LAT-deficient environments (Zhu et al. 2005; Hundt et al. 2009). Therefore C26 and C29 of LAT were necessary for plasma membrane localization and for the in vivo function of LAT, but raft localization was not necessary for T-cell development.

In vitro experiments have shown the importance of the distal four phosphotyrosines for the function of LAT. A "knock-in" LAT mutant in which the endogenous wild type LAT gene was replaced with a mutant form of LAT containing Y→F substitutions at the four distal tyrosines revealed that those four distal tyrosines were required for T-cell development (Sommers et al. 2001). Knock-in mutants of Y136→ F (corresponding to human Y132→ F) showed an incomplete block in early αβ T-cell development and a fatal lymphoproliferative disease involving marked lymphadenopathy, splenomegaly, and multiorgan lymphocyte infiltration (Fig. 4) (Aguado et al. 2002; Sommers et al. 2002). A knock-in LAT mutant containing Y→F mutations at Y175, Y195, and Y235, (corresponding to human Y171, Y191, and Y226) also developed lymphoproliferative disease (Nunez-Cruz et al. 2003). Although the lymphoproliferative phenotypes of these two mouse models were similar, the intrathymic T-cell development differed between these two models. In LAT Y136F knock-in mice, some αβ T cells developed and expanded, whereas in LAT Y175/195/235F knock-in mice, cells of the γδ T cells lineage developed and expanded.

In addition to the developmental block in LAT Y136F mice, thymocyte selection was profoundly affected (Sommers et al. 2005) and Treg

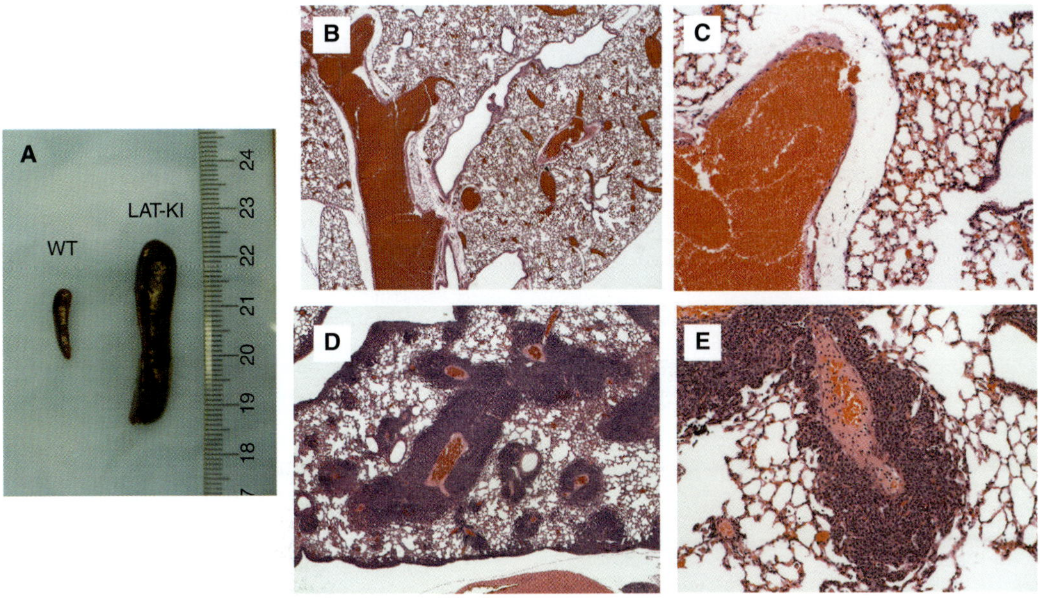

Figure 4. Lymphoproliferative disease in LAT Y136F KI mice. (*A*) Spleens from 10-week-old wild type C57BL/6 and LAT Y136F KI mice are shown. The ruler depicts centimeters. (*B–E*) Lymphocyte infiltration into lung. H&E-stained sections of lung from wild type C57BL/6 (*B,C*) and LAT Y136F KI mice (*D,E*) are shown. Images were photographed using a 2X (*B,D*) or 10X (*C,E*) objective.

cells (Lu and Rudensky 2009) did not develop (Koonpaew et al. 2006; Wang et al. 2008). Abnormal cytokine production was observed in these mice, especially elevated levels of IL-4. In fact the mice showed many aspects of Th2 lymphoproliferative disease including IgG1 and IgE hypergammaglobulinemia, elevated serum IL-4, eosinophilia, and lymphocyte proliferation in lungs (Aguado et al. 2002; Sommers et al. 2002; Genton et al. 2006; Miyaji et al. 2009). The contributing factors leading to development of lymphoproliferative disease are still under investigation, but some are known. Altered or crippled signaling leads to a block in T-cell development and a lymphopenic environment. Altered signaling in the T cells developing in this lymphopenic environment results in T-cell hyperproliferation, defective activation-induced cell death and altered cytokine production (↑IL-4,↓IL-2). Lack of development and function of Treg cells may be secondary to decreased IL-2 production. Altered signaling in LAT Y136F T cells consists of decreased TCR-induced calcium flux and other events downstream of PLC-γ1 (Sommers et al. 2002). Reports of Erk1/2 activation in LAT Y136F T cells have been variable (Sommers et al. 2002; Mingueneau et al. 2009; Miyaji et al. 2009), most likely because of different methods of in vitro activation and age of the animals from which the T cells were derived. However, a report by Miyaji et al. showed that Erk activation played an important role in lymphoproliferative disease in LAT Y136F mice (Miyaji et al. 2009). Precisely how other aspects of altered signaling (especially altered calcium signaling) lead to hyperproliferation, decreased cell death and altered cytokine production in LAT Y136F T cells is a subject for future investigations.

LAT-independent TCR Signaling

A recent report by Mingueneau et al. highlights a mouse model in which LAT-independent T cell signaling was demonstrated (Mingueneau et al. 2009). In these experiments, $CD4^+$ T cells that had developed in a LAT-sufficient environment were rendered LAT-deficient following expression of Cre recombinase, which was accomplished by retroviral infection. After additional culturing with IL-2, the cells were stimulated through the TCR and various responses were assessed or the T cells were adoptively transferred to T cell-deficient hosts. Although phosphorylation of several key signaling molecules was not observed in the LAT-deficient mature T cells (including PLC-γ1 and Erk 1/2), phosphorylation of some signaling molecules was observed (e.g., SLP-76 and Akt). TCR-induced calcium flux and interferon-γ production were also profoundly diminished in these cells. However, these experiments lend credence to the hypothesis that although LAT is critical for TCR signaling, some TCR-induced, LAT-independent signaling can occur in mature T cells. Furthermore, LAT-deficient mature T cells that had developed in a LAT-sufficient environment could initiate lymphoproliferative disease when transferred into T-cell-deficient (CD3ε knockout) hosts. This raises the fascinating possibility that TCR-induced, LAT-independent signaling mechanisms (in a lymphopenic context) can lead to lymphoproliferative disease. It will be interesting to compare the LAT-independent signaling mechanisms from this situation with LAT-independent signaling described in T-cell clones (Chau and Madrenas 1999) and in Jurkat T cells (Ku et al. 2001; Shan et al. 2001).

CONCLUDING REMARKS

As described in this review, identification of the critical adapter molecule LAT over a decade ago has led to a multitude of studies. Our understanding of TCR-mediated signaling and T cell biology in general has been thereby enriched. We expect that continued application of state-of-the-art approaches will lead to further insight into how LAT serves as the focal point of TCR-mediated activation.

ACKNOWLEDGMENTS

The authors would like to thank Robert Kortum, Valarie Barr and Ronald Wange for helpful suggestions. The authors' research was supported by the Intramural Research Program of the NIH, NCI, CCR.

REFERENCES

Aguado E, Richelme S, Nunez-Cruz S, Miazek A, Mura AM, Richelme M, Guo XJ, Sainty D, He HT, Malissen B, et al. 2002. Induction of T helper type 2 immunity by a point mutation in the LAT adaptor. *Science* **296:** 2036–2040.

Baker RG, Hsu CJ, Lee D, Jordan MS, Maltzman JS, Hammer DA, Baumgart T, Koretzky GA. 2009. The adapter protein SLP-76 mediates "outside-in" integrin signaling and function in T cells. *Mol Cell Biol* **29:** 5578–5589.

Baker JE, Majeti R, Tangye SG, Weiss A. 2001. Protein tyrosine phosphatase CD148-mediated inhibition of T-cell receptor signal transduction is associated with reduced LAT and phospholipase Cγ1 phosphorylation. *Mol Cell Biol* **21:** 2393–2403.

Balagopalan L, Barr VA, Samelson LE. 2009. Endocytic events in TCR signaling: Focus on adapters in microclusters. *Immunol Rev* **232:** 84–98.

Balagopalan L, Barr VA, Sommers CL, Barda-Saad M, Goyal A, Isakowitz MS, Samelson LE. 2007. c-Cbl-mediated regulation of LAT-nucleated signaling complexes. *Mol Cell Biol* **27:** 8622–8636.

Barda-Saad M, Braiman A, Titerence R, Bunnell SC, Barr VA, Samelson LE. 2005. Dynamic molecular interactions linking the T cell antigen receptor to the actin cytoskeleton. *Nat Immunol* **6:** 80–89.

Barr VA, Balagopalan L, Barda-Saad M, Polishchuk R, Boukari H, Bunnell SC, Bernot KM, Toda Y, Nossal R, Samelson LE. 2006. T-cell antigen receptor-induced signaling complexes: Internalization via a cholesterol-dependent endocytic pathway. *Traffic* **7:** 1143–1162.

Beach D, Gonen R, Bogin Y, Reischl IG, Yablonski D. 2007. Dual role of SLP-76 in mediating T cell receptor-induced activation of phospholipase C-γ1. *J Biol Chem* **282:** 2937–2946.

Berry DM, Nash P, Liu SK, Pawson T, McGlade CJ. 2002. A high-affinity Arg-X-X-Lys SH3 binding motif confers specificity for the interaction between Gads and SLP-76 in T cell signaling. *Curr Biol* **12:** 1336–1341.

Bonello G, Blanchard N, Montoya MC, Aguado E, Langlet C, He HT, Nunez-Cruz S, Malissen M, Sanchez-Madrid F, Olive D, et al. 2004. Dynamic recruitment of the adaptor protein LAT: LAT exists in two distinct intracellular pools and controls its own recruitment. *J Cell Sci* **117:** 1009–1016.

Boussiotis VA, Freeman GJ, Berezovskaya A, Barber DL, Nadler LM. 1997. Maintenance of human T cell anergy: Blocking of IL-2 gene transcription by activated Rap1. *Science* **278:** 124–128.

Braiman A, Barda-Saad M, Sommers CL, Samelson LE. 2006. Recruitment and activation of PLCγ1 in T cells: A new insight into old domains. *EMBO J* **25:** 774–784.

Brdicka T, Cerny J, Horejsi V. 1998. T cell receptor signalling results in rapid tyrosine phosphorylation of the linker protein LAT present in detergent-resistant membrane microdomains. *Biochem Biophys Res Commun* **248:** 356–360.

Brignatz C, Restouin A, Bonello G, Olive D, Collette Y. 2005. Evidences for ubiquitination and intracellular trafficking of LAT, the linker of activated T cells. *Biochim Biophys Acta* **1746:** 108–115.

Brown DA. 2006. Lipid rafts, detergent-resistant membranes, and raft targeting signals. *Physiology (Bethesda)* **21:** 430–439.

Brown DA, London E. 1998. Structure and origin of ordered lipid domains in biological membranes. *J Membr Biol* **164:** 103–114.

Buday L, Egan SE, Rodriguez Viciana P, Cantrell DA, Downward J. 1994. A complex of Grb2 adaptor protein, Sos exchange factor, and a 36-kDa membrane-bound tyrosine phosphoprotein is implicated in ras activation in T cells. *J Biol Chem* **269:** 9019–9023.

Bunnell SC, Diehn M, Yaffe MB, Findell PR, Cantley LC, Berg LJ. 2000. Biochemical interactions integrating Itk with the T cell receptor-initiated signaling cascade. *J Biol Chem* **275:** 2219–2230.

Bunnell SC, Hong DI, Kardon JR, Yamazaki T, McGlade CJ, Barr VA, Samelson LE. 2002. T cell receptor ligation induces the formation of dynamically regulated signaling assemblies. *J Cell Biol* **158:** 1263–1275.

Bunnell SC, Kapoor V, Trible RP, Zhang W, Samelson LE. 2001. Dynamic actin polymerization drives T cell receptor-induced spreading: A role for the signal transduction adaptor LAT. *Immunity* **14:** 315–329.

Bunnell SC, Singer AL, Hong DI, Jacque BH, Jordan MS, Seminario MC, Barr VA, Koretzky GA, Samelson LE. 2006. Persistence of cooperatively stabilized signaling clusters drives T-cell activation. *Mol Cell Biol* **26:** 7155–7166.

Campi G, Varma R, Dustin ML. 2005. Actin and agonist MHC-peptide complex-dependent T cell receptor microclusters as scaffolds for signaling. *J Exp Med* **202:** 1031–1036.

Cemerski S, Das J, Giurisato E, Markiewicz MA, Allen PM, Chakraborty AK, Shaw AS. 2008. The balance between T cell receptor signaling and degradation at the center of the immunological synapse is determined by antigen quality. *Immunity* **29:** 414–422.

Chau LA, Madrenas J. 1999. Phospho-LAT-independent activation of the ras-mitogen-activated protein kinase pathway: A differential recruitment model of TCR partial agonist signaling. *J Immunol* **163:** 1853–1858.

Chiang YJ, Sommers CL, Jordan MS, Gu H, Samelson LE, Koretzky GA, Hodes RJ. 2004. Inactivation of c-Cbl reverses neonatal lethality and T cell developmental arrest of SLP-76-deficient mice. *J Exp Med* **200:** 25–34.

Cruz-Orcutt N, Houtman JC. 2009. PI3 kinase function is vital for the function but not formation of LAT-mediated signaling complexes. *Mol Immunol* **46:** 2274–2283.

da Silva AJ, Li Z, de Vera C, Canto E, Findell P, Rudd CE. 1997. Cloning of a novel T-cell protein FYB that binds FYN and SH2-domain-containing leukocyte protein 76 and modulates interleukin 2 production. *Proc Natl Acad Sci* **94:** 7493–7498.

Daeron M, Lesourne R. 2006. Negative signaling in Fc receptor complexes. *Adv Immunol* **89:** 39–86.

Delon J, Kaibuchi K, Germain RN. 2001. Exclusion of CD43 from the immunological synapse is mediated by phosphorylation-regulated relocation of the cytoskeletal adaptor moesin. *Immunity* **15:** 691–701.

Di Bartolo V, Montagne B, Salek M, Jungwirth B, Carrette F, Fourtane J, Sol-Foulon N, Michel F, Schwartz O, Lehmann WD, et al. 2007. A novel pathway down-modulating T cell activation involves HPK-1-dependent recruitment of 14-3-3 proteins on SLP-76. *J Exp Med* **204:** 681–691.

Diakowski W, Grzybek M, Sikorski AF. 2006. Protein 4.1, a component of the erythrocyte membrane skeleton and its related homologue proteins forming the protein 4.1/FERM superfamily. *Folia Histochem Cytobiol* **44:** 231–248.

Dong S, Corre B, Foulon E, Dufour E, Veillette A, Acuto O, Michel F. 2006. T cell receptor for antigen induces linker for activation of T cell-dependent activation of a negative signaling complex involving Dok-2, SHIP-1, and Grb-2. *J Exp Med* **203:** 2509–2518.

Donovan JA, Wange RL, Langdon WY, Samelson LE. 1994. The protein product of the c-cbl protooncogene is the 120-kDa tyrosine-phosphorylated protein in Jurkat cells activated via the T cell antigen receptor. *J Biol Chem* **269:** 22921–22924.

Douglass AD, Vale RD. 2005. Single-molecule microscopy reveals plasma membrane microdomains created by protein-protein networks that exclude or trap signaling molecules in T cells. *Cell* **121:** 937–950.

Duan L, Reddi AL, Ghosh A, Dimri M, Band H. 2004. The Cbl family and other ubiquitin ligases: Destructive forces in control of antigen receptor signaling. *Immunity* **21:** 7–17.

Facchetti F, Chan JK, Zhang W, Tironi A, Chilosi M, Parolini S, Notarangelo LD, Samelson LE. 1999. Linker for activation of T cells (LAT), a novel immunohistochemical marker for T cells, NK cells, mast cells, and megakaryocytes: Evaluation in normal and pathological conditions. *Am J Pathol* **154:** 1037–1046.

Finco TS, Kadlecek T, Zhang W, Samelson LE, Weiss A. 1998. LAT is required for TCR-mediated activation of PLCγ1 and the Ras pathway. *Immunity* **9:** 617–626.

Freiberg BA, Kupfer H, Maslanik W, Delli J, Kappler J, Zaller DM, Kupfer A. 2002. Staging and resetting T cell activation in SMACs. *Nat Immunol* **3:** 911–917.

Fruman DA, Bismuth G. 2009. Fine tuning the immune response with PI3K. *Immunol Rev* **228:** 253–272.

Fruman DA, Rameh LE, Cantley LC. 1999. Phosphoinositide binding domains: Embracing 3-phosphate. *Cell* **97:** 817–820.

Fukazawa T, Reedquist KA, Panchamoorthy G, Soltoff S, Trub T, Druker B, Cantley L, Shoelson SE, Band H. 1995a. T cell activation-dependent association between the p85 subunit of the phosphatidylinositol 3-kinase and Grb2/phospholipase C-γ 1-binding phosphotyrosyl protein pp36/38. *J Biol Chem* **270:** 20177–20182.

Fukazawa T, Reedquist KA, Trub T, Soltoff S, Panchamoorthy G, Druker B, Cantley L, Shoelson SE, Band H. 1995b. The SH3 domain-binding T cell tyrosyl phosphoprotein p120. Demonstration of its identity with the c-cbl protooncogene product and in vivo complexes with Fyn, Grb2, and phosphatidylinositol 3-kinase. *J Biol Chem* **270:** 19141–19150.

Gascoigne NR, Ampudia J, Clamme JP, Fu G, Lotz C, Mallaun M, Niederberger N, Palmer E, Rybakin V, Yachi PP, et al. 2009. Visualizing intermolecular interactions in T cells. *Curr Top Microbiol Immunol* **334:** 31–46.

Genton C, Wang Y, Izui S, Malissen B, Delsol G, Fournie GJ, Malissen M, Acha-Orbea H. 2006. The Th2 lymphoproliferation developing in LatY136F mutant mice triggers polyclonal B cell activation and systemic autoimmunity. *J Immunol* **177:** 2285–2293.

Gilliland LK, Schieven GL, Norris NA, Kanner SB, Aruffo A, Ledbetter JA. 1992. Lymphocyte lineage-restricted tyrosine-phosphorylated proteins that bind PLC γ 1 SH2 domains. *J Biol Chem* **267:** 13610–13616.

Grakoui A, Bromley SK, Sumen C, Davis MM, Shaw AS, Allen PM, Dustin ML. 1999. The immunological synapse: A molecular machine controlling T cell activation. *Science* **285:** 221–227.

Gringhuis SI, Leow A, Papendrecht-Van Der Voort EA, Remans PH, Breedveld FC, Verweij CL. 2000. Displacement of linker for activation of T cells from the plasma membrane due to redox balance alterations results in hyporesponsiveness of synovial fluid T lymphocytes in rheumatoid arthritis. *J Immunol* **164:** 2170–2179.

Harder T, Kuhn M. 2000. Selective accumulation of raft-associated membrane protein LAT in T cell receptor signaling assemblies. *J Cell Biol* **151:** 199–208.

Harder T, Simons K. 1999. Clusters of glycolipid and glycosylphosphatidylinositol-anchored proteins in lymphoid cells: Accumulation of actin regulated by local tyrosine phosphorylation. *Eur J Immunol* **29:** 556–562.

Harkiolaki M, Lewitzky M, Gilbert RJ, Jones EY, Bourette RP, Mouchiroud G, Sondermann H, Moarefi I, Feller SM. 2003. Structural basis for SH3 domain-mediated high-affinity binding between Mona/Gads and SLP-76. *EMBO J* **22:** 2571–2582.

Hartgroves LC, Lin J, Langen H, Zech T, Weiss A, Harder T. 2003. Synergistic assembly of linker for activation of T cells signaling protein complexes in T cell plasma membrane domains. *J Biol Chem* **278:** 20389–20394.

He HT, Marguet D. 2008. T-cell antigen receptor triggering and lipid rafts: A matter of space and time scales. Talking Point on the involvement of lipid rafts in T-cell activation. *EMBO Rep* **9:** 525–530.

Horn J, Wang X, Reichardt P, Stradal TE, Warnecke N, Simeoni L, Gunzer M, Yablonski D, Schraven B, Kliche S. 2009. Src homology 2-domain containing leukocyte-specific phosphoprotein of 76 kDa Is mandatory for TCR-mediated inside-out signaling, but dispensable for CXCR4-mediated LFA-1 activation, adhesion, and migration of T cells. *J Immunol* **183:** 5756–5767.

Houtman JC, Brown PH, Bowden B, Yamaguchi H, Appella E, Samelson LE, Schuck P. 2007. Studying multisite binary and ternary protein interactions by global analysis of isothermal titration calorimetry data in SEDPHAT: Application to adaptor protein complexes in cell signaling. *Protein Sci* **16:** 30–42.

Houtman JC, Higashimoto Y, Dimasi N, Cho S, Yamaguchi H, Bowden B, Regan C, Malchiodi EL, Mariuzza R, Schuck P, et al. 2004. Binding specificity of multiprotein signaling complexes is determined by both cooperative interactions and affinity preferences. *Biochemistry* **43:** 4170–4178.

Houtman JC, Houghtling RA, Barda-Saad M, Toda Y, Samelson LE. 2005. Early phosphorylation kinetics of proteins involved in proximal TCR-mediated signaling pathways. *J Immunol* **175:** 2449–2458.

Houtman JC, Yamaguchi H, Barda-Saad M, Braiman A, Bowden B, Appella E, Schuck P, Samelson LE. 2006. Oligomerization of signaling complexes by the multipoint binding of GRB2 to both LAT and SOS1. *Nat Struct Mol Biol* **13:** 798–805.

Hu MC, Qiu WR, Wang X, Meyer CF, Tan TH. 1996. Human HPK1, a novel human hematopoietic progenitor kinase that activates the JNK/SAPK kinase cascade. *Genes Dev* **10:** 2251–2264.

Hundt M, Harada Y, De Giorgio L, Tanimura N, Zhang W, Altman A. 2009. Palmitoylation-dependent plasma membrane transport but lipid raft-independent signaling by linker for activation of T cells. *J Immunol* **183:** 1685–1694.

Hundt M, Tabata H, Jeon MS, Hayashi K, Tanaka Y, Krishna R, De Giorgio L, Liu YC, Fukata M, Altman A. 2006. Impaired activation and localization of LAT in anergic T cells as a consequence of a selective palmitoylation defect. *Immunity* **24:** 513–522.

Hunter T. 2009. Tyrosine phosphorylation: Thirty years and counting. *Curr Opin Cell Biol* **21:** 140–146.

Huse M, Klein LO, Girvin AT, Faraj JM, Li QJ, Kuhns MS, Davis MM. 2007. Spatial and temporal dynamics of T cell receptor signaling with a photoactivatable agonist. *Immunity* **27:** 76–88.

Ilani T, Vasiliver-Shamis G, Vardhana S, Bretscher A, Dustin ML. 2009. T cell antigen receptor signaling and immunological synapse stability require myosin IIA. *Nat Immunol* **10:** 531–539.

Irvin BJ, Williams BL, Nilson AE, Maynor HO, Abraham RT. 2000. Pleiotropic contributions of phospholipase C-γ1 (PLC-γ1) to T-cell antigen receptor-mediated signaling: Reconstitution studies of a PLC-γ1-deficient Jurkat T-cell line. *Mol Cell Biol* **20:** 9149–9161.

Janes PW, Ley SC, Magee AI. 1999. Aggregation of lipid rafts accompanies signaling via the T cell antigen receptor. *J Cell Biol* **147:** 447–461.

Jiang Y, Cheng H. 2007. Evidence of LAT as a dual substrate for Lck and Syk in T lymphocytes. *Leuk Res* **31:** 541–545.

June CH, Fletcher MC, Ledbetter JA, Samelson LE. 1990. Increases in tyrosine phosphorylation are detectable before phospholipase C activation after T cell receptor stimulation. *J Immunol* **144:** 1591–1599.

Kaizuka Y, Douglass AD, Vardhana S, Dustin ML, Vale RD. 2009. The coreceptor CD2 uses plasma membrane microdomains to transduce signals in T cells. *J Cell Biol* **185:** 521–534.

Kaizuka Y, Douglass AD, Varma R, Dustin ML, Vale RD. 2007. Mechanisms for segregating T cell receptor and adhesion molecules during immunological synapse formation in Jurkat T cells. *Proc Natl Acad Sci* **104:** 20296–20301.

Kane LP, Lin J, Weiss A. 2000. Signal transduction by the TCR for antigen. *Curr Opin Immunol* **12:** 242–249.

Kang Q, Yu Y, Pei X, Hughes R, Heck S, Zhang X, Guo X, Halverson G, Mohandas N, An X. 2009. Cytoskeletal protein 4.1R negatively regulates T-cell activation by inhibiting the phosphorylation of LAT. *Blood* **113:** 6128–6137.

Kenworthy AK. 2008. Have we become overly reliant on lipid rafts? Talking point on the involvement of lipid rafts in T-cell activation. *EMBO Rep* **9:** 531–535.

Kiefer F, Tibbles LA, Anafi M, Janssen A, Zanke BW, Lassam N, Pawson T, Woodgett JR, Iscove NN. 1996. HPK1, a hematopoietic protein kinase activating the SAPK/JNK pathway. *EMBO J* **15:** 7013–7025.

Koonpaew S, Shen S, Flowers L, Zhang W. 2006. LAT-mediated signaling in CD4+CD25+ regulatory T cell development. *J Exp Med* **203:** 119–129.

Koretzky GA, Abtahian F, Silverman MA. 2006. SLP76 and SLP65: Complex regulation of signalling in lymphocytes and beyond. *Nat Rev Immunol* **6:** 67–78.

Krummel MF, Sjaastad MD, Wulfing C, Davis MM. 2000. Differential clustering of CD4 and CD3zeta during T cell recognition. *Science* **289:** 1349–1352.

Ku GM, Yablonski D, Manser E, Lim L, Weiss A. 2001. A PAK1-PIX-PKL complex is activated by the T-cell receptor independent of Nck, Slp-76 and LAT. *EMBO J* **20:** 457–465.

Lahesmaa R, Allsup A, Soderberg C, Jackman J, Findell P, Peltz G. 1995. Modulation of the Grb2-associated protein complex in human CD4+ T cells by receptor activation. *J Immunol* **155:** 3815–3822.

Lee KH, Dinner AR, Tu C, Campi G, Raychaudhuri S, Varma R, Sims TN, Burack WR, Wu H, Wang J, et al. 2003. The immunological synapse balances T cell receptor signaling and degradation. *Science* **302:** 1218–1222.

Lillemeier BF, Mortelmaier MA, Forstner MB, Huppa JB, Groves JT, Davis MM. 2010. TCR and Lat are expressed on separate protein islands on T cell membranes and concatenate during activation. *Nat Immunol* **11:** 90–96.

Lillemeier BF, Pfeiffer JR, Surviladze Z, Wilson BS, Davis MM. 2006. Plasma membrane-associated proteins are clustered into islands attached to the cytoskeleton. *Proc Natl Acad Sci* **103:** 18992–18997.

Lin J, Weiss A. 2001. Identification of the minimal tyrosine residues required for linker for activation of T cell function. *J Biol Chem* **276:** 29588–29595.

Lin J, Weiss A, Finco TS. 1999. Localization of LAT in glycolipid-enriched microdomains is required for T cell activation. *J Biol Chem* **274:** 28861–28864.

Lindholm CK, Gylfe E, Zhang W, Samelson LE, Welsh M. 1999. Requirement of the Src homology 2 domain protein Shb for T cell receptor-dependent activation of the interleukin-2 gene nuclear factor for activation of T cells element in Jurkat T cells. *J Biol Chem* **274:** 28050–28057.

Lindholm CK, Henriksson ML, Hallberg B, Welsh M. 2002. Shb links SLP-76 and Vav with the CD3 complex in Jurkat T cells. *Eur J Biochem* **269:** 3279–3288.

Liou J, Kiefer F, Dang A, Hashimoto A, Cobb MH, Kurosaki T, Weiss A. 2000. HPK1 is activated by lymphocyte antigen receptors and negatively regulates AP-1. *Immunity* **12:** 399–408.

Liu SK, Fang N, Koretzky GA, McGlade CJ. 1999. The hematopoietic-specific adaptor protein Gads functions in T-cell signaling via interactions with the SLP-76 and LAT adaptors. *Curr Biol* **9:** 67–75.

Lorenz U. 2009. SHP-1 and SHP-2 in T cells: Two phosphatases functioning at many levels. *Immunol Rev* **228:** 342–359.

Lu LF, Rudensky A. 2009. Molecular orchestration of differentiation and function of regulatory T cells. *Genes Dev* **23:** 1270–1282.

Mingueneau M, Roncagalli R, Gregoire C, Kissenpfennig A, Miazek A, Archambaud C, Wang Y, Perrin P, Bertosio E, Sansoni A, et al. 2009. Loss of the LAT adaptor converts antigen-responsive T cells into pathogenic effectors that function independently of the T cell receptor. *Immunity* **31:** 197–208.

Miyaji M, Kortum RL, Surana R, Li W, Woolard KD, Simpson RM, Samelson LE, Sommers CL. 2009. Genetic evidence for the role of Erk activation in a lymphoproliferative disease of mice. *Proc Natl Acad Sci* **106:** 14502–14507.

Monks CR, Freiberg BA, Kupfer H, Sciaky N, Kupfer A. 1998. Three-dimensional segregation of supramolecular activation clusters in T cells. *Nature* **395:** 82–86.

Montixi C, Langlet C, Bernard AM, Thimonier J, Dubois C, Wurbel MA, Chauvin JP, Pierres M, He HT. 1998. Engagement of T cell receptor triggers its recruitment to low-density detergent-insoluble membrane domains. *EMBO J* **17:** 5334–5348.

Mossman KD, Campi G, Groves JT, Dustin ML. 2005. Altered TCR signaling from geometrically repatterned immunological synapses. *Science* **310:** 1191–1193.

Munro S. 2003. Lipid rafts: Elusive or illusive? *Cell* **115:** 377–388.

Murphy MA, Schnall RG, Venter DJ, Barnett L, Bertoncello I, Thien CB, Langdon WY, Bowtell DD. 1998. Tissue hyperplasia and enhanced T-cell signalling via ZAP-70 in c-Cbl-deficient mice. *Mol Cell Biol* **18:** 4872–4882.

Musci MA, Hendricks-Taylor LR, Motto DG, Paskind M, Kamens J, Turck CW, Koretzky GA. 1997. Molecular cloning of SLAP-130, an SLP-76-associated substrate of the T cell antigen receptor-stimulated protein tyrosine kinases. *J Biol Chem* **272:** 11674–11677.

Myers MD, Sosinowski T, Dragone LL, White C, Band H, Gu H, Weiss A. 2006. Src-like adaptor protein regulates TCR expression on thymocytes by linking the ubiquitin ligase c-Cbl to the TCR complex. *Nat Immunol* **7:** 57–66.

Nguyen K, Sylvain NR, Bunnell SC. 2008. T cell costimulation via the integrin VLA-4 inhibits the actin-dependent centralization of signaling microclusters containing the adaptor SLP-76. *Immunity* **28:** 810–821.

Nunez-Cruz S, Aguado E, Richelme S, Chetaille B, Mura AM, Richelme M, Pouyet L, Jouvin-Marche E, Xerri L, Malissen B, et al. LAT regulates gammadelta T cell homeostasis and differentiation. *Nat Immunol* **4:** 999–1008.

Oh-hora M. 2009. Calcium signaling in the development and function of T-lineage cells. *Immunol Rev* **231:** 210–224.

Ott VL, Tamir I, Niki M, Pandolfi PP, Cambier JC. 2002. Downstream of kinase, p62(dok), is a mediator of Fc γ IIB inhibition of Fc epsilon RI signaling. *J Immunol* **168:** 4430–4439.

Oya K, Wang J, Watanabe Y, Koga R, Watanabe T. 2003. Appearance of the LAT protein at an early stage of

B-cell development and its possible role. *Immunology* **109:** 351–359.

Paz PE, Wang S, Clarke H, Lu X, Stokoe D, Abo A. 2001. Mapping the Zap-70 phosphorylation sites on LAT (linker for activation of T cells) required for recruitment and activation of signalling proteins in T cells. *Biochem J* **356:** 461–471.

Perez-Villar JJ, Whitney GS, Sitnick MT, Dunn RJ, Venkatesan S, O'Day K, Schieven GL, Lin TA, Kanner SB. 2002. Phosphorylation of the linker for activation of T-cells by Itk promotes recruitment of Vav. *Biochemistry* **41:** 10732–10740.

Quilliam LA. 2007. New insights into the mechanisms of SOS activation. *Sci STKE* **2007:** pe67.

Readinger JA, Mueller KL, Venegas AM, Horai R, Schwartzberg PL. 2009. Tec kinases regulate T-lymphocyte development and function: New insights into the roles of Itk and Rlk/Txk. *Immunol Rev* **228:** 93–114.

Rellahan BL, Graham LJ, Tysgankov AY, DeBell KE, Veri MC, Noviello C, Bonvini E. 2003. A dynamic constitutive and inducible binding of c-Cbl by PLCγ1 SH3 and SH2 domains (negatively) regulates antigen receptor-induced PLCγ1 activation in lymphocytes. *Exp Cell Res* **289:** 184–194.

Reynolds LF, Smyth LA, Norton T, Freshney N, Downward J, Kioussis D, Tybulewicz VL. 2002. Vav1 transduces T cell receptor signals to the activation of phospholipase C-γ1 via phosphoinositide 3-kinase-dependent and -independent pathways. *J Exp Med* **195:** 1103–1114.

Rhee SG, Bae YS. 1997. Regulation of phosphoinositide-specific phospholipase C isozymes. *J Biol Chem* **272:** 15045–15048.

Rivero-Lezcano OM, Sameshima JH, Marcilla A, Robbins KC. 1994. Physical association between Src homology 3 elements and the protein product of the c-cbl proto-oncogene. *J Biol Chem* **269:** 17363–17366.

Rohatgi R, Nollau P, Ho HY, Kirschner MW, Mayer BJ. 2001. Nck and phosphatidylinositol 4,5-bisphosphate synergistically activate actin polymerization through the N-WASP-Arp2/3 pathway. *J Biol Chem* **276:** 26448–26452.

Samelson LE. 2002. Signal transduction mediated by the T cell antigen receptor: The role of adapter proteins. *Annu Rev Immunol* **20:** 371–394.

Sangani D, Venien-Bryan C, Harder T. 2009. Phosphotyrosine-dependent in vitro reconstitution of recombinant LAT-nucleated multiprotein signalling complexes on liposomes. *Mol Membr Biol* **26:** 159–170.

Seminario MC, Bunnell SC. 2008. Signal initiation in T-cell receptor microclusters. *Immunol Rev* **221:** 90–106.

Serrano CJ, Graham L, DeBell K, Rawat R, Veri MC, Bonvini E, Rellahan BL, Reischl IG. 2005. A new tyrosine phosphorylation site in PLC γ 1: The role of tyrosine 775 in immune receptor signaling. *J Immunol* **174:** 6233–6237.

Shan X, Balakir R, Criado G, Wood JS, Seminario MC, Madrenas J, Wange RL. 2001. Zap-70-independent Ca(2+) mobilization and Erk activation in Jurkat T cells in response to T-cell antigen receptor ligation. *Mol Cell Biol* **21:** 7137–7149.

Shen R, Ouyang YB, Qu CK, Alonso A, Sperzel L, Mustelin T, Kaplan MH, Feng GS. 2002. Grap negatively regulates T-cell receptor-elicited lymphocyte proliferation and interleukin-2 induction. *Mol Cell Biol* **22:** 3230–3236.

Shen S, Zhu M, Lau J, Chuck M, Zhang W. 2009. The essential role of LAT in thymocyte development during transition from the double-positive to single-positive stage. *J Immunol* **182:** 5596–5604.

Shim EK, Moon CS, Lee GY, Ha YJ, Chae SK, Lee JR. 2004. Association of the Src homology 2 domain-containing leukocyte phosphoprotein of 76 kD (SLP-76) with the p85 subunit of phosphoinositide 3-kinase. *FEBS Lett* **575:** 35–40.

Shui JW, Boomer JS, Han J, Xu J, Dement GA, Zhou G, Tan TH. 2007. Hematopoietic progenitor kinase 1 negatively regulates T cell receptor signaling and T cell-mediated immune responses. *Nat Immunol* **8:** 84–91.

Sieh M, Batzer A, Schlessinger J, Weiss A. 1994. GRB2 and phospholipase C-γ 1 associate with a 36- to 38-kilodalton phosphotyrosine protein after T-cell receptor stimulation. *Mol Cell Biol* **14:** 4435–4442.

Sigismund S, Woelk T, Puri C, Maspero E, Tacchetti C, Transidico P, Di Fiore PP, Polo S. 2005. Clathrin-independent endocytosis of ubiquitinated cargos. *Proc Natl Acad Sci* **102:** 2760–2765.

Singer AL, Bunnell SC, Obstfeld AE, Jordan MS, Wu JN, Myung PS, Samelson LE, Koretzky GA. 2004. Roles of the proline-rich domain in SLP-76 subcellular localization and T cell function. *J Biol Chem* **279:** 15481–15490.

Sommers CL, Lee J, Steiner KL, Gurson JM, DePersis CL, El-Khoury D, Fuller CL, Shores EW, Love PE, Samelson LE. 2005. Mutation of the phospholipase C-γ1-binding site of LAT affects both positive and negative thymocyte selection. *J Exp Med* **201:** 1125–1134.

Sommers CL, Samelson LE, Love PE. 2004. LAT: A T lymphocyte adapter protein that couples the antigen receptor to downstream signaling pathways. *Bioessays* **26:** 61–67.

Sommers CL, Menon RK, Grinberg A, Zhang W, Samelson LE, Love PE. 2001. Knock-in mutation of the distal four tyrosines of linker for activation of T cells blocks murine T cell development. *J Exp Med* **194:** 135–142.

Sommers CL, Park CS, Lee J, Feng C, Fuller CL, Grinberg A, Hildebrand JA, Lacana E, Menon RK, Shores EW, et al. 2002. A LAT mutation that inhibits T cell development yet induces lymphoproliferation. *Science* **296:** 2040–2043.

Stoica B, DeBell KE, Graham L, Rellahan BL, Alava MA, Laborda J, Bonvini E. 1998. The amino-terminal Src homology 2 domain of phospholipase C γ 1 is essential for TCR-induced tyrosine phosphorylation of phospholipase C γ 1. *J Immunol* **160:** 1059–1066.

Tamir I, Stolpa JC, Helgason CD, Nakamura K, Bruhns P, Daeron M, Cambier JC. 2000. The RasGAP-binding protein p62dok is a mediator of inhibitory FcγRIIB signals in B cells. *Immunity* **12:** 347–358.

Tanimura N, Nagafuku M, Liddicoat DR, Hamaoka T, Kosugi A. 2003a. Analysis of the mobility of signaling molecules in lymphocytes using fluorescence photobleaching techniques. *Sci STKE* **2003:** pl10.

Tanimura N, Nagafuku M, Minaki Y, Umeda Y, Hayashi F, Sakakura J, Kato A, Liddicoat DR, Ogata M, Hamaoka T, et al. 2003b. Dynamic changes in the mobility of LAT in aggregated lipid rafts upon T cell activation. *J Cell Biol* **160:** 125–135.

Tanimura N, Saitoh S, Kawano S, Kosugi A, Miyake K. 2006. Palmitoylation of LAT contributes to its subcellular localization and stability. *Biochem Biophys Res Commun* **341:** 1177–1183.

Trub T, Frantz JD, Miyazaki M, Band H, Shoelson SE. 1997. The role of a lymphoid-restricted, Grb2-like SH3-SH2-SH3 protein in T cell receptor signaling. *J Biol Chem* **272:** 894–902.

Varma R, Campi G, Yokosuka T, Saito T, Dustin ML. 2006. T cell receptor-proximal signals are sustained in peripheral microclusters and terminated in the central supramolecular activation cluster. *Immunity* **25:** 117–127.

Wang X. 2009. Cre transgenic mouse lines. *Methods Mol Biol* **561:** 265–273.

Wang H, Wei B, Bismuth G, Rudd CE. 2009. SLP-76-ADAP adaptor module regulates LFA-1 mediated costimulation and T cell motility. *Proc Natl Acad Sci* **106:** 12436–12441.

Wang Y, Kissenpfennig A, Mingueneau M, Richelme S, Perrin P, Chevrier S, Genton C, Lucas B, DiSanto JP, Acha-Orbea H, et al. 2008. Th2 lymphoproliferative disorder of LatY136F mutant mice unfolds independently of TCR-MHC engagement and is insensitive to the action of Foxp3+ regulatory T cells. *J Immunol* **180:** 1565–1575.

Wange RL. 2000. LAT, the linker for activation of T cells: A bridge between T cell-specific and general signaling pathways. *Sci STKE* **2000:** RE1.

Weber JR, Bell GM, Han MY, Pawson T, Imboden JB. 1992. Association of the tyrosine kinase LCK with phospholipase C-γ 1 after stimulation of the T cell antigen receptor. *J Exp Med* **176:** 373–379.

Weber JR, Orstavik S, Torgersen KM, Danbolt NC, Berg SF, Ryan JC, Tasken K, Imboden JB, Vaage JT. 1998. Molecular cloning of the cDNA encoding p36, a tyrosine-phosphorylated adaptor protein selectively expressed by T cells and natural killer cells. *J Exp Med* **187:** 1157–1161.

Weissman AM. 2001. Themes and variations on ubiquitylation. *Nat Rev Mol Cell Biol* **2:** 169–178.

Wilson BS, Pfeiffer JR, Surviladze Z, Gaudet EA, Oliver JM. 2001. High resolution mapping of mast cell membranes reveals primary and secondary domains of Fc(epsilon)RI and LAT. *J Cell Biol* **154:** 645–658.

Wittekind M, Mapelli C, Farmer BT 2nd, Suen KL, Goldfarb V, Tsao J, Lavoie T, Barbacid M, Meyers CA, Mueller L. 1994. Orientation of peptide fragments from Sos proteins bound to the N-terminal SH3 domain of Grb2 determined by NMR spectroscopy. *Biochemistry* **33:** 13531–13539.

Wulfing C, Davis MM. 1998. A receptor/cytoskeletal movement triggered by costimulation during T cell activation. *Science* **282:** 2266–2269.

Wunderlich L, Farago A, Downward J, Buday L. 1999. Association of Nck with tyrosine-phosphorylated SLP-76 in activated T lymphocytes. *Eur J Immunol* **29:** 1068–1075.

Yablonski D, Kadlecek T, Weiss A. 2001. Identification of a phospholipase C-γ1 (PLC-γ1) SH3 domain-binding site in SLP-76 required for T-cell receptor-mediated activation of PLC-γ1 and NFAT. *Mol Cell Biol* **21:** 4208–4218.

Yablonski D, Kuhne MR, Kadlecek T, Weiss A. 1998. Uncoupling of nonreceptor tyrosine kinases from PLC-γ1 in an SLP-76-deficient T cell. *Science* **281:** 413–416.

Yamasaki S, Nishida K, Hibi M, Sakuma M, Shiina R, Takeuchi A, Ohnishi H, Hirano T, Saito T. 2001. Docking protein Gab2 is phosphorylated by ZAP-70 and negatively regulates T cell receptor signaling by recruitment of inhibitory molecules. *J Biol Chem* **276:** 45175–45183.

Yamasaki S, Nishida K, Sakuma M, Berry D, McGlade CJ, Hirano T, Saito T. 2003. Gads/Grb2-mediated association with LAT is critical for the inhibitory function of Gab2 in T cells. *Mol Cell Biol* **23:** 2515–2529.

Yasuda T, Bundo K, Hino A, Honda K, Inoue A, Shirakata M, Osawa M, Tamura T, Nariuchi H, Oda H, et al. 2007. Dok-1 and Dok-2 are negative regulators of T cell receptor signaling. *Int Immunol* **19:** 487–495.

Yokosuka T, Kobayashi W, Sakata-Sogawa K, Takamatsu M, Hashimoto-Tane A, Dustin ML, Tokunaga M, Saito T. 2008. Spatiotemporal regulation of T cell costimulation by TCR-CD28 microclusters and protein kinase C theta translocation. *Immunity* **29:** 589–601.

Yokosuka T, Sakata-Sogawa K, Kobayashi W, Hiroshima M, Hashimoto-Tane A, Tokunaga M, Dustin ML, Saito T. 2005. Newly generated T cell receptor microclusters initiate and sustain T cell activation by recruitment of Zap70 and SLP-76. *Nat Immunol* **6:** 1253–1262.

Zeng R, Cannon JL, Abraham RT, Way M, Billadeau DD, Bubeck-Wardenberg J, Burkhardt JK. 2003. SLP-76 coordinates Nck-dependent Wiskott-Aldrich syndrome protein recruitment with Vav-1/Cdc42-dependent Wiskott-Aldrich syndrome protein activation at the T cell-APC contact site. *J Immunol* **171:** 1360–1368.

Zhang W, Irvin BJ, Trible RP, Abraham RT, Samelson LE. 1999a. Functional analysis of LAT in TCR-mediated signaling pathways using a LAT-deficient Jurkat cell line. *Int Immunol* **11:** 943–950.

Zhang W, Sloan-Lancaster J, Kitchen J, Trible RP, Samelson LE. 1998a. LAT: The ZAP-70 tyrosine kinase substrate that links T cell receptor to cellular activation. *Cell* **92:** 83–92.

Zhang W, Sommers CL, Burshtyn DN, Stebbins CC, DeJarnette JB, Trible RP, Grinberg A, Tsay HC, Jacobs HM, Kessler CM, et al. 1999b. Essential role of LAT in T cell development. *Immunity* **10:** 323–332.

Zhang W, Trible RP, Samelson LE. 1998b. LAT palmitoylation: Its essential role in membrane microdomain targeting and tyrosine phosphorylation during T cell activation. *Immunity* **9:** 239–246.

Zhang W, Trible RP, Zhu M, Liu SK, McGlade CJ, Samelson LE. 2000. Association of Grb2, Gads, and phospholipase C-γ 1 with phosphorylated LAT tyrosine residues. Effect of LAT tyrosine mutations on T cell antigen receptor-mediated signaling. *J Biol Chem* **275:** 23355–23361.

Zhu M, Janssen E, Zhang W. 2003. Minimal requirement of tyrosine residues of Linker for Activation of T Cells in TCR signaling and thymocyte development. *J Immunol* **170:** 325–333.

Zhu M, Shen S, Liu Y, Granillo O, Zhang W. 2005. Cutting Edge: Localization of linker for activation of T cells to lipid rafts is not essential in T cell activation and development. *J Immunol* **174:** 31–35.

Coordination of Receptor Signaling in Multiple Hematopoietic Cell Lineages by the Adaptor Protein SLP-76

Martha S. Jordan[1,2] and Gary A. Koretzky[1,2,3]

[1]Abramson Family Cancer Institute, University of Pennsylvania, Philadelphia, Pennsylvania 19104

[2]Department of Pathology and Laboratory Medicine, University of Pennsylvania, Philadelphia, Pennsylvania 19104

[3]Department of Medicine, University of Pennsylvania, Philadelphia, Pennsylvania 19104

Correspondence: koretzky@mail.med.upenn.edu

The adaptor protein SLP-76 is expressed in multiple hematopoietic lineages including T cells, platelets, and neutrophils. SLP-76 mediated signaling is dependent on its multiple protein interaction domains, as it creates a scaffold on which key signaling complexes are built. SLP-76 is critical for supporting signaling downstream of both immunoreceptors and integrins. The signaling molecules used both upstream and downstream of SLP-76 are similar among these receptors and across cell types; however, important differences exist. Appreciating how SLP-76 coordinates signal transduction across different cell and receptor types provides insights into the complex interplay of pathways critical for activation of cells of the immune system that are essential for host defense.

Adaptor proteins are an important component of many signaling systems both within and beyond the immune system. Unlike enzymes that catalyze the chemical reactions required for signal propagation, adaptor proteins are molecular platforms on which other proteins assemble. Although lacking enzymatic function, adaptor proteins regulate signaling by stabilizing or restricting molecular interactions required for proper enzyme activation and for localizing these key effector molecules appropriately within the cell.

Src homology 2 (SH2) domain-containing leukocyte protein of 76 kDa (SLP-76) is an adaptor present in a number of hematopoietic cell lineages including T cells, platelets, neutrophils, mast cells, macrophages, and NK cells. There are two SLP-76 homologs: SLP-65/B cell linker protein (BLNK), and cytokine-dependent hematopoietic cell linker (CLNK), also known as mast cell immunoreceptor signal transducer (MIST). BLNK is expressed in B cells and monocytes, and CLNK is expressed in mast cells, activated T cells, and NK cells. SLP-76 was

 doi: 10.1101/cshperspect.a002501
Cite this article as *Cold Spring Harb Perspect Biol* 2010;2:a002501

first identified in T cells as a substrate of T-cell receptor (TCR) stimulated protein tyrosine kinases (PTKs) through "pull down" experiments using immobilized growth factor receptor-bound protein 2 (Grb2) (Jackman et al. 1995). Since then, SLP-76 has been shown to be critically important in the development of T cells and for propagating signals downstream of not only the TCR but also additional receptors present on hematopoietic cells.

A MODEL FOR SLP-76 FUNCTION FOLLOWING IMMUNORECEPTOR ENGAGEMENT

TCR stimulation results in activation of Lck, a member of the Src family of PTKs, leading to phosphorylation of immunoreceptor tyrosine-based activation motifs (ITAMs) present within the CD3 chains of the TCR. Phosphorylated ITAMS recruit the Syk PTK ZAP-70, which is responsible for tyrosine phosphorylation of several proteins including the adaptors linker for activation of T cells (LAT) (Zhang et al. 1998) and SLP-76 (Bubeck Wardenburg et al. 1996). Phosphorylation of these adaptors is a critical step in the assembly of the LAT/SLP-76 signaling complex and is necessary for the recruitment and activation of additional proteins including phospholipase C γ1 (PLCγ1), the phospholipase responsible for hydrolysis of phosphatidylinositol 4,5-bisphosphate (PIP2) and generation of the critical second messengers inositol trisphosphate (IP3) and diacylglycerol (DAG).

Initial insights into the role of SLP-76 in TCR signaling arose from overexpression studies performed in the Jurkat T-cell line, demonstrating that increased expression of SLP-76 led to enhanced TCR signaling. The complementary loss-of-function studies including the generation of SLP-76 deficient Jurkat cells (J14) (Yablonski et al. 1998) and SLP-76 deficient mice (Clements et al. 1998; Pivniouk et al. 1998) revealed the essential role of SLP-76 in T-cell biology. Although TCR stimulation of J14 cells still results in Lck and ZAP-70 activation, TCR ligation fails to result in activation of key signaling components including PLCγ1 (Yablonski et al. 1998), resulting in diminished Ca^{2+} flux, protein kinase C-θ (PKCθ) activation, Erk phosphorylation, and decreased activity of several transcriptional reporter constructs, including nuclear factor of activated T cells (NFAT) and activator protein-1 (AP-1) (Yablonski et al. 1998; Herndon et al. 2001). In vivo, SLP-76 deficiency results in a complete failure to generate mature T cells. T-cell development in mice deficient for SLP-76 is blocked at the transition from the DN3 ($CD4^-CD8^-$, double negative) to DN4 stage (Clements et al. 1998; Pivniouk et al. 1998), a transition that requires signaling from the pre-TCR. Many of the biochemical defects described in J14 cells have now been shown in T cells from mice that generate a normal T-cell repertoire before excision of endogenous SLP-76 through cre-lox technology (Maltzman et al. 2005).

SLP-76 was defined initially as having three functional domains: the amino-terminal acidic domain, the central proline rich region (PPR), and the carboxy-terminal SH2 domain (Fig. 1). The amino terminus contains three tyrosines at amino acids 112 (113 in humans), 128, and 145 that become phosphorylated after TCR stimulation. When mutated at all three tyrosine residues (Y3F mutant), SLP-76 loses nearly all of its ability to support PLCγ1 and PKCθ activation, Ca^{2+} flux, and Erk phosphorylation (Musci et al. 1997b; Yablonski et al. 1998; Herndon et al. 2001). Mutation of this domain also leads to the dysregulation of actin polymerization and polarized actin cap formation following TCR stimulation (Bubeck Wardenburg et al. 1998). Loss of these functions has been associated with the inability of the Y3F SLP-76 mutant to associate with Vav1, the adaptor non-catalytic region of tyrosine kinase (Nck), and IL-2-inducible T-cell kinase (Itk), the proteins shown to bind to these phosphorylated tyrosines. In cell lines, mutation of tyrosines 112 and/or 128 abrogates the inducible association of SLP-76 with the guanine nucleotide exchange factor (GEF) Vav1 and Nck, two proteins involved in actin reorganization (Wu et al. 1996; Raab et al. 1997; Bubeck Wardenburg et al. 1998). Mutation of Y145 results in the loss of binding to Itk, a Tec family tyrosine kinase

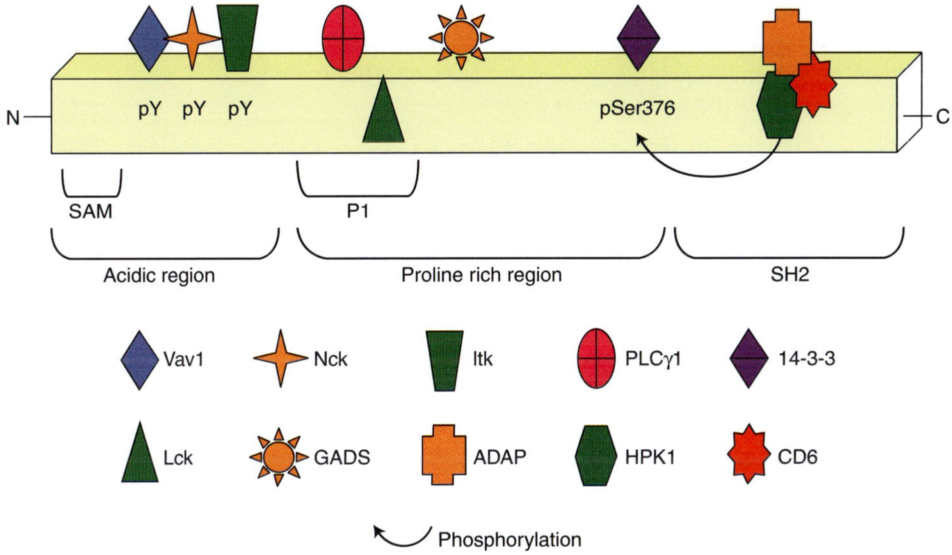

Figure 1. The domains of SLP-76 regulate its function. SLP-76 contains multiple regions that dictate its localization and association with other signaling molecules (adaptors in orange and kinases in green). The amino terminus contains three tyrosines that become phosphorylated after TCR ligation and function together to regulate the activity of Itk and activation of PLCγ1. This region also contains a critical SAM domain. The proline-rich region constitutively binds Gads as well as PLCγ1 and Lck, which lie within a PRR subdomain termed P1. Serine 376 in the PRR is a target for HPK1 phosphorylation that appears to allow for the subsequent binding of 14-3-3 family members. The carboxyl terminus of SLP-76 harbors an SH2 domain that associates with phosphorylated ADAP, CD6, and HPK1. The interaction of the SH2 domain with HPK1 regulates its kinase activity.

important for PLCγ1 activation (Su et al. 1999; Bunnell et al. 2000). Based on these protein/protein associations, original models posited that loss of Vav1 and Nck binding primarily result in decreased actin reorganization, whereas loss of Itk association leads to defective PLCγ1 activation. As discussed later, further investigations have shown that the role of these proteins is more complex than originally described.

To determine the in vivo consequence of mutating these three tyrosines, SLP-$76^{-/-}$ mice were reconstituted with wild-type (WT) or a Y3F mutant SLP-76 (Myung et al. 2001; Kumar et al. 2002). Expression of the Y3F SLP-76 transgene partially relieves the DN3 thymic block, enabling double positive ($CD4^+$ $CD8^+$, DP) thymocytes to develop; however, progression to the mature single positive ($CD4^+CD8^-$ or $CD4^-CD8^+$, SP) stage is dramatically impaired. The T cells that do arise fail to proliferate or form actin caps in response to TCR stimulation in addition to those biochemical defects observed in Y3F-expressing J14 cells (Myung et al. 2001; Kumar et al. 2002).

The central PRR of SLP-76 contains a novel SH3-domain binding motif (RxxK) that is responsible for the constitutive association of SLP-76 with the carboxy-terminal SH3 domain of the adaptor Gads (Asada et al. 1999; Liu et al. 1999; Berry et al. 2002). Gads also contains an SH2 domain that mediates its interaction with phosphorylated LAT (Asada et al. 1999; Liu et al. 1999). LAT is constitutively located in glycolipid-enriched membrane (GEM) microdomains present within the plasma membrane, and the relocalization of SLP-76 from the cytosol to the GEMs and into the LAT-nucleated complex requires its interaction with Gads (Singer et al. 2004; Bunnell et al. 2006). Disruption of the SLP-76/Gads interaction results in diminished Ca^{2+} flux, IP3 generation, and activation of Erk, PLCγ1, and NFAT (Musci et al.

1997b; Yablonski et al. 2001; Singer et al. 2004). In vivo, transgenic expression of a Gads-binding mutant allows for DP thymocyte development and progression of some thymocytes to the SP stage, but development is far from normal, as the thymic size is only 20% of that seen in WT SLP-76 reconstituted mice (Myung et al. 2001; Kumar et al. 2002). The smaller thymic size is at least in part because of the requirement of this region for proper positive selection (Jordan et al. 2007). Peripheral T cells in Gads-binding mutant mice are only weakly activated downstream of the TCR, as shown by their decreased ability to up-regulate activation markers, flux Ca^{2+}, and proliferate (Myung et al. 2001; Kumar et al. 2002). The ability to retain SLP-76 function above that seen in the Y3F mutant and in the face of disrupted SLP-76 localization may be because of retention of SLP-76 phosphorylation under these circumstances (Yablonski et al. 2001).

The third primary region of SLP-76 is its carboxy-terminal SH2 domain, which inducibly associates with adhesion and degranulation-promoting adapter protein (ADAP) and the serine-threonine kinase hematopoietic progenitor kinase 1 (HPK1) (Musci et al. 1997a; Sauer et al. 2001). Early studies in Jurkat cells gave little insight into the function of the SH2 domain, other than the finding that its absence led to only modest decreases in PLCγ1 and Erk activation (Yablonski et al. 2001). Additional studies showed a requirement for this domain in the activation of HPK1, although the cellular consequences of this function are unclear (Liou et al. 2000; Sauer et al. 2001). Transgenic expression of an SH2 SLP-76 mutant on a SLP-$76^{-/-}$ background allows for the generation of mature T cells that show nearly normal Ca^{2+} flux and Erk activation but diminished proliferation (Myung et al. 2001; Kumar et al. 2002). These characteristics are similar to those seen in mice deficient in ADAP, which is known to be required for the up-regulation and activation of integrins (Griffiths et al. 2001; Peterson et al. 2001). Based on these similarities, the primary role of this domain was assumed to be that of supporting of ADAP function in integrin activation.

NEW INSIGHTS INTO HOW THE STRUCTURE OF SLP-76 MEDIATES ITS FUNCTION

Over the past several years, further investigation into the structure of SLP-76 has led to more sophisticated models of how SLP-76 operates. These studies include analysis of mice expressing double or single mutations in the amino-terminal tyrosines of SLP-76, identification of additional regions within the amino terminus and PPR, and further in vivo analysis of the SH2 domain (Table 1).

The Amino Terminus: SLP-76 Tyrosines and the SAM Domain

To understand how the individual amino-terminal tyrosines of SLP-76 contribute to its function, knockin mice bearing mutations at both tyrosines 112 and 128 or tyrosine 145 alone were generated (Jordan et al. 2008). Unlike Y3F transgenic mice that have a paucity of mature thymocytes, both Y112/128F and Y145F mice produce mature thymocytes and peripheral T cells, although positive and negative selection is severely defective, as revealed by crossing the tyrosine mutants to mice expressing TCR transgenes. Biochemically, T cells from both SLP-76 mutant strains show impaired but not absent Ca^{2+} flux, actin polymerization and phosphorylation of PLCγ1 and Erk, with Y145F thymocytes consistently demonstrating a more severe phenotype both biochemically and developmentally. The similarity between Y112/128F and Y145F mice clearly demonstrate that there is coordinate regulation of multiple signaling pathways downstream of the amino-terminal tyrosines of SLP-76 in terms of PLCγ1 activation as well as actin reorganization.

This in vivo observation of coordinate regulation of actin polymerization and PLCγ1 function is largely supported by mechanistic studies performed in cell lines. In Jurkat cells, the amino-terminal tyrosines of SLP-76 are important for the relocalization of Nck and its binding partner WASP, an actin regulatory protein, to the T cell:APC interface. These tyrosines also

Table 1. Summary of signaling and developmental defects reported for primary T cells expressing mutant forms of SLP-76.

Domain	Mutation name	Defective signaling	Developmental defects	Defects in peripheral T-cell function
Amino terminus	ΔSAM-SLP-76[1]	Ca^{2+} flux, pErk	Thymic size, DP → SP, (+) and (−) selection	Proliferation, IL-2 production, SLP-76 clustering
	Y3F[2]	pSLP-76, pPLCγ1, Ca^{2+} flux	Thymic size, DN3 → DN4, DP → SP	CD69/CD25 up-regulation, proliferation, TCR clustering
	Y112/128F[3]	pSLP-76, pVav1, pPLCγ1, Ca^{2+}, pErk, actin	DP → SP, (+) and (−) selection, Nur77 upregulation	Not reported
	Y145F[3]	pPLCγ1, Ca^{2+}, pErk, actin	DP → SP, (+) and (−) selection, Nur77 up-regulation, ↑CD8SP with memory/innate markers	Not reported
PRR	Gads Binding Mutant[2,4]	pPLCγ1, Ca^{2+}	Thymic size, DP → SP, (+) selection	CD69/CD25 up-regulation, proliferation, TCR clustering
	P1[5*]	pPLCγ1, Ca^{2+}, pErk	n.d.	n.d.
	Lck[6]	pPLCγ1, Ca^{2+}, pErk	Thymic size, DN3 → DN4, DP → SP	CD25 up-regulation, proliferation, IL-2 production
SH2	R448K[2,7]	pSLP-76 Ser376, pPKD,	DN3 → DN4, DP → SP, (+) selection	CD69/CD25 up-regulation, proliferation, IL-2 production

1. Shen et al. 2009.
2. Myung et al. 2001.
3. Jordan et al. 2008.
4. Jordan et al. 2007.
*Data reported for cell lines.
5. Yablonski et al. 2001.
6. Kumar et al. 2005.
7. M. S. J. unpublished data.

direct the membrane recruitment of Vav1 and activated Cdc42, a GTPase required for WASP activation and a target of Vav1 GEF activity (Zeng et al. 2003). Such studies suggested that the coordinated recruitment of Vav1 and Nck and the subsequent activation of their targets or binding partners were sufficient for actin regulation. However, studies in $Itk^{-/-}$ T cells revealed that recruitment of activated WASP, Cdc42, and Vav1 to the T cell:APC interface is also dependent on Itk and specifically its SH2 domain, which interacts with Y145 of SLP-76. It was further shown that Itk is important for establishing/stabilizing the interaction between SLP-76 and Vav1 (Labno et al. 2003; Dombroski et al. 2005). These studies support a model in which all three tyrosines of SLP-76 and their respective binding partners function together to properly regulate TCR-induced actin polymerization.

Similar to actin regulation, cell line studies also support the notion that PLCγ1 regulation is controlled by multiple SLP-76 binding partners. Such studies revealed that the amino-terminal tyrosines of SLP-76 are required for Itk kinase activity and that the majority of catalytically active Itk is associated with SLP-76 (Bogin et al. 2007). However, Vav1 also regulates Itk and PLCγ1 activation as well as the interaction between SLP-76 and PLCγ1 (Reynolds et al. 2002). These data, when considered with the role of Itk in mediating Vav1/SLP-76 interactions, suggest that these molecules must be involved in coordinate regulation of one another and that they function together to support many if not most of the signaling pathways following TCR engagement. In generating a unified model of SLP-76 function using data from primary T cells and cell lines, it is important to acknowledge that unlike in Jurkat cells where mutation of the amino-terminal tyrosines result in loss of SLP-76 associations with Vav1, Nck, and Itk, introduction of these mutations into primary T cells does not (Jordan et al. 2008). Thus, it appears that formation of the signaling complex in primary cells results from numerous intermolecular interactions. It remains likely, however, that the SLP-76 tyrosines are still critical for binding to Vav1, Nck, and Itk, and that these associations are essential for optimal function of each of the SLP-76-associated proteins.

Although several similarities exist between Y112/128F and Y145F mutant mice, there are also distinct differences. TCR stimulation of thymocytes from Y112/128F but not Y145F mice fails to result in phosphorylation of Vav1 following TCR stimulation, whereas mutation of Y145 but not Y112/128 results in the selection of CD8SP thymocytes that have characteristics shared by memory and/or innate-like T cells (Jordan et al. 2008). Two distinct features of these CD8SP cells are their capacity to produce copious amounts of IFNγ directly ex vivo and their expression of high levels of eomesodermin, a transcription factor found at elevated levels in memory $CD8^+$ T cells. Because Itk deficiency also leads to increased development of this thymic subset, it is likely that SLP-76-dependent activation of Itk regulates their development (Atherly et al. 2006; Broussard et al. 2006). How tyrosine 145 of SLP-76 plays a role in the regulation of eomesodermin is unclear, but insight into this question will likely be informative in understanding T cell fate decisions both in the thymus and in memory T-cell development.

The generation of SLP-76 tyrosine mutant mice led to experiments addressing whether the two mutations could complement one another functionally in vivo when coexpressed. Double-mutant mice (mice expressing one Y112/128F and one Y145F allele) were analyzed for T-cell development and for signaling downstream of the TCR (Jordan et al. 2008). Thymocyte development in double-mutant mice is phenotypically and functionally normal, with no increase in CD8SP memory/innate-like T cells, as seen selectively in Y145F mice. Biochemically, expression of both mutant alleles restores Vav1 phosphorylation, a selective defect seen in Y112/128F thymocytes, and PLCγ1 phosphorylation, a defect shared by both lineages. These data demonstrate that downstream of the TCR, active Vav1, Nck, and Itk do not have to be present on the same SLP-76 molecule to support wild-type signal transduction. However, given the plethora of data indicating that these proteins coordinately regulate one another, it is likely that formation of a signaling complex consisting of at least two SLP-76 molecules forms following TCR ligation. Such a complex may be dependent on LAT, which has been shown to oligomerize after TCR stimulation because of its inducible association with Grb2 (Houtman et al. 2006). The precise stoichiometry of the LAT/SLP-76 complex has yet to be defined.

The amino terminus of SLP-76 also contains a sterile-α motif (SAM) located just proximal to tyrosines 112, 128, and 145 at amino acids 12–78. This motif was identified several years ago, but only recently has its function begun to be appreciated. SAM domains typically mediate protein/protein interactions that may or may not involve other SAM domains. Currently, it is not known what interaction(s) the SAM domain of SLP-76 mediates, but knockin mice with a mutation in this motif demonstrate its importance in thymocyte development (Shen et al. 2009). Thymocytes from ΔSAM-SLP-76 mice transition through the DN stages of development relatively normally, but diminished numbers are present at the DP and SP stages, consistent with defects in both positive and negative selection. Following TCR engagement, T cells from these mice show decreased Ca^{2+} flux and Erk activation accompanied by weak clustering of SLP-76. Mechanistically, the underlying cause of these observed defects is not known, but they are at least partially independent of the SLP-76/Gads interaction, as mice bearing mutations in both of these domains show severe developmental defects of thymic subsets (Shen et al. 2009).

The SLP-76 PRR: The P1 Domain and Ser376 Phosphorylation

In addition to the Gads-binding site, other important regions within the PRR of SLP-76 have recently been described. Defined as a region necessary for the basal association of SLP-76 with PLCγ1, the P1 domain (amino acids 157–223) was shown to be important for activating several pathways downstream of SLP-76, including PLCγ1 and Erk phosphorylation, IP3 production, and NFAT activation (Yablonski et al. 2001). Further studies showed that the critical aspect of this region probably lies more in its role as a "molecular spacer" rather than mediating an essential protein/protein interaction (Gonen et al. 2005). However, manipulation of this domain can lead to loss of SLP-76/Vav1 associations and/or SLP-76-associated Itk kinase activity, both of which are associated with decreased PLCγ1 activation (Beach et al. 2007).

Other residues within the PRR have been identified as either additional regions of protein/protein interaction or sites of phosphorylation. Amino acids 185–194, just proximal to the Gads-binding domain, mediate a constitutive interaction with the SH3 domain of Lck (Sanzenbacher et al. 1999). The importance of this region was recognized through in vivo expression of a SLP-76Δ185-194 mutant on a SLP-$76^{-/-}$ background. Mice bearing this mutation have a severe DN3 block and very few peripheral T cells that show only residual function (Kumar et al. 2005). Despite these dramatic defects, it is unclear how the loss of Lck binding to SLP-76 leads to this phenotype. It is possible that disruption of SLP-76/Lck prevents proper phosphorylation of Itk by Lck (Heyeck et al. 1997). Recently, serine 376 within the PRR was shown to be phosphorylated by the kinase HPK1, one of the binding partners of the SLP-76 SH2 domain. Phosphorylation at this site results in decreased TCR signal transduction. Although the mechanism has yet to be defined, early studies suggest that 14-3-3 family members, which are recruited to SLP-76 following Ser376 phosphorylation, may play a role (Di Bartolo et al. 2007; Shui et al. 2007).

The SLP-76 SH2 Domain: ADAP Dependent and Independent Functions

Because of its association with ADAP, the SLP-76 SH2 domain has been implicated in TCR-induced integrin activation. For full activation, integrins must undergo both affinity and avidity changes. These changes are governed by engagement of other cell surface receptors, such as the TCR, that lead to conformational changes in integrins and integrin clustering. These changes increase integrin affinity and avidity, respectively, for ligands and are referred to as "inside-out" signaling. Support for a role for SLP-76 in inside-out signaling comes from studies performed in cell lines, which showed that mutation of ADAP at sites of SLP-76/ADAP interaction diminish the ability of T cells to adhere to integrin-coated surfaces (Wang et al. 2004; Kliche et al. 2006). Until recently, data linking the SLP-76 SH2 domain was based on mutational analysis of ADAP. New data in primary T cells now reveal that mutation of the SLP-76 SH2 domain itself inhibits TCR-induced inside-out signaling (Baker et al. 2009). These data also reveal a role for this domain in signaling directly downstream of integrin engagement (discussed later).

The molecular mechanism for SLP-76 mediated inside-out signaling to integrins is thus likely ADAP dependent. In addition to its inducible interaction with SLP-76, ADAP is constitutively present in a trimolecular complex with Src kinase-associated phosphoprotein of 55 kDa (SKAP55), via a direct interaction, and Ras-proximity-1 (Rap1)-GTP-interacting adapter molecule (RIAM), via a tertiary interaction. On TCR stimulation, this ADAP/RIAM/SKAP55 complex associates with the small GTPase Rap1 and is responsible for the relocalization of activated Rap1 to the membrane, where it aids in promoting clustering and affinity modulation of integrins, a step necessary for their full activation (Kliche et al. 2006; Menasche et al. 2007). It has been shown that neither ADAP (Kliche et al. 2006) nor the SLP-76 SH2 domain (unpubl. data) regulates Rap1 activation. However, recent data indicate that SLP-76 is involved in the activation of this GTPase, although the domain requirements have not been identified (Horn et al. 2009).

Recent in vivo studies comparing the function of primary T cells harboring a SLP-76 SH2 domain mutation to ADAP-deficient T cells suggest that some SH2 domain functions are ADAP independent. Specifically, PKD phosphorylation, a target of PKCθ, and phosphorylation of SLP-76 on Ser376, an HPK1 target, is dependent on the SLP-76 SH2 domain but not ADAP (M.S.J. unpubl. data). These findings indicate that the pathways downstream of the SLP-76 SH2 domain and ADAP are not linear, and they may help explain why activation of NFκB targets is more defective in the absence of a functional SLP-76 SH2 domain versus in the deficiency of ADAP (M.S.J. unpubl. data).

The full function of the SLP-76 SH2 domain may not be solely based on ADAP and HPK1 interactions. This domain can also associate with the cytoplasmic tail of the T-cell surface receptor CD6 (Hassan et al. 2006). Studies have shown that engagement of CD6 costimulates TCR-driven IL-2 production and that mutation of the tyrosine on CD6 that mediates its interaction with SLP-76 abolishes CD6 costimulation (Hassan et al. 2006). How this mutation affects in vivo function awaits further investigations.

SLP-76 SIGNALING IN OTHER HEMATOPOIETIC CELLS

Initial insight into the function of SLP-76 was based solely on its role in TCR signaling. Subsequent evaluation of signal transduction in platelets and neutrophils has shown that SLP-76 is important for supporting signals downstream of immunoreceptors present on non-T cells and provided the first evidence for the role of SLP-76 downstream of integrins. Experiments examining the molecular basis of SLP-76 function in various lineages allow for interesting comparisons between the requirements of signal propagation downstream of a variety of receptor systems.

GPVI and FcγR Signaling in Platelets and Neutrophils

Platelets use multiple cell surface receptors to maintain hemostasis, including the collagen receptor GPVI and integrins of the β1 and β2 families. Acting as a first line of defense, neutrophils rely on FcγRs for recognition of IgG-opsonized particles or soluble IgG immune complexes and integrins, including the β2 integrin Mac-1 (CD18/CD11b), for adhesion to extracellular matrix proteins. GPVI associates with the Fcγ chain and thus propagates signals in a manner similar to that of the FcγR present on neutrophils. Like TCR signaling, the FcγR cascade relies on phosphorylation of ITAMs present within the Fcγ chain by Src family members (Lyn and/or Fyn, Hgr, and Fgr for platelets and neutrophils and Lck for T cells) followed by recruitment of Syk (ZAP-70 in T cells) to these phosphorylated motifs. Activated Syk is required for subsequent phosphorylation of PLCγ2, a critical activator of downstream effector functions. Ligation of GPVI in platelets and FcγR in neutrophils results in actin reorganization, cell spreading, degranulation, aggregation in the case of platelets, and production of reactive oxygen intermediates (ROI) in the case of neutrophils. All of these functions are dependent on SLP-76, as platelets and neutrophils from SLP-76-deficient mice either fail completely or show greatly reduced responses when these events are interrogated (Clements et al. 1999; Newbrough et al. 2003). Although the molecular mechanism defining how SLP-76 contributes to PLCγ activation and several other SLP-76-dependent pathways have been best described in T cells, many of the same effector molecules, including Vav and Tec family members, have also been shown to be important for signaling downstream of platelet and neutrophil FcγRs (reviewed in Bezman and Koretzky 2007). Similar to signaling downstream of the TCR, LAT is phosphorylated following GPVI stimulation and is required for optimal platelet degranulation following stimulation with convulxin, a selective GPVI agonist (Gibbins et al. 1998; Judd et al. 2002). In T cells, Gads is largely required for development and peripheral T-cell function; however, in platelets, the role for Gads is more complex. For some GPV1-mediated functions such as convulxin-induced degranulation, Gads is completely dispensable (Judd et al. 2002). In contrast, Gads is required for optimal GPVI-induced aggregation (Hughes et al. 2008). Therefore, although signaling through a SLP-76/LAT axis is common between the TCR and GPVI, their differential requirement for Gads in certain functions indicates that downstream integration of signals for these receptors can be distinct.

Integrin Signaling in Platelets and Neutrophils

Signaling downstream of integrins is also dependent on SLP-76 and has been described best in platelets and neutrophils (Fig. 2). Similar to the TCR, engagement of FcγRs can induce the integrin affinity and avidity changes required for full integrin activation.

Once activated, ligation of integrins such as αIIbβIIIa on platelets and Mac-1 on neutrophils by fibrinogen or polyRGD (an engineered tri-peptide integrin binding motif) results in "outside-in" signaling from the integrin. Similar to the TCR and FcγR, integrin outside-in signaling uses Src and Syk protein tyrosine kinases as well as SLP-76 and many of its associated molecules for signal propagation. Exactly how integrins, which lack ITAMs, couple to Syk is an area of active investigation. Coupling may be through a direct ITAM-independent interaction (Woodside et al. 2001) or may rely on associations with ITAM-bearing receptors, such as the Fcγ chain or DAP12, as suggested from recent studies performed in mice deficient in these ITAM-bearing proteins (Mocsai et al. 2006). Under normal circumstances, integrin ligation results in cell spreading, ROI production, and degranulation of intracellular components; however, these functions are greatly impaired in platelets and/or neutrophils lacking SLP-76 (Judd et al. 2000; Judd et al. 2002; Newbrough et al. 2003). The decreased functions observed in vitro correlate with altered neutrophil and platelet-dependent functions in vivo, as mice lacking SLP-76 within the myeloid compartment show defective integrin-dependent accumulation and degranulation of neutrophils as measured by a model of LPS-induced dermal inflammation (Clemens et al. 2007). Moreover, mice bearing particular SLP-76 mutations within their platelets fail to form stable thrombi in a model of arterial thrombosis (Bezman et al. 2008). These data demonstrate the importance of SLP-76 in these non-T-cell compartments.

Structure/Function Studies of SLP-76 in Platelets and Neutrophils

Mice reconstituted with bone marrow progenitor cells from SLP-76-deficient mice expressing mutant forms of SLP-76 have been critically important in determining the domain requirements of SLP-76 in non-T-cell immunoreceptor and integrin function. Using this system, it was shown, similar to their essential role in TCR signal transduction, that the amino-terminal tyrosines of SLP-76 are critical for downstream propagation of GPVI as well as integrin initiated signaling (Abtahian et al. 2006). Y3F-expressing platelets fail to degranulate in response to GPVI, and diminished platelet spreading and neutrophil ROI production is observed in Y3F-expressing cells stimulated through their integrin receptors. Studies in individual amino-terminal tyrosine knockin mice revealed that αIIbβIIIa signaling, as measured by platelet spreading, is diminished in mice harboring

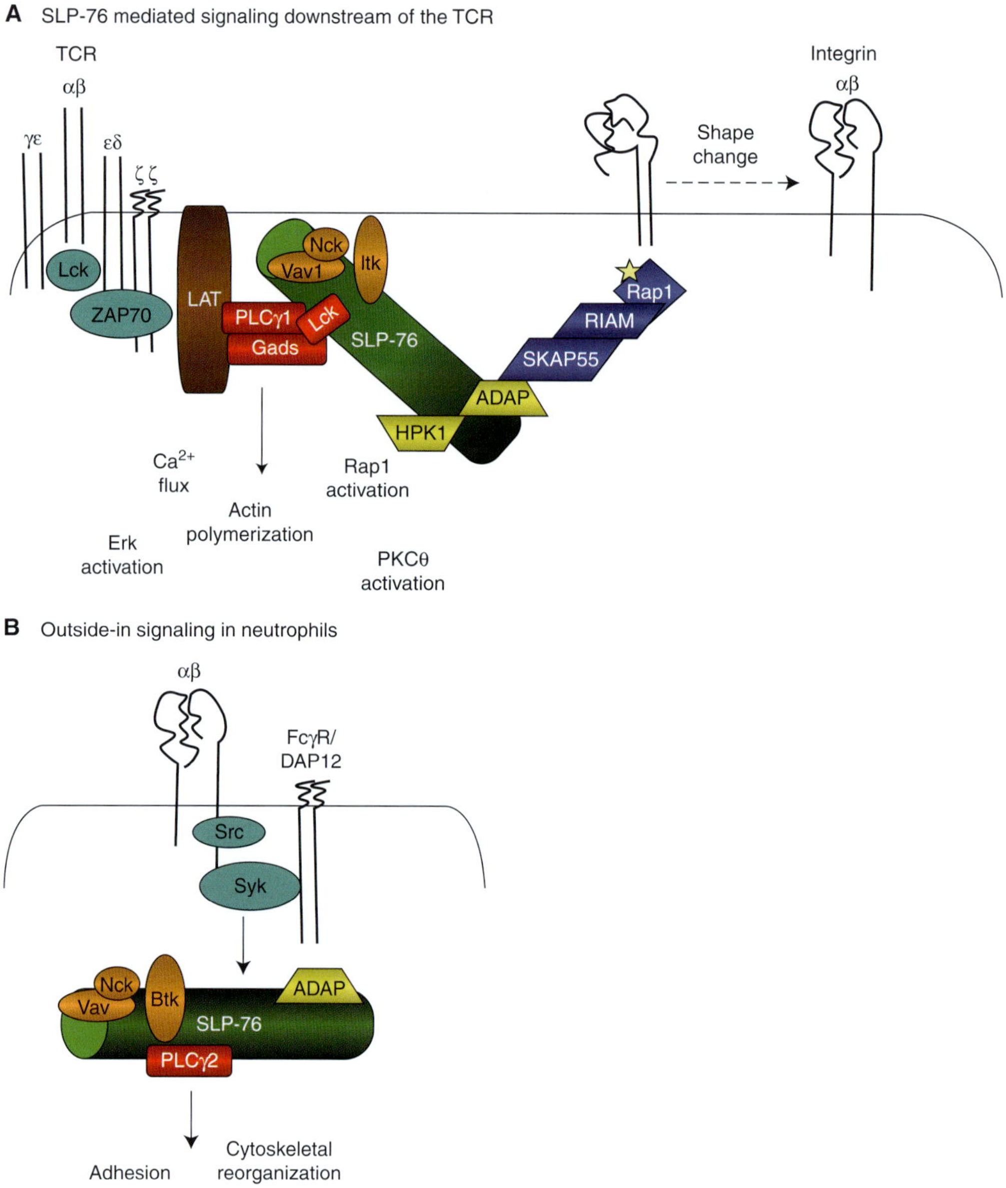

Figure 2. SLP-76 supported TCR and integrin signaling. (*A*) SLP-76 mediated signaling downstream of the TCR. TCR ligation initiates activation of Lck that phosphorylates ITAMs present on the CD3 chains of the TCR. ZAP-70 is recruited, activated, and phosphorylates LAT and SLP-76. Phosphorylated LAT associates with PLCγ1 and Gads and in doing so, recruits SLP-76 to the membrane. Phosphorylated SLP-76 associates with Vav1, Nck, and Itk, which together activate PLCγ1 and initiate Ca^{2+} flux and actin polymerization. SLP-76 also regulates the activation of Erk, PKCθ, and Rap1. Association of SLP-76 with the ADAP/SKAP55/RIAM complex promotes inside-out signaling by localizing active Rap1. Conformational changes mediated by inside-out signaling enable high affinity binding of integrins to their receptors. (*B*) Outside-in signaling in neutrophils. Integrin ligation activates associated Src kinases in neutrophils. Syk may be recruited to an integrin-associated ITAM-bearing receptor or may directly associate with the integrin β chain tail. Activated Syk phosphorylates SLP-76 and initiates PLCγ2 activation, leading to cytoskeletal reorganization and cellular adhesion in a LAT- and Gads-independent manner.

mutations in tyrosines 112 and 128 of SLP-76, whereas platelet spreading in Y145F mutant mice is normal (Bezman et al. 2008). In contrast, GPVI signaling is strikingly more dependent on tyrosine 145 than on tyrosines 112 and 128 for most functions except cell spreading, where both mutants show similar defects. This latter observation of defective Y112/128F spreading may reflect a role for integrin outside-in signaling, which follows GPVI inside-out signaling. It is interesting to note that Vav phosphorylation is defective in Y145F platelets and preserved in Y112/128F platelets, the opposite of what is observed in T cells bearing these mutations (Jordan et al. 2008). Therefore, although signals transduced downstream of different receptors may use identical or similar proteins, the exact details of how transduction pathways are integrated appear to differ between cell types and receptor classes.

Another intriguing difference between the requirements for the amino-terminal tyrosines of SLP-76 in T cells versus platelets was revealed through complementation experiments. As noted earlier, mice expressing one Y145F and one Y112/128F allele functionally complement one another in *trans* in T cells (Jordan et al. 2008). This is not the case for GPVI signaling, as dual expression of each mutant allele results in a functional response that is only marginally better than the response from Y145F platelets (Bezman et al. 2008). Similar results are seen in neutrophils from these mice. Both Y145F and Y112/128F neutrophils display defective FcγR and integrin signal transduction, and coexpression of each mutant allele does little to rescue these defects (Lenox et al. 2009). One clue as to why complementation occurs in T cells and not platelets and neutrophils may lie in the differential requirement for Gads in signaling downstream of the TCR versus GPVI. LAT oligomerizes following TCR ligation (Houtman et al. 2006). Thus, the association of SLP-76 with LAT via Gads may allow for trans-activation of those molecules associated with SLP-76 due to their ability to be in close proximity to one another. The fact that Gads is not required for all GPVI functions suggests that the association of SLP-76 with LAT can be regulated differently in T cells versus platelets. It is possible that this difference somehow precludes complementation. Additionally, whether LAT oligomerizes in platelets and whether SLP-76 is affected by this oligomerization remain to be determined. A related possibility that has been posited to explain the discrepancy in complementation is the difference in the number of ITAMs present in the TCR versus the GPVI/FcγR signaling complex (Bezman et al. 2008). The TCR/CD3 complex consists of ten ITAMs compared with the two associated with the Fcγ chain. An increased number of ITAMs may result in the assembly of a larger molecular signaling complex, allowing for greater opportunity for *trans*-complementation.

Analysis of the role of the PRR region of SLP-76 in non-T cells has also revealed differences in the requirements of this domain for downstream signal transduction of various receptors, specifically integrins versus GPVI/FcγRs. In T cells, mutation of the residues of the PRR that mediate the interaction of SLP-76 with Gads results in greatly diminished but not absent TCR-mediated functions. GPVI signaling in platelets is also dependent on the SLP-76/Gads interaction domain, although Gads itself can be dispensable under some circumstances, as discussed above (Hughes et al. 2008). In contrast, studies on platelets and neutrophils from chimeric mice expressing mutations in the Gads-binding site have shown that integrin function appears normal in the absence of a SLP-76/Gads interaction, as the mutant platelets spread normally on integrin ligand-coated surfaces and mutant neutrophils produce copious amounts of ROI following integrin stimulation (Abtahian et al. 2006). How SLP-76 functions when it is unable to bind to Gads is unknown. It has been suggested that either SLP-76 does not need to relocalize to the membrane after integrin stimulation or that SLP-76 utilizes other protein interactions to achieve proper localization (Abtahian et al. 2006). Recent studies of integrin outside-in signaling in T cells have begun to help answer some of these questions.

The integrin studies performed in platelets and neutrophils have paved the way for subsequent studies in T cells. Interestingly, however, findings in T cells may also provide new insights to foster our understanding of integrin signaling in myeloid derived lineages. Similar to what is seen following TCR stimulation, integrin stimulation of T cells leads to the formation of SLP-76 microclusters that localize with phosphorylated signaling proteins such as ZAP-70 and Lck (Nguyen et al. 2008; Baker et al. 2009). Unlike TCR stimulation, integrin-induced SLP-76 clustering does not require LAT or the Gads-binding region of SLP-76 (Baker et al. 2009). These results are consistent with the requirements of SLP-76 domains for integrin activation in platelets and extend our understanding, indicating that SLP-76 clustering downstream of integrins is mediated by proteins other than Gads or by domains outside of the Gads-binding region.

Additional studies investigating the role of the SLP-76 SH2 domain in integrin-mediated signals show that this domain is required for both integrin-induced SLP-76 clustering and outside-in mediated integrin adhesion. Studies in ADAP-deficient T cells reveal that ADAP is required for these same functions, suggesting that perhaps ADAP plays a role in SLP-76 membrane localization following integrin stimulation. Interestingly, ADAP is not required for SLP-76 clustering post-TCR ligation (Baker et al. 2009). Whether these findings are applicable to platelets and neutrophils remains to be determined, but these data do provide yet another example of differential regulation of SLP-76 and the signaling pathways it supports.

ADDITIONAL ROLES OF SLP-76

In this article, we discussed only a subset of cell types and receptor systems in which SLP-76 plays a role. In addition to the lineages considered above, mast cells use SLP-76 to transduce signals downstream of FcεRI. SLP-76-deficient mast cells have impaired cytokine release, degranulation, and Ca^{2+} flux, and these mice are resistant to IgE-induced passive systemic anaphylaxis (Pivniouk et al. 1999). Structure/function analysis has revealed that the functional hierarchy of the domains of SLP-76 in mast cells is similar to that seen in T cells (Wu et al. 2004). Very recently, mice deficient in SLP-76 were also shown to be resistant to serum-induced arthritis, a disease that is dependent on functional mast cells and neutrophils, which highlights the in vivo importance of SLP-76 in these lineages (Lenox et al. 2009).

SLP-76 also impacts systems beyond hematopoietic cells. The subcutaneous hemorrhaging observed in most SLP-76-deficient fetuses and the blood-filled peritoneum seen in surviving pups was initially attributed to defective platelet function. However, this phenotype was shown subsequently to be the result of improper vascular development with a failure of blood and lymphatic vessels to separate normally (Abtahian et al. 2003). The exact means by which SLP-76 controls this separation has yet to be determined.

In the years since the identification and initial characterization of SLP-76 as a substrate of TCR-stimulated PTKs, much has been learned about the details of how this protein functions, both in T cells and other immune lineages. It is likely that this key adaptor protein has yet other roles, perhaps as a regulator of signals initiated by additional cell surface receptors. Insights gleaned from studies of how SLP-76 regulates immune cell development and function have provided important information about the complexities of signaling in these lineages, and they have also taught us a great deal about more general principles of how signal transduction cascades are integrated by adaptor proteins. We anticipate that future studies of SLP-76, using mice in which expression of mutant variants of this protein can be more tightly regulated temporally and in a lineage specific manner, will teach us much more about the intricate processes of cell differentiation and fate decisions and more globally about how the immune system uses various strategies for optimal host defense.

ACKNOWLEDGMENTS

We thank Justina Stadanlick for excellent editorial assistance.

REFERENCES

Abtahian F, Bezman N, Clemens R, Sebzda E, Cheng L, Shattil SJ, Kahn ML, Koretzky GA. 2006. Evidence for the requirement of ITAM domains but not SLP-76/Gads interaction for integrin signaling in hematopoietic cells. *Mol Cell Biol* **26:** 6936–6949.

Abtahian F, Guerriero A, Sebzda E, Lu MM, Zhou R, Mocsai A, Myers EE, Huang B, Jackson DG, Ferrari VA, et al. 2003. Regulation of blood and lymphatic vascular separation by signaling proteins SLP-76 and Syk. *Science* **299:** 247–251.

Asada H, Ishii N, Sasaki Y, Endo K, Kasai H, Tanaka N, Takeshita T, Tsuchiya S, Konno T, Sugamura K. 1999. Grf40, A novel Grb2 family member is involved in T cell signaling through interaction with SLP-76 and LAT. *J Exp Med* **189:** 1383–1390.

Atherly LO, Lucas JA, Felices M, Yin CC, Reiner SL, Berg LJ. 2006. The Tec family tyrosine kinases Itk and Rlk regulate the development of conventional CD8+ T cells. *Immunity* **25:** 79–91.

Baker RG, Hsu CJ, Lee D, Jordan MS, Maltzman JS, Hammer DA, Baumgart T, Koretzky GA. 2009. The adapter protein SLP-76 mediates "outside-in" integrin signaling and function in T cells. *Mol Cell Biol* doi: 10.1128/MCB.00283-09.

Beach D, Gonen R, Bogin Y, Reischl IG, Yablonski D. 2007. Dual role of SLP-76 in mediating T cell receptor-induced activation of phospholipase C-gamma1. *J Biol Chem* **282:** 2937–2946.

Berry DM, Nash P, Liu SK, Pawson T, McGlade CJ. 2002. A high-affinity Arg-X-X-Lys SH3 binding motif confers specificity for the interaction between Gads and SLP-76 in T cell signaling. *Curr Biol* **12:** 1336–1341.

Bezman N, Koretzky GA. 2007. Compartmentalization of ITAM and integrin signaling by adapter molecules. *Immunol Rev* **218:** 9–28.

Bezman NA, Lian L, Abrams CS, Brass LF, Kahn ML, Jordan MS, Koretzky GA. 2008. Requirements of SLP76 tyrosines in ITAM and integrin receptor signaling and in platelet function in vivo. *J Exp Med* **205:** 1775–1788.

Bogin Y, Ainey C, Beach D, Yablonski D. 2007. SLP-76 mediates and maintains activation of the Tec family kinase ITK via the T cell antigen receptor-induced association between SLP-76 and ITK. *Proc Natl Acad Sci* **104:** 6638–6643.

Broussard C, Fleischacker C, Horai R, Chetana M, Venegas AM, Sharp LL, Hedrick SM, Fowlkes BJ, Schwartzberg PL. 2006. Altered development of CD8+ T cell lineages in mice deficient for the Tec kinases Itk and Rlk. *Immunity* **25:** 93–104.

Bubeck Wardenburg J, Fu C, Jackman JK, Flotow H, Wilkinson SE, Williams DH, Johnson R, Kong G, Chan AC, Findell PR. 1996. Phosphorylation of SLP-76 by the ZAP-70 protein-tyrosine kinase is required for T-cell receptor function. *J Biol Chem* **271:** 19641–19644.

Bubeck Wardenburg J, Pappu R, Bu JY, Mayer B, Chernoff J, Straus D, Chan AC. 1998. Regulation of PAK activation and the T cell cytoskeleton by the linker protein SLP-76. *Immunity* **9:** 607–616.

Bunnell SC, Diehn M, Yaffe MB, Findell PR, Cantley LC, Berg LJ. 2000. Biochemical interactions integrating Itk with the T cell receptor-initiated signaling cascade. *J Biol Chem* **275:** 2219–2230.

Bunnell SC, Singer AL, Hong DI, Jacque BH, Jordan MS, Seminario MC, Barr VA, Koretzky GA, Samelson LE. 2006. Persistence of cooperatively stabilized signaling clusters drives T-cell activation. *Mol Cell Biol* **26:** 7155–7166.

Clemens RA, Lenox LE, Kambayashi T, Bezman N, Maltzman JS, Nichols KE, Koretzky GA. 2007. Loss of SLP-76 expression within myeloid cells confers resistance to neutrophil-mediated tissue damage while maintaining effective bacterial killing. *J Immunol* **178:** 4606–4614.

Clements JL, Lee JR, Gross B, Yang B, Olson JD, Sandra A, Watson SP, Lentz SR, Koretzky GA. 1999. Fetal hemorrhage and platelet dysfunction in SLP-76-deficient mice. *The Journal of clinical investigation* **103:** 19–25.

Clements JL, Yang B, Ross-Barta SE, Eliason SL, Hrstka RF, Williamson RA, Koretzky GA. 1998. Requirement for the leukocyte-specific adapter protein SLP-76 for normal T cell development. *Science* **281:** 416–419.

Di Bartolo V, Montagne B, Salek M, Jungwirth B, Carrette F, Fourtane J, Sol-Foulon N, Michel F, Schwartz O, Lehmann WD, et al. 2007. A novel pathway down-modulating T cell activation involves HPK-1-dependent recruitment of 14-3-3 proteins on SLP-76. *J Exp Med* **204:** 681–691.

Dombroski D, Houghtling RA, Labno CM, Precht P, Takesono A, Caplen NJ, Billadeau DD, Wange RL, Burkhardt JK, Schwartzberg PL. 2005. Kinase-independent functions for Itk in TCR-induced regulation of Vav and the actin cytoskeleton. *J Immunol* **174:** 1385–1392.

Gibbins JM, Briddon S, Shutes A, van Vugt MJ, van de Winkel JG, Saito T, Watson SP. 1998. The p85 subunit of phosphatidylinositol 3-kinase associates with the Fc receptor gamma-chain and linker for activitor of T cells (LAT) in platelets stimulated by collagen and convulxin. *J Biol Chem* **273:** 34437–34443.

Gonen R, Beach D, Ainey C, Yablonski D. 2005. T cell receptor-induced activation of phospholipase C-gamma1 depends on a sequence-independent function of the P-I region of SLP-76. *J Biol Chem* **280:** 8364–8370.

Griffiths EK, Krawczyk C, Kong YY, Raab M, Hyduk SJ, Bouchard D, Chan VS, Kozieradzki I, Oliveira-Dos-Santos AJ, Wakeham A, et al. 2001. Positive regulation of T cell activation and integrin adhesion by the adapter Fyb/Slap. *Science* **293:** 2260–2263.

Hassan NJ, Simmonds SJ, Clarkson NG, Hanrahan S, Puklavec MJ, Bomb M, Barclay AN, Brown MH. 2006. CD6 regulates T-cell responses through activation-dependent recruitment of the positive regulator SLP-76. *Mol Cell Biol* **26:** 6727–6738.

Herndon TM, Shan XC, Tsokos GC, Wange RL. 2001. ZAP-70 and SLP-76 regulate protein kinase C-theta and NF-kappa B activation in response to engagement of CD3 and CD28. *J Immunol* **166:** 5654–5664.

Heyeck SD, Wilcox HM, Bunnell SC, Berg LJ. 1997. Lck phosphorylates the activation loop tyrosine of the Itk kinase domain and activates Itk kinase activity. *J Biol Chem* **272:** 25401–25408.

Horn J, Wang X, Reichardt P, Stradal TE, Warnecke N, Simeoni L, Gunzer M, Yablonski D, Schraven B, Kliche S. 2009. Src homology 2-domain containing leukocyte-specific phosphoprotein of 76 kDa Is mandatory for TCR-mediated inside-out signaling, but dispensable for CXCR4-mediated LFA-1 activation, adhesion, and migration of T cells. *J Immunol* **183:** 5756–5767.

Houtman JC, Yamaguchi H, Barda-Saad M, Braiman A, Bowden B, Appella E, Schuck P, Samelson LE. 2006. Oligomerization of signaling complexes by the multipoint binding of GRB2 to both LAT and SOS1. *Nat Structural Mol Biol* **13:** 798–805.

Hughes CE, Auger JM, McGlade J, Eble JA, Pearce AC, Watson SP. 2008. Differential roles for the adapters Gads and LAT in platelet activation by GPVI and CLEC-2. *J Thromb Haemost* **6:** 2152–2159.

Jackman JK, Motto DG, Sun Q, Tanemoto M, Turck CW, Peltz GA, Koretzky GA, Findell PR. 1995. Molecular cloning of SLP-76, a 76-kDa tyrosine phosphoprotein associated with Grb2 in T cells. *J Biol Chem* **270:** 7029–7032.

Jordan MS, Maltzman JS, Kliche S, Shabason J, Smith JE, Obstfeld A, Schraven B, Koretzky GA. 2007. In vivo disruption of T cell development by expression of a dominant-negative polypeptide designed to abolish the SLP-76/Gads interaction. *Eur J Immunol* **37:** 2961–2972.

Jordan MS, Smith JE, Burns JC, Austin JE, Nichols KE, Aschenbrenner AC, Koretzky GA. 2008. Complementation in trans of altered thymocyte development in mice expressing mutant forms of the adaptor molecule SLP76. *Immunity* **28:** 359–369.

Judd BA, Myung PS, Leng L, Obergfell A, Pear WS, Shattil SJ, Koretzky GA. 2000. Hematopoietic reconstitution of SLP-76 corrects hemostasis and platelet signaling through alpha IIb beta 3 and collagen receptors. *Proceedings of the National Academy of Sciences of the United States of America* **97:** 12056–12061.

Judd BA, Myung PS, Obergfell A, Myers EE, Cheng AM, Watson SP, Pear WS, Allman D, Shattil SJ, Koretzky GA. 2002. Differential requirement for LAT and SLP-76 in GPVI versus T cell receptor signaling. *J Exp Med* **195:** 705–717.

Kliche S, Breitling D, Togni M, Pusch R, Heuer K, Wang X, Freund C, Kasirer-Friede A, Menasche G, Koretzky GA, et al. 2006. The ADAP/SKAP55 signaling module regulates T-cell receptor-mediated integrin activation through plasma membrane targeting of Rap1. *Mol Cell Biol* **26:** 7130–7144.

Kumar L, Feske S, Rao A, Geha RS. 2005. A 10-aa-long sequence in SLP-76 upstream of the Gads binding site is essential for T cell development and function. *Proc Natl Acad Sci* **102:** 19063–19068.

Kumar L, Pivniouk V, de la Fuente MA, Laouini D, Geha RS. 2002. Differential role of SLP-76 domains in T cell development and function. *Proc Natl Acad Sci* **99:** 884–889.

Labno CM, Lewis CM, You D, Leung DW, Takesono A, Kamberos N, Seth A, Finkelstein LD, Rosen MK, Schwartzberg PL, et al. 2003. Itk functions to control actin polymerization at the immune synapse through localized activation of Cdc42 and WASP. *Curr Biol* **13:** 1619–1624.

Lenox LE, Kambayashi T, Okumura M, Prieto C, Sauer K, Bunte RM, Jordan MS, Koretzky GA, Nichols KE. 2009. Mutation of tyrosine 145 of lymphocyte cytosolic protein 2 protects mice from anaphylaxis and arthritis. *J Allergy Clin Immunol* **124:** 1088–1098.

Liou J, Kiefer F, Dang A, Hashimoto A, Cobb MH, Kurosaki T, Weiss A. 2000. HPK1 is activated by lymphocyte antigen receptors and negatively regulates AP-1. *Immunity* **12:** 399–408.

Liu SK, Fang N, Koretzky GA, McGlade CJ. 1999. The hematopoietic-specific adaptor protein gads functions in T-cell signaling via interactions with the SLP-76 and LAT adaptors. *Curr Biol* **9:** 67–75.

Maltzman JS, Kovoor L, Clements JL, Koretzky GA. 2005. Conditional deletion reveals a cell-autonomous requirement of SLP-76 for thymocyte selection. *J Exp Med* **202:** 893–900.

Menasche G, Kliche S, Chen EJ, Stradal TE, Schraven B, Koretzky G. 2007. RIAM links the ADAP/SKAP-55 signaling module to Rap1, facilitating T-cell-receptor-mediated integrin activation. *Mol Cell Biol* **27:** 4070–4081.

Mocsai A, Abram CL, Jakus Z, Hu Y, Lanier LL, Lowell CA. 2006. Integrin signaling in neutrophils and macrophages uses adaptors containing immunoreceptor tyrosine-based activation motifs. *Nat Immunol* **7:** 1326–1333.

Musci MA, Hendricks-Taylor LR, Motto DG, Paskind M, Kamens J, Turck CW, Koretzky GA. 1997a. Molecular cloning of SLAP-130, an SLP-76-associated substrate of the T cell antigen receptor-stimulated protein tyrosine kinases. *J Biol Chem* **272:** 11674–11677.

Musci MA, Motto DG, Ross SE, Fang N, Koretzky GA. 1997b. Three domains of SLP-76 are required for its optimal function in a T cell line. *J Immunol* **159:** 1639–1647.

Myung PS, Derimanov GS, Jordan MS, Punt JA, Liu QH, Judd BA, Meyers EE, Sigmund CD, Freedman BD, Koretzky GA. 2001. Differential requirement for SLP-76 domains in T cell development and function. *Immunity* **15:** 1011–1026.

Newbrough SA, Mocsai A, Clemens RA, Wu JN, Silverman MA, Singer AL, Lowell CA, Koretzky GA. 2003. SLP-76 regulates Fcgamma receptor and integrin signaling in neutrophils. *Immunity* **19:** 761–769.

Nguyen K, Sylvain NR, Bunnell SC 2008. T cell costimulation via the integrin VLA-4 inhibits the actin-dependent centralization of signaling microclusters containing the adaptor SLP-76. *Immunity* **28:** 810–821.

Peterson EJ, Woods ML, Dmowski SA, Derimanov G, Jordan MS, Wu JN, Myung PS, Liu QH, Pribila JT, Freedman BD, et al. 2001. Coupling of the TCR to integrin activation by Slap-130/Fyb. *Science* **293:** 2263–2265.

Pivniouk V, Tsitsikov E, Swinton P, Rathbun G, Alt FW, Geha RS. 1998. Impaired viability and profound block in thymocyte development in mice lacking the adaptor protein SLP-76. *Cell* **94:** 229–238.

Pivniouk VI, Martin TR, Lu-Kuo JM, Katz HR, Oettgen HC, Geha RS. 1999. SLP-76 deficiency impairs signaling via the high-affinity IgE receptor in mast cells. *J Clin Invest* **103:** 1737–1743.

Raab M, da Silva AJ, Findell PR, Rudd CE. 1997. Regulation of Vav-SLP-76 binding by ZAP-70 and its relevance to TCR zeta/CD3 induction of interleukin-2. *Immunity* **6:** 155–164.

Reynolds LF, Smyth LA, Norton T, Freshney N, Downward J, Kioussis D, Tybulewicz VL. 2002. Vav1 transduces T cell receptor signals to the activation of phospholipase

C-gamma1 via phosphoinositide 3-kinase-dependent and -independent pathways. *J Exp Med* **195:** 1103–1114.

Sanzenbacher R, Kabelitz D, Janssen O. 1999. SLP-76 binding to p56lck: a role for SLP-76 in CD4-induced desensitization of the TCR/CD3 signaling complex. *J Immunol* **163:** 3143–3152.

Sauer K, Liou J, Singh SB, Yablonski D, Weiss A, Perlmutter RM. 2001. Hematopoietic progenitor kinase 1 associates physically and functionally with the adaptor proteins B cell linker protein and SLP-76 in lymphocytes. *J Biol Chem* **276:** 45207–45216.

Shen S, Lau J, Zhu M, Zou J, Fuller D, Li QJ, Zhang W. 2009. The importance of Src homology 2 domain-containing leukocyte phosphoprotein of 76 kilodaltons sterile-alpha motif domain in thymic selection and T-cell activation. *Blood* **114:** 74–84.

Shui JW, Boomer JS, Han J, Xu J, Dement GA, Zhou G, Tan TH. 2007. Hematopoietic progenitor kinase 1 negatively regulates T cell receptor signaling and T cell-mediated immune responses. *Nat Immunol* **8:** 84–91.

Singer AL, Bunnell SC, Obstfeld AE, Jordan MS, Wu JN, Myung PS, Samelson LE, Koretzky GA. 2004. Roles of the proline-rich domain in SLP-76 subcellular localization and T cell function. *J Biol Chem* **279:** 15481–15490.

Su YW, Zhang Y, Schweikert J, Koretzky GA, Reth M, Wienands J. 1999. Interaction of SLP adaptors with the SH2 domain of Tec family kinases. *Eur J Immunol* **29:** 3702–3711.

Wang H, McCann FE, Gordan JD, Wu X, Raab M, Malik TH, Davis DM, Rudd CE. 2004. ADAP-SLP-76 binding differentially regulates supramolecular activation cluster (SMAC) formation relative to T cell-APC conjugation. *The Journal of experimental medicine* **200:** 1063–1074.

Woodside DG, Obergfell A, Leng L, Wilsbacher JL, Miranti CK, Brugge JS, Shattil SJ, Ginsberg MH. 2001. Activation of Syk protein tyrosine kinase through interaction with integrin beta cytoplasmic domains. *Curr Biol* **11:** 1799–1804.

Wu J, Motto DG, Koretzky GA, Weiss A. 1996. Vav and SLP-76 interact and functionally cooperate in IL-2 gene activation. *Immunity* **4:** 593–602.

Wu JN, Jordan MS, Silverman MA, Peterson EJ, Koretzky GA 2004. Differential requirement for adapter proteins Src homology 2 domain-containing leukocyte phosphoprotein of 76 kDa and adhesion- and degranulation-promoting adapter protein in FcepsilonRI signaling and mast cell function. *J Immunol* **172:** 6768–6774.

Yablonski D, Kadlecek T, Weiss A. 2001. Identification of a phospholipase C-gamma1 (PLC-gamma1) SH3 domain-binding site in SLP-76 required for T-cell receptor-mediated activation of PLC-gamma1 and NFAT. *Molecular and cellular biology* **21:** 4208–4218.

Yablonski D, Kuhne MR, Kadlecek T, Weiss A. 1998. Uncoupling of nonreceptor tyrosine kinases from PLC-gamma1 in an SLP-76-deficient T cell. *Science (New York, NY)* **281:** 413–416.

Zeng R, Cannon JL, Abraham RT, Way M, Billadeau DD, Bubeck-Wardenberg J, Burkhardt JK. 2003. SLP-76 coordinates Nck-dependent Wiskott-Aldrich syndrome protein recruitment with Vav-1/Cdc42-dependent Wiskott-Aldrich syndrome protein activation at the T cell-APC contact site. *J Immunol* **171:** 1360–1368.

Zhang W, Sloan-Lancaster J, Kitchen J, Trible RP, Samelson LE. 1998. LAT: the ZAP-70 tyrosine kinase substrate that links T cell receptor to cellular activation. *Cell* **92:** 83–92.

T-Cell Signaling Regulated by the Tec Family Kinase, Itk

Amy H. Andreotti[1], Pamela L. Schwartzberg[2], Raji E. Joseph[1], and Leslie J. Berg[3]

[1]Department of Biochemistry, Biophysics and Molecular Biology, Iowa State University, Ames, Iowa 50011
[2]National Human Genome Research Institute, National Institutes of Health, Bethesda, Maryland 20814
[3]Department of Pathology, University of Massachusetts Medical School, Worcester, Massachusetts 01655
Correspondence: amyand@iastate.edu

The Tec family tyrosine kinases regulate lymphocyte development, activation, and differentiation. In T cells, the predominant Tec kinase is Itk, which functions downstream of the T-cell receptor to regulate phospholipase C-γ. This review highlights recent advances in our understanding of Itk kinase structure and enzymatic regulation, focusing on Itk protein domain interactions and mechanisms of substrate recognition. We also discuss the role of Itk in the development of conventional versus innate T-cell lineages, including both $\alpha\beta$ and $\gamma\delta$ T-cell subsets. Finally, we describe the complex role of Itk signaling in effector T-cell differentiation and the regulation of cytokine gene expression. Together, these data implicate Itk as an important modulator of T-cell signaling and function.

The Tec family nonreceptor tyrosine kinases, Tec, Btk, Itk/Emt/Tsk, Rlk/Txk, and Bmx/Etk, are expressed primarily in hematopoietic cells and serve as important mediators of antigen receptor signaling in lymphocytes (Berg et al. 2005; Felices et al. 2007; Readinger et al. 2009). The demonstration that the human B-cell immunodeficiency, X-linked agammaglobulinemia (XLA), is caused by mutations in Btk first underscored the importance of this tyrosine kinase family in lymphocyte development and antigen receptor signaling (Rawlings et al. 1993; Thomas et al. 1993; Tsukada et al. 1993; Vetrie et al. 1993). T lymphocytes express three Tec kinases: Itk, Rlk and Tec. To date, only Itk has been found to have a clearly defined function in T cells, leading to the conclusion that Itk is the predominant Tec kinase in T cells. In this review, we will cover recent findings that highlight the critical role of Itk in T-cell signaling and function.

STRUCTURE AND REGULATION OF ITK ACTIVITY

Tec Kinase Domain Structure

Itk (inducible T-cell kinase; also known as Emt, expressed in mast cells and T lymphocytes, and Tsk, T-cell-specific kinase) was cloned in the early 1990s (Siliciano et al. 1992; Gibson et al. 1993; Heyeck and Berg 1993; Tanaka et al. 1993; Yamada et al. 1993). The domain organization of Itk and related Tec kinase family

 doi: 10.1101/cshperspect.a002287
Cite this article as *Cold Spring Harb Perspect Biol* 2010;2:a002287

members shares similarities with other tyrosine kinase families but also reveals unique features (Fig. 1A). Shared features include the common SH3-SH2-kinase cassette also found in the Src, Csk, and Abl kinases (Williams et al. 1998; Tsygankov 2003). Four of the five Tec tyrosine kinases (Itk, Btk, Tec, and Bmx) contain an amino-terminal region that includes a pleckstrin homology (PH) domain, followed by a Zn^{++} binding region termed the Btk homology (BH) motif, and (except for Bmx) a proline rich region (PRR) that conforms to the consensus sequence of an SH3 ligand. Instead of the PH-BH region, Txk (also known as Rlk) contains a cysteine string motif at its amino terminus.

Itk Activation Downstream of the TCR

The Tec family kinases are involved in multiple receptor-mediated signaling pathways in many cell types; several recent reviews have covered these in depth (Berg et al. 2005; Felices et al. 2007; Gomez-Rodriguez et al. 2007; Mihara

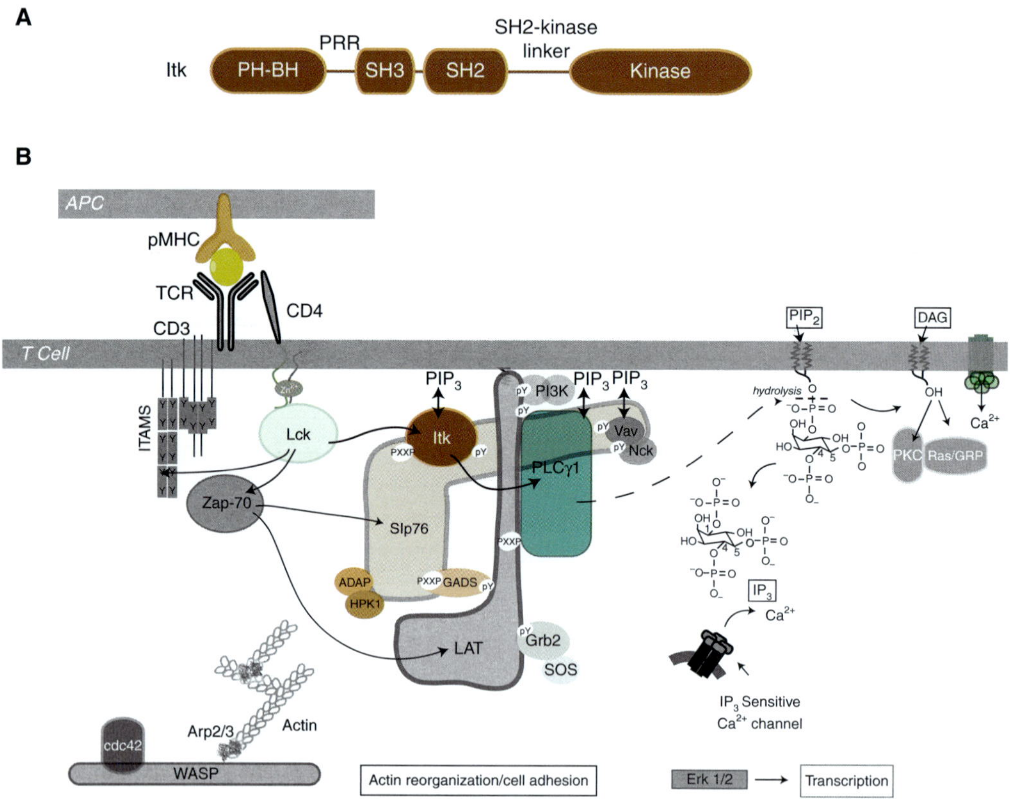

Figure 1. Itk domain organization and it's role in T cell signaling. (*A*) The domain structure of Itk. (*B*) Schematic of T-cell signaling emphasizing the proximal signaling complex downstream of the TCR as described in the text. The trio of tyrosine kinases, Zap-70, Lck, and Itk that initiate signaling following TCR engagement are shown as gray, green, and brown circles, respectively. Arrows indicate early phosphorylation events on TCR/pMHC recognition. The amino-terminal region of Lck binds to the cytoplasmic tail of CD4 via coordination of Zn^{++} (Huse et al. 1998). Active Lck phosphorylates the ITAMs, Zap-70, and Itk, which in turn phosphorylate their targets Slp-76, LAT (by Zap-70) and PLCγ1 (by Itk). Activated PLCγ1 then hydrolyzes PIP_2 (dashed arrow) to produce DAG and IP_3 leading to increased calcium flux. Multiple copies of these signaling molecules likely cooperate in the phosphorylation cascade and so stoichiometry as shown is overly simplified. Moreover, negative regulation of the signaling cascade in the form of phosphatases, kinases and other types of molecules also plays a significant role in T-cell function but are not represented here.

and Suzuki 2007; Gilfillan and Rivera 2009; Koprulu and Ellmeier 2009; Prince et al. 2009; Readinger et al. 2009). For Itk, significant attention has been given to activation via T-cell receptor (TCR) stimulation (see Fig. 1B). Interaction of the TCR with peptide-MHC complexes on antigen presenting cells (APC) activates the Src kinase Lck, leading to phosphorylation of the CD3 immunoreceptor tyrosine-based activation motifs (ITAMs). Zap-70 then binds to the phosphorylated ITAMs and is phosphorylated by Lck, resulting in Zap-70 activation and subsequent phosphorylation of the adaptors LAT and SLP-76 (Pitcher and van Oers 2003; Houtman et al. 2005; Au-Yeung et al. 2009; Smith-Garvin et al. 2009). Following activation of PI3K and accumulation of phosphatidylinositol (3,4,5) trisphosphate (PIP_3) in the plasma membrane, Itk is recruited to the membrane via its PH domain (August et al. 1997; Ching et al. 1999; Yang et al. 2001). There, Itk interacts with the phosphorylated SLP-76/LAT adapter complex via Itk's Src homology (SH3 and SH2) domains (Bubeck Wardenburg et al. 1996; Zhang et al. 1998; Shan and Wange 1999; Su et al. 1999; Bunnell et al. 2000; Ching et al. 2000; Sommers et al. 2004; Koretzky et al. 2006), permitting phosphorylation of Itk on its activation loop (Y511) by Lck (Heyeck et al. 1997).

Once activated, Itk undergoes cis autophosphorylation on Y180 in its SH3 domain (Wilcox and Berg 2003; Joseph et al. 2007a). Itk also interacts with and directly phosphorylates its downstream target, PLCγ1 (Perez-Villar and Kanner 1999; Joseph et al. 2007c), resulting in phospholipase activation (Liu et al. 1998; Schaeffer et al. 1999; Wilcox and Berg 2003; Bogin et al. 2007). The kinetics characterizing many of these early phosphorylation events have been delineated, providing an understanding of the sequential nature of signaling following TCR engagement (Houtman et al. 2005). It is not known, however, whether Itk autophosphorylation occurs prior to, or following, phosphorylation of PLCγ1. Because autophosphorylation on Y180 does not affect Itk catalytic activity but instead modulates binding of the Itk SH3 domain to different targets (Wilcox and Berg 2003; Joseph et al. 2007a), Itk autophosphorylation may serve as a trigger to alter protein interaction partners or localization either before or after PLCγ1 activation. PLCγ1, once activated, hydrolyzes PIP_2 to produce the second messengers IP_3 and DAG (Rhee 2001). Downstream consequences include Ca^{2+} flux (Lewis and Cahalan 1995), Erk activation (Schaeffer et al. 1999; Schaeffer et al. 2000; Miller and Berg 2002), transcription (Crabtree and Olson 2002), cytokine release (Liao and Littman 1995; Liu et al. 1998), and actin reorganization (Labno et al. 2003), all of which are impaired in the absence of Itk. Importantly, these downstream consequences of TCR activation are not ablated by a deficiency in Itk, but instead, are substantially reduced, leading to altered development and differentiation of distinct T-cell lineages, as discussed later.

The Itk Signaling Complex

The Itk/SLP-76 interaction likely involves the SH3 and SH2 domains of Itk targeting a proline-rich stretch within SLP-76 (residues 184-195) and an adjacent phosphorylated tyrosine (Y145) in SLP-76, respectively (Su et al. 1999; Bunnell et al. 2000). In an elegant knockin mouse study, tyrosine 145 of SLP-76 was replaced with the nonphosphorylatable phenylalanine to probe the contribution of the ItkSH2/pY145 interaction to signaling (Jordan et al. 2008). Mutation of Y145 resulted in markedly diminished PLCγ1 phosphorylation, decreased Ca^{2+} flux, lowered phospho-ERK levels and gave rise to T cells that resembled those lacking Itk, as discussed later. This observation (Jordan et al. 2008) is consistent with findings that the Itk/SLP-76 interaction is required to activate Itk kinase activity in Jurkat cells (Bogin et al. 2007). Intriguingly, the majority of the catalytically active Itk in the cell consists of only the small fraction of Itk that is bound to SLP-76 (Bogin et al. 2007). Surprisingly, the SLP-76 Y145F mutation did not lead to loss of the Itk/SLP-76 association (Jordan et al. 2008) despite altered signaling. This finding suggests that cooperative multivalent interactions preserve Itk localization in this signaling complex even when a single contact site is disrupted.

The importance of the Itk PH domain/PIP_3 interaction (Fukuda et al. 1996; August et al. 1997) was also recently probed by inhibition of phosphatidylinositol-3 kinase (PI3K) activity in human peripheral blood T cells (Cruz-Orcutt and Houtman 2009). In a manner similar to SLP-76 Y145F, loss of the Itk PH domain/PIP_3 interaction resulted in impaired downstream signaling but did not affect the ability of Itk and other PH domain containing signaling proteins to assemble into the signaling complex (Cruz-Orcutt and Houtman 2009). In a separate study, soluble IP_4 was found to promote binding of the Itk PH domain to PIP_3 (Huang et al. 2007). Reminiscent of the Itk/SLP-76 interaction (Bogin et al. 2007), these studies also showed that only a limited pool of Itk in the cell is capable of binding PIP_3 even in the presence of excess PIP_3 (Huang et al. 2007). These separate findings suggest that multiple interaction sites exist to properly localize Itk following TCR stimulation (Graef et al. 1997; Garcon and Nunes 2006; Beach et al. 2007). Furthermore, part of the function of Itk and other proteins in the LAT/SLP-76 complex may be to function as adaptors and stabilize proteins in the complex. Kinase-independent functions for Itk have been observed in the regulation of actin polymerization (Dombroski et al. 2005) and activation of the transcription factor, SRF (Hao et al. 2006); these functions may be secondary to effects on recruitment of the guanine-nucleotide exchange factor Vav to this complex (Labno et al. 2003).

Regulation of Itk Activity

The regulatory mechanisms that control the activity of Lck and Zap-70 during TCR signaling have been the subject of intense investigation, leading to detailed insights into function (Palacios and Weiss 2004; Au-Yeung et al. 2009). In contrast, a mechanistic understanding of Itk regulation remains elusive, as Itk and other Tec kinases have, to date, resisted crystallization in their full-length form. Comparing the Tec and Src kinase families, it is evident that regulatory differences must exist. In spite of the shared SH3-SH2-Kinase cassette, the Tec kinases lack the carboxy-terminal regulatory tail that plays an important role in Src regulation (Brown and Cooper 1996; Xu et al. 1997; Xu et al. 1999; Cowan-Jacob et al. 2005). Moreover, removal of the noncatalytic domains from the Src kinase domain yields an active enzyme, whereas the isolated Itk and Btk kinase domains show poor catalytic activity (Fig. 2A) (LaFevre-Bernt et al. 1998; Hawkins and Marcy 2001; Guo et al. 2006; Joseph et al. 2007b). Mutation of a well-conserved tryptophan residue in the Itk and Btk SH2-kinase linkers (W355 in Itk, W395 in Btk) results in a significant decrease in kinase activity (Lin et al. 2005; Guo et al. 2006; Joseph et al. 2007b) whereas mutation of the corresponding tryptophan in the Src kinase Hck leads to a dramatic increase in the activity of that kinase (LaFevre-Bernt et al. 1998). In this regard, Itk is similar to Csk (carboxy-terminal Src kinase), another tyrosine kinase comprised of the common SH3-SH2-Kinase cassette and regulated in a manner opposite that of Src (Huang et al. 2009). The isolated Csk kinase domain shows negligible catalytic activity (Fig. 2A) and mutation of the conserved tryptophan (W188) decreases Csk activity (Lin et al. 2005). Based on similar biochemical profiles, a model for active Itk can be developed using the available structure of active Csk (Fig. 2B).

In agreement with the established interaction between active Itk and SLP-76 (Jordan et al. 2008; Bogin et al. 2007), the distance between the SH3 and SH2 binding pockets in the high-resolution structure of Csk closely matches the distance between the phosphotyrosine motif and proline-rich region in SLP-76 (residues 143–195). Moreover, the Itk PRR region adjacent to the Itk SH3 domain in the model of active Itk (Fig. 2B) could be available for binding to an SH3 domain from an exogenous signaling partner. The precise binding partner of the Itk PRR is not yet clear but candidates include the SH3 domains of Vav, Fyn, Src, PLCγ1 or Grb2 (Yang et al. 1995; Perez-Villar and Kanner 1999; Bunnell et al. 2000; Chamorro et al. 2001). Csk lacks the PH-BH region of the Tec kinases and so comparative model building for this region of Itk becomes

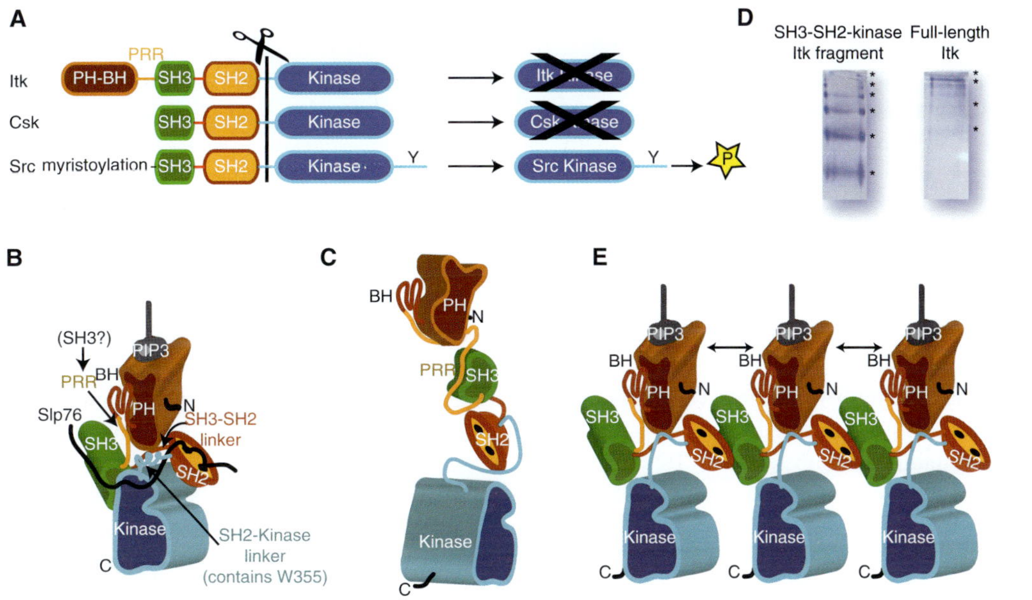

Figure 2. Activation/inactivation of Itk. (*A*) Comparison of the domain structures of Itk, Csk and Src kinases. The domains are colored as follows: PH-BH is brown, PRR is yellow, SH3 is green, SH2 is orange, SH2-Kinase linker and Kinase domain are blue. All three kinases share the SH3-SH2-kinase cassette but differ with respect to resulting catalytic activity of the isolated kinase domain. Itk and Csk kinase domains alone show very poor catalytic activity whereas the Src kinase domain readily phosphorylates substrates. (*B*) Model of active Itk based on biochemical data and the high resolution structure of Csk (PDB code: 1K9A). The SH2-Kinase linker and SH3-SH2 linker are both arranged along the back of the small kinase lobe. Activating interactions between the kinase domain and the SH2-kinase linker in particular are also supported by a recent structure of Btk (Marcotte et al. 2010). SLP-76 is depicted by the black line and is shown binding to both Itk SH3 and SH2. The PH-BH domain is placed so that its carboxy-terminus is close in space to the amino-terminus of the SH3 domain. The intervening sequence is the PRR, which would be available in this model for binding an SH3 domain from another protein. PH domain is shown bound to its PIP_3 ligand. (*C*) Model of Itk adopting the extended conformation seen for Btk in SAXS studies (Marquez et al. 2003). Of note is the fact that the intramolecular PRR/SH3 interaction is sterically allowed and that the SH2-Kinase linker region is likely in an extended conformation and not the active conformation shown in (*B*). (*D*) Native gel electrophoresis for the SH3-SH2-Kinase fragment of Itk and for full length Itk showing ladders indicative of mulitmerization. Sample concentration for both purified proteins is approximately 5 micromolar. (*E*) Model of Itk multimerization based on the structure of the binary Itk SH3/SH2 complex (PDB code: 2K79). Doubled headed arrows indicate PH domain self-association. Whether Itk multimerization occurs exclusively at the membrane remains to be determined. It is of note that the SH3/SH2 interface defined in PDB 2K79 is accessible in the models shown in both (*B*) and (*C*) suggesting that either configuration could multimerize. In (*B*), (*C*) and (*E*) the amino and carboxy-termimi of full length Itk are indicated by N and C, respectively.

speculative. It is interesting to note however, that the PH-BH domain of Itk extends from the amino-terminus of the PRR region and, within the context of the Csk-based model, the PH-BH domain could pack between the SH3 and SH2 domains making contact with the small lobe of the kinase domain and/or the SH3-SH2/SH2-kinase linker regions potentially affecting kinase activity (Joseph et al. 2007b) (Fig. 2B).

Mechanism of Itk Substrate Recognition

Once activated and assembled within the SLP-76/LAT signaling complex, Itk meets its downstream substrate, PLCγ1 (Liu et al. 1998;

Schaeffer et al. 1999; Wilcox and Berg 2003; Houtman et al. 2005; Bogin et al. 2007). Itk mediated phosphorylation of Y783 in PLCγ1 requires a remote specificity-determining element within the second SH2 domain of PLCγ1 (Joseph et al. 2007c). Mutation in the PLCγ1 SH2 domain that disrupts the Itk/PLCγ1 interaction results in decreased PLCγ1 activation as measured by calcium flux in Jurkat T cells (Braiman et al. 2006; Min et al. 2009). Earlier work also identified a TCR-regulated association between Itk and PLCγ1 involving the SH3 domain of PLCγ1 (Perez-Villar and Kanner 1999). Although PLCγ1 is the most well established substrate of Itk, additional candidate substrates have been described and include TFII-I (August 2009; Sacristan et al. 2009), Tim-3 (van de Weyer et al. 2006) and T-bet (Hwang et al. 2005); the molecular details of Itk recognition of these substrates has not been elucidated.

Downregulation of Itk Kinase Activity

Inactivation of kinase activity is an important regulatory feature that must appropriately balance activating signals. Although few negative regulators of Itk have been described, expression of PTEN, a lipid phosphatase, can decrease Itk activity by reducing PIP_3 levels (Shan et al. 2000). Negative regulation of Itk by the peptidyl prolyl isomerase, cyclophilin A, has been described (Brazin et al. 2002; Colgan et al. 2004). Interestingly, another peptidyl prolyl isomerase, Pin1 (protein interacting with NIMA1), regulates phosphorylation and steady state levels of Btk (Yu et al. 2006). However, the mechanism of action remains poorly understood for both Pin1 and CypA (Joseph and Andreotti 2009; Mohamed et al. 2009).

As already described, the regulatory domains of the Tec kinases exert a positive influence on the kinase domain in a manner similar to Csk (Fig. 2B) and so conformational rearrangements that uncouple the regulatory domains (particularly the SH2-kinase linker region) from the catalytic domain would down-regulate Itk activity. In this context, the only structural data available for any full length Tec kinase comes from small-angle X-ray scattering (SAXS) analysis of Btk (Marquez et al. 2003). These data indicate that an inactive preparation of Btk (Marquez et al. 2003; Lin et al. 2009) adopts an extended conformation in solution with negligible contacts between domains (Fig. 2C). Whether the fully extended configuration derived from SAXS data represent a physiologically relevant state remains to be determined but the available data suggest that a conformational shift of the regulatory domains away from the kinase domain could provide one layer of negative regulation.

In addition to negatively regulating the catalytic domain, the configuration of the inactive kinase might also serve to sequester its ligand binding sites (Pawson and Kofler 2009). The SH3/PXXP interaction, in particular, is constitutive and not controlled by transient phosphorylation/dephosphorylation events. In down-regulated Itk, the SH3 domain-binding site might be masked in an intramolecular fashion by the adjacent PRR (Andreotti et al. 1997; Laederach et al. 2003) (Fig. 2C). However, it is experimentally difficult to probe the precise role of the PRR in kinase regulation because it is also likely involved in activating interactions with SH3 domain-containing signaling partners (Yang et al. 1995; Bunnell et al. 2000). Efforts that use deletion of the PRR region to probe its regulatory role (Hao and August 2002) are therefore complicated by the dual nature of the PRR.

Regulation of Itk Activity by Multimerization

Structural studies using different Tec kinase fragments indicate that the noncatalytic regulatory domains interact in an intermolecular fashion driving multimerization in solution (Andreotti et al. 1997; Brazin et al. 2000; Hansson et al. 2001; Laederach et al. 2002; Okoh and Vihinen 2002; Pursglove et al. 2002; Laederach et al. 2003; Severin et al. 2009). For Itk, multimerization is mediated by self-association of the PH domain (Huang et al. 2007) and intermolecular interactions between the SH3 and SH2 domains of different Itk molecules (Brazin et al. 2000; Severin et al. 2009). Native gel electrophoresis indicates that the full-length Itk molecule, the SH3-SH2-kinase fragment, and the Itk SH3-SH2 fragment all form multimers (Severin et al. 2009) (Fig. 2D). The high resolution structure of the Itk SH3/SH2 complex has been solved and provides insight into how full length Itk might self-associate (Severin et al. 2009) (Fig. 2E). Functionally, disruption of the SH3/SH2 interface by mutation reduced oligomerization and increased Itk kinase activity both in vitro and in T cells (Severin et al. 2009; Min et al. 2010). Moreover, both binding experiments and structural data indicate that intermolecular association of Itk is mutually exclusive with binding of SLP-76 to the SH3 and SH2 domains (Brazin et al. 2000; Pletneva et al. 2006; Severin et al. 2009). Thus, down-regulation of Itk activity might involve intermolecular self-association whereas multimerization is disfavored for active, SLP-76 bound Itk. The interplay between specific clustering of signaling molecules (Houtman et al. 2006), the formation of TCR microclusters (Bunnell et al. 2006; Seminario and Bunnell 2008) and the specific role of Itk multimerization during T-cell signaling certainly deserve continued attention.

A split YFP/fluorescence complementation approach has been used to probe the different conformational and oligomeric states of Itk in cells (Qi et al. 2006; Qi and August 2009). Itk that is doubly tagged at the amino- and carboxy-termini results in fluorescence complementation in the cytosol and at the membrane. The authors suggest that the observed fluorescence complementation within a single Itk molecule is indicative of a folded compact structure. Taking the linker lengths between Itk and each of the split YFPs (two linkers each ~40 Å) and the diameter of YFP itself (~30 Å) into account, fluorescence complementation in this system would occur for distances of approximately 110 Å between the Itk termini. It should be noted that the distance between the amino- and carboxy-termini of Btk in the fully extended model derived from SAXS data is 125 Å and given the flexibility inherent in the extended model (Marquez et al. 2003), it seems likely that fluorescence complementation would be observed for almost any domain arrangement of Itk; compact or extended. Thus, the fluorescence "ruler" used in these experiments is unlikely to discriminate between different domain arrangements.

Intermolecular fluorescence complementation between differentially tagged Itk molecules is consistently observed only at the cell membrane and not in the cytosol (Qi et al. 2006; Qi and August 2009). Although this data alone does not rule out multimerization in the cytosol, the spatial segregation of the stronger fluorescence signal to the membrane is certainly interesting and suggests that Itk clusters are at least stabilized upon or after membrane localization. Notably, Itk autophosphorylation enhances the affinity between the SH3 and SH2 domains (Joseph et al. 2007a), possibly triggering sequestration of Itk into clusters and down-regulation of kinase activity. Release of Itk from the membrane, promoted by dephosphorylation of PIP_3 and the SLP-76/LAT complex (Shan et al. 2000; Baker et al. 2001; Yang et al. 2001; Tomlinson et al. 2004), would then result in decreased local Itk concentration, which could cause Itk oligomers to disassociate into inactive, monomeric species in the cytosol. Regardless of aggregation state, the inactive species would adopt a conformation that decouples the regulatory region (in particular the SH2-kinase linker) from the Itk kinase domain.

ITK REGULATION OF T-CELL FUNCTION

Naïve T-Cell Activation is Impaired in the Absence of Itk

The central role of Itk in the LAT-SLP-76 complex suggests that Itk is an important component of TCR signaling. Early studies using primary murine *Itk*$^{-/-}$ T cells showed that Itk is required for robust T-cell activation in response to TCR plus costimulatory receptor signaling. However, these studies also revealed that Itk is not absolutely required for all responses downstream of the TCR (Liao and Littman 1995; Liu et al. 1998; Schaeffer et al. 1999). One aspect of this modulatory role for Itk is the partial redundancy between Itk and a second Tec kinase family member, Rlk. Thus, a combined deficiency in both Itk and Rlk results in TCR signaling defects that are significantly more profound than is seen in T cells lacking Itk alone (Schaeffer et al. 1999). Overall, the message from these experiments was that Itk functions to amplify TCR signaling, rather than as an all-or-nothing component of the TCR signaling pathway.

T-Cell Differentiation and Effector Functions are Regulated by Itk

In spite of the relatively modest impact of the Itk deficiency on naïve T-cell activation, numerous studies have shown that Itk signaling plays a critical role in regulating T-cell differentiation and T-cell effector functions (Fig. 3). The first of these studies showed that *Itk*$^{-/-}$ Balb/c mice were unable to generate the T_H2 response to *Leishmania major* infection typically seen in wild-type Balb/c mice and instead, mounted a protective T_H1 response that cleared the infection (Fowell et al. 1999). These results highlighted the importance of Itk signaling in the generation of T_H2 responses, a function of Itk that has been studied in more detail both in vitro and in vivo.

Itk Promotes T_H2 Differentiation and Cytokine Production

The initial findings that Itk-deficient mice were unable to mount a protective T_H2 response to infection were verified in a number of distinct infectious disease systems including *Leishmania major* in *Itk*$^{-/-}$ Balb/c mice, as well as *Nippostrongylus brasiliensis* and *Schistosoma mansoni*, two parasites that are eliminated by T_H2-dependent recruitment and activation of granulocytes

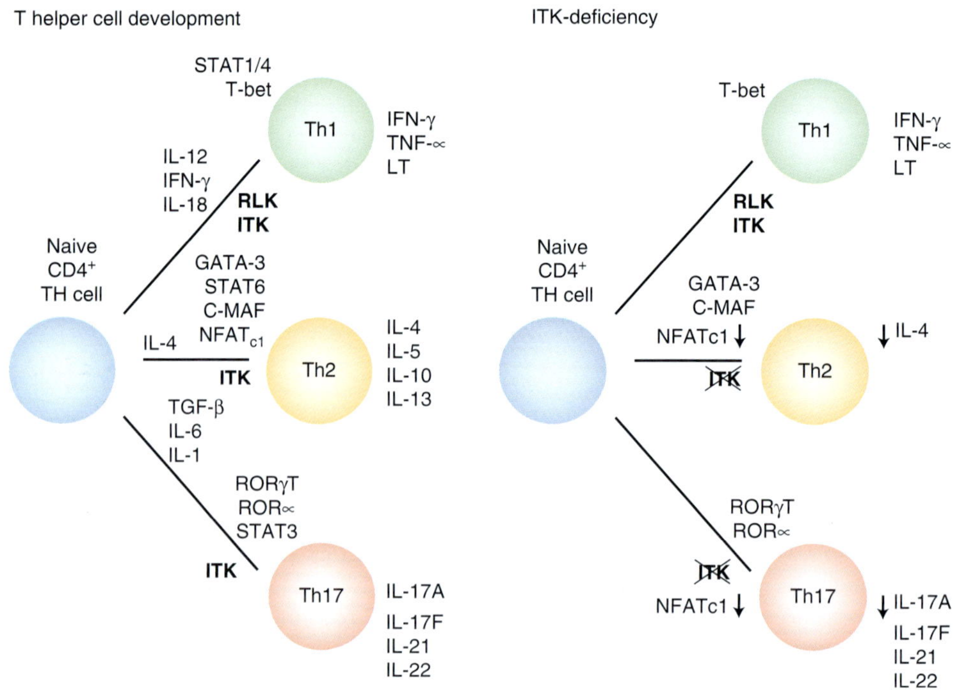

Figure 3. Itk function in T helper cell differentiation. At the *left*, the three major lineages of effector T cells are shown, along with the cytokine signals and transcription factors that regulate their differentiation from naïve $CD4^+$ T cells. Unlike T_H1 cells, which coexpress Itk and Rlk, T_H2 cells express only Itk. The status of Rlk expression in T_H17 cells is currently unknown. In the absence of Itk (*right* panel), $NFAT_{c1}$ activity is greatly reduced, leading to impaired production of IL-4 by T_H2 cells and IL-17A by T_H17 cells. *Itk*$^{-/-}$ T_H1 cells, which continue to express Rlk, have a more modest defect in effector cytokine production.

(Fowell et al. 1999; Schaeffer et al. 2001). Analyses of cytokine responses of the parasite-specific T cells from these infected mice indicated that *Itk*$^{-/-}$ T cells produced decreased IL-4, IL-5, and IL-10 (Schaeffer et al. 2001). Similarly, in vitro experiments showed that *Itk*$^{-/-}$ Balb/c CD4^{+} T cells stimulated under T_H2-polarizing conditions produced much less IL-4 than their wild-type counterparts (Fowell et al. 1999). These cytokine responses correlated with defective TCR-induced activation of the calcium-dependent transcription factor, NFAT, by *Itk*$^{-/-}$ T cells, providing a potential explanation for the reduced *Il4* transcription (Fig. 3).

However, it is difficult to infer from these experiments the precise nature of the T-cell defect resulting in impaired protection to parasitic infections. Reductions in T-cell activation, T-cell differentiation, T-cell migration, or elicitation of recall responses at the site of infection could each contribute to the poor protection observed in *Itk*$^{-/-}$ mice. An additional complication is the presence of a large population of previously activated/memory CD4^{+} T cells in *Itk*$^{-/-}$ mice; global gene expression analysis has confirmed that *Itk*$^{-/-}$ CD4^{+} T cells are not comparable to wild-type CD4^{+} T cells prior to their activation (Blomberg et al. 2009), and thus, might respond differently than naïve T cells to parasite infections.

Nonetheless, useful information has been gleaned from direct infection of *Itk*$^{-/-}$ mice with pathogenic microorganisms. One elegant approach to evaluating T cell defects in the absence of Itk was taken by Au-Yeung and colleagues, who crossed Balb/c *Itk*$^{-/-}$ mice to reporter mice that express GFP under the control of the IL-4 promoter (Au-Yeung et al. 2006). Following infection of these mice with *Leishmania major*, equal numbers of CD4^{+}GFP^{+} T cells were found at the sites of infection in wild-type versus *Itk*$^{-/-}$ mice, indicating that Itk was not required for the initiation of the T_H2 response or for the migration of IL-4-competent T cells to these sites. Instead, the effector T-cell recall response at the sites of infection were impaired in the absence of Itk, leading to reduced IL-4 production by *Itk*$^{-/-}$ T cells following restimulation. However, these studies are not in complete agreement with an independent set of experiments examining T-cell-mediated airway hyperresponsiveness in *Itk*$^{-/-}$ mice (Mueller and August 2003). In this system, in addition to impaired T_H2 cytokine production, T-cell recruitment to the lungs was also impaired following inhaled antigen challenge in *Itk*$^{-/-}$ mice compared to controls. Interestingly, a partial inhibition of Itk kinase activity was able to block T-cell migration to the lung in this model system, but had no effect on the priming of T_H2 effector responses in secondary lymphoid organs (Sahu et al. 2008a). These findings indicate roles for Itk in both T-cell priming and T-cell migration, findings that are supported by studies showing a role for Itk in signaling downstream of chemokine receptors (Fischer et al. 2004; Takesono et al. 2004).

These in vivo experiments have been complemented by a series of in vitro experiments performed with naïve *Itk*$^{-/-}$ CD4^{+} T cells. All of these studies agreed that *Itk*$^{-/-}$ CD4^{+} T cells produced substantially less IL-4, IL-5, and IL-13 than wild-type T cells, even after differentiation in a powerful T_H2 polarizing environment (Fowell et al. 1999; Schaeffer et al. 2001; Miller et al. 2004; Au-Yeung et al. 2006). In these studies, the defects in effector cytokine production were not because of impaired T_H2 differentiation, as molecular analyses indicated that *Itk*$^{-/-}$ T cells polarized into the T_H2 lineage showed all of the hallmarks of *bona fide* T_H2 effector cells (Schaeffer et al. 2001; Miller et al. 2004; Au-Yeung et al. 2006). Interestingly, the requirement for Itk in effector cytokine production is shared by αβTCR^{+} NKT cells specific for α-galactosyl ceramide bound to CD1d; in the absence of Itk, NKT cells stimulated in vivo or in vitro produce substantially less IL-4 or IFNγ than wild-type NKT cells (Au-Yeung and Fowell 2007; Felices and Berg 2008).

T_H1 Differentiation and Effector Functions Are Modestly Impaired in the Absence of Itk

Unlike the essential role for Itk in protective immunity to helminthic parasites, Itk signaling in T cells is largely dispensible for protective

$CD4^+$ T_H1 responses to intracellular pathogens. Thus, *Itk*$^{-/-}$ mice clear infections of *Leishmania major*, and are only modestly more susceptible to sublethal doses of *Toxoplasma gondii* than wild-type mice (Fowell et al. 1999; Schaeffer et al. 1999). These observations may reflect the fact that IFNγ production by *Itk*$^{-/-}$ $CD4^+$ T cells is not reduced to the same extent as production of T_H2 cytokines (Fowell et al. 1999; Schaeffer et al. 2001; Miller et al. 2004; Au-Yeung et al. 2006). Indeed, under certain conditions that normally lead to T_H2 cytokine production, such as stimulation with altered peptide ligands, *Itk*$^{-/-}$ $CD4^+$ T cells produce IFN-γ (Miller et al. 2004). A second aspect of this differential requirement for Itk in T_H1 versus T_H2 effector responses is the interesting expression pattern of the two predominant Tec kinases in T cells, Itk and Rlk (Fig. 3). Unlike naïve T cells which express both Itk and Rlk, T_H2 effector cells lose Rlk and express only Itk; in fact, the levels of Itk are several-fold increased in T_H2 cells compared to naïve T cells (Miller et al. 2004). In contrast, T_H1 effector cells continue to express both of these Tec kinases. As a consequence, whereas *Itk*$^{-/-}$ T_H1 cells continue to express Rlk, *Itk*$^{-/-}$ T_H2 cells lack both Itk and Rlk, thus accounting for the more severe functional defect observed in *Itk*$^{-/-}$ T_H2 responses. Corroborating this notion, ectopic expression of Rlk in effector T_H2 cells restores wild-type T cell responses in *Itk*$^{-/-}$ mice injected with *Schistosoma mansoni* eggs or immunized to induce allergic airway hypersensitivity (Sahu et al. 2008b).

Itk Is Required for IL-17A Production in T_H17 Effector Cells

In addition to the global defects in effector T-cell responses described above for T_H2, and to a lesser extent, T_H1 responses, T cells lacking Itk display more selective alterations in cytokine production that highlight the complex nature of the signaling pathways that regulate T-cell effector responses. One recent example is the role of Itk in the regulation of cytokine production by T_H17 cells, a recently recognized CD4 effector T-cell lineage that differentiates in response to IL-6 and TGF-β and expresses the pro-inflammatory cytokines IL-17A, IL-17F, IL-21 and IL-22 (Korn et al. 2009). T_H17 cells are important for antimicrobial activity, particularly in the gastrointestinal system, but are also hallmarks of inflammation involved in multiple autoimmune disorders.

Although the requirements for cytokine signals for T_H17 differentiation have been extensively examined, the contribution of TCR signaling to the regulation of T_H17 cytokine production was unknown. To evaluate this question, naïve $CD4^+$ T cells from *Itk*$^{-/-}$ and *Rlk*$^{-/-}$*Itk*$^{-/-}$ mice were differentiated under T_H1 and T_H17 polarizing conditions, and cytokine production evaluated by intracellular staining (Gomez-Rodriguez et al. 2009). Although both cell types could generate large percentages of IFNγ producing T_H1 cells, *Itk*$^{-/-}$ and even more dramatically *Rlk*$^{-/-}$*Itk*$^{-/-}$ T cells showed reduced IL-17A production. These defects did not appear to be because of altered thymic development, because they could be reversed by re-introduction of Itk by retroviral transduction before T_H17 differentiation. Further analyses showed decreased *Il17a* message in *Itk*$^{-/-}$ T cells (Fig. 3). However, surprisingly, expression of other T_H17 cytokines including the closely linked *Il17f* gene, was relatively intact, as was the expression of genes encoding the master regulators RORγt and RORα. Although phosphorylation of STAT3 in response to IL-6 was slightly attenuated in Itk-deficient T cells, re-expression of a constitutively activated STAT3 construct did not restore normal IL-17A production.

The selective role for Itk in IL-17A production resulted from a requirement for robust calcium signaling, leading to NFAT activation, in the transcription of the *Il17a* gene (Fig. 3). Thus, differentiation of *Itk*$^{-/-}$ T cells in the presence of anti-CD3 plus ionomycin, which can rescue Ca^{2+} mobilization in Itk-deficient T cells (Liu et al. 1998), did restore IL-17A production, directly implicating defective TCR-driven Ca^{2+} mobilization in this phenotype (Gomez-Rodriguez et al. 2009). Supporting this idea, stimulation of wild-type T cells with decreasing amounts of anti-CD3 during differentiation preferentially reduced expression of IL-17A, whereas leaving IL-17F relatively spared.

Similarly, differentiation of wild-type T cells in the presence of low-dose CyclosporinA or FK506, two inhibitors of Calcineurin that prevent NFAT activation (Winslow et al. 2003), also preferentially decreased IL-17A expression. Conversely, expression of a constitutively active mutant of $NFAT_{c1}$ (Neal and Clipstone 2003) restored IL-17A production by $Itk^{-/-}$ T_H17 cells, suggesting that Itk-mediated activation of NFAT contributes to the regulation of IL-17A. Sequence analyses showed a conserved NFAT binding site located 3.5 kb upstream of the start site of the *Il17a* gene and ChIP analyses showed that this site was occupied in wild-type but not Itk-deficient cells differentiated under T_H17 conditions. Although potential NFAT binding sites were found upstream of the *Il17f* gene, none were conserved cross-species. Nonetheless, conserved noncoding sequences in the region around both the linked *Il17a* and *Il17f* genes showed epigenetic modifications suggestive of open chromatin conformations (Wilson et al. 2009).

These results suggest that a TCR-driven pathway involving Itk-mediated regulation of Ca^{2+} and NFAT is required for full expression of IL-17A and that this pathway distinguishes expression of different T_H17 cytokines. These data support the growing view that there are distinct subclasses of T_H17 cells that produce overlapping sets of cytokines based upon distinct inputs. Thus, TCR signaling may be particularly important for expression of IL-17A, which is more strongly proinflammatory than IL-17F (Yang et al. 2008; Ishigame et al. 2009). This regulation operates during the differentiation of T_H17 cells, because the defect in IL-17A production is seen even upon restimulation of cells with PMA and ionomycin (Gomez-Rodriguez et al. 2009). $NFAT_{c1}$ autoregulates its own expression and previous data showed defective $NFAT_{c1}$ induction in Itk-deficient cells (Nurieva et al. 2007), perhaps accounting for the requirement for Itk during T_H17 differentiation. Alternatively, other chromatin remodeling factors that are recruited by transcription factors, including the BRG proteins, may not assemble properly on the IL-17A locus in the absence of efficient NFAT activation (Berg 2009). Although the exact mechanism of this defect awaits further study, these detailed investigations of Itk-deficient T lymphocytes have revealed another feature of how TCR signaling can affect the differentiation of distinct cytokine producing cells.

TCR Signaling in $CD8^+$ T Cells Requires Itk

Although fewer studies have addressed the role of Itk in $CD8^+$ T-cell responses, the conclusions thus far indicate that primary responses to viral infection are largely intact in the absence of Itk, or Itk and Rlk (Bachmann et al. 1997; Atherly et al. 2006). Biochemical studies show that $Itk^{-/-}$ and $Rlk^{-/-}Itk^{-/-}$ $CD8^+$ T cells share a similar deficiency in TCR-induced PLC-γ1 activation, calcium mobilization, ERK activation, and cytokine production as $CD4^+$ T cells lacking the Tec kinases (Atherly et al. 2006). Nonetheless, $Itk^{-/-}$ mice do mount protective immune responses to LCMV, vaccinia virus, and VSV, and $Rlk^{-/-}Itk^{-/-}$ mice to LCMV, although kinetics of viral clearance were often slightly delayed. Tracking of virus-specific $CD8^+$ T cells indicated that $Itk^{-/-}$ and $Rlk^{-/-}Itk^{-/-}$ T cells were unable to expand to the extent seen with wild-type T cells, resulting in an overall deficit in the magnitude of the $CD8^+$ T-cell response. This reduction in T-cell expansion could not be accounted for by impaired $CD4^+$ T cell help in $Itk^{-/-}$ mice, as adoptive transfer of wild-type LCMV-immune T cells ($CD4^+$ and $CD8^+$) into $Itk^{-/-}$ mice prior to LCMV challenge was unable to rescue the $CD8^+$ T-cell defects. In addition, the impaired response of $Itk^{-/-}$ $CD8^+$ T-cells was not because of the predominant population of "innate-like" $CD8^+$ T-cells present in these mice (see the following), as OT-1 TCR transgenic $Itk^{-/-}$ $CD8^+$ T-cells, which develop as conventional naïve T-cells, also shared the defect seen with polyclonal $Itk^{-/-}$ $CD8^+$ T-cells in non-TCR transgenic mice. It still remains to be determined whether $Itk^{-/-}$ $CD8^+$ T-cells can generate functional memory cells. Although virus-specific T-cells persist in $Itk^{-/-}$ mice well after virus has been cleared from the animals, the ability of these cells to provide protection, and to mount an effective recall response, has not yet been tested.

This issue is complex, as defects in *Itk*$^{-/-}$ CD4^{+} T-cells could impact the generation of robust memory CD8^{+} T-cells; thus, appropriate studies will be required to distinguish CD4^{+} T-cell-intrinsic from CD8^{+} T-cell-intrinsic contributions to antiviral memory responses.

TEC KINASES IN T-CELL DEVELOPMENT

Role of Itk and Rlk in Conventional T-Cell Development

TCR signaling plays a critical role in the development of T cells within the thymus, where the strength or duration of TCR signaling helps determine the survival, maturation, and differentiation of thymocytes into mature T cells. Double negative CD4^{-}CD8^{-} (DN) cells that have rearranged their TCRβ chains must receive competent signals through the pre-TCR to progress through the DN3 stage and expand in the DN4 to double positive (DP) transition (Ciofani and Zuniga-Pflucker 2006). At the double positive stage, TCR signaling regulates cell survival and differentiation of conventional T cells (Jameson et al. 1995). Thymocytes that receive very poor or no TCR signals undergo "death by neglect." Thymocytes that receive strong TCR signals, as from agonist peptides, particularly in the context of costimulation, undergo negative selection and also die within the thymus. Only thymocytes that receive weak but adequate signals will progress onto CD4/CD8 selection, where the duration of TCR signaling, in large part dictated by the expression patterns of the co-receptors CD4 and CD8 upon TCR engagement, will determine their ultimate lineage as CD4 or CD8 single positive (SP) cells (Bosselut 2004).

It is therefore not surprising that *Itk*$^{-/-}$ and *Rlk*$^{-/-}$*Itk*$^{-/-}$ T cells show altered thymocyte development (Fig. 4). *Itk*$^{-/-}$ mice have smaller thymi and show phenotypes consistent with impaired pre-TCR signaling. Although these defects are relatively subtle, competitive bone marrow transfers using mixtures of WT and *Itk*$^{-/-}$ bone marrow indicated that both *Itk*$^{-/-}$ and *Rlk*$^{-/-}$*Itk*$^{-/-}$ cells had a selective disadvantage at the DN-DP transition (Lucas et al. 2007). However, *Rlk*$^{-/-}$*Itk*$^{-/-}$ T cells also showed a competitive advantage compared to WT cells at earlier stages (prior to DN3), perhaps reflecting a negative signaling role for Rlk in cell survival/cytokine signaling pathways.

Defects in positive selection of thymocytes lacking Itk have been shown using a variety of MHC Class I and Class II-restricted TCR transgenes (Liao and Littman 1995; Schaeffer et al. 2000; Lucas et al. 2002; Lucas et al. 2003). The extent of these defects depended on the selecting TCR, with the weakest selecting TCRs showing the largest effects (Lucas et al. 2002). Defects in positive selection were exacerbated in *Rlk*$^{-/-}$*Itk*$^{-/-}$ mice, providing evidence that these genes are partially redundant, and supporting the idea that decreased TCR signaling reduces positive selection of conventional T cells (Schaeffer et al. 2000). Supporting partial redundancy of these genes, overexpression of Txk/Rlk partially rescued positive selection defects in *Itk*$^{-/-}$ mice (Sommers et al. 1999).

Itk-deficient animals also show reduced or delayed negative selection in multiple TCR transgenic systems (Liao and Littman 1995; Schaeffer et al. 2000; Lucas et al. 2002). In *Rlk*$^{-/-}$*Itk*$^{-/-}$ mice, negative selection is more clearly impaired, leading to positive selection of normally deleted HY TCR transgenic cells in male *Rlk*$^{-/-}$*Itk*$^{-/-}$ mice (Schaeffer et al. 2000). These data support a model where negative selection is converted to positive selection in the context of impaired TCR signaling and may account for the increased SP cells in *Rlk*$^{-/-}$*Itk*$^{-/-}$ mice.

Development of Innate T-Lymphocyte Populations

Recent data have revealed surprising new findings on thymic selection in mice deficient in Itk and Rlk (Atherly et al. 2006; Broussard et al. 2006; Dubois et al. 2006; Berg 2007; Hu et al. 2007). Although *Itk*$^{-/-}$ and *Rlk*$^{-/-}$*Itk*$^{-/-}$ mice show decreased positive selection of conventional T cells using both MHC Class I and Class II-restricted TCR transgenes, surprisingly, *Itk*$^{-/-}$ and *Rlk*$^{-/-}$*Itk*$^{-/-}$ mice have increased numbers of CD8 SP cells (Liao and Littman 1995;

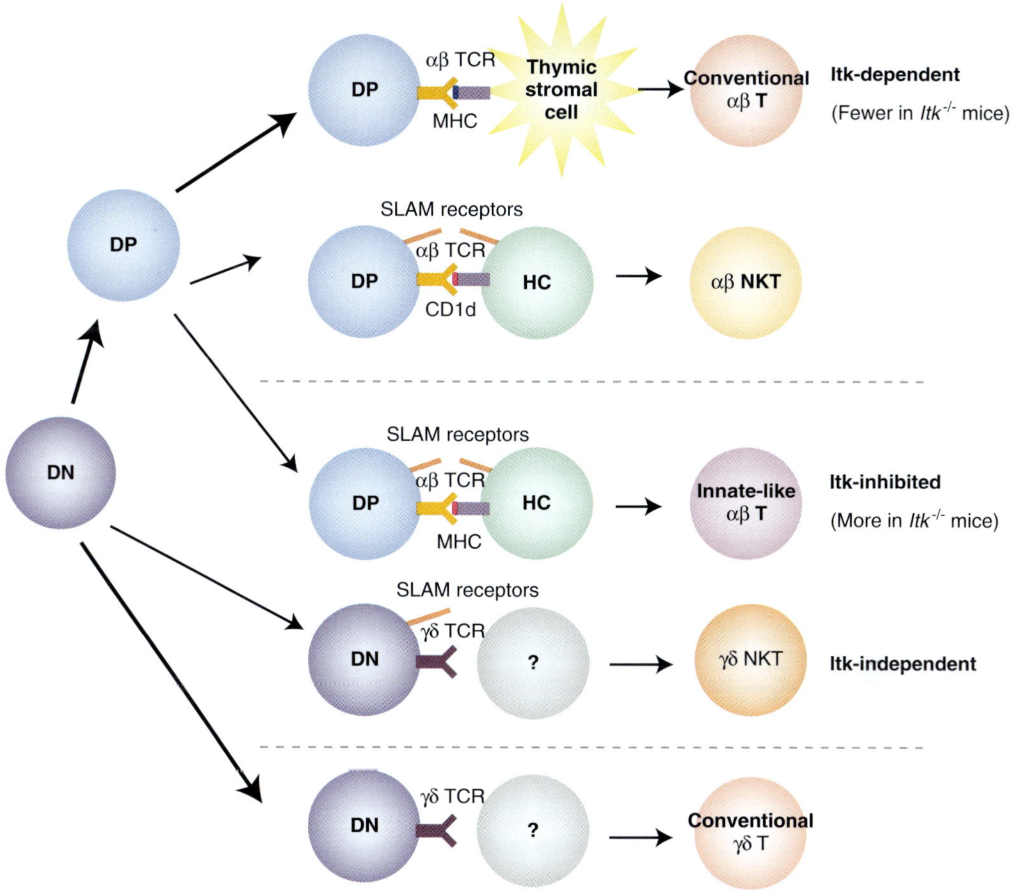

Figure 4. Regulation of T-cell lineages by Itk. In the thymus, progenitor T cells ($CD4^-CD8^-CD3^-$; DN) develop into γδ T cells and αβ T cells. Cells developing in the αβ T-cell lineage progress through an intermediate $CD4^+CD8^+$ (DP) stage prior to their maturation into mature conventional αβ T cells, CD1d-restricted αβ NKT cells and other subsets of innate-like αβ T cells. Conventional αβ T cells, which are selected on classical MHC molecules expressed on thymic stromal cells, and αβ NKT cells, which are selected on the MHC class IB molecule, CD1d, expressed on hematopoietic cells (HC) and dependent on SLAM receptor signaling, are greatly reduced in the absence of Itk. In contrast, innate-like αβ T cells, which are also selected by recognition of MHC molecules on HC and require SLAM receptor signaling, and "innate-like" γδ T cells (γδ NKT) are substantially increased in number in *Itk*$^{-/-}$ mice. A third subset of cells, illustrated by conventional γδ T cells, appear unaffected by the presence versus the absence of Itk. Currently, the cell–cell interactions required for γδ T-cell development are unknown.

Schaeffer et al. 2000; Lucas et al. 2002). Further examination of these CD8 SP populations suggested that these cells do not resemble conventional naive CD8 SP thymocytes. Instead, virtually all of the CD8 SP thymocytes in these mice showed markers seen on memory or activated T cells, including CD44 and CD122 (the IL-2Rβ chain), and rapidly expressed cytokines upon activation (Atherly et al. 2006; Broussard et al. 2006; Dubois et al. 2006; Hu et al. 2007). These phenotypes are characteristic of memory cells in the periphery—however, developmental studies and fetal thymic organ cultures provided evidence that Itk-deficient CD8 cells developed these characteristics within the thymus (Atherly et al. 2006; Broussard et al. 2006; Dubois et al.

2006). Although genetic crosses showed a requirement for MHC Class I for their development, transfers of bone marrow from *Itk*$^{-/-}$ or *Rlk*$^{-/-}$*Itk*$^{-/-}$ mice into irradiated β2m-deficient mice, which lack MHC-Class I on their radioresistant thymic stroma, showed that *Itk*$^{-/-}$ and *Rlk*$^{-/-}$*Itk*$^{-/-}$ CD8$^+$ thymocytes could be selected on hematopoietically derived cells (Broussard et al. 2006). Although these observations are atypical for conventional thymocytes, these features are all reminiscent of NKT cells and other related innate T-cell lineages; thus, many innate T-cell lineages have been shown to be selected on double-positive thymocytes, to express activation or memory cell markers, and to rapidly produce cytokines even in the thymus (Berg 2007). Like NKT cells, *Itk*$^{-/-}$ and *Rlk*$^{-/-}$*Itk*$^{-/-}$ CD8$^+$ cells also expressed NK cell markers and high levels of the transcription factor Eosmesodermin, and were dependent on IL-15 (Atherly et al. 2006; Dubois et al. 2006). Moreover, like NKT cells, development of these CD8 cells was dependent on SAP, a small adaptor molecule that is required for signaling from the SLAM family of receptors that are expressed on hematopoietic cells, but not on the thymic epithelium (Horai et al. 2007). Together these characteristics suggested that Itk-deficiency led to the preferential selection of an innate CD8 cell population on hematopoietic cells within the thymus (Fig. 4).

Although this phenotype is primarily seen in CD8 cells, there is also a small population of CD4 cells that show these properties in *Itk*$^{-/-}$ mice. These CD44^{hi}CD4$^+$ cells are also dependent on SAP and express PLZF, a transcription factor responsible for the innate characteristics of NKT cells (PLS and LJB, unpublished data and Broussard et al. 2006; Hu and August 2008; Raberger et al. 2008). Although the selecting cell population for the innate type T cells in *Itk*$^{-/-}$ mice is yet undetermined, the large number of CD8 cells with this characteristic (and relatively low numbers of CD4 cells) suggests that it is likely to be DP thymocytes, which express MHC Class I but not Class II, and are the selecting cells for NKT cells (Bendelac 1995).

Why loss of Itk leads to the preferential development of CD8 cells with this phenotype is not clear. Reciprocal transfers of *Itk*$^{-/-}$β2m$^{-/-}$ bone marrow provided evidence that the development of these innate phenotypes required selection on hematopoietic cells (Horai et al. 2007). One possibility is that poor selection of conventional thymocytes could permit selection or expansion of cells selected on hematopoietic cells. However, transfer of WT bone marrow into irradiated β2m-deficient hosts prevents selection of conventional CD8$^+$ T cells and yet does not give rise to a large number of innate CD8 cells, suggesting that development of these cells is normally suppressed. Given the rapid production of cytokines showed by these cells, it is likely that the generation of these cells may be under tight control.

Several theories could account for the development of this population in *Itk*$^{-/-}$ mice. Itk could play a negative role in signaling pathways required for the development of these lineages, such those downstream of SLAM family receptors. Alternatively, these cells might normally be deleted in wild-type animals—negative selection can occur on hematopoietic cells, which express costimulatory molecules, such as CD28, thought to be important in this process (Punt et al. 1994). Alternatively, defective TCR signaling may make these cells more susceptible to signals from costimulatory molecules, such as SLAM family members, or cytokines that may be required either for selection or development of these phenotypes (Atherly et al. 2006; Dubois et al. 2006; Horai et al. 2007). Further analyses will provide clues to the regulation of these interesting lineages that appear to be at the boundary of innate and adaptive immune responses (Berg 2007).

Itk Signaling Regulates αβ and γδ NKT-Cell Maturation

In addition to defects in the development and selection of conventional naïve CD4$^+$ and CD8$^+$ T-cells, αβTCR$^+$ NKT-cell maturation is also impaired in the absence of Itk and Rlk (Gadue and Stein 2002; Au-Yeung and Fowell 2007; Felices and Berg 2008; and Fig. 5). In these Tec kinase-deficient mice, αβ NKT cell numbers are substantially reduced; furthermore, of the cells that remain, few represent the most mature αβ NKT subset characterized by high expression of NK1.1 and CD122, and down-regulation of CD4. *Itk*$^{-/-}$ and *Itk*$^{-/-}$*Rlk*$^{-/-}$ NKT cells also show poor responses to TCR stimulation, producing little to no IL-4 and IFNγ following in vivo activation. Thus, similar to conventional αβ T cells, αβ NKT cells require Tec kinases for their development and function.

In stark contrast to the requirements for αβTCR$^+$ T cells, the development of γδ T cells does not require Itk signaling (Fig. 4). The major γδ T-cell populations that arise in the fetus and adult are largely unchanged in *Itk*$^{-/-}$ mice (Felices et al. 2009; C. Yin and LJB, unpub. observ.). Indeed, overall γδT cell numbers in *Itk*$^{-/-}$ mice are significantly increased compared to wild-type mice, because of dramatic increases in one particular γδ T-cell subset, the γδ NKT cells (Felices et al. 2009; Qi et al. 2009). This subset expresses the Vγ1.1/Vδ6.3 TCR; however, in all other regards, these γδ T cells share characteristics of CD1d-specific αβ NKT cells rather than traits associated with most adult-derived γδ T cells (Vicari et al. 1996; Azuara et al. 1997; Lees et al. 2001). For instance, the majority of γδ NKT cells express CD4, NK1.1, and the signature NKT cell transcription factor, PLZF (Felices et al. 2009; Kreslavsky et al. 2009). As a result, γδ NKT cells, like their αβ NKT cell counterparts, produce both IFNγ and T_H2 cytokines. In addition, these cells preferentially home to the liver, rather than to lymphoid organs such as the spleen (Lees et al. 2001; Felices et al. 2009).

Surprisingly, this expanded population of γδ NKT cells is responsible for a spontaneous hyper-IgE syndrome observed in *Itk*$^{-/-}$ mice (Felices et al. 2009; Qi et al. 2009). When *Itk*$^{-/-}$ mice are crossed to mice lacking γδ T cells, the elevated serum IgE and germinal center hyperplasia seen in *Itk*$^{-/-}$ mice disappear. *Itk*$^{-/-}$ splenic γδ NKT cells constitutively express ICOS, and readily upregulate CD40-ligand and OX40 following γδ TCR stimulation in vitro (Felices et al. 2009). These findings suggest that *Itk*$^{-/-}$ γδ NKT cells are being activated in vivo, allowing them to provide T cell help to B cells and secrete cytokines that promote switching to IgE. As with the majority of γδ T cells, the ligand recognized by the Vγ1.1/Vδ6.3$^+$ TCR is unknown, nor is it known whether activation of these cells requires signaling through the γδ TCR.

These findings raise several interesting issues regarding the development and signaling properties of γδ T cells. In contrast to αβ T cells, γδ T cells are not reduced in the absence of Itk; further, the γδ NKT cell population is, in fact, greatly expanded. These data indicate that the signals regulating γδ T-cell development do not require Itk, a situation comparable to that seen for innate CD8$^+$ αβ T cells. The expansion of both the γδ NKT and the innate CD8$^+$ population in *Itk*$^{-/-}$ mice may reflect the fact that the majority of these cells normally undergo negative selection, but that their deletion is impaired in the absence of Itk.

A second interesting issue is the apparently robust TCR signaling in *Itk*$^{-/-}$ γδ T cells. αβ T cells lacking Itk have substantially reduced responses to TCR signaling, particularly in their ability to synthesize effector cytokines. *Itk*$^{-/-}$ γδ T cells, however, produce copious amounts of cytokines in response to γδ TCR stimulation. These data suggest that the biochemical pathways downstream of the γδ TCR are likely distinct from those in αβ T cells. Consistent with this idea, mice carrying a mutant version of the adapter protein LAT also have few αβ T cells but greatly increased numbers of γδ T cells that overproduce T_H2 cytokines; further, these mice share the hyper-IgE syndrome seen in *Itk*$^{-/-}$ mice (Nunez-Cruz et al. 2003). This LAT mutant is lacking a critical tyrosine in the cytoplasmic tail that is essential for efficient TCR signaling in αβ T cells, reinforcing the notion that αβ and γδ TCR signaling have distinct biochemical requirements.

ITK AND HUMAN DISEASE

Although mutations affecting Btk were identified as the cause of X-linked Agammaglobulinemia 17 years ago (Rawlings et al. 1993; Thomas et al. 1993; Tsukada et al. 1993; Vetrie et al. 1993), alterations in Itk have only recently been associated with human disorders. These include Itk promoter polymorphisms associated with atopy (Graves et al. 2005) and the expression of an Itk-Syk fusion protein generated by a chromosomal translocation in T-cell lymphomas (Streubel et al. 2006); in this latter case, the Itk PH domain becomes fused to the kinase domain of Syk, leading to constitutive activation of the kinase. Most recently, a mutation affecting Itk was found in 2 siblings who died from fatal Epstein-Barr Virus infection (Huck et al. 2009). This mutation changes a single amino acid in the Itk SH2 domain and destabilizes the Itk protein. Although the basis for this particular immunodeficiency is unknown, it may be related to Itk function in cytotoxic $CD8^+$ T cells (Bachmann et al. 1997; Atherly et al. 2006). Importantly, the implication of Itk in a human primary immunodeficiency has validated its role as a critical regulator of T lymphocyte function. Given the depth of information on lymphocyte biology provided through the study of Tec kinases, it is clear that their study will continue to provide important insights into the signaling pathways involved in immune cell development, homeostasis and function.

ACKNOWLEDGMENTS

This work was supported by grants from the National Institutes of Health (AI37584 and AI46564 to LJB and AI043957 and AI075150 to AHA) and by intramural funds from the NHGRI/NIH (PLS). We thank Julia Fekecs for assistance with graphic illustrations.

REFERENCES

Andreotti AH, Bunnell SC, Feng S, Berg LJ, Schreiber SL. 1997. Regulatory intramolecular association in a tyrosine kinase of the Tec family. *Nature* **385:** 93–97.

Atherly LO, Brehm MA, Welsh RM, Berg LJ. 2006. Tec kinases Itk and Rlk are required for CD8+ T cell responses to virus infection independent of their role in CD4+ T cell help. *J Immunol* **176:** 1571–1581.

Au-Yeung BB, Deindl S, Hsu LY, Palacios EH, Levin SE, Kuriyan J, Weiss A. 2009. The structure, regulation, and function of ZAP-70. *Immunol Rev* **228:** 41–57.

Au-Yeung BB, Fowell DJ. 2007. A key role for Itk in both IFN gamma and IL-4 production by NKT cells. *J Immunol* **179:** 111–119.

Au-Yeung BB, Katzman SD, Fowell DJ. 2006. Cutting edge: Itk-dependent signals required for CD4+ T cells to exert, but not gain, Th2 effector function. *J Immunol* **176:** 3895–3899.

August A. 2009. IL-2-inducible T-cell kinase (ITK) finds another (dance) partner ... TFII-I. *Eur J Immunol* **39:** 2354–2357.

August A, Sadra A, Dupont B, Hanafusa H. 1997. Src-induced activation of inducible T cell kinase (ITK) requires phosphatidylinositol 3-kinase activity and the Pleckstrin homology domain of inducible T cell kinase. *Proc Natl Acad Sci U S A* **94:** 11227–11232.

Azuara V, Levraud JP, Lembezat MP, Pereira P. 1997. A novel subset of adult gamma delta thymocytes that secretes a distinct pattern of cytokines and expresses a very restricted T cell receptor repertoire. *Eur J Immunol* **27:** 544–553.

Bachmann MF, Littman DR, Liao XC. 1997. Antiviral immune responses in Itk-deficient mice. *J Virol* **71:** 7253–7257.

Baker JE, Majeti R, Tangye SG, Weiss A. 2001. Protein tyrosine phosphatase CD148-mediated inhibition of T-cell receptor signal transduction is associated with reduced LAT and phospholipase Cgamma1 phosphorylation. *Mol Cell Biol* **21:** 2393–2403.

Beach D, Gonen R, Bogin Y, Reischl IG, Yablonski D. 2007. Dual role of SLP-76 in mediating T cell receptor-induced activation of phospholipase C-gamma1. *J Biol Chem* **282:** 2937–2946.

Bendelac A. 1995. Positive selection of mouse NK1+ T cells by CD1-expressing cortical thymocytes. *J Exp Med* **182:** 2091–2096.

Berg LJ. 2007. Signalling through TEC kinases regulates conventional versus innate CD8(+) T-cell development. *Nat Rev Immunol* **7:** 479–485.

Berg LJ. 2009. Strength of T cell receptor signaling strikes again. *Immunity* **31:** 529–531.

Berg LJ, Finkelstein LD, Lucas JA, Schwartzberg PL. 2005. Tec family kinases in T lymphocyte development and function. *Annu Rev Immunol* **23:** 549–600.

Blomberg KE, Boucheron N, Lindvall JM, Yu L, Raberger J, Berglof A, Ellmeier W, Smith CE. 2009. Transcriptional signatures of Itk-deficient CD3+, CD4+ and CD8+ T-cells. *BMC Genomics* **10:** 233.

Bogin Y, Ainey C, Beach D, Yablonski D. 2007. SLP-76 mediates and maintains activation of the Tec family kinase ITK via the T cell antigen receptor-induced association between SLP-76 and ITK. *Proc Natl Acad Sci U S A* **104:** 6638–6643.

Bosselut R. 2004. CD4/CD8-lineage differentiation in the thymus: from nuclear effectors to membrane signals. *Nat Rev Immunol* **4:** 529–540.

Braiman A, Barda-Saad M, Sommers CL, Samelson LE. 2006. Recruitment and activation of PLCgamma1 in T cells: a new insight into old domains. *EMBO J* **25:** 774–784.

Brazin KN, Fulton DB, Andreotti AH. 2000. A specific intermolecular association between the regulatory domains of a Tec family kinase. *J Mol Biol* **302:** 607–623.

Brazin KN, Mallis RJ, Fulton DB, Andreotti AH. 2002. Regulation of the tyrosine kinase Itk by the peptidyl-prolyl isomerase cyclophilin A. *Proc Natl Acad Sci U S A* **99:** 1899–1904.

Broussard C, Fleischacker C, Horai R, Chetana M, Venegas AM, Sharp LL, Hedrick SM, Fowlkes BJ, Schwartzberg PL. 2006. Altered development of CD8+ T cell lineages in mice deficient for the Tec kinases Itk and Rlk. *Immunity* **25:** 93–104.

Brown MT, Cooper JA. 1996. Regulation, substrates and functions of src. *Biochim Biophys Acta* **1287:** 121–149.

Bubeck Wardenburg J, Fu C, Jackman JK, Flotow H, Wilkinson SE, Williams DH, Johnson R, Kong G, Chan AC, Findell PR. 1996. Phosphorylation of SLP-76 by the ZAP-70 protein-tyrosine kinase is required for T-cell receptor function. *J Biol Chem* **271:** 19641–19644.

Bunnell SC, Diehn M, Yaffe MB, Findell PR, Cantley LC, Berg LJ. 2000. Biochemical interactions integrating Itk with the T cell receptor-initiated signaling cascade. *J Biol Chem* **275:** 2219–2230.

Bunnell SC, Singer AL, Hong DI, Jacque BH, Jordan MS, Seminario MC, Barr VA, Koretzky GA, Samelson LE. 2006. Persistence of cooperatively stabilized signaling clusters drives T-cell activation. *Mol Cell Biol* **26:** 7155–7166.

Chamorro M, Czar MJ, Debnath J, Cheng G, Lenardo MJ, Varmus HE, Schwartzberg PL. 2001. Requirements for activation and RAFT localization of the T-lymphocyte kinase Rlk/Txk. *BMC Immunol* **2:** 3.

Ching KA, Grasis JA, Tailor P, Kawakami Y, Kawakami T, Tsoukas CD. 2000. TCR/CD3-Induced activation and binding of Emt/Itk to linker of activated T cell complexes: requirement for the Src homology 2 domain. *J Immunol* **165:** 256–262.

Ching KA, Kawakami Y, Kawakami T, Tsoukas CD. 1999. Emt/Itk associates with activated TCR complexes: role of the pleckstrin homology domain. *J Immunol* **163:** 6006–6013.

Ciofani M, Zuniga-Pflucker JC. 2006. A survival guide to early T cell development. *Immunol Res* **34:** 117–132.

Colgan J, Asmal M, Neagu M, Yu B, Schneidkraut J, Lee Y, Sokolskaja E, Andreotti A, Luban J. 2004. Cyclophilin A regulates TCR signal strength in CD4+ T cells via a proline-directed conformational switch in Itk. *Immunity* **21:** 189–201.

Cowan-Jacob SW, Fendrich G, Manley PW, Jahnke W, Fabbro D, Liebetanz J, Meyer T. 2005. The crystal structure of a c-Src complex in an active conformation suggests possible steps in c-Src activation. *Structure* **13:** 861–871.

Crabtree GR, Olson EN. 2002. NFAT signaling: choreographing the social lives of cells. *Cell* **109 Suppl:** S67–79.

Cruz-Orcutt N, Houtman JC. 2009. PI3 kinase function is vital for the function but not formation of LAT-mediated signaling complexes. *Mol Immunol* **46:** 2274–2283.

Dombroski D, Houghtling RA, Labno CM, Precht P, Takesono A, Caplen NJ, Billadeau DD, Wange RL, Burkhardt JK, Schwartzberg PL. 2005. Kinase-independent functions for Itk in TCR-induced regulation of Vav and the actin cytoskeleton. *J Immunol* **174:** 1385–1392.

Dubois S, Waldmann TA, Muller JR. 2006. ITK and IL-15 support two distinct subsets of CD8+ T cells. *Proc Natl Acad Sci U S A* **103:** 12075–12080.

Felices M, Berg LJ. 2008. The Tec kinases Itk and Rlk regulate NKT cell maturation, cytokine production, and survival. *J Immunol* **180:** 3007–3018.

Felices M, Falk M, Kosaka Y, Berg LJ. 2007. Tec kinases in T cell and mast cell signaling. *Adv Immunol* **93:** 145–184.

Felices M, Yin CC, Kosaka Y, Kang J, Berg LJ. 2009. Tec kinase Itk in γδ T cells is pivotal for controlling IgE production in vivo. *Proc Natl Acad Sci* **106:** 8308–8313.

Fischer AM, Mercer JC, Iyer A, Ragin MJ, August A. 2004. Regulation of CXC chemokine receptor 4-mediated migration by the Tec family tyrosine kinase ITK. *J Biol Chem* **279:** 29816–29820.

Fowell DJ, Shinkai K, Liao XC, Beebe AM, Coffman RL, Littman DR, Locksley RM. 1999. Impaired NFATc translocation and failure of Th2 development in Itk-deficient CD4+ T cells. *Immunity* **11:** 399–409.

Fukuda M, Kojima T, Kabayama H, Mikoshiba K. 1996. Mutation of the pleckstrin homology domain of Bruton's tyrosine kinase in immunodeficiency impaired inositol 1,3,4,5-tetrakisphosphate binding capacity. *J Biol Chem* **271:** 30303–30306.

Gadue P, Stein PL. 2002. NK T cell precursors exhibit differential cytokine regulation and require Itk for efficient maturation. *J Immunol* **169:** 2397–2406.

Garcon F, Nunes JA. 2006. Travel informations on the Tec kinases during lymphocyte activation. *Adv Exp Med Biol* **584:** 15–27.

Gibson S, Leung B, Squire JA, Hill M, Arima N, Goss P, Hogg D, Mills GB. 1993. Identification, cloning, and characterization of a novel human T-cell-specific tyrosine kinase located at the hematopoietin complex on chromosome 5q. *Blood* **82:** 1561–1572.

Gilfillan AM, Rivera J. 2009. The tyrosine kinase network regulating mast cell activation. *Immunol Rev* **228:** 149–169.

Gomez-Rodriguez J, Readinger JA, Viorritto IC, Mueller KL, Houghtling RA, Schwartzberg PL. 2007. Tec kinases, actin, and cell adhesion. *Immunol Rev* **218:** 45–64.

Gomez-Rodriguez J, Sahu N, Handon R, Davidson TS, Anderson SM, Kirby MR, August A, Schwartzberg PL. 2009. Differential expression of interleukin-17A and -17F is coupled to T cell receptor signaling via inducible T cell kinase. *Immunity* **31:** 587–597.

Graef IA, Holsinger LJ, Diver S, Schreiber SL, Crabtree GR. 1997. Proximity and orientation underlie signaling by the non-receptor tyrosine kinase ZAP70. *EMBO J* **16:** 5618–5628.

Graves PE, Siroux V, Guerra S, Klimecki WT, Martinez FD. 2005. Association of atopy and eczema with polymorphisms in T-cell immunoglobulin domain and mucin domain-IL-2-inducible T-cell kinase gene cluster in chromosome 5 q 33. *J Allergy Clin Immunol* **116:** 650–656.

Guo S, Wahl MI, Witte ON. 2006. Mutational analysis of the SH2-kinase linker region of Bruton's tyrosine kinase defines alternative modes of regulation for cytoplasmic tyrosine kinase families. *Int Immunol* **18:** 79–87.

Hansson H, Okoh MP, Smith CI, Vihinen M, Hard T. 2001. Intermolecular interactions between the SH3 domain and the proline-rich TH region of Bruton's tyrosine kinase. *FEBS Lett* **489:** 67–70.

Hao S, August A. 2002. The proline rich region of the Tec homology domain of ITK regulates its activity. *FEBS Lett* **525:** 53–58.

Hao S, Qi Q, Hu J, August A. 2006. A kinase independent function for Tec kinase ITK in regulating antigen receptor induced serum response factor activation. *FEBS Lett* **580:** 2691–2697.

Hawkins J, Marcy A. 2001. Characterization of Itk tyrosine kinase: contribution of noncatalytic domains to enzymatic activity. *Protein Expr Purif* **22:** 211–219.

Heyeck SD, Berg LJ. 1993. Developmental regulation of a murine T-cell-specific tyrosine kinase gene, Tsk. *Proc Natl Acad Sci U S A* **90:** 669–673.

Heyeck SD, Wilcox HM, Bunnell SC, Berg LJ. 1997. Lck phosphorylates the activation loop tyrosine of the Itk kinase domain and activates Itk kinase activity. *J Biol Chem* **272:** 25401–25408.

Horai R, Mueller KL, Handon RA, Cannons JL, Anderson SM, Kirby MR, Schwartzberg PL. 2007. Requirements for selection of conventional and innate T lymphocyte lineages. *Immunity* **27:** 775–785.

Houtman JC, Houghtling RA, Barda-Saad M, Toda Y, Samelson LE. 2005. Early phosphorylation kinetics of proteins involved in proximal TCR-mediated signaling pathways. *J Immunol* **175:** 2449–2458.

Houtman JC, Yamaguchi H, Barda-Saad M, Braiman A, Bowden B, Appella E, Schuck P, Samelson LE. 2006. Oligomerization of signaling complexes by the multipoint binding of GRB2 to both LAT and SOS1. *Nat Struct Mol Biol* **13:** 798–805.

Hu J, August A. 2008. Naive and innate memory phenotype CD4+ T cells have different requirements for active Itk for their development. *J Immunol* **180:** 6544–6552.

Hu J, Sahu N, Walsh E, August A. 2007. Memory phenotype CD8+ T cells with innate function selectively develop in the absence of active Itk. *Eur J Immunol* **37:** 2892–2899.

Huang K, Wang YH, Brown A, Sun G. 2009. Identification of N-terminal lobe motifs that determine the kinase activity of the catalytic domains and regulatory strategies of Src and Csk protein tyrosine kinases. *J Mol Biol* **386:** 1066–1077.

Huang YH, Grasis JA, Miller AT, Xu R, Soonthornvacharin S, Andreotti AH, Tsoukas CD, Cooke MP, Sauer K. 2007. Positive regulation of Itk PH domain function by soluble IP4. *Science* **316:** 886–889.

Huck K, Feyen O, Niehues T, Ruschendorf F, Hubner N, Laws HJ, Telieps T, Knapp S, Wacker HH, Meindl A, et al. 2009. Girls homozygous for an IL-2-inducible T cell kinase mutation that leads to protein deficiency develop fatal EBV-associated lymphoproliferation. *J Clin Invest* **119:** 1350–1358.

Huse M, Eck MJ, Harrison SC. 1998. A Zn2+ ion links the cytoplasmic tail of CD4 and the N-terminal region of Lck. *J Biol Chem* **273:** 18729–18733.

Hwang ES, Szabo SJ, Schwartzberg PL, Glimcher LH. 2005. T helper cell fate specified by kinase-mediated interaction of T-bet with GATA-3. *Science* **307:** 430–433.

Ishigame H, Kakuta S, Nagai T, Kadoki M, Nambu A, Komiyama Y, Fujikado N, Tanahashi Y, Akitsu A, Kotaki H, et al. 2009. Differential roles of interleukin-17A and -17F in host defense against mucoepithelial bacterial infection and allergic responses. *Immunity* **30:** 108–119.

Jameson SC, Hogquist KA, Bevan MJ. 1995. Positive selection of thymocytes. *Annu Rev Immunol* **13:** 93–126.

Jordan MS, Smith JE, Burns JC, Austin JE, Nichols KE, Aschenbrenner AC, Koretzky GA. 2008. Complementation in trans of altered thymocyte development in mice expressing mutant forms of the adaptor molecule SLP76. *Immunity* **28:** 359–369.

Joseph RE, Andreotti AH. 2009. Conformational snapshots of Tec kinases during signaling. *Immunol Rev* **228:** 74–92.

Joseph RE, Fulton DB, Andreotti AH. 2007a. Mechanism and functional significance of Itk autophosphorylation. *J Mol Biol* **373:** 1281–1292.

Joseph RE, Min L, Andreotti AH. 2007b. The linker between SH2 and kinase domains positively regulates catalysis of the Tec family kinases. *Biochemistry* **46:** 5455–5462.

Joseph RE, Min L, Xu R, Musselman ED, Andreotti AH. 2007c. A remote substrate docking mechanism for the tec family tyrosine kinases. *Biochemistry* **46:** 5595–5603.

Koprulu AD, Ellmeier W. 2009. The role of Tec family kinases in mononuclear phagocytes. *Crit Rev Immunol* **29:** 317–333.

Koretzky GA, Abtahian F, Silverman MA. 2006. SLP76 and SLP65: complex regulation of signalling in lymphocytes and beyond. *Nat Rev Immunol* **6:** 67–78.

Korn T, Bettelli E, Oukka M, Kuchroo VK. 2009. IL-17 and Th17 Cells. *Annu Rev Immunol* **27:** 485–517.

Kreslavsky T, Savage AK, Hobbs R, Gounari F, Bronson R, Pereira P, Pandolfi PP, Bendelac A, von Boehmer H. 2009. TCR-inducible PLZF transcription factor required for innate phenotype of a subset of gammadelta T cells with restricted TCR diversity. *Proc Natl Acad Sci U S A* **106:** 12453–12458.

Labno CM, Lewis CM, You D, Leung DW, Takesono A, Kamberos N, Seth A, Finkelstein LD, Rosen MK, Schwartzberg PL, et al. 2003. Itk functions to control actin polymerization at the immune synapse through localized activation of Cdc42 and WASP. *Curr Biol* **13:** 1619–1624.

Laederach A, Cradic KW, Brazin KN, Zamoon J, Fulton DB, Huang XY, Andreotti AH. 2002. Competing modes of self-association in the regulatory domains of Bruton's tyrosine kinase: intramolecular contact versus asymmetric homodimerization. *Protein Sci* **11:** 36–45.

Laederach A, Cradic KW, Fulton DB, Andreotti AH. 2003. Determinants of intra versus intermolecular self-association within the regulatory domains of Rlk and Itk. *J Mol Biol* **329:** 1011–1020.

LaFevre-Bernt M, Sicheri F, Pico A, Porter M, Kuriyan J, Miller WT. 1998. Intramolecular regulatory interactions in the Src family kinase Hck probed by mutagenesis of a conserved tryptophan residue. *J Biol Chem* **273:** 32129–32134.

Lees RK, Ferrero I, MacDonald HR. 2001. Tissue-specific segregation of TCRgamma delta+ NKT cells according to phenotype TCR repertoire and activation status: parallels with TCR alphabeta+NKT cells. *Eur J Immunol* **31:** 2901–2909.

Lewis RS, Cahalan MD. 1995. Potassium and calcium channels in lymphocytes. *Annu Rev Immunol* **13:** 623–653.

Liao XC, Littman DR. 1995. Altered T cell receptor signaling and disrupted T cell development in mice lacking Itk. *Immunity* **3:** 757–769.

Lin L, Czerwinski R, Kelleher K, Siegel MM, Wu P, Kriz R, Aulabaugh A, Stahl M. 2009. Activation loop phosphorylation modulates Bruton's tyrosine kinase (Btk) kinase domain activity. *Biochemistry* **48:** 2021–2032.

Lin X, Ayrapetov MK, Lee S, Parang K, Sun G. 2005. Probing the communication between the regulatory and catalytic domains of a protein tyrosine kinase, Csk. *Biochemistry* **44:** 1561–1567.

Liu KQ, Bunnell SC, Gurniak CB, Berg LJ. 1998. T cell receptor-initiated calcium release is uncoupled from capacitative calcium entry in Itk-deficient T cells. *J Exp Med* **187:** 1721–1727.

Lucas JA, Atherly LO, Berg LJ. 2002. The absence of Itk inhibits positive selection without changing lineage commitment. *J Immunol* **168:** 6142–6151.

Lucas JA, Felices M, Evans JW, Berg LJ. 2007. Subtle defects in pre-TCR signaling in the absence of the Tec kinase Itk. *J Immunol* **179:** 7561–7567.

Lucas JA, Miller AT, Atherly LO, Berg LJ. 2003. The role of Tec family kinases in T cell development and function. *Immunol Rev* **191:** 119–138.

Marcotte DJ, Liu YT, Arduini RM, Hession CA, Miatkowski K, Wildes CP, Cullen PF, Hong V, Hopkins BT, Mertsching E, et al. 2010. Structures of human Bruton's tyrosine kinase in active and inactive conformations suggest a mechanism of activation for TEC family kinases. *Protein Sci* **19:** 429–439.

Marquez JA, Smith CI, Petoukhov MV, Lo Surdo P, Mattsson PT, Knekt M, Westlund A, Scheffzek K, Saraste M, Svergun DI. 2003. Conformation of full-length Bruton tyrosine kinase (Btk) from synchrotron X-ray solution scattering. *EMBO J* **22:** 4616–4624.

Mihara S, Suzuki N. 2007. Role of Txk, a member of the Tec family of tyrosine kinases, in immune-inflammatory diseases. *Int Rev Immunol* **26:** 333–348.

Miller AT, Berg LJ. 2002. Defective Fas ligand expression and activation-induced cell death in the absence of IL-2-inducible T cell kinase. *J Immunol* **168:** 2163–2172.

Miller AT, Wilcox HM, Lai Z, Berg LJ. 2004. Signaling through Itk promotes T helper 2 differentiation via negative regulation of T-bet. *Immunity* **21:** 67–80.

Min L, Joseph RE, Fulton DB, Andreotti AH. 2009. Itk tyrosine kinase substrate docking is mediated by a nonclassical SH2 domain surface of PLCgamma1. *Proc Natl Acad Sci U S A* **106:** 21143–21148.

Min L, Wu W, Joseph RE, Fulton DB, Berg L, Andreotti AH. 2010. Disrupting the intermolecular self-association of Itk enhances T cell signaling. *J Immunology* **184:** 4228–4235.

Mohamed AJ, Yu L, Backesjo CM, Vargas L, Faryal R, Aints A, Christensson B, Berglof A, Vihinen M, Nore BF, et al. 2009. Bruton's tyrosine kinase (Btk): function, regulation, and transformation with special emphasis on the PH domain. *Immunol Rev* **228:** 58–73.

Mueller C, August A. 2003. Attenuation of Immunological Symptoms of Allergic Asthma in Mice Lacking the Tyrosine Kinase ITK. *J Immunol* **170:** 5056–5063.

Neal JW, Clipstone NA. 2003. A constitutively active NFATc1 mutant induces a transformed phenotype in 3T3-L1 fibroblasts. *J Biol Chem* **278:** 17246–17254.

Nunez-Cruz S, Aguado E, Richelme S, Chetaille B, Mura AM, Richelme M, Pouyet L, Jouvin-Marche E, Xerri L, Malissen B, et al. 2003. LAT regulates gammadelta T cell homeostasis and differentiation. *Nat Immunol* **4:** 999–1008.

Nurieva RI, Chuvpilo S, Wieder ED, Elkon KB, Locksley R, Serfling E, Dong C. 2007. A costimulation-initiated signaling pathway regulates NFATc1 transcription in T lymphocytes. *J Immunol* **179:** 1096–1103.

Okoh MP, Vihinen M. 2002. Interaction between Btk TH and SH3 domain. *Biopolymers* **63:** 325–334.

Palacios EH, Weiss A. 2004. Function of the Src-family kinases, Lck and Fyn, in T-cell development and activation. *Oncogene* **23:** 7990–8000.

Pawson T, Kofler M. 2009. Kinome signaling through regulated protein-protein interactions in normal and cancer cells. *Curr Opin Cell Biol* **21:** 147–153.

Perez-Villar JJ, Kanner SB. 1999. Regulated association between the tyrosine kinase Emt/Itk/Tsk and phospholipase-C gamma 1 in human T lymphocytes. *J Immunol* **163:** 6435–6441.

Pitcher LA, van Oers NS. 2003. T-cell receptor signal transmission: who gives an ITAM? *Trends Immunol* **24:** 554–560.

Pletneva EV, Sundd M, Fulton DB, Andreotti AH. 2006. Molecular details of Itk activation by prolyl isomerization and phospholigand binding: the NMR structure of the Itk SH2 domain bound to a phosphopeptide. *J Mol Biol* **357:** 550–561.

Prince AL, Yin CC, Enos ME, Felices M, Berg LJ. 2009. The Tec kinases Itk and Rlk regulate conventional versus innate T-cell development. *Immunol Rev* **228:** 115–131.

Punt JA, Osborne BA, Takahama Y, Sharrow SO, Singer A. 1994. Negative selection of CD4+CD8+ thymocytes by T cell receptor-induced apoptosis requires a costimulatory signal that can be provided by CD28. *J Exp Med* **179:** 709–713.

Pursglove SE, Mulhern TD, Mackay JP, Hinds MG, Booker GW. 2002. The solution structure and intramolecular associations of the Tec kinase SRC homology 3 domain. *J Biol Chem* **277:** 755–762.

Qi Q, August A. 2009. The tec family kinase itk exists as a folded monomer in vivo. *J Biol Chem* **284:** 29882–29892.

Qi Q, Sahu N, August A. 2006. Tec kinase Itk forms membrane clusters specifically in the vicinity of recruiting receptors. *J Biol Chem* **281:** 38529–38534.

Qi Q, Xia M, Hu J, Hicks E, Iyer A, Xiong N, August A. 2009. Enhanced development of CD4+ gammadelta T cells in the absence of Itk results in elevated IgE production. *Blood* **114:** 564–571.

Raberger J, Schebesta A, Sakaguchi S, Boucheron N, Blomberg KE, Berglof A, Kolbe T, Smith CI, Rulicke T, Ellmeier W. 2008. The transcriptional regulator PLZF induces the development of CD44 high memory phenotype T cells. *Proc Natl Acad Sci U S A* **105:** 17919–17924.

Rawlings DJ, Saffran DC, Tsukada S, Largaespada DA, Grimaldi JC, Cohen L, Mohr RN, Bazan JF, Howard M,

Copeland NG, et al. 1993. Mutation of unique region of Bruton's tyrosine kinase in immunodeficient XID mice. *Science* **261:** 358–361.

Readinger JA, Mueller KL, Venegas AM, Horai R, Schwartzberg PL. 2009. Tec kinases regulate T-lymphocyte development and function: new insights into the roles of Itk and Rlk/Txk. *Immunol Rev* **228:** 93–114.

Rhee SG. 2001. Regulation of phosphoinositide-specific phospholipase C. *Annu Rev Biochem* **70:** 281–312.

Sacristan C, Schattgen SA, Berg LJ, Bunnell SC, Roy AL, Rosenstein Y. 2009. Characterization of a novel interaction between transcription factor TFII-I and the inducible tyrosine kinase in T cells. *Eur J Immunol* **39:** 2584–2595.

Sahu N, Mueller C, Fischer A, August A. 2008a. Differential sensitivity to Itk kinase signals for T helper 2 cytokine production and chemokine-mediated migration. *J Immunol* **180:** 3833–3838.

Sahu N, Venegas AM, Jankovic D, Mitzner W, Gomez-Rodriguez J, Cannons JL, Sommers C, Love P, Sher A, Schwartzberg PL, et al. 2008b. Selective expression rather than specific function of Txk and Itk regulate Th1 and Th2 responses. *J Immunol* **181:** 6125–6131.

Schaeffer EM, Broussard C, Debnath J, Anderson S, McVicar DW, Schwartzberg PL. 2000. Tec family kinases modulate thresholds for thymocyte development and selection. *J Exp Med* **192:** 987–1000.

Schaeffer EM, Debnath J, Yap G, McVicar D, Liao XC, Littman DR, Sher A, Varmus HE, Lenardo MJ, Schwartzberg PL. 1999. Requirement for Tec kinases Rlk and Itk in T cell receptor signaling and immunity. *Science* **284:** 638–641.

Schaeffer EM, Yap GS, Lewis CM, Czar MJ, McVicar DW, Cheever AW, Sher A, Schwartzberg PL. 2001. Mutation of Tec family kinases alters T helper cell differentiation. *Nat Immunol* **2:** 1183–1188.

Seminario MC, Bunnell SC. 2008. Signal initiation in T-cell receptor microclusters. *Immunol Rev* **221:** 90–106.

Severin A, Joseph RE, Boyken S, Fulton DB, Andreotti AH. 2009. Proline isomerization preorganizes the Itk SH2 domain for binding to the Itk SH3 domain. *J Mol Biol* **387:** 726–743.

Shan X, Czar MJ, Bunnell SC, Liu P, Liu Y, Schwartzberg PL, Wange RL. 2000. Deficiency of PTEN in Jurkat T cells causes constitutive localization of Itk to the plasma membrane and hyperresponsiveness to CD3 stimulation. *Mol Cell Biol* **20:** 6945–6957.

Shan X, Wange RL. 1999. Itk/Emt/Tsk activation in response to CD3 cross-linking in Jurkat T cells requires ZAP-70 and Lat and is independent of membrane recruitment. *J Biol Chem* **274:** 29323–29330.

Siliciano JD, Morrow TA, Desiderio SV. 1992. itk, a T-cell-specific tyrosine kinase gene inducible by interleukin 2. *Proc Natl Acad Sci U S A* **89:** 11194–11198.

Smith-Garvin JE, Koretzky GA, Jordan MS. 2009. T cell activation. *Annu Rev Immunol* **27:** 591–619.

Sommers CL, Rabin RL, Grinberg A, Tsay HC, Farber J, Love PE. 1999. A role for the Tec family tyrosine kinase Txk in T cell activation and thymocyte selection. *J Exp Med* **190:** 1427–1438.

Sommers CL, Samelson LE, Love PE. 2004. LAT: a T lymphocyte adapter protein that couples the antigen receptor to downstream signaling pathways. *Bioessays* **26:** 61–67.

Streubel B, Vinatzer U, Willheim M, Raderer M, Chott A. 2006. Novel t(5;9)(q33;q22) fuses ITK to SYK in unspecified peripheral T-cell lymphoma. *Leukemia* **20:** 313–318.

Su YW, Zhang Y, Schweikert J, Koretzky GA, Reth M, Wienands J. 1999. Interaction of SLP adaptors with the SH2 domain of Tec family kinases. *Eur J Immunol* **29:** 3702–3711.

Takesono A, Horai R, Mandai M, Dombroski D, Schwartzberg PL. 2004. Requirement for Tec kinases in chemokine-induced migration and activation of Cdc42 and Rac. *Curr Biol* **14:** 917–922.

Tanaka N, Asao H, Ohtani K, Nakamura M, Sugamura K. 1993. A novel human tyrosine kinase gene inducible in T cells by interleukin 2. *FEBS Lett* **324:** 1–5.

Thomas JD, Sideras P, Smith CI, Vorechovsky I, Chapman V, Paul WE. 1993. Colocalization of X-linked agammaglobulinemia and X-linked immunodeficiency genes. *Science* **261:** 355–358.

Tomlinson MG, Heath VL, Turck CW, Watson SP, Weiss A. 2004. SHIP family inositol phosphatases interact with and negatively regulate the Tec tyrosine kinase. *J Biol Chem* **279:** 55089–55096.

Tsukada S, Saffran DC, Rawlings DJ, Parolini O, Allen RC, Klisak I, Sparkes RS, Kubagawa H, Mohandas T, Quan S, et al. 1993. Deficient expression of a B cell cytoplasmic tyrosine kinase in human X-linked agammaglobulinemia. *Cell* **72:** 279–290.

Tsygankov AY. 2003. Non-receptor protein tyrosine kinases. *Front Biosci* **8:** 595–635.

van de Weyer PS, Muehlfeit M, Klose C, Bonventre JV, Walz G, Kuehn EW. 2006. A highly conserved tyrosine of Tim-3 is phosphorylated upon stimulation by its ligand galectin-9. *Biochem Biophys Res Commun* **351:** 571–576.

Vetrie D, Vorechovsky I, Sideras P, Holland J, Davies A, Flinter F, Hammarstrom L, Kinnon C, Levinsky R, Bobrow M, et al. 1993. The gene involved in X-linked agammaglobulinaemia is a member of the src family of protein-tyrosine kinases. *Nature* **361:** 226–233.

Vicari AP, Mocci S, Openshaw P, O'Garra A, Zlotnik A. 1996. Mouse gamma delta TCR+NK1.1+ thymocytes specifically produce interleukin-4, are major histocompatibility complex class I independent, and are developmentally related to alpha beta TCR+NK1.1+ thymocytes. *Eur J Immunol* **26:** 1424–1429.

Wilcox HM, Berg LJ. 2003. Itk phosphorylation sites are required for functional activity in primary T cells. *J Biol Chem* **278:** 37112–37121.

Williams JC, Wierenga RK, Saraste M. 1998. Insights into Src kinase functions: structural comparisons. *Trends Biochem Sci* **23:** 179–184.

Wilson CB, Rowell E, Sekimata M. 2009. Epigenetic control of T-helper-cell differentiation. *Nat Rev Immunol* **9:** 91–105.

Winslow MM, Neilson JR, Crabtree GR. 2003. Calcium signalling in lymphocytes. *Curr Opin Immunol* **15:** 299–307.

Xu W, Doshi A, Lei M, Eck MJ, Harrison SC. 1999. Crystal structures of c-Src reveal features of its autoinhibitory mechanism. *Mol Cell* **3:** 629–638.

Xu W, Harrison SC, Eck MJ. 1997. Three-dimensional structure of the tyrosine kinase c-Src. *Nature* **385:** 595–602.

Yamada N, Kawakami Y, Kimura H, Fukamachi H, Baier G, Altman A, Kato T, Inagaki Y, Kawakami T. 1993. Structure and expression of novel protein-tyrosine kinases, Emb and Emt, in hematopoietic cells. *Biochem Biophys Res Commun* **192:** 231–240.

Yang W, Malek SN, Desiderio S. 1995. An SH3-binding site conserved in Bruton's tyrosine kinase and related tyrosine kinases mediates specific protein interactions in vitro and in vivo. *J Biol Chem* **270:** 20832–20840.

Yang WC, Ching KA, Tsoukas CD, Berg LJ. 2001. Tec kinase signaling in T cells is regulated by phosphatidylinositol 3-kinase and the Tec pleckstrin homology domain. *J Immunol* **166:** 387–395.

Yang XO, Chang SH, Park H, Nurieva R, Shah B, Acero L, Wang YH, Schluns KS, Broaddus RR, Zhu Z, et al. 2008. Regulation of inflammatory responses by IL-17F. *J Exp Med* **205:** 1063–1075.

Yu L, Mohamed AJ, Vargas L, Berglof A, Finn G, Lu KP, Smith CI. 2006. Regulation of Bruton tyrosine kinase by the peptidylprolyl isomerase Pin1. *J Biol Chem* **281:** 18201–18207.

Zhang W, Sloan-Lancaster J, Kitchen J, Trible RP, Samelson LE. 1998. LAT: the ZAP-70 tyrosine kinase substrate that links T cell receptor to cellular activation. *Cell* **92:** 83–92.

An Enigmatic Tail of CD28 Signaling

Jonathan S. Boomer and Jonathan M. Green

Department of Internal Medicine, Washington University School of Medicine, St Louis, Missouri 63110

Correspondence: jgreen@im.wustl.edu

CD28 costimulation regulates a wide range of cellular processes, from proliferation and survival to promoting the differentiation of specialized T-cell subsets. Since first being identified over 20 years ago, CD28 has remained a subject of intense study because of its profound consequences on T cell function and its potential for therapeutic manipulation. In this review we highlight the signaling cascades initiated by the major signaling motifs in CD28, focusing on PI-3 kinase-dependent and -independent pathways and how these are linked to specific cellular outcomes. Recent studies using gene targeted knockin mice have clarified the relative importance of these motifs on in vivo immune responses; however, much remains to be elucidated. Understanding the mechanism behind costimulation holds great potential for development of new clinically relevant reagents, a fact beginning to be realized with the advent of drugs that prevent CD28 ligation and signaling.

OVERVIEW OF CD28-MEDIATED COSTIMULATION

A critical component of the immunological response to foreign protein is the activation of T lymphocytes. T-cell receptor (TCR) recognition of peptide loaded MHC molecules provides antigen specificity and initiates the required steps for T-cell activation, although additional signals are needed for complete T-cell activation. A two signal model for lymphocyte activation was first proposed by Bretscher and Cohn in 1970 as a mechanism for self:nonself discrimination in B cells (Bretscher and Cohn 1970) and was subsequently expanded to incorporate CD8 T-cell activation by Lafferty et al (Lafferty and Cunningham 1975). Using T-cell clones, Schwartz and Jenkins showed that engagement of the TCR by peptide loaded MHC in the absence of other signals is insufficient to activate the T cell, and in fact may render it unresponsive to further antigenic stimulation, a condition termed anergy (Quill and Schwartz 1987; Mueller et al. 1989; Schwartz et al. 1989; Jenkins et al. 1990; Linsley and Ledbetter 1993). They further showed that provision of additional, non-MHC restricted signals provided by professional APC, resulted in the T cell becoming fully activated leading to clonal expansion and development of effector function. Thus, the two-signal model, where signal one is defined as the binding of peptide/MHC complexes and signal two refers to the additional, costimulatory signal(s), provided a potential mechanism for peripheral tolerance (June et al. 1990; June et al. 1994; Bluestone 1995). The most studied and well characterized costimulatory receptor/ligand

 doi: 10.1101/cshperspect.a002436
Cite this article as *Cold Spring Harb Perspect Biol* 2010;2:a002436

complex is the CD28/B7 interaction (as reviewed in Rudd and Schneider 2003; Miller et al. 2009; Paterson et al. 2009; Rudd et al. 2009; Sharpe 2009).

CD28 was first identified by antibodies recognizing a 44 kDa protein on the surface of human T cells that synergized with PHA in the induction of a proliferative response (Hansen et al. 1980; Gmünder and Lesslauer 1984). The cDNA was expression cloned in 1987 by Aruffo and Seed and shown to encode a 23 kD type 1 transmembrane protein (Aruffo and Seed 1987). CD28 is expressed on the cell surface as a glycosylated, disulfide-linked homodimer (Aruffo and Seed 1987). In humans, CD28 is expressed on approximately 80% of CD4+ T cells and 50% of CD8+ T cells and has been detected on plasma cells, neutrophils, and eosinophils, although the function in these latter cell types is unclear (Lee et al. 1990; Venuprasad et al. 2001; Woerly et al. 2002). In contrast to humans, in mouse 100% of both CD4 and CD8 T cells express CD28 (Gross et al. 1990).

Most important in the initial activation of naïve T cells, ligation of CD28 on CD4+ T cells has diverse and profound consequences. CD28 signaling increases the sensitivity of the T cell to antigen receptor engagement, and as a result proliferation is induced at otherwise submitogenic concentrations of antigen (Gmünder and Lesslauer 1984; Damle et al. 1988; Damle and Doyle 1989; Shahinian et al. 1993b; Bachmann et al. 1996; Bachmann et al. 1997). Cytokine production is greatly increased, most significantly IL-2. CD28 costimulation results in a 50-fold increase in IL-2 secretion by both transcriptional and posttranscriptional regulation of expression (Lindsten et al. 1989; Fraser et al. 1991). Interestingly, under some conditions, this has been found to be resistant to suppression by cyclosporine, suggesting engagement of a novel signal transduction cascade (Thompson et al. 1989; Ledbetter et al. 1990). In addition, cell survival is enhanced by CD28 costimulation, in part by inducing expression of anti-apoptotic proteins including Bcl-X_L (Boise et al. 1995; Radvanyi et al. 1996; Sperling et al. 1996). Thus, by a variety of mechanisms, CD28 mediated costimulation greatly enhances the T cells effector response to antigen

Whether CD28 can activate signaling independent of TCR engagement or function only in a TCR dependent manner has been controversial. The ability of certain "superagonistic" anti-CD28 antibodies to independently activate T cells suggests that in some circumstances CD28 can function in the absence of a TCR derived signal (Tacke et al. 1997; Luhder et al. 2003). Furthermore, microarray analysis of T cells stimulated only through CD28 supported TCR independent signaling; however, this effect appeared to be relatively short lived (Riley et al. 2002; Marinari et al. 2004). Importantly, the existence of both TCR dependent and independent pathways is not mutually exclusive, and it is likely that both are operative and important.

CD28 is the founding member of the immunoglobulin (Ig) family of costimulatory receptors (Hansen et al. 1980; Gmünder and Lesslauer 1984), that now includes the receptors: cytotoxic T lymphocyte-associated antigen 4 (CTLA-4, CD152) (Dariavach et al. 1988), inducible costimulator (ICOS) (Hutloff et al. 1999; Tezuka et al. 2000), programmed death receptor 1 (PD-1) (Ishida et al. 1992), and B- and T-lymphocyte attenuator (BTLA) (Watanabe et al. 2003). Although CD28 and CTLA-4 both bind the ligands B7-1 (CD80) and B7-2 (CD86) (Freeman et al. 1989; Balzano et al. 1992; Azuma et al. 1993; Freeman et al. 1993) and are located on human chromosome 2q33 (Naluai et al. 2000) (mouse chromosome 1 (Gross et al. 1990; Howard et al. 1991)), CTLA-4 forms bivalent homodimers between CD80 molecules which contributes to its higher affinity for CD80 (12 compared to a 200 nM Kda value for CD28) (van der Merwe et al. 1997; Evans et al. 2005). CD28 and ICOS predominately enhance T-cell activation unlike CTLA-4, PD-1 and BTLA which are inhibitory (Rudd and Schneider 2003; Miller et al. 2009; Paterson et al. 2009; Rudd et al. 2009; Sharpe 2009). This review will cover the diverse functions of the CD28-B7 interaction highlighting the complex signaling events that define CD28 function.

THE MOTIFS BEHIND CD28 SIGNALING

The extracellular domain of CD28 binds to B7 proteins using a MYPPPY motif and this interaction initiates the costimulatory signal transduction cascade (Freeman et al. 1989; Balzano et al. 1992; Azuma et al. 1993; Freeman et al. 1993; Rudd and Schneider 2003). CD28 has a highly conserved, relatively short cytoplasmic tail (41 aa in human, 38 aa in mouse) that has no intrinsic enzymatic activity. However, several motifs have been identified including: four tyrosine residues, four serine and two threonine residues, two PxxP motifs, and two lysine residues each of which may be important in function (Rudd and Schneider 2003; Rudd et al. 2009). Phosphorylation of the tyrosine residues provides docking sites for src homology-2 (SH2) domain containing proteins whereas the proline-rich motifs can bind src homology-3 (SH3) domain containing proteins. The serine and threonine residues can also be phosphorylated and finally the lysine residues provide potential sites for ubiquitination. Interestingly CD28, CTLA-4, and ICOS share an YxxM motif (Rudd and Schneider 2003) a consensus binding site for the p85 subunit of the lipid kinase phosphatidyl-inositol 3-kinase (PI3K) (August and Dupont 1994; Pagès et al. 1994; Prasad et al. 1994). CD28 contains the sequence YMNM, where the Asn residue in the +2 position confers additional specificity for Grb2/GADS binding (Songyang et al. 1993; Schneider et al. 1995b), whereas the methionine at the +3 position confers p85 specificity (Takeda et al. 2008). The ability to bind Grb2 is absent from both ICOS and CTLA-4, which may in part explain the signaling and functional differences between these coreceptors (Rudd and Schneider 2003; Rudd et al. 2009). Early work has shown that the cytoplasmic tail of CD28 is tyrosine phosphorylated presumably by the Src family kinases Fyn and Lck (Raab et al. 1995; King et al. 1997), as well as phosphorylated on serine/threonine residues (Hutchcroft and Bierer 1994; Hutchcroft et al. 1996; Parry et al. 1997a). These events initiate specific protein–protein interactions and subsequent activation of downstream signaling cascades that define the costimulatory functions of CD28.

Motif Specific Protein–Protein Interactions

In the simplest format, CD28 signaling is initiated by activation of two dominant signaling cascades, one that requires phosphorylation of a tyrosine residue within the membrane proximal YMNM motif and subsequent binding of the p85 subunit of PI3K whereas another pathway is initiated by the more distal proline-rich regions (Rudd and Schneider 2003; Miller et al. 2009; Paterson et al. 2009; Rudd et al. 2009; Sharpe 2009) (Fig. 1). Each of these bind distinct sets of proteins, although there is

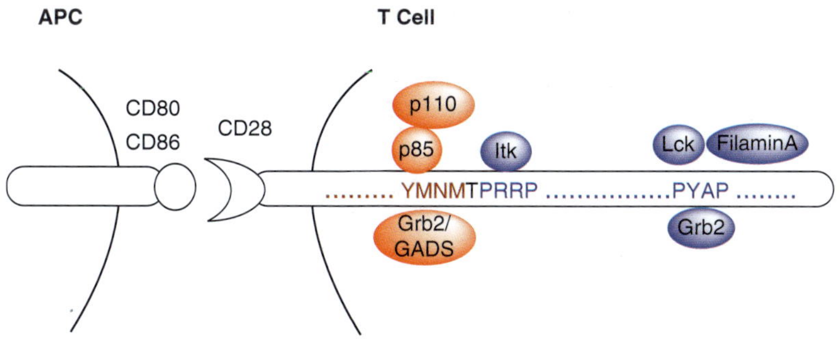

Figure 1. Motif specific protein:protein interactions with the cytoplasmic tail of CD28. Engagement of CD28 by CD80 or CD86 expressed on the antigen presenting cell (APC), initiates signal transduction cascades dependent on specific association of proteins with the cytoplasmic tail of CD28. The proximal YMNM motif when phosphorylated via Src family kinases binds the SH2 containing (indicated in red) proteins p85, a subunit of PI3K, and Grb2 or GADS via their SH2 domain. The distal proline-rich motifs bind the SH3 containing (indicated in blue) proteins Itk, at the sequence PRRP, and Grb2 (via its SH3 domain), filamin-A and Lck at the sequence PYAP.

significant overlap and potential for interaction. The observation that PI3K activation was enhanced via CD28 signaling led to the discovery that the p85 subunit of PI3K directly binds the YMNM motif via an SH2 interaction following phosphorylation of the tyrosine residue (Pagès et al. 1994; Truitt et al. 1994; Cai et al. 1995; Raab et al. 1995). Although PI3K-dependent signaling is likely the major pathway initiated by the proximal motif, its relative importance in CD28 function has been highly controversial. Initial in vitro CD28 signaling results should be interpreted with caution as the Jurkat T-cell line lacks both the PTEN and SHIP-1 phosphatases that function to inactivate the products of PI3K; therefore, the threshold to fully activate PI3K is reduced (Shan et al. 2000; Freeburn et al. 2002). Other proteins that can bind this region include Grb2 and GADS via their SH2 domain (Raab et al. 1995; Schneider et al. 1995a; Kim et al. 1998; Ellis et al. 2000). Competition between Grb2, GADS and PI3K for binding to the YMNM motif may be important in determining downstream CD28-dependent signaling events

Signaling by the distal PYAP motif is likely initiated by binding and activation of the Src family kinase Lck (King et al. 1997; Holdorf et al. 1999) although Fyn, Grb2, and GADS (via their SH3 domains) can also bind this region (Okkenhaug and Rottapel 1998; Ellis et al. 2000). The association of IL-2-inducible T-cell kinase (Itk) directly with the proximal proline-rich region (PRRP) of CD28 via its SH3 domain (Marengere et al. 1997) has been shown; however, if this event requires PI3K-dependent production of D3-lipids remains poorly understood. The PYAP motif also associates with filamin-A, an actin binding protein that functions as a scaffold for lipid raft formation (Tavano et al. 2006).

The generation of two specific knockin mice by our laboratory, the CD28 Y170F and CD28 AYAA mice, has clarified the role(s) of these two motifs in CD28 function (Friend et al. 2006; Dodson et al. 2009). Mutation of the proximal YMNM motif, although abrogating PI3K binding and subsequent protein kinase B (PKB)/Akt activation has no discernible in vivo effect (Dodson et al. 2009). On the other hand, mutation of the distal proline motif (PYAP) resulted in impaired CD28-dependent functions that included proliferation, IL-2 and other cytokine secretion, and adaptive immune system defects (Friend et al. 2006; Dodson et al. 2009). These data provide evidence that the distal proline motif initiates a critical, nonredundant signaling pathway required for CD28 function, whereas the proximal tyrosine based motif may be dispensable.

PI3K-DEPENDENT SIGNALING PATHWAYS

Activation of PI3K induces the production of the D3-lipids, phosphatidylinositol (3,4)-bisphosphate (PIP2) and phosphatidylinositol (3,4,5)-triphosphate (PIP3) where PIP3 is the main TCR/CD28 induced D3-lipid (Sasaki et al. 2000; Wang and Rudd 2008) (Fig. 2). These D3-lipids recruit pleckstrin homology (PH) domain containing proteins including phosphoinositide-dependent kinase 1 (PDK1), PKB/Akt, and possibly Wiskott Aldrich Syndrome Protein (WASP), a sorting nexin (SNX9), Itk and Vav (Parry et al. 1997b; Costello et al. 2002; Harriague and Bismuth 2002; Badour et al. 2007). PKB can be phosphorylated and activated by PDK1 and is a particularly important PI3K-dependent effector protein in CD28 signaling (Vanhaesebroeck and Alessi 2000). Downstream targets of PKB are diverse and include glycogen synthase kinase 3 (GSK3), mTOR, cAMP responsive element binding protein-1 (CREB), Bcl-2 antagonist of cell death (BAD), Bcl-X_L, FOXO family of transcription factors, and inhibitor of nuclear factor-κB (IκB) (Vanhaesebroeck and Alessi 2000). Thus, the PI3K-dependent signaling pathway has the ability to regulate a diverse array of CD28 functions that include cell cycle progression, apoptosis, cellular metabolism, and IL-2 transcription.

SIGNALING PATHWAYS INITIATED BY BINDING OF ADAPTOR PROTEINS

In addition to the PI3K-dependent pathway, binding of adaptor proteins (i.e., Grb2 or GADS) to the proximal motif can initiate

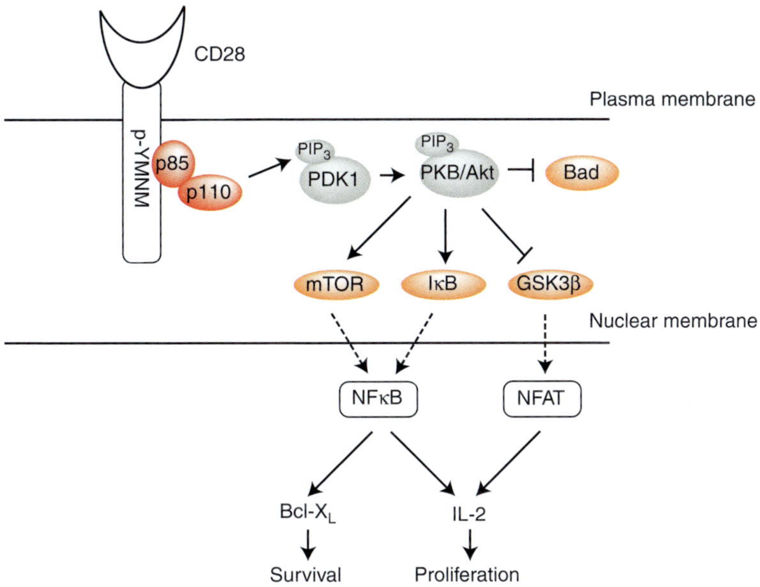

Figure 2. PI3K dependent signal pathway. Tyrosine phosphorylation of the YMNM motif via Src family kinases initiates the binding of the p85 subunit of PI3K. PI3K activity leads to the production of D-3 lipids, which recruit proteins via their pleckstrin homology domain (PH), including PDK1 and PKB/Akt. Once PKB is phosphorylated by PDK1, PKB phosphorylates its downstream targets including mTOR, IκB, GSK3β and Bad. Active mTOR and IκB result in increased NF-κB transcriptional activity whereas the phosphorylation of Bad and GSK3β results in increased survival and NFAT transcriptional regulation, respectively. The activation of NF-κB and NFAT (indicated by the dotted lines) induces the transcription of both Bcl-X_L, a prosurvival factor, and IL-2, an important T-cell cytokine required for proliferation as well as other genes.

signaling (Fig. 3). Grb2 recruits Son of Sevenless (Sos), a guanine nucleotide exchange factor and activator of GTPase $p21^{ras}$, and Vav, also an exchange factor (Schneider et al. 1995a; Collins et al. 1997; Bustelo 2000). A mechanism by which CD28 via Grb2 binding to the PxxP motif mediates JNK activation is plausible where Grb2 recruits Sos and Vav which in turn phosphorylate and activate Ras-related C3 botulinum toxin substrate (Rac1), PKCθ and cell division cycle 42 (CDC42) (Collins et al. 1997; Bustelo 2000). Rac1 and CDC42 activate the MEK4/MEK7 to MEKK1 signaling cascade culminating in JNK activation (Su et al. 1994). In support of this model, CD28 induces a GTP-RAC complex, possibly through the phosphorylation of Vav by ZAP-70 (Salojin et al. 1999), and activates MEKK1 in a Vav-dependent manner (Marinari et al. 2002). CD28-dependent activation of JNK (Su et al. 1994) is inhibited by mutant forms of Vav that lack functional GTP-GDP exchange activity and a dominant negative mutant of MEKK1 (Marinari et al. 2002; Wood et al. 2006). GADS potentially interacts with SLP76 and LAT in a signaling complex thereby providing CD28 a means to modulate TCR mediated signaling (Rudd and Schneider 2003; Rudd et al. 2009). Although these studies imply Grb2/GADS, Vav, PKCθ, and downstream MAPK signaling are linked to the PxxP motifs in CD28, this cascade has yet to be fully defined.

REGULATION OF IL-2 PRODUCTION BY CD28

Transcriptional Regulation of the IL-2 Gene by CD28

The regulation of IL-2 by CD28 is complex involving both increased gene transcription and post-transcriptional stabilization of mRNA

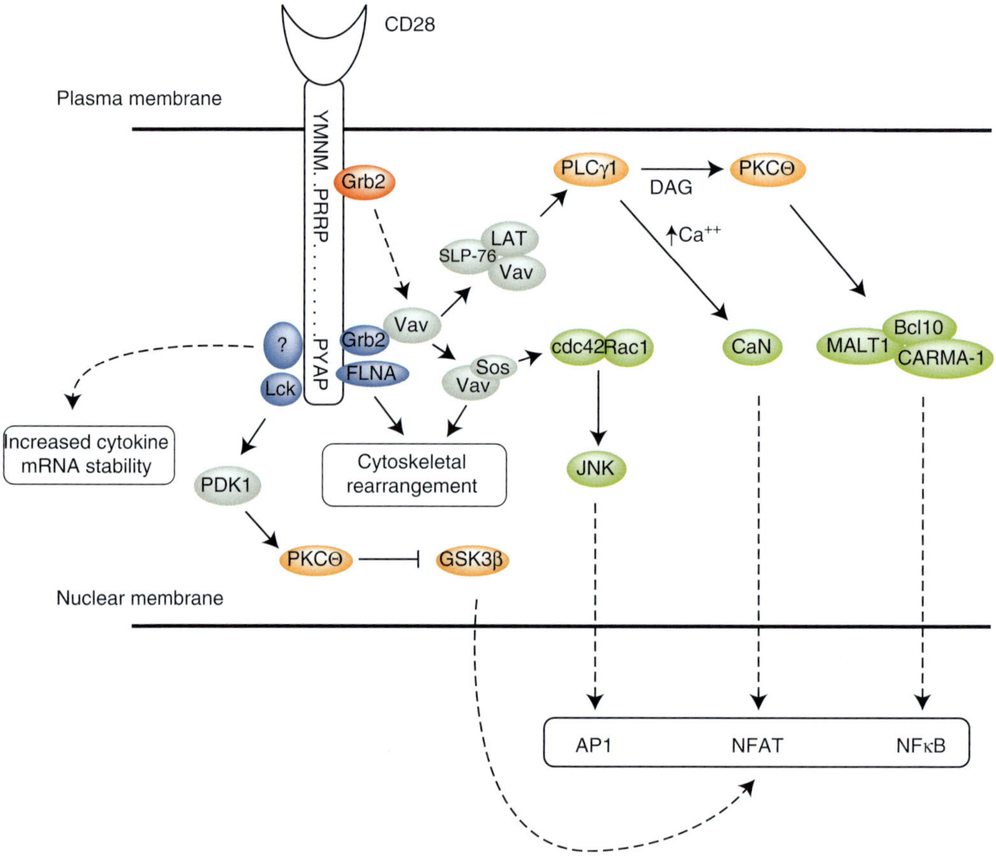

Figure 3. Adaptor protein intiated signaling events. The binding of the adaptor protein Grb2 to CD28 occurs via its SH2 domain at the proximal YMNM motif or its SH3 domain at the distal PYAP motif. Grb2 through an SH3 interaction may subsequently bind Vav which has the potential to initiate two signaling complexes. The Vav-Sos complex results in cdc42/Rac1 activation an initiator of cytoskeletal rearrangement and downstream MAPK activation (JNK) that induces the formation of the AP1 transcriptional complex. Vav also binds the SLP-76-LAT complex, which activates PLCγ1 a kinase that increases intracellular Ca^{2+}, through IP3 production, and activates PKCθ via DAG. The increase in intracellular Ca^{2+} results in calcineurin activation, a phosphatase that acts on NFAT allowing for its nuclear translocation. PKCθ activation leads to the formation of a multi-protein complex with Bcl10, MALT1 and CARMA-1 that induces NF-κB transcriptional activation. The distal PYAP motif also binds Lck a Src kinase that phosphorylates PDK1 which phosphorylates and activates PKCθ and subsequently inactivates GSK3β via phosphorylation resulting in enhanced transcription of NFAT–dependent genes. Stabilization of cytokine mRNA is also dependent on signals originating at the PYAP through a mechanism that may involve Lck binding (indicated via the dotted arrow) or an unknown protein whereas cytoskeletal rearrangement at the PYAP motif is filamin-A (FLNA) dependent.

(Lindsten et al. 1989; Bohjanen et al. 1992). The IL-2 promoter has binding elements for members of the NFAT, NF-κB, Oct-1, and AP-1 (c-fos/Jun) families of transcription factors and CD28 regulates the expression and activity of NFAT, NF-κB, and AP-1 (Fraser et al. 1991; Granelli-Piperno and Nolan 1991; Shapiro et al. 1997; Miller et al. 2009). Research pioneered by the Weiss laboratory identified a specific region in the IL-2 promoter, termed the CD28-response element (CD28RE) that binds transcription factors in a CD28-dependent manner thus providing a site for signal integration (Fraser et al. 1991; Shapiro et al. 1997). Analyses of the CD28RE/AP-1 complex determined that CD28 signaling activated both IκBα and IκBβ resulting in increased transcription of IL-2, which was inhibited by dominant negative mutants of these IκB proteins (Lai and Tan 1994; Harhaj et al. 1996; Harhaj and Sun 1998; Zhou et al. 2002). The phosphorylation of the IκB complex, which targets IκB for degradation through ubiquitination, via upstream serine/threonine kinases called the IκB kinases (IKKs), results in NF-κB translocation to the nucleus (Lai and Tan 1994; Harhaj et al. 1996; Harhaj and Sun 1998). CD28-dependent IKK activation was inhibited by a dominant negative MEKK1 and IKK deficient cells have defective IL-2 production implicating CD28 as an upstream mediator of the IKK complex and NF-κB activation (Harhaj et al. 2000; Tuosto et al. 2000). Through a mutational screen, mice identified as defective in antigen receptor activation of JNK and NF-κB were subsequently determined to also be defective in caspase recruitment domain-containing membrane associated guanylate kinase protein-1 (CARMA-1) activation (Jun et al. 2003). CARMA-1 binds Bcl10 via CARD-CARD binding domains and activation of CARMA-1 is enhanced by PKCθ phosphorylation (Gaide et al. 2002; Takeda et al. 2008). Bcl10 deficient T cells also show impaired NF-κB activation; therefore, a CD28-dependent signaling cascade can be predicted for NF-κB that includes PKCθ, Bcl10 and CARMA-1 proteins (Gaide et al. 2002; Takeda et al. 2008). The series of events that initiate this complex is not well understood, but the oligimerization of the CARMA-1 protein through its coiled-coil domain (CC2) and its localization via its CC1 domain are involved (Tanner et al. 2007) along with a lysine residue at 808 of CARMA-1 (Wang et al. 2004).

Analyses using CD28 deficient T cells determined that CD28 is required for proper segregation of PKCθ into the immunological synapse and the failure to localize PKCθ correctly results in poor activation of NF-κB and IL-2 gene transcription (Huang et al. 2002; Sanchez-Lockhart et al. 2004; Sanchez-Lockhart and Miller 2006). Mice or T cell lines deficient in PKCθ have a pronounced defect in NF-κB activation through TCR-CD28 signals (Sun et al. 2000; Matsumoto et al. 2005). In Jurkat T cells Vav has been shown to bind PKCθ and acts synergistically to enhance CD28-dependent activation of NF-κB (Coudronniere et al. 2000; Dienz et al. 2000; Hehner et al. 2000a; Hehner et al. 2000b; Villalba et al. 2000a). Results indicate the CARMA-1, Bcl10, and mucosa-associated lymphoid tissue (MALT) lymphoma translocation gene 1 (MALT1) complex is dependent on CD28-dependent PKCθ translocation and activation, possibly through Vav and Grb2, to regulate NF-κB activation (Wang et al. 2002; Hara et al. 2003; Jun et al. 2003; Wang et al. 2004; Tanner et al. 2007). CD28 activation of NF-κB may involve RIP-2, a CARD containing adaptor also called CARDIAK or RICK, as RIP-2 deficient T cells have impaired IL-2 production, proliferation, and NF-κB transcription (Kobayashi et al. 2002). CD28 mediated activation of NF-κB may include the role of PKB, NIK, and Cot; however, this remains unclear as these experiments used overexpression of mutant proteins or resulted in partial inhibition of CD28 functions (Kane et al. 1999; Lin et al. 1999; Hehner et al. 2000b; Yamada et al. 2000; Bauer et al. 2001).

Regulation of NFAT by CD28 is another costimulatory pathway that augments IL-2 expression (Fraser et al. 1991). TCR and CD28-dependent signaling pathways cooperate in the activation of PLCγ-1 leading to increased intracellular calcium levels, PKC phosphorylation and activation of calcineurin (Ca^{2+}-dependent Ser/Thr phosphatase, CaN), which then dephosphorylates NFAT enhancing its nuclear translocation (Fraser et al. 1991). The immunosuppressant drugs cyclosporine and tacrolimus block CaN activation and thereby prevent NFAT translocation (Fraser et al. 1991). In contrast, CD28 derived signals induce the phosphorylation and inactivation of GSK3β, a kinase that promotes the nuclear export of NFAT (Diehn et al. 2002; Pan et al. 2007). By this mechanism, CD28 leads to NFAT nuclear trapping and increased transcription of NFAT-dependent genes, including IL-2 (Diehn et al. 2002; Pan et al. 2007).

Postranscriptional Regulation of IL-2 mRNA Expression

In addition to transcriptional control of the IL-2 gene, CD28 regulates the stability of IL-2 and other cytokine mRNA (Lindsten et al. 1989; Miller et al. 2009). Specific AUUA sequences in the 3′ untranslated region destabilize the message (Bohjanen et al. 1992). Briefly, there are two general models of mRNA stability 1) microRNA and 2) AU-rich elements (ARE) within the 3′ UTR of the mRNA that affect degradation of the RNA especially that of cytokine genes including IL-2 (Lindsten et al. 1989). In unstimulated cells AU-binding proteins bind the 3′ UTR and recruit a multicomponent exosome that deadenylates and digests the 3′ UTR of the mRNA via exonucleases (Ogilvie et al. 2005). T-cell stimulation induces TTP activation, an AU-binding protein, and subsequent binding to the ARE within the 3′ UTR of IL-2 mRNA that induces degradation (Ogilvie et al. 2005). The absence of TTP results in increased IL-2 expression after T-cell activation (Ogilvie et al. 2005). To counter TTP induced degradation, a CD28-dependent mechanism may require HuR translocation to the cytoplasm (Seko et al. 2004) and/or elements of MAPK activation (i.e., JNK and p38) (Chen et al. 1998; Winzen et al. 1999; Chen et al. 2000). As CD28 regulates JNK activation (Su et al. 1994; Marinari et al. 2002; Wood et al. 2006), YB-1 and nucleolin, JNK induced proteins, stabilize the IL-2 transcript at the 5′ end; however, these proteins are not sufficient and may require another protein that binds to the 3′ UTR (Chen et al. 2000). One possible candidate, NF90, a transcription factor associated with NFAT and IL-2 activation (Corthesy and Kao 1994), has been shown to compete with TTP for binding at the ARE of IL-2 mRNA (Shim et al. 2002). NF90 knockout T cells have decreased IL-2 secretion but whether this effect is on the regulation of IL-2 transcription or mRNA stability, as well as, how CD28 signaling affects NF90 function is unclear (Shi et al. 2007). T-cell activation shuttles NF90 from the nucleus to the cytoplasm that dissociates TTP from the ARE and results in IL-2 mRNA stabilization (Shim et al. 2002; Shi et al. 2007).

Attempts to dissect the signaling pathway from CD28 to IL-2 using a mutational approach have yielded inconsistent results. Early experiments in Jurkat cells showed an association of PI3K with CD28 but the dependency of IL-2 secretion on this association were conflicting (Stein et al. 1994; Truitt et al. 1994; Crooks et al. 1995; Lu et al. 1995; Truitt et al. 1996; Sadra et al. 1999). A role for the YMNM motif in IL-2 secretion was suggested using murine cell lines but whether this was mediated through Grb2 or PI3K was not clear (Pagès et al. 1994; Cai et al. 1995; Kim et al. 1998; Watanabe et al. 2006). The role of the YMNM motif in IL-2 secretion remained unclear from studies in transgenic mice or retroviral reconstitution on the CD28-deficient background (Burr et al. 2001b; Harada et al. 2001; Okkenhaug et al. 2001; Andres et al. 2004b). These data showed either no effect on proliferation and IL-2 secretion (Okkenhaug et al. 2001; Andres et al. 2004b), a partial inhibition (Burr et al. 2001b), or inhibition of proliferation and IL-2 secretion at early time points (Harada et al. 2001). A caveat of these experiments is that CD28 expression is not under endogenous control but that of a heterologous promoter.

Recent data has supported the PYAP motif as the dominant signaling motif determining the net output of IL-2 following CD28 costimulation. In a retroviral reconstitution system, the proline to alanine mutation within the PYAP motif resulted in either decreased CD28-dependent proliferation and IL-2 production (Burr et al. 2001b) or no effect unless this mutation is combined with the tyrosine mutation in the YMNM motif (Andres et al. 2004b). In transgenic mice, the PYAP motif had impaired CD28-dependent proliferation whereas the Itk binding proline motif (PRRP) did not (Gogishvili et al. 2008). Interestingly, in a PI3K mutant mouse ($p110\delta^{D910A/D910A}$) CD28-dependent proliferation was maintained whereas Vav deficient primary T cells had defective proliferation (Gogishvili et al. 2008). These results predict that the PYAP motif mediates CD28-dependent proliferation through Vav and Lck independently of PI3K, although the role of Grb2 or GADS (Watanabe et al. 2006) was not directly addressed.

We directly tested the relative importance of the YMNM and the PYAP motifs in regulating CD28-dependent IL-2 production using gene targeted knockin mice. This approach benefits from the study of truly naïve T cells in which CD28 expression is controlled by endogenous elements. Under these conditions the PYAP motif has a critical role in CD28 mediated proliferation and IL-2 secretion (Friend et al. 2006; Dodson et al. 2009). Importantly, these studies also showed that CD28 expression level is an important determinant of the outcome of engagement as mice with a single AYAA allele had greater impairment in both proliferation and IL-2 secretion than mice with two AYAA alleles. In contrast, the Y170F mutation had little appreciable effect independent of gene dosage (Dodson et al. 2009). There is segregation of transcriptional control of IL-2 expression and regulation of mRNA stability by the YMNM and PYAP motifs. The Y170F mutation impairs IL-2 transcription but has little effect on net IL-2 secretion and no effect on mRNA stability whereas mutation of the PYAP motif has the reciprocal effect (Sanchez-Lockhart et al. 2004; Miller et al. 2009) (X. Wang, J. Green and J. Miller, unpublished data). The PYAP motif can mediate Grb2 binding and downstream JNK activation; therefore, these data give credence to the theory that CD28-dependent JNK activation is the dominant pathway resulting in IL-2 mRNA stability.

The CD28 knockin mice also provided an opportunity for us to directly examine components of the signaling cascade leading to IL-2 production in naïve, primary T cells. Previous studies showed that the YMNM motif was required for CD28-dependent activation of PKB, which in turn led to phosphorylation and inactivation of GSK3β resulting in enhanced IL-2 gene transcription (Ohteki et al. 2000; Appleman et al. 2002; Diehn et al. 2002; Wood et al. 2006). However, data from the knockin mice showed the phosphorylation of both PKCθ and GSK3β were dependent on the PYAP and not the YMNM motif (Dodson et al. 2009). An alternative mechanism of the phosphorylation of GSK3β has been proposed where PKC isoforms directly phosphorylate GSK3β possibly through Vav or PDK1 (Lee et al. 2005) independent of PKB (Goode et al. 1992; Fang et al. 2002; Vilimek and Duronio 2006), a model supported by the knockin data.

CD28 SIGNALING AND THE IMMUNOLOGICAL SYNAPSE

The organization of specific proteins within the contact site of the APC and T cell during T-cell activation is referred to as the immunological synapse (Grakoui et al. 1999; Dustin and Chan 2000; van der Merwe 2002). Proteins segregate into supramolecular activation clusters (SMAC), which contain three distinct domains: the cSMAC ("c" indicates central), enriched in TCR, CD4, CD28, and signaling components such as PKCθ and Lck, the pSMAC ("p" indicates peripheral), enriched in LFA-1 and other cytoskeletal components and an outer region that contains mostly CD45 (Grakoui et al. 1999; Bromley et al. 2001; Bunnell et al. 2002; Freiberg et al. 2002). Initially, TCR and CD4 form small microclusters that also contain CD28 within the cSMAC whereas LFA-1 moves into the pSMAC (Grakoui et al. 1999; Krummel et al. 2000; Andres et al. 2004a). Although how the activation of Lck and its role in CD28 signaling is poorly understood, the colocalization of CD45, a phosphatase that activates Lck by removing an inhibitory phosphate (Y505), with CD4 associated Lck and CD28 in these microclusters provides a possible model of Lck activation and association with CD28 (Johnson et al. 2000; Freiberg et al. 2002).

CD28 is required for the localization of PKCθ to the immunologic synapse (Monks et al. 1998; Bromley et al. 2001; Andres et al. 2004a). Mutational analysis mapped this requirement to the YMNM motif, suggesting a PI3K-dependent mechanism (Sanchez-Lockhart et al. 2004; Sanchez-Lockhart and Miller 2006; Yokosuka et al. 2008). In recent experiments, the ability of CD28 to recruit PKCθ to the immunological synapse was subsequently determined to be independent of PIP3 generation (Garcon et al. 2008). Interestingly PKCθ colocalizes with filamin-A, and when filamin-A is knocked down via siRNA, PKCθ activity, as well as, NF-κB, NFAT/AP-1, and IL-2 transcription are reduced (Hayashi and Altman 2006). As CD28 has the ability to modulate the cytoskeletal components, Rho (Kaga et al. 1998) SLP76, Rac1 (Michel et al. 2000; Raab et al. 2001), and PKCθ (Villalba et al. 2000a) in a Vav-dependent mechanism (Villalba et al. 2000b) and Vav has been shown to enable the TCR/CD3 complex to cluster (Fischer et al. 1998; Holsinger et al. 1998; Krawczyk et al. 2002), a CD28-dependent molecular signaling scaffold may be postulated at the immunological synapse. CD28 is tethered to the lipid raft via binding of filamin-A (Tavano et al. 2006) where CD28, through the YMNM motif, recruits PKCθ via PI3K activation (Sanchez-Lockhart et al. 2008). Thus, CD28 actin rearrangement functions through the binding of filamin-A, the downstream activation of Vav and PKCθ and the Vav effectors CDC42 and Rho-GTPases (Tavano et al. 2006). These data predict the movement of CD28 to the immunological synapse occurs before the majority of signaling events, through its PYAP motif, and is critical for CD28's ability to function.

After the PYAP motif has recruited CD28 to the immunological synapse, the binding of PI3K to the YMNM motif also modulates CD28 receptor mediated endocytosis. CD28 undergoes internalization, an event that requires the YMNM motif and binding of PI3K, which results in degradation via the lysosome or a recycling back to the cell surface (Cefai et al. 1998). CD28 internalization also requires the recruitment of WASP, via its SH3 domain, and a sorting nexin (SNX9) (Snapper and Rosen 1999; Badour et al. 2007). An SNX9 mutant that lacks the phagocyte oxidase homology (PX) domain that interacts with the p85 subunit of PI3K and PIP3 failed to induce CD28 internalization and NFAT transcription, whereas the overexpression of SNX9 increased these processes (Badour et al. 2007). These results indicate that CD28 recruitment and signaling is functionally linked to the active signaling environment of the immunological synapse.

CD28 ENHANCES T-CELL SURVIVAL

A major consequence of CD28 signaling is protection from cell death (Boise et al. 1995; Noel et al. 1996; Radvanyi et al. 1996; Sperling et al. 1996). The two dominant pathways of apoptosis in T cells are (1) receptor mediated (i.e., TNF family) or (2) mitochondrial-associated proteins (i.e., the Bcl family) both of which are modulated via CD28 signaling (Rudd et al. 2009). CD28 costimulation prevents activation induced cell death (AICD) by inhibiting the expression of CD95L expression (*cis* mediated death) on the T cell and by decreasing the formation of the death inducing signaling complex (DISC) (Kirchhoff et al. 2000). CD28 stimulation increases the protein expression of c-FLIPs which compete with procaspase 8 for binding to the death domains of CD95 and FADD thereby preventing the conversion of procaspase 8 to active caspase 8 and the induction of cell death (Kirchhoff et al. 2000). CD28-deficient cells are more sensitive to Fas-mediated cell death, which could be reversed by the overexpression of PKB (Jones et al. 2002). The overexpression of PKB prevented the binding and downstream conversion of procaspase 8 to caspase 8 through inhibition of the DISC complex (Jones et al. 2002). CD28 also enhances targets of glycolysis, including Glut1 expression and glucose uptake, all required for increased metabolism during cellular proliferation through the activation of PKB (Frauwirth et al. 2002).

CD28 stimulation also induces the up-regulation of antiapoptotic factors including Bcl-X_L (Boise et al. 1995). The overexpression of Bcl-X_L in CD28-deficient mice restores cellular survival but not proliferation (Dahl et al. 2000); thereby, separating these two CD28-dependent functions. Bcl-X_L, an NF-κB inducible gene, requires CD28-dependent IKK activation through the activation of PKCθ and formation of the CARMA-1, Bcl10, and MALT1 complex (Wang et al. 2002; Hara et al. 2003; Jun et al. 2003;

Wang et al. 2004; Tanner et al. 2007; Takeda et al. 2008). CD28-deficient mice reconstituted with a disrupted YMNM motif failed to induce Bcl-X_L whereas CD28-deficient mice reconstituted with disrupted proline rich motifs of CD28 retained the capacity to induce Bcl-X_L (Burr et al. 2001b; Harada et al. 2001; Okkenhaug et al. 2001). The overexpression of PKB led to constitutively elevated amounts of Bcl-X_L protein and inhibited FAS mediated apoptosis (Jones et al. 2000; Jones et al. 2002). These data suggested a central role for the YMNM motif, via PKB activation, in the regulation of cell survival by CD28. These results held true until the generation and analysis of CD28 knockin mice, which surprisingly indicated that neither mutation in the YMNM or PYAP motifs alone prevented CD28-dependent up-regulation of Bcl-X_L (Dodson et al. 2009). However, the expression of Bcl-X_L is dependent on both PKCθ and NF-κB activation, events that are regulated by CD28 and may not be exclusively downstream of the YMNM motif (Khoshnan et al. 2000; Marinari et al. 2004; Li et al. 2005; Manicassamy et al. 2006).

CD28 REGULATES T-CELL SUBTYPE DEVELOPMENT AND DIFFERENTIATION

Although the peripheral T-cell repertoire is not markedly altered in the absence of CD28, a role in negative selection has been suggested (Shahinian et al. 1993b; Lucas et al. 1995; Sperling et al. 1996; Noel et al. 1998). However a more dramatic phenotype has been uncovered with the identification of regulatory T cells (Tregs) (Boden et al. 2003; Tang et al. 2003; Tai et al. 2005; Guo et al. 2008). CD28-deficient mice have drastically reduced numbers of thymically derived Tregs (Salomon et al. 2000; Tang et al. 2003; Lohr et al. 2004; Tai et al. 2005). CD28 could contribute to Treg development enhancing IL-2 production (Guo et al. 2008) or by a cell intrinsic mechanism (Tai et al. 2005; Tao et al. 2005). However, the failure of wild-type cells to support the generation of CD28-deficient Tregs suggests that exogenous IL-2 is not sufficient in the absence of other CD28-dependent signals (Tai et al. 2005). Additional data recently obtained using the knockin mice supports that CD28, partially via the distal carboxy-terminal proline motif, provides an intrinsic signal necessary for the generation of cytokine responsive Treg precursors (C Lio, C Hsieh, and J Green, unpublished data).

As the differentiation of T cells into distinctive populations (TH0, TFh, TH1, TH2, and TH17) are defined in part by their cytokine secretion profiles, a role for CD28 in the generation and maintenance of these populations has been implicated. Although CD28 regulates multiple cytokines, it is neither absolutely required nor sufficient for the generation of any distinct T-cell subset. CD28 is important for TH2 generation but not the secretion of TH2 cytokines (King et al. 1995; Rulifson et al. 1997); however, TH1 generation does not require CD28 but the secretion of IL-2 by these cells requires CD28 (Shahinian et al. 1993a; Lucas et al. 1995; Villegas et al. 1999). Transgenic overexpression of PKB in CD28-deficient mice restored the production of IL-2 and IFN-γ but not IL-4 and IL-5 the cytokines required for the generation of TH2 cells (Kane et al. 2001). CD28-deficient mice show a defect in IL-4 production on receptor stimulation, which is partially restored by the overexpression of PDK1 (Andres et al. 2004b; Nirula et al. 2006). Although PDK1 is an upstream activator of PKB, in this case PDK1 was determined to be dependent on PKA and NFAT (Nirula et al. 2006). CD28-deficient and IL-4 deficient mice fail to induce germinal center formation and upregulation of CD40L in an Ova model of stimulation (Reiter and Pfeffer 2002). Interestingly, AYAA knockin mice also have defects in IL-4 production, germinal center formation, and decreased activation of PDK1 (Friend et al. 2006; Dodson et al. 2009). These data illustrate the complex role of the YMNM and PYAP motifs in the activation of PKB and PKA in TH2 type T cells and in the secondary immune response.

The development of a specialized cell type called NKT cells, as characterized by the expression of both a T-cell marker, invariant TCR (Vα14-Jβ18), and an NK marker (NK1.1) that respond to glycolipid α-galactosylceramide when presented by CD1d, is also dependent on CD28 signals (Williams et al. 2008). Through the use of CD28-deficient and B7-deficient mice, a distinctive role of CD28 in the thymic generation of these NKT cells was identified, as the absence of either CD28 or its ligand B7 resulted in severely decreased numbers of NKT cells (Williams et al. 2008). Paradoxically, the overexpression of B71 in the thymus of transgenic mice also resulted in decreased NKT cells (Williams et al. 2008). These results suggest that not only the presence of CD28 but the strength of signaling through CD28 coordinate in the development and differentiation of cell types.

THE ROLE OF CD28 COSTIMULATION IN DISEASE

Perhaps surprisingly, mutations in CD28 have not been identified as the genetic basis for any human disease. However, mouse models of disease have been shown to depend on CD28 function. The use of CD28- and B7-deficient cells has implicated this pathway in the generation of a TH2 response required for the destruction of islet cells (autoimmune diabetes in the NOD mouse model) (Lenschow et al. 1996) and in the activation of newly recruited T cells responsible for epitope spreading (relapsing EAE in the EAE mouse model) (Miller et al. 1995). CD28 has been shown to be required in a model of allergic airway inflammation (Harris et al. 1997a; Harris et al. 1997b; Keane-Myers et al. 1997; Mathur et al. 1999; Burr et al. 2001a). A variety of transplantation studies showed that a lack of CD28 could prolong graft acceptance, but in most cases the organs were eventually rejected (Turka et al. 1992; Sayegh and Turka 1998; Harada et al. 2001). However, when combined with blockade of CD40:CD40L interactions, indefinite graft acceptance was achieved (Larsen et al. 1996).

Examination of the role for the specific motifs of CD28 in disease models are limited; however, the PYAP motif was important in a model of autoimmune disease using CTLA-4-deficient mice (Tai et al. 2007), a functional YMNM motif is needed to generate a graft versus host response (Harada et al. 2001) whereas disease severity in experimental allergic encephalomyelitis is reduced in CD28-deficient mice and the CD28-AYAA knockin mice, but not the CD28-Y170F knockin mice (Friend et al. 2006; Dodson et al. 2009). In the absence of CD28 or in the CD28-AYAA knockin mice, antibody production and germinal center formation is markedly reduced (Friend et al. 2006; Dodson et al. 2009). Thus, in a number of model systems, CD28 plays a critical role.

An understanding of costimulation, and the mechanism by which it regulates immune responses provides new avenues for the development of therapeutics. As CD28 decreases the threshold required to activate naïve T cells, as well as increasing the survival of activated T cells, one can envision development of drugs that either potentiate or down-regulate an immune response. Although global inhibition of CD28 function would be anticipated to profoundly inhibit T-cell function, more specific targeting of specific regions of the extracellular or intracellular regions of CD28 (i.e., the B7 binding domain or the signaling motifs YMNM or PYAP) via antibodies, small molecule mimetics or siRNAs could potentially alter some aspects of CD28 effects whereas leaving others intact. This is beginning to be realized with the approval and marketing of a humanized version of CTLA4Ig (abatacept, Orencia™) for the treatment of rheumatoid arthritis in adults and Stills disease in children (Genovese et al. 2005) and is being explored for other conditions (Linsley and Nadler 2009; Podojil and Miller 2009). Second generation agents have shown promise in the prevention of transplant rejection (Vincenti et al. 2005). The potential for the manipulation of costimulatory signals as an approach for cancer immunotherapy is currently an area of active study (Waldmann and Morris 2006). Although the exact mechanism of action of CTLA4Ig and related compounds is not fully understood these agents have promise for providing greater specificity in the manipulation of the immune response with less toxicity than many currently used drugs.

ACKNOWLEDGMENTS

Supported in part by National Institutes of Health (NIH) grant HL062683 to JMG. We thank J. Miller, X. Wang, C. Lio and Chyi-Song Hsieh for willingness to share unpublished data.

REFERENCES

Andres PG, Howland KC, Dresnek D, Edmondson S, Abbas AK, Krummel MF. 2004a. CD28 signals in the immature immunological synapse. *J Immunol* **172:** 5880–5886.

Andres PG, Howland KC, Nirula A, Kane LP, Barron L, Dresnek D, Sadra A, Imboden J, Weiss A, Abbas AK. 2004b. Distinct regions in the CD28 cytoplasmic domain are required for T helper type 2 differentiation. *Nat Immunol* **5:** 435–442.

Appleman LJ, van Puijenbroek AA, Shu KM, Nadler LM, Boussiotis VA. 2002. CD28 costimulation mediates down-regulation of p27kip1 and cell cycle progression by activation of the PI3K/PKB signaling pathway in primary human T cells. *J Immunol* **168:** 2729–2736.

Aruffo A, Seed B. 1987. Molecular cloning of a CD28 cDNA by a high-efficiency COS cell expression system. *Proc Nath Acad Sci* **84:** 8573–8577.

August A, Dupont B. 1994. CD28 of T lymphocytes associates with phosphatidylinositol 3-kinase. *Int Immunol* **6:** 769–774.

Azuma M, Ito D, Yagita H, Okumura K, Phillips JH, Lanier LL, Somoza C. 1993. B70 antigen is a second ligand for CTLA-4 and CD28. *Nature* **366:** 76–79.

Bachmann MF, McKall-Faienza K, Schmits R, Bouchard D, Beach J, Speiser DE, Mak TW, Ohashi PS. 1997. Distinct roles for LFA-1 and CD28 during the activation of naive T cells: Adhesion versus costimulation. *Immunity* **7:** 549–557.

Bachmann MF, Sebzda E, Kündig TM, Shahinian A, Speiser DE, Mak TW, Ohashi PS. 1996. T cell responses are governed by avidity and co-stimulatory thresholds. *Eur J Immunol* **26:** 2017–2022.

Badour K, McGavin MK, Zhang J, Freeman S, Vieira C, Filipp D, Julius M, Mills GB, Siminovitch KA. 2007. Interaction of the Wiskott-Aldrich syndrome protein with sorting nexin 9 is required for CD28 endocytosis and cosignaling in T cells. *Proc Natl Acad Sci* **104:** 1593–1598.

Balzano C, Buonavista N, Rouvier E, Golstein P. 1992. CTLA-4 and CD28: similar proteins, neighbouring genes. *Int J Cancer Suppl* **7:** 28–32.

Bauer B, Krumbock N, Fresser F, Hochholdinger F, Spitaler M, Simm A, Uberall F, Schraven B, Baier G. 2001. Complex Formation and Cooperation of Protein Kinase Cθ and Akt1/Protein Kinase Bα in the NF-κ B Transactivation Cascade in Jurkat T Cells. *J Biol Chem* **276:** 31627–31634.

Bluestone JA. 1995. New perspectives of CD28-B7-mediated T cell costimulation. *Immunity* **2:** 555–559.

Boden E, Tang Q, Bour-Jordan H, Bluestone JA. 2003. The role of CD28 and CTLA4 in the function and homeostasis of CD4+CD25+ regulatory T cells. *Novartis Found Symp* **252:** 55–63; discussion 63-56, 106–114.

Bohjanen PR, Petryniak B, June CH, Thompson CB, Lindsten T. 1992. AU RNA-binding factors differ in their binding specificities and affinities. *J Biol Chem* **267:** 6302–6309.

Boise LH, Minn AJ, Noel PJ, June CH, Accavitti MA, Lindsten T, Thompson CB. 1995. CD28 costimulation can promote T cell survival by enhancing the expression of Bcl-x_L. *Immunity* **3:** 87–98.

Bretscher P, Cohn M. 1970. A theory of self-nonself discrimination. *Science* **169:** 1042–1049.

Bromley SK, Iaboni A, Davis SJ, Whitty A, Green JM, Shaw AS, Weiss A, Dustin ML. 2001. The immunological synapse and CD28-CD80 interactions. *Nat Immunol* **2:** 1159–1166.

Bunnell SC, Hong DI, Kardon JR, Yamazaki T, McGlade CJ, Barr VA, Samelson LE. 2002. T cell receptor ligation induces the formation of dynamically regulated signaling assemblies. *J Cell Biol* **158:** 1263–1275.

Burr JS, Kimzey SL, Randolph DR, Green JM. 2001a. CD28 and CTLA4 coordinately regulate airway inflammatory cell recruitment and T-helper cell differentiation after inhaled allergen. *Am J Respir Cell Mol Biol* **24:** 563–568.

Burr JS, Savage NDL, Messah GE, Kimzey SL, Shaw AS, Arch RH, Green JM. 2001b. Cutting Edge: Distinct Motifs Within CD28 Regulate T Cell Proliferation and Induction of Bcl-XL. *J Immunol* **166:** 5331–5335.

Bustelo XR. 2000. Regulatory and signaling properties of the Vav family. *Mol Cell Biol* **20:** 1461–1477.

Cai YC, Cefai D, Schneider H, Raab M, Nabavi N, Rudd CE. 1995. Selective CD28pYMNM mutations implicate phosphatidylinositol 3-kinase in CD86-CD28-mediated costimulation. *Immunity* **3:** 417–426.

Cefai D, Schneider H, Matangkasombut O, Kang H, Brody J, Rudd CE. 1998. CD28 receptor endocytosis is targeted by mutations that disrupt phosphatidylinositol 3-kinase binding and costimulation. *J Immunol* **160:** 2223–2230.

Chen CY, Del Gatto-Konczak F, Wu Z, Karin M. 1998. Stabilization of interleukin-2 mRNA by the c-Jun NH2-terminal kinase pathway. *Science* **280:** 1945–1949.

Chen CY, Gherzi R, Andersen JS, Gaietta G, Jurchott K, Royer HD, Mann M, Karin M. 2000. Nucleolin and YB-1 are required for JNK-mediated interleukin-2 mRNA stabilization during T-cell activation. *Genes Dev* **14:** 1236–1248.

Collins TL, Deckert M, Altman A. 1997. Views on Vav. *Immunol Today* **18:** 221–225.

Corthesy B, Kao PN. 1994. Purification by DNA affinity chromatography of two polypeptides that contact the NF-AT DNA binding site in the interleukin 2 promoter. *J Biol Chem* **269:** 20682–20690.

Costello PS, Gallagher M, Cantrell DA. 2002. Sustained and dynamic inositol lipid metabolism inside and outside the immunological synapse. *Nat Immunol* **3:** 1082–1089.

Coudronniere N, Villalba M, Englund N, Altman A. 2000. NF-κ B activation induced by T cell receptor/CD28 costimulation is mediated by protein kinase C-θ. *Proc Natl Acad Sci* **97:** 3394–3399.

Crooks ME, Littman DR, Carter RH, Fearon DT, Weiss A, Stein PH. 1995. CD28-mediated costimulation in the absence of phosphatidylinositol 3- kinase association and activation. *Mol Cell Biol* **15:** 6820–6828.

Dahl AM, Klein C, Andres PG, London CA, Lodge MP, Mulligan RC, Abbas AK. 2000. Expression of Bcl-XL Restores Cell Survival, but Not Proliferation and Effector Differentiation, in CD28-deficient T Lymphocytes. *J Exp Med* **191:** 2031–2038.

Damle NK, Doyle LV. 1989. Stimulation via the CD3 and CD28 molecules induces responsiveness to IL- 4 in CD4+CD29+CD45R- memory T lymphocytes. *J Immunol* **143:** 1761–1767.

Damle NK, Doyle LV, Grosmaire LS, Ledbetter JA. 1988. Differential regulatory signals delivered by antibody binding to the CD28 (Tp44) molecule during the activation of human T lymphocytes. *J Immunol* **140:** 1753–1761.

Dariavach P, Mattei MG, Golstein P, Lefranc MP. 1988. Human Ig superfamily CTLA-4 gene: Chromosomal localization and identity of protein sequence between murine and human CTLA-4 cytoplasmic domains. *Eur J Immunol* **18:** 1901–1905.

Diehn M, Alizadeh AA, Rando OJ, Liu CL, Stankunas K, Botstein D, Crabtree GR, Brown PO. 2002. Genomic expression programs and the integration of the CD28 costimulatory signal in T cell activation. *PNAS* **99:** 11796–11801.

Dienz O, Hehner SP, Droge W, Schmitz ML. 2000. Synergistic activation of NF-κ B by functional cooperation between vav and PKCθ in T lymphocytes. *J Biol Chem* **275:** 24547–24551.

Dodson LF, Boomer JS, Deppong CM, Shah DD, Sim J, Bricker TL, Russell JH, Green JM. 2009. Targeted knockin mice expressing mutations of CD28 reveal an essential pathway for costimulation. *MCB* **29:** 3710–3721.

Dustin ML, Chan AC. 2000. Signaling takes shape in the immune system. *Cell* **103:** 283–294.

Ellis JH, Ashman C, Burden MN, Kilpatrick KE, Morse MA, Hamblin PA. 2000. GRID: a novel Grb-2-related adapter protein that interacts with the activated T cell costimulatory receptor CD28. *J Immunol* **164:** 5805–5814.

Evans EJ, Esnouf RM, Manso-Sancho R, Gilbert RJ, James JR, Yu C, Fennelly JA, Vowles C, Hanke T, Walse B et al. 2005. Crystal structure of a soluble CD28-Fab complex. *Nat Immunol* **6:** 271–279.

Fang X, Yu S, Tanyi JL, Lu Y, Woodgett JR, Mills GB. 2002. Convergence of multiple signaling cascades at glycogen synthase kinase 3: Edg receptor-mediated phosphorylation and inactivation by lysophosphatidic acid through a protein kinase C-dependent intracellular pathway. *Mol Cell Biol* **22:** 2099–2110.

Fischer KD, Tedford K, Penninger JM. 1998. Vav links antigen-receptor signaling to the actin cytoskeleton. *Semin Immunol* **10:** 317–327.

Fraser JD, Irving BA, Crabtree GR, Weiss A. 1991. Regulation of interleukin-2 gene enhancer activity by the T cell accessory molecule CD28. *Science* **251:** 313–316.

Frauwirth KA, Riley JL, Harris MH, Parry RV, Rathmell JC, Plas DR, Elstrom RL, June CH, Thompson CB. 2002. The CD28 signaling pathway regulates glucose metabolism. *Immunity* **16:** 769–777.

Freeburn RW, Wright KL, Burgess SJ, Astoul E, Cantrell DA, Ward SG. 2002. Evidence that SHIP-1 contributes to phosphatidylinositol 3,4,5-trisphosphate metabolism in T lymphocytes and can regulate novel phosphoinositide 3-kinase effectors. *J Immunol* **169:** 5441–5450.

Freeman GJ, Freedman AS, Segil JM, Lee G, Whitman JF, Nadler LM. 1989. B7, a new member of the Ig superfamily with unique expression on activated and neoplastic B cells. *J Immunol* **143:** 2714–2722.

Freeman GJ, Gribben JG, Boussiotis VA, Ng JW, Restivo VA Jr, Lombard LA, Gray GS, Nadler LM. 1993. Cloning of B7-2: A CTLA-4 counter-receptor that costimulates human T cell proliferation. *Science* **262:** 909–911.

Freiberg BA, Kupfer H, Maslanik W, Delli J, Kappler J, Zaller DM, Kupfer A. 2002. Staging and resetting T cell activation in SMACs. *Nat Immunol* **3:** 911–917.

Friend LD, Shah DD, Deppong C, Lin J, Bricker TL, Juehne TI, Rose CM, Green JM. 2006. A dose-dependent requirement for the proline motif of CD28 in cellular and humoral immunity revealed by a targeted knockin mutant. *J Exp Med* **203:** 2121–2133.

Gaide O, Favier B, Legler DF, Bonnet D, Brissoni B, Valitutti S, Bron C, Tschopp J, Thome M. 2002. CARMA1 is a critical lipid raft-associated regulator of TCR-induced NF-κ B activation. *Nat Immunol* **3:** 836–843.

Garcon F, Patton DT, Emery JL, Hirsch E, Rottapel R, Sasaki T, Okkenhaug K. 2008. CD28 provides T-cell costimulation and enhances PI3K activity at the immune synapse independently of its capacity to interact with the p85/p110 heterodimer. *Blood* **111:** 1464–1471.

Genovese MC, Becker JC, Schiff M, Luggen M, Sherrer Y, Kremer J, Birbara C, Box J, Natarajan K, Nuamah I et al. 2005. Abatacept for rheumatoid arthritis refractory to tumor necrosis factor α inhibition. *N Engl J Med* **353:** 1114–1123.

Gmünder H, Lesslauer W. 1984. A 45-kDa human T-cell membrane glycoprotein functions in the regulation of cell proliferative responses. *Eur J Biochem* **142:** 153–160.

Gogishvili T, Elias F, Emery JL, McPherson K, Okkenhaug K, Hunig T, Dennehy KM. 2008. Proliferative signals mediated by CD28 superagonists require the exchange factor Vav1 but not phosphoinositide 3-kinase in primary peripheral T cells. *Eur J Immunol* **38:** 2528–2533.

Goode N, Hughes K, Woodgett JR, Parker PJ. 1992. Differential regulation of glycogen synthase kinase-3 β by protein kinase C isotypes. *J Biol Chem* **267:** 16878–16882.

Grakoui A, Bromley SK, Sumen C, Davis MM, Shaw AS, Allen PM, Dustin ML. 1999. The immunological synapse: A molecular machine controlling T cell activation. *Science* **285:** 221–227.

Granelli-Piperno A, Nolan P. 1991. Nuclear transcription factors that bind to elements of the IL-2 promoter. Induction requirements in primary human T cells. *J Immunol* **147:** 2734–2739.

Gross JA, St John T, Allison JP. 1990. The murine homologue of the T lymphocyte antigen CD28. Molecular cloning and cell surface expression. *J Immunol* **144:** 3201–3210.

Guo F, Iclozan C, Suh WK, Anasetti C, Yu XZ. 2008. CD28 controls differentiation of regulatory T cells from naive CD4 T cells. *J Immunol* **181:** 2285–2291.

Hansen JA, Martin PJ, Nowinski RC. 1980. Monoclonal antibodies identifying a novel T-cell antigen and Ia antigens of human lymphocytes. *Immunogenetics* **10:** 247–260.

Hara H, Wada T, Bakal C, Kozieradzki I, Suzuki S, Suzuki N, Nghiem M, Griffiths EK, Krawczyk C, Bauer B et al. 2003. The MAGUK family protein CARD11 is essential for lymphocyte activation. *Immunity* **18:** 763–775.

Harada Y, Tokushima M, Matsumoto Y, Ogawa S, Otsuka M, Hayashi K, Weiss BD, June CH, Abe R. 2001. Critical requirement for the membrane-proximal cytosolic tyrosine residue for CD28-mediated costimulation in vivo. *J Immunol* **166:** 3797–3803.

Harhaj EW, Sun SC. 1998. IκB kinases serve as a target of CD28 signaling. *J Biol Chem* **273:** 25185–25190.

Harhaj EW, Good L, Xiao G, Uhlik M, Cvijic ME, Rivera-Walsh I, Sun SC. 2000. Somatic mutagenesis studies of NF-κ B signaling in human T cells: Evidence for an essential role of IKK γ in NF-γ B activation by T-cell costimulatory signals and HTLV-I Tax protein. *Oncogene* **19:** 1448–1456.

Harhaj EW, Maggirwar SB, Good L, Sun SC. 1996. CD28 mediates a potent costimulatory signal for rapid degradation of IγBβ which is associated with accelerated activation of various NF-κB/Rel heterodimers. *Mol Cell Biol* **16:** 6736–6743.

Harriague J, Bismuth G. 2002. Imaging antigen-induced PI3K activation in T cells. *Nat Immunol* **3:** 1090–1096.

Harris N, Campbell C, Le Gros G, Ronchese F. 1997a. Blockade of CD28/B7 co-stimulation by mCTLA4-Hγ1 inhibits antigen- induced lung eosinophilia but not Th2 cell development or recruitment in the lung. *Eur J Immunol* **27:** 155–161.

Harris N, Peach R, Naemura J, Linsley PS, Le Gros G, Ronchese F. 1997b. CD80 costimulation is essential for the induction of airway eosinophilia. *J Exp Med* **185:** 177–182.

Hayashi K, Altman A. 2006. Filamin A is required for T cell activation mediated by protein kinase C-θ. *J Immunol* **177:** 1721–1728.

Hehner SP, Hofmann TG, Dienz O, Droge W, Schmitz ML. 2000a. Tyrosine-phosphorylated Vav1 as a point of integration for T-cell receptor- and CD28-mediated activation of JNK, p38, and interleukin-2 transcription. *J Biol Chem* **275:** 18160–18171.

Hehner SP, Hofmann TG, Ushmorov A, Dienz O, Wing-Lan Leung I, Lassam N, Scheidereit C, Droge W, Schmitz ML. 2000b. Mixed-lineage kinase 3 delivers CD3/CD28-derived signals into the IκB kinase complex. *Mol Cell Biol* **20:** 2556–2568.

Holdorf AD, Green JM, Levin SD, Denny MF, Straus DB, Link V, Changelian PS, Allen PM, Shaw AS. 1999. Proline residues in CD28 and the Src homology (SH)3 domain of Lck are required for T cell costimulation. *J Exp Med* **190:** 375–384.

Holsinger LJ, Graef IA, Swat W, Chi T, Bautista DM, Davidson L, Lewis RS, Alt FW, Crabtree GR. 1998. Defects in actin-cap formation in Vav-deficient mice implicate an actin requirement for lymphocyte signal transduction. *Curr Biol* **8:** 563–572.

Howard TA, Rochelle JM, Seldin MF. 1991. Cd28 and Ctla-4, two related members of the Ig supergene family, are tightly linked on proximal mouse chromosome 1. *Immunogenetics* **33:** 74–76.

Huang J, Lo P-F, Zal T, Gascoigne NRJ, Smith BA, Levin SD, Grey HM. 2002. CD28 plays a critical role in the segregation of PKCθ within the immunologic synapse. *PNAS* **99:** 9369–9373.

Hutchcroft JE, Bierer BE. 1994. Activation-dependent phosphorylation of the T-lymphocyte surface receptor CD28 and associated proteins. *Proc Natl Acad Sci* **91:** 3260–3264.

Hutchcroft JE, Tsai B, Bierer BE. 1996. Differential phosphorylation of the T lymphocyte costimulatory receptor CD28. Activation-dependent changes and regulation by protein kinase C. *J Biol Chem* **271:** 13362–13370.

Hutloff A, Dittrich AM, Beier KC, Eljaschewitsch B, Kraft R, Anagnostopoulos I, Kroczek RA. 1999. ICOS is an inducible T-cell co-stimulator structurally and functionally related to CD28. *Nature* **397:** 263–266.

Ishida Y, Agata Y, Shibahara K, Honjo T. 1992. Induced expression of PD-1, a novel member of the immunoglobulin gene superfamily, upon programmed cell death. *Embo J* **11:** 3887–3895.

Jenkins MK, Chen CA, Jung G, Mueller DL, Schwartz RH. 1990. Inhibition of antigen-specific proliferation of type 1 murine T cell clones after stimulation with immobilized anti-CD3 monoclonal antibody. *J Immunol* **144:** 16–22.

Johnson KG, Bromley SK, Dustin ML, Thomas ML. 2000. A supramolecular basis for CD45 tyrosine phosphatase regulation in sustained T cell activation. *Proc Natl Acad Sci* **97:** 10138–10143.

Jones RG, Elford AR, Parsons MJ, Wu L, Krawczyk CM, Yeh WC, Hakem R, Rottapel R, Woodgett JR, Ohashi PS. 2002. CD28-dependent activation of protein kinase B/Akt blocks Fas-mediated apoptosis by preventing death-inducing signaling complex assembly. *J Exp Med* **196:** 335–348.

Jones RG, Parsons M, Bonnard M, Chan VS, Yeh WC, Woodgett JR, Ohashi PS. 2000. Protein kinase B regulates T lymphocyte survival, nuclear factor κB activation, and Bcl-X(L) levels in vivo. *J Exp Med* **191:** 1721–1734.

Jun JE, Wilson LE, Vinuesa CG, Lesage S, Blery M, Miosge LA, Cook MC, Kucharska EM, Hara H, Penninger JM et al. 2003. Identifying the MAGUK protein Carma-1 as a central regulator of humoral immune responses and atopy by genome-wide mouse mutagenesis. *Immunity* **18:** 751–762.

June CH, Bluestone JA, Nadler LM, Thompson CB. 1994. The B7 and CD28 receptor families. *Immunol Today* **15:** 321–330.

June CH, Ledbetter JA, Linsley PS, Thompson CB. 1990. Role of the CD28 receptor in T-cell activation. *Immunol Today* **11:** 211–216.

Kaga S, Ragg S, Rogers KA, Ochi A. 1998. Stimulation of CD28 with B7-2 promotes focal adhesion-like cell contacts where Rho family small G proteins accumulate in T cells. *J Immunol* **160:** 24–27.

Kane LP, Andres PG, Howland KC, Abbas AK, Weiss A. 2001. Akt provides the CD28 costimulatory signal for up-regulation of IL-2 and IFN-γ but not TH2 cytokines. *Nat Immunol* **2:** 37–44.

Kane LP, Shapiro VS, Stokoe D, Weiss A. 1999. Induction of NF-κB by the Akt/PKB kinase. *Curr Biol* **9:** 601–604.

Keane-Myers A, Gause WC, Linsley PS, Chen SJ, Wills-Karp M. 1997. B7-CD28/CTLA-4 costimulatory pathways are required for the development of T helper cell 2-mediated allergic airway responses to inhaled antigens. *J Immunol* **158:** 2042–2049.

Khoshnan A, Tindell C, Laux I, Bae D, Bennett B, Nel AE. 2000. The NF-κ B cascade is important in Bcl-xL expression and for the anti-apoptotic effects of the CD28 receptor in primary human CD4+ lymphocytes. *J Immunol* **165:** 1743–1754.

Kim H-H, Tharayil M, Rudd CE. 1998. Growth factor receptor-bound protein 2 SH2/SH3 domain binding to CD28 and its role in co-signaling. *J Biol Chem* **273:** 296–301.

King PD, Sadra A, Teng JM, Xiao-Rong L, Han A, Selvakumar A, August A, Dupont B. 1997. Analysis of CD28 cytoplasmic tail tyrosine residues as regulators and substrates for the protein tyrosine kinases, EMT and LCK. *J Immunol* **158:** 580–590.

King CL, Stupi RJ, Craighead N, June CH, Thyphronitis G. 1995. CD28 activation promotes Th2 subset differentiation by human CD4+ cells. *Eur J Immunol* **25:** 587–595.

Kirchhoff S, Muller WW, Li-Weber M, Krammer PH. 2000. Up-regulation of c-FLIPshort and reduction of activation-induced cell death in CD28-costimulated human T cells. *Eur J Immunol* **30:** 2765–2774.

Kobayashi K, Inohara N, Hernandez LD, Galan JE, Nunez G, Janeway CA, Medzhitov R, Flavell RA. 2002. RICK/Rip2/CARDIAK mediates signalling for receptors of the innate and adaptive immune systems. *Nature* **416:** 194–199.

Krawczyk C, Oliveira-dos-Santos A, Sasaki T, Griffiths E, Ohashi PS, Snapper S, Alt F, Penninger JM. 2002. Vav1 controls integrin clustering and MHC/peptide-specific cell adhesion to antigen-presenting cells. *Immunity* **16:** 331–343.

Krummel MF, Sjaastad MD, Wulfing C, Davis MM. 2000. Differential clustering of CD4 and CD3zeta during T cell recognition. *Science* **289:** 1349–1352.

Lafferty KJ, Cunningham AJ. 1975. A new analysis of allogeneic interactions. *Aust J Exp Biol Med Sci* **53:** 27–42.

Lai JH, Tan TH. 1994. CD28 signaling causes a sustained down-regulation of I κ B α which can be prevented by the immunosuppressant rapamycin. *J Biol Chem* **269:** 30077–30080.

Larsen CP, Elwood ET, Alexander DZ, Ritchie SC, Hendrix R, Tucker-Burden C, Cho HR, Aruffo A, Hollenbaugh D, Linsley PS et al. 1996. Long-term acceptance of skin and cardiac allografts after blocking CD40 and CD28 pathways. *Nature* **381:** 434–438.

Ledbetter JA, Imboden JB, Schieven GL, Grosmaire LS, Rabinovitch PS, Lindsten T, Thompson CB, June CH. 1990. CD28 ligation in T-cell activation: Evidence for two signal transduction pathways. *Blood* **75:** 1531–1539.

Lee KY, D'Acquisto F, Hayden MS, Shim JH, Ghosh S. 2005. PDK1 nucleates T cell receptor-induced signaling complex for NF-κB activation. *Science* **308:** 114–118.

Lee KP, Taylor C, Petryniak B, Turka LA, June CH, Thompson CB. 1990. The genomic organization of the CD28 gene. Implications for the regulation of CD28 mRNA expression and heterogeneity. *J Immunol* **145:** 344–352.

Lenschow DJ, Herold KC, Rhee L, Patel B, Koons A, Kin H-Y, Fuchs E, Singh B, Thompson CB, Bluestone JA. 1996. CD28/B7 regulation of Th1 and Th2 subsets in the development of autoimmune diabetes. *Immunity* **5:** 285–293.

Li Y, Sedwick CE, Hu J, Altman A. 2005. Role for protein kinase Cθ (PKCθ) in TCR/CD28-mediated signaling through the canonical but not the non-canonical pathway for NF-κB activation. *J Biol Chem* **280:** 1217–1223.

Lin X, Cunningham ETJr, Mu Y, Geleziunas R, Greene WC. 1999. The proto-oncogene Cot kinase participates in CD3/CD28 induction of NF-κB acting through the NF-κB-inducing kinase and IκB kinases. *Immunity* **10:** 271–280.

Lindsten T, June CH, Ledbetter JA, Stella G, Thompson CB. 1989. Regulation of lymphokine messenger RNA stability by a surface-mediated T cell activation pathway. *Science* **244:** 339–343.

Linsley PS, Ledbetter JA. 1993. The role of the CD28 receptor during T cell responses to antigen. *Annu Rev Immunol* **11:** 191–212.

Linsley PS, Nadler SG. 2009. The clinical utility of inhibiting CD28-mediated costimulation. *Immunol Rev* **229:** 307–321.

Lohr J, Knoechel B, Kahn EC, Abbas AK. 2004. Role of B7 in T cell tolerance. *J Immunol* **173:** 5028–5035.

Lu Y, Phillips CA, Trevillyan JM. 1995. Phosphatidylinositol 3-kinase activity is not essential for CD28 costimulatory activity in Jurkat T cells: Studies with a selective inhibitor, wortmannin. *Eur J Immunol* **25:** 533–537.

Lucas PJ, Negishi I, Nakayama K, Fields LE, Loh DY. 1995. Naive CD28-deficient T cells can initiate but not sustain an in vitro antigen-specific immune response. *J Immunol* **154:** 5757–5768.

Luhder F, Huang Y, Dennehy KM, Guntermann C, Muller I, Winkler E, Kerkau T, Ikemizu S, Davis SJ, Hanke T et al. 2003. Topological requirements and signaling properties of T cell-activating, anti-CD28 antibody superagonists. *J Exp Med* **197:** 955–966.

Manicassamy S, Gupta S, Huang Z, Sun Z. 2006. Protein Kinase C-{θ}-Mediated Signals Enhance CD4+ T Cell Survival by Up-Regulating Bcl-xL. *J Immunol* **176:** 6709–6716.

Marengere LE, Okkenhaug K, Clavreul A, Couez D, Gibson S, Mills GB, Mak TW, Rottapel R. 1997. The SH3 domain of Itk/Emt binds to proline-rich sequences in the cytoplasmic domain of the T cell costimulatory receptor CD28. *J Immunol* **159:** 3220–3229.

Marinari B, Costanzo A, Marzano V, Piccolella E, Tuosto L. 2004. CD28 delivers a unique signal leading to the selective recruitment of RelA and p52 NF-κB subunits on IL-8 and Bcl-xL gene promoters. *Proc Natl Acad Sci* **101:** 6098–6103.

Marinari B, Costanzo A, Viola A, Michel F, Mangino G, Acuto O, Levrero M, Piccolella E, Tuosto L. 2002. Vav cooperates with CD28 to induce NF-κB activation via a pathway involving Rac-1 and mitogen-activated kinase kinase 1. *Eur J Immunol* **32:** 447–456.

Mathur M, Herrmann K, Qin Y, Gulmen F, Li X, Krimins R, Weinstock J, Elliott D, Bluestone JA, Padrid P. 1999. CD28 interactions with either CD80 or CD86 are sufficient to induce allergic airway inflammation in mice. *Am J Respir Cell Mol Biol* **21:** 498–509.

Matsumoto R, Wang D, Blonska M, Li H, Kobayashi M, Pappu B, Chen Y, Lin X. 2005. Phosphorylation of CARMA1 plays a critical role in T Cell receptor-mediated NF-κB activation. *Immunity* **23:** 575–585.

Michel F, Mangino G, Attal-Bonnefoy G, Tuosto L, Alcover A, Roumier A, Olive D, Acuto O. 2000. CD28 utilizes Vav-1 to enhance TCR-proximal signaling and NF-AT activation. *J Immunol* **165:** 3820–3829.

Miller J, Baker C, Cook K, Graf B, Sanchez-Lockhart M, Sharp K, Wang X, Yang B, Yoshida T. 2009. Two pathways of costimulation through CD28. *Immunol Res*

Miller SD, Vanderlugt CL, Lenschow DJ, Pope JG, Karandikar NJ, Dal Canto MC, Bluestone JA. 1995. Blockade of CD28/B7-1 interaction prevents epitope spreading and clinical relapses of murine EAE. *Immunity* **3:** 739–745.

Monks CR, Freiberg BA, Kupfer H, Sciaky N, Kupfer A. 1998. Three-dimensional segregation of supramolecular activation clusters in T cells. *Nature* **395:** 82–86.

Mueller D, Jenkins MK, Schwartz RH. 1989. An accessory cell-derived costimulatory signal acts independently of protein kinase C activation to allow T cell proliferation and prevent the induction of unresponsiveness. *J Immunol* **142:** 2617–2628.

Naluai AT, Nilsson S, Samuelsson L, Gudjonsdottir AH, Ascher H, Ek J, Hallberg B, Kristiansson B, Martinsson T, Nerman O et al. 2000. The CTLA4/CD28 gene region on chromosome 2q33 confers susceptibility to celiac disease in a way possibly distinct from that of type 1 diabetes and other chronic inflammatory disorders. *Tissue Antigens* **56:** 350–355.

Nirula A, Ho M, Phee H, Roose J, Weiss A. 2006. Phosphoinositide-dependent kinase 1 targets protein kinase A in a pathway that regulates interleukin 4. *J Exp Med* **203:** 1733–1744.

Noel PJ, Alegre ML, Reiner SL, Thompson CB. 1998. Impaired negative selection in CD28-deficient mice. *Cell Immunol* **187:** 131–138.

Noel PJ, Boise LH, Green JM, Thompson CB. 1996. CD28 costimulation prevents cell death during primary T cell activation. *J Immunol* **157:** 636–642.

Ogilvie RL, Abelson M, Hau HH, Vlasova I, Blackshear PJ, Bohjanen PR. 2005. Tristetraprolin down-regulates IL-2 gene expression through AU-rich element-mediated mRNA decay. *J Immunol* **174:** 953–961.

Ohteki T, Parsons M, Zakarian A, Jones RG, Nguyen LT, Woodgett JR, Ohashi PS. 2000. Negative regulation of T cell proliferation and interleukin 2 production by the serine threonine kinase GSK-3. *J Exp Med* **192:** 99–104.

Okkenhaug K, Rottapel R. 1998. Grb2 forms an inducible protein complex with CD28 through a Src homology 3 domain-proline interaction. *J Biol Chem* **273:** 21194–21202.

Okkenhaug K, Wu L, Garza KM, La Rose J, Khoo W, Odermatt B, Mak TW, Ohashi PS, Rottapel R. 2001. A point mutation in CD28 distinguishes proliferative signals from survival signals. *Nat Immunol* **2:** 325–332.

Pagès F, Ragueneau M, Rottapel R, Truneh A, Nunes J, Imbert J, Olive D. 1994. Binding of phosphatidylinositol-3-OH kinase to CD28 is required for T- cell signalling. *Nature* **369:** 327–329.

Pan M, Winslow MM, Chen L, Kuo A, Felsher D, Crabtree GR. 2007. Enhanced NFATc1 Nuclear Occupancy Causes T Cell Activation Independent of CD28 Costimulation. *J Immunol* **178:** 4315–4321.

Parry RV, Olive D, Westwick J, Sansom DM, Ward SG. 1997a. Evidence that a kinase distinct from protein kinase C and phosphatidylinositol 3-kinase mediates ligation-dependent serine/threonine phosphorylation of the T-lymphocyte co-stimulatory molecule CD28. *Biochem J* **326:** 249–257.

Parry RV, Reif K, Smith G, Sansom DM, Hemmings BA, Ward SG. 1997b. Ligation of the T cell co-stimulatory receptor CD28 activates the serine-threonine protein kinase protein kinase B. *Eur J Immunol* **27:** 2495–2501.

Paterson AM, Vanguri VK, Sharpe AH. 2009. SnapShot: B7/CD28 costimulation. *Cell* **137:** 974–974 e971.

Podojil JR, Miller SD. 2009. Molecular mechanisms of T-cell receptor and costimulatory molecule ligation/blockade in autoimmune disease therapy. *Immunol Rev* **229:** 337–355.

Prasad KV, Cai YC, Raab M, Duckworth B, Cantley L, Shoelson SE, Rudd CE. 1994. T-cell antigen CD28 interacts with the lipid kinase phosphatidylinositol 3-kinase by a cytoplasmic Tyr(P)-Met-Xaa-Met motif. *Proc Natl Acad Sci* **91:** 2834–2838.

Quill H, Schwartz RH. 1987. Stimulation of normal inducer T cell clones with antigen presented by purified Ia molecules in planar lipid membranes: Specific induction of a long-lived state of proliferative nonresponsiveness. *J Immunol* **138:** 3704–3712.

Raab M, Pfister S, Rudd CE. 2001. CD28 signaling via VAV/SLP-76 adaptors: Regulation of cytokine transcription independent of TCR ligation. *Immunity* **15:** 921–933.

Raab M, Cai YC, Bunnell SC, Heyeck SD, Berg LJ, Rudd CE. 1995. p56Lck and p59Fyn regulate CD28 binding to phosphatidylinositol 3- kinase, growth factor receptor-bound protein GRB-2, and T cell specific protein-tyrosine kinase ITK: Implications for T-cell costimulation. *Proc Natl Acad Sci* **92:** 8891–8895.

Radvanyi LG, Shi Y, Vaziri H, Sharma A, Dhala R, Mills GB, Miller RG. 1996. CD28 costimulation inhibits TCR-induced apoptosis during a primary T cell response. *J Immunol* **156:** 1788–1798.

Reiter R, Pfeffer K. 2002. Impaired germinal centre formation and humoral immune response in the absence of CD28 and interleukin-4. *Immunol* **106:** 222–228.

Riley JL, Mao M, Kobayashi S, Biery M, Burchard J, Cavet G, Gregson BP, June CH, Linsley PS. 2002. Modulation of TCR-induced transcriptional profiles by ligation of CD28, ICOS, and CTLA-4 receptors. *Proc Natl Acad Sci* **99:** 11790–11795.

Rudd CE, Schneider H. 2003. Unifying concepts in CD28, ICOS and CTLA4 co-receptor signalling. *Nat Rev Immunol* **3:** 544–556.

Rudd CE, Taylor A, Schneider H. 2009. CD28 and CTLA-4 coreceptor expression and signal transduction. *Immunol Rev* **229:** 12–26.

Rulifson IC, Sperling AI, Fields PE, Fitch FW, Bluestone JA. 1997. CD28 costimulation promotes the production of Th2 cytokines. *J Immunol* **158:** 658–665.

Sadra A, Cinek T, Arellano JL, Shi J, Truitt KE, Imboden JB. 1999. Identification of Tyrosine Phosphorylation Sites in the CD28 Cytoplasmic Domain and Their Role in the Costimulation of Jurkat T Cells. *J Immunol* **162:** 1966–1973.

Salojin KV, Zhang J, Delovitch TL. 1999. TCR and CD28 are coupled via ZAP-70 to the activation of the Vav/Rac-1-/PAK-1/p38 MAPK signaling pathway. *J Immunol* **163:** 844–853.

Salomon B, Lenschow DJ, Rhee L, Ashourian N, Singh B, Sharpe A, Bluestone JA. 2000. B7/CD28 costimulation is essential for the homeostasis of the CD4+CD25+ immunoregulatory T cells that control autoimmune diabetes. *Immunity* **12:** 431–440.

Sanchez-Lockhart M, Miller J. 2006. Engagement of CD28 Outside of the Immunological Synapse Results in Up-Regulation of IL-2 mRNA Stability but Not IL-2 Transcription. *J Immunol* **176:** 4778–4784.

Sanchez-Lockhart M, Graf B, Miller J. 2008. Signals and sequences that control CD28 localization to the central region of the immunological synapse. *J Immunol* **181:** 7639–7648.

Sanchez-Lockhart M, Marin E, Graf B, Abe R, Harada Y, Sedwick CE, Miller J. 2004. Cutting edge: CD28-mediated transcriptional and posttranscriptional regulation of IL-2 expression are controlled through different signaling pathways. *J Immunol* **173:** 7120–7124.

Sasaki T, Irie-Sasaki J, Jones RG, Oliveira-dos-Santos AJ, Stanford WL, Bolon B, Wakeham A, Itie A, Bouchard D, Kozieradzki I et al. 2000. Function of PI3Kγ in thymocyte development, T cell activation, and neutrophil migration. *Science* **287:** 1040–1046.

Sayegh MH, Turka LA. 1998. The role of T-cell costimulatory activation pathways in transplant rejection. *N Engl J Med* **338:** 1813–1821.

Schneider H, Cai Y-C, Prasad KVS, Shoelson SE, Rudd CE. 1995a. T cell antigen CD28 binds to the GRB-2/SOS complex, regulators of p21ras. *Eur J Immunol* **25:** 1044–1050.

Schneider H, Cai YC, Prasad KV, Shoelson SE, Rudd CE. 1995b. T cell antigen CD28 binds to the GRB-2/SOS complex, regulators of p21ras. *Eur J Immunol* **25:** 1044–1050.

Schwartz RH, Mueller DL, Jenkins MK, Quill H. 1989. T-cell clonal anergy. *Cold Spring Harb Symp Quant Biol* **54:** 605–610.

Seko Y, Azmi H, Fariss R, Ragheb JA. 2004. Selective cytoplasmic translocation of HuR and site-specific binding to the interleukin-2 mRNA are not sufficient for CD28-mediated stabilization of the mRNA. *J Biol Chem* **279:** 33359–33367.

Shahinian A, Pfeffer K, Lee KP, Kundig TM, Kishihara K, Wakeham A, Kawai K, Ohashi PS, Thompson CB, Mak TW. 1993a. Differential T cell costimulatory requirements in CD28-deficient mice. *Science* **261:** 609–612.

Shahinian A, Pfeffer K, Lee KP, Kündig TM, Kishihara K, Wakeham A, Kawai K, Ohashi PS, Thompson CB, Mak TW. 1993b. Differential T cell costimulatory requirements in CD28-deficient mice. *Science* **261:** 609–612.

Shan X, Czar MJ, Bunnell SC, Liu P, Liu Y, Schwartzberg PL, Wange RL. 2000. Deficiency of PTEN in Jurkat T cells causes constitutive localization of Itk to the plasma membrane and hyperresponsiveness to CD3 stimulation. *Mol Cell Biol* **20:** 6945–6957.

Shapiro VS, Truitt KE, Imboden JB, Weiss A. 1997. CD28 mediates transcriptional upregulation of the interleukin-2 (IL-2) promoter through a composite element containing the CD28RE and NF-IL-2B AP-1 sites. *Mol Cel Biol* **17:** 4051–4058.

Sharpe AH. 2009. Mechanisms of costimulation. *Immunol Rev* **229:** 5–11.

Shi L, Godfrey WR, Lin J, Zhao G, Kao PN. 2007. NF90 regulates inducible IL-2 gene expression in T cells. *J Exp Med* **204:** 971–977.

Shim J, Lim H, R Yates J, Karin M. 2002. Nuclear export of NF90 is required for interleukin-2 mRNA stabilization. *Mol Cell* **10:** 1331–1344.

Snapper SB, Rosen FS. 1999. The Wiskott-Aldrich syndrome protein (WASP): Roles in signaling and cytoskeletal organization. *Annu Rev Immunol* **17:** 905–929.

Songyang Z, Shoelson SE, Chaudhuri M, Gish G, Pawson T, Haser WG, King F, Roberts T, Ratnofsky S, Lechleider RJ et al. 1993. SH2 domains recognize specific phosphopeptide sequences. *Cell* **72:** 767–778.

Sperling AI, Auger JA, Ehst BD, Rulifson IC, Thompson CB, Bluestone JA. 1996. CD28/B7 interactions deliver a unique signal to naive T cells that regulates cell survival but not early proliferation. *J Immunol* **157:** 3909–3917.

Stein PH, Fraser JD, Weiss A. 1994. The cytoplasmic domain of CD28 is both necessary and sufficient for costimulation of interleukin-2 secretion and association with phosphatidylinositol 3'-kinase. *Mol Cell Biol* **14:** 3392–3402.

Su B, Jacinto E, Hibi M, Kallunki T, Karin M, Ben-Neriah Y. 1994. JNK is involved in signal integration during costimulation of T lymphocytes. *Cell* **77:** 727–736.

Sun Z, Arendt CW, Ellmeier W, Schaeffer EM, Sunshine MJ, Gandhi L, Annes J, Petrzilka D, Kupfer A, Schwartzberg PL et al. 2000. PKC-θ is required for TCR-induced NF-κB activation in mature but not immature T lymphocytes. *Nature* **404:** 402–407.

Tacke M, Hanke G, Hanke T, Hunig T. 1997. CD28-mediated induction of proliferation in resting T cells in vitro and in vivo without engagement of the T cell receptor: Evidence for functionally distinct forms of CD28. *Eur J Immunol* **27:** 239–247.

Tai X, Cowan M, Feigenbaum L, Singer A. 2005. CD28 costimulation of developing thymocytes induces Foxp3 expression and regulatory T cell differentiation independently of interleukin 2. *Nat Immunol* **6:** 152–162.

Tai X, Van Laethem F, Sharpe AH, Singer A. 2007. Induction of autoimmune disease in CTLA-4-/- mice depends on a specific CD28 motif that is required for in vivo costimulation. *Proc Natl Acad Sci* **104:** 13756–13761.

Takeda K, Harada Y, Watanabe R, Inutake Y, Ogawa S, Onuki K, Kagaya S, Tanabe K, Kishimoto H, Abe R. 2008. CD28 stimulation triggers NF-κB activation through the CARMA1-PKCθ-Grb2/Gads axis. *Int Immunol* **20:** 1507–1515.

Tang Q, Henriksen KJ, Boden EK, Tooley AJ, Ye J, Subudhi SK, Zheng XX, Strom TB, Bluestone JA. 2003. Cutting edge: CD28 controls peripheral homeostasis of CD4+ CD25+ regulatory T cells. *J Immunol* **171:** 3348–3352.

Tanner MJ, Hanel W, Gaffen SL, Lin X. 2007. CARMA1 coiled-coil domain is involved in the oligomerization and subcellular localization of CARMA1 and is required for T cell receptor-induced NF-κB activation. *J Biol Chem* **282:** 17141–17147.

Tao R, Wang L, Han R, Wang T, Ye Q, Honjo T, Murphy TL, Murphy KM, Hancock WW. 2005. Differential effects of B and T lymphocyte attenuator and programmed death-1 on acceptance of partially versus fully MHC-mismatched cardiac allografts. *J Immunol* **175:** 5774–5782.

Tavano R, Contento RL, Baranda SJ, Soligo M, Tuosto L, Manes S, Viola A. 2006. CD28 interaction with filamin-A controls lipid raft accumulation at the T-cell immunological synapse. *Nat Cell Biol* **8:** 1270–1276.

Tezuka K, Tsuji T, Hirano D, Tamatani T, Sakamaki K, Kobayashi Y, Kamada M. 2000. Identification and characterization of rat AILIM/ICOS, a novel T-cell costimulatory molecule, related to the CD28/CTLA4 family. *Biochem Biophys Res Commun* **276:** 335–345.

Thompson CB, Lindsten T, Ledbetter JA, Kunkel SL, Young HA, Emerson SG, Leiden JM, June CH. 1989. CD28 activation pathway regulates the production of multiple T-cell-derived lymphokines/cytokines. *Proc Natl Acad Sci* **86:** 1333–1337.

Truitt KE, Hicks CM, Imboden JB. 1994. Stimulation of CD28 triggers an association between CD28 and phosphatidylinositol 3-kinase in Jurkat T cells. *J Exp Med* **179:** 1071–1076.

Truitt KE, Nagel T, Suen LF, Imboden JB. 1996. Structural requirements for CD28-mediated costimulation of IL-2 production in Jurkat T cells. *J Immunol* **156:** 4539–4541.

Tuosto L, Costanzo A, Guido F, Marinari B, Vossio S, Moretti F, Levrero M, Piccolella E. 2000. Mitogen-activated kinase kinase kinase 1 regulates T cell receptor- and CD28-mediated signaling events which lead to NF-κB activation. *Eur J Immunol* **30:** 2445–2454.

Turka LA, Linsley PS, Lin H, Brady W, Leiden JM, Wei R-Q, Gibson ML, Zheng X-G, Myrdal S, Gordon D et al. 1992. T-cell activation by the CD28 ligand B7 is required for cardiac allograft rejection in vivo. *Proc Natl Acad Sci* **89:** 11102–11105.

van der Merwe PA. 2002. Formation and function of the immunological synapse. *Curr Opin Immunol* **14:** 293–298.

van der Merwe PA, Bodian DL, Daenke S, Linsley P, Davis SJ. 1997. CD80 (B7-1) binds both CD28 and CTLA-4 with a low affinity and very fast kinetics. *J Exp Med* **185:** 393–403.

Vanhaesebroeck B, Alessi DR. 2000. The PI3K-PDK1 connection: more than just a road to PKB. *Biochem J* **346:** 561–576.

Venuprasad K, Parab P, Prasad DV, Sharma S, Banerjee PR, Deshpande M, Mitra DK, Pal S, Bhadra R, Mitra D et al. 2001. Immunobiology of CD28 expression on human neutrophils. I. CD28 regulates neutrophil migration by modulating CXCR-1 expression. *Eur J Immunol* **31:** 1536–1543.

Vilimek D, Duronio V. 2006. Cytokine-stimulated phosphorylation of GSK-3 is primarily dependent upon PKCs, not PKB. *Biochem Cell Biol* **84:** 20–29.

Villalba M, Coudronniere N, Deckert M, Teixeiro E, Mas P, Altman A. 2000a. A novel functional interaction between Vav and PKCθ is required for TCR-induced T cell activation. *Immunity* **12:** 151–160.

Villalba M, Hernandez J, Deckert M, Tanaka Y, Altman A. 2000b. Vav modulation of the Ras/MEK/ERK signaling pathway plays a role in NFAT activation and CD69 up-regulation. *Eur J Immunol* **30:** 1587–1596.

Villegas EN, Elloso MM, Reichmann G, Peach R, Hunter CA. 1999. Role of CD28 in the generation of effector and memory responses required for resistance to Toxoplasma gondii. *J Immunol* **163:** 3344–3353.

Vincenti F, Larsen C, Durrbach A, Wekerle T, Nashan B, Blancho G, Lang P, Grinyo J, Halloran PF, Solez K et al. 2005. Costimulation blockade with belatacept in renal transplantation. *N Engl J Med* **353:** 770–781.

Waldmann TA, Morris JC. 2006. Development of antibodies and chimeric molecules for cancer immunotherapy. *Adv Immunol* **90:** 83–131.

Wang H, Rudd CE. 2008. SKAP-55, SKAP-55-related and ADAP adaptors modulate integrin-mediated immune-cell adhesion. *Trends Cell Biol* **18:** 486–493.

Wang D, Matsumoto R, You Y, Che T, Lin XY, Gaffen SL, Lin X. 2004. CD3/CD28 costimulation-induced NF-κB activation is mediated by recruitment of protein kinase C-θ, Bcl10, and IκB kinase β to the immunological synapse through CARMA1. *Mol Cell Biol* **24:** 164–171.

Wang D, You Y, Case SM, McAllister-Lucas LM, Wang L, DiStefano PS, Nunez G, Bertin J, Lin X. 2002. A requirement for CARMA1 in TCR-induced NF-κ B activation. *Nat Immunol* **3:** 830–835.

Watanabe N, Gavrieli M, Sedy JR, Yang J, Fallarino F, Loftin SK, Hurchla MA, Zimmerman N, Sim J, Zang X et al. 2003. BTLA is a lymphocyte inhibitory receptor with similarities to CTLA-4 and PD-1. *Nat Immunol* **4:** 670–679.

Watanabe R, Harada Y, Takeda K, Takahashi J, Ohnuki K, Ogawa S, Ohgai D, Kaibara N, Koiwai O, Tanabe K et al. 2006. Grb2 and Gads exhibit different interactions with CD28 and play distinct roles in CD28-mediated costimulation. *J Immunol* **177:** 1085–1091.

Williams JA, Lumsden JM, Yu X, Feigenbaum L, Zhang J, Steinberg SM, Hodes RJ. 2008. Regulation of thymic NKT cell development by the B7-CD28 costimulatory pathway. *J Immunol* **181:** 907–917.

Winzen R, Kracht M, Ritter B, Wilhelm A, Chen CY, Shyu AB, Muller M, Gaestel M, Resch K, Holtmann H. 1999. The p38 MAP kinase pathway signals for cytokine-induced mRNA stabilization via MAP kinase-activated protein kinase 2 and an AU-rich region-targeted mechanism. *Embo J* **18:** 4969–4980.

Woerly G, Lacy P, Younes AB, Roger N, Loiseau S, Moqbel R, Capron M. 2002. Human eosinophils express and release IL-13 following CD28-dependent activation. *J Leukoc Biol* **72:** 769–779.

Wood JE, Schneider H, Rudd CE. 2006. TcR and TcR-CD28 Engagement of Protein Kinase B (PKB/AKT) and Glycogen Synthase Kinase-3 (GSK-3) Operates Independently of Guanine Nucleotide Exchange Factor VAV-1. *J Biol Chem* **281:** 32385–32394.

alternative activities are controlled by whether or not SLAM-related receptors are coexpressed with members of the SLAM-associated protein (SAP) family of intracellular adaptor molecules. The functions and mechanisms of action of the SLAM and SAP families are reviewed herein.

THE SLAM FAMILY OF RECEPTORS

Properties

The SLAM family of receptors is a group of type I transmembrane receptors that includes SLAM (CD150; SLAMF1), 2B4 (CD244; SLAMF4), Ly-9 (CD229; SLAMF3), natural killer, T- and B-cell antigen (NTB-A) or Ly108 (in the mouse) (SLAMF6), CD84 (SLAMF5), and CD2-like receptor activating cytotoxic cells (CRACC; CD319; and SLAMF7) (Table 1) (Schwartzberg et al. 2009; Ma et al. 2007; Veillette et al. 2007; Veillette 2006b; Veillette 2006a). SLAM-family receptors are broadly expressed in immune cells, and are not found in nonimmune cells. They possess an extracellular segment usually composed of two immunoglobulin (Ig)-like domains—one variable (V)-like domain and one constant 2 (C2)-like domain—in addition to a transmembrane region and a cytoplasmic domain bearing multiple tyrosine-based motifs. The one exception to this structural organization is Ly-9, which has four Ig-like domains in its extracellular domain (two tandem repeats of the basic V-like and C2-like organization).

One of the distinguishing features of SLAM-family members is that, unlike most other receptors expressed in immune cells, SLAM-related receptors are typically self-ligands. There is only one exception to this property, that is 2B4, which interacts with CD48, another Ig superfamily member found on nearly all hematopoietic cells (Latchman et al. 1998; Brown et al. 1998). Thus, SLAM-family receptors can be triggered in the context of homotypic or heterotypic cell–cell interactions (Fig. 1). SLAM is also the receptor for a variety of morbilliviruses (Tatsuo et al. 2000). These include measles virus, which binds and enters immune cells via SLAM in humans. In all cases examined by crystallography (NTB-A, 2B4, and CD84), the association of SLAM-family receptors with their physiological ligand was observed to be mediated via the amino-terminal V-like domain (Cao et al. 2006; Velikovsky et al. 2007; Yan et al. 2007).

Table 1. SLAM-family receptors

Receptor	Physiological ligand	Expression pattern	Interaction with SAP	Interaction with EAT-2 (ERT)	Phenotypes knockout mice	Genetic background knockout mice
SLAM (SLAMF1)	SLAM	T, B, DC, Mφ, plat.	+	+	T, Mφ, plat., NK-T (with Ly108)	mixed 129-B6
2B4 (SLAMF4)	CD48	NK, T (CD8$^+$), DC, Mφ, eos.	+	+	NK	pure B6
NTB-A/Ly108 (SLAMF6)	NTB-A/Ly108	T, B, NK, DC, neutro.	+	+	T, neutro., NK-T	mixed 129-B6
Ly-9 (SLAMF3)	Ly-9	T, B, NK, DC, Mφ	+	+	T	mixed 129-B6
CD84 (SLAMF5)	CD84	T, B, NK, DC, Mφ, plat., mast, eos.	+	+	Not reported	Not reported
CRACC (SLAMF7)	CRACC	T (activated), B, NK, DC, Mφ	−	+	NK	pure 129

The salient features of SLAM-family receptors are depicted. T, T cells; B, B cells; DC, dendritic cells, Mφ, macrophages; plat., platelets; NK, natural killer cells; eos., eosinophils; neutro., neutrophils; mast, mast cells; NK-T, natural killer-T cells; B6, C57BL/6.

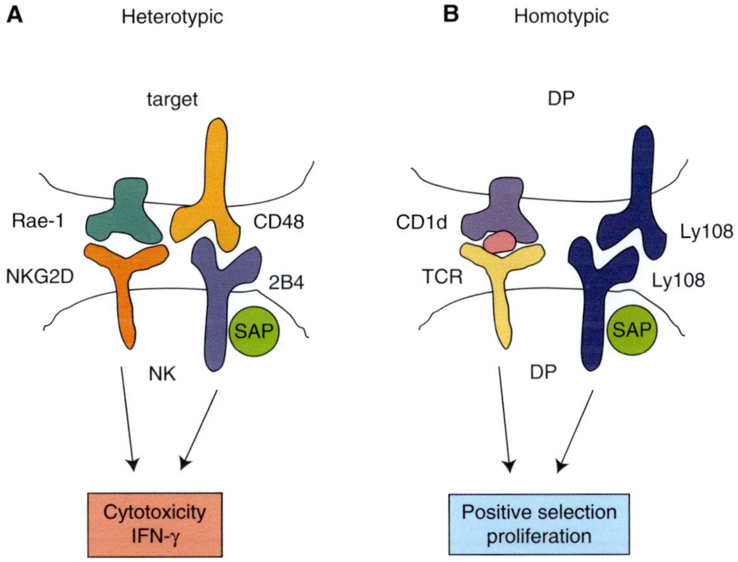

Figure 1. Triggering of SLAM-family receptors by heterotypic or homotypic cell–cell interactions. All SLAM-family receptors, except 2B4, are self-ligands. In the case of 2B4, the ligand is CD48, another receptor expressed on hematopoietic cells. Consequently, SLAM-family receptors can be triggered during heterotypic or homotypic cell–cell interactions. (*A*) Heterotypic cell–cell interactions. Triggering of 2B4 on natural killer cells (NK) by CD48 on target cells results in SAP family adaptor-dependent signals, which enhance cytotoxicity and interferon (IFN)-γ secretion induced through stimulation of the primary activating receptor NKG2D by its ligand Rae-1. (*B*) Homotypic cell–cell interactions. Triggering of Ly108 by Ly108 on two double-positive thymocytes (DP) results in SAP-dependent signals, which promote positive selection and/or proliferation of DP expressing the appropriate T cell receptor (TCR) and engaged by self-lipid-loaded CD1d.

The genes coding from SLAM-family receptors are located within a ~400 kilobase (kb)-cluster on chromosome 1, in humans and mice (Morra et al. 2001). This observation, coupled with the conserved exon-intron structure of *SLAM*-related genes, implies that the *SLAM* family was generated by sequential duplication of a single ancestor gene. Sequence polymorphisms were documented in multiple members of the SLAM family. In some cases, these polymorphisms alter protein sequence. Interestingly, polymorphisms of *SLAM-family* genes among mouse strains, in particular of the Ly108-encoding gene, were found to correlate with susceptibility to the auto-immune disease systemic lupus erythematosus (SLE) (Wandstrat et al. 2004). Likewise, variations in the Ly-9- and 2B4-encoding genes were linked to susceptibility to SLE and rheumatoid arthritis (RA), respectively, in humans (Cunninghame Graham et al. 2008; Suzuki et al. 2008).

Functions of SLAM-Family Receptors

Several lines of evidence support the notion that SLAM-family receptors carry out important functions in immune cells (Schwartzberg et al. 2009; Ma et al. 2007; Veillette et al. 2007; Veillette 2006b; Veillette 2006a). These include antibody stimulation experiments, studies in which SLAM-family receptors or their ligands are ectopically expressed on immune cells or target cells, genetic linkage analyses, and the creation of genetically modified mouse strains. In addition, SLAM-family receptors physically associate with SAP-family adaptors, which have been firmly implicated in normal immunity.

Antibody Stimulation Experiments

The first indication that the SLAM family is involved in immune regulation was provided for 2B4 (Garni-Wagner et al. 1993). It was noted that a monoclonal antibody directed against mouse 2B4 was able to stimulate cytotoxicity and interferon (IFN)-γ secretion by lymphokine-activated killer (LAK) cells. It also promoted that ability of some T cells to mediate nonmajor histocompatibility complex (MHC)-restricted killing of target cells. Similar effects were later seen with antibodies directed against human 2B4, as well as human NTB-A and CRACC, on natural killer (NK) cells (Nakajima et al. 1999; Bottino et al. 2001; Bouchon et al. 2001). Thus, these findings suggested that 2B4, NTB-A, and CRACC are activating receptors in NK cells and T cells.

Subsequently, it was reported that antibody-mediated engagement of SLAM potentiated the proliferation of human B cells in response to interleukin-4 (IL-4) and anti-CD40 (Sidorenko and Clark 1993). Another group also showed that stimulation of previously activated human T cells (T-cell blasts) or human T-cell clones by a monoclonal antibody against SLAM was able to induce production of IFN-γ, but not IL-4, even in the absence of concomitant engagement of the T-cell receptor (TCR) (Cocks et al. 1995). A similar antibody raised against mouse SLAM augmented IFN-γ production by mouse T cells stimulated by CD3-specific monoclonal antibodies (Castro et al. 1999). However, unlike human T cells, activation of mouse T cells by anti-SLAM antibodies did not occur in the absence of TCR coengagement. In other studies, antibody-mediated ligation of SLAM was shown to promote cytotoxicity by $CD8^{+}$ T cells, secretion of IL-12 and IL-8 by dendritic cells (DCs), and production of IL-12 by macrophages (Wang et al. 2004; Henning et al. 2001; Rethi et al. 2006; Bleharski et al. 2001; Kruse et al. 2001). Collectively, these data hinted that SLAM is a positive regulator of activation of T cells, B cells, DCs, and macrophages.

Ectopic Expression Studies

Further substantiation of the regulatory activity of SLAM-related receptors was obtained through studies in which ligands for SLAM-family receptors are ectopically expressed on targets for NK cells. Several groups reported that expression of CD48, the ligand of 2B4, on target cells enhanced the cytotoxicity mediated by human NK cells (Nakajima et al. 1999). This effect was subsequently extended to NTB-A, CRACC, and, at least in the mouse, Ly-9 (Bottino et al. 2001; Bouchon et al. 2001; Dong et al. 2009). Yet other studies raised the possibility that, in the mouse, 2B4 may carry out an inhibitory function in NK cells. Indeed, two groups found that expression of CD48 on target cells inhibited, rather than enhanced, mouse NK-cell functions (Lee et al. 2004; Mooney et al. 2004). Nonetheless, two additional laboratories found that expression of CD48 on targets augmented mouse NK-cell cytotoxicity (Dong et al. 2009; Bloch-Queyrat et al. 2005). Although the basis for this disparity is not clarified, these observations nonetheless supported the idea that 2B4 regulates NK-cell functions.

Other data also suggested that 2B4, as well as other SLAM-related receptors, can inhibit immune cell functions in some settings. Chief among these was the finding that expression of CD48 on targets inhibited the ability of normal immature or uterine human NK cells to mediate killing (Vacca et al. 2006; Sivori et al. 2002). Likewise, expression of CD84 on the rat basophilic leukemia cell line RBL-2H3 suppressed IgE-mediated degranulation (Oliver-Vila et al. 2008). Furthermore, expression of CRACC on a fibroblast cell line bearing class II MHC molecules inhibited antigen-triggered activation of mouse T cells (Cruz-Munoz et al. 2009).

Hence, in combination, these findings raised the possibility that SLAM-family receptors have pleiotropic effects in immune cells, and that they can be at times activating and at times inhibitory.

Genetic Linkage Analyses

Genetic data cemented the proposal that SLAM-family receptors have pivotal roles in immune regulation. It was found that the *Sle1b* locus, which confers enhanced susceptibility to SLE in mouse strains such as NZM2410 and 129Sv compared with C57BL/6, corresponds to the *Slam-family* locus on mouse chromosome 1

(Wandstrat et al. 2004). Polymorphisms in the gene encoding Ly108 were found to correlate most robustly with disease susceptibility, although the genes coding for other members of the SLAM family (in particular Ly-9) may also participate. These polymorphisms were reported to regulate lupus susceptibility by altering T-cell and B-cell functions (Kumar et al. 2006; Wandstrat et al. 2004).

Likewise, the *Nkt1* locus, which influences NK-T cell numbers and functions in autoimmune nonobese diabetes (NOD) mice, was also observed to correspond to the *Slam* locus (Jordan et al. 2007). In particular, the genes coding for SLAM and Ly108 were identified as likely susceptibility genes. The *Nkt1* locus in NOD mice was shown to result in reduced SLAM expression on double-positive thymocytes, possibly explaining the reduced NK-T cell numbers in these animals.

Lastly, in humans, a polymorphism in the gene coding for 2B4 was reported to increase the susceptibility to rheumatoid arthritis in a Japanese population (Suzuki et al. 2008). Moreover, a polymorphism in the gene encoding Ly-9 affected susceptibility to familial SLE in Canada and the United Kingdom (Cunninghame Graham et al. 2008). It also influenced the representation of $CD8^+$ and $CD4^+$ T-cell subsets.

Combined, these observations provided strong correlative evidence that, through sequence variations that affect protein function or expression, *SLAM*-related genes are implicated in the pathophysiology of auto-immune diseases in humans and mice.

Analyses of Genetically Modified Mice

A major limitation of antibody stimulation experiments and overexpression studies is that these manipulations can evoke nonphysiological effects. There are many examples highlighting this caveat. Similarly, genetic linkage analyses provide only correlative evidence of involvement of gene products in a function. For these reasons, studies of genetically modified mice are crucial for formally determining the physiological activities of molecules, although conventional gene ablation approaches do not address well the possibilities of functional redundancy and compensation by other molecules.

Mutant mice were engineered for several SLAM-family receptors (Table 1). Mice lacking SLAM were found to show a defect in TCR-mediated IL-4 secretion (Wang et al. 2004; Davidson et al. 2004). Moreover, their macrophages had altered production of IL-12, tumor necrosis factor-α (TNF-α), IL-6, and nitric oxide (NO) in response to lipopolysaccharide (LPS). Lastly, SLAM-deficient platelets showed defective aggregation in response to collagen or thrombin receptor stimulation (Nanda et al. 2005).

2B4-deficient mice were observed to have a complex NK-cell phenotype, which was not expected from the earlier antibody stimulation experiments (Vaidya et al. 2005; McNerney et al. 2005; Lee et al. 2004). NK cells lacking 2B4 had augmented, instead of diminished, cytotoxicity toward target cells. Intriguingly, this effect was restricted to target cells expressing CD48. In contrast, an enhancement of cytotoxicity was seen with target cells lacking CD48. Whereas these genetic data undoubtedly implicated 2B4 in the regulation of NK-cell-mediated cytotoxicity, they also implied that 2B4-CD48 interactions can have two types of effect. Seemingly, 2B4 is inhibitory when engaged by CD48 located on target cells. However, it becomes activating when triggered by CD48 on other NK cells. Additional evaluations are needed to explain these findings, especially in the context of the other data indicating that 2B4 is primarily an activating receptor in NK cells.

Mice devoid of CRACC were also found to have an NK-cell defect (Cruz-Munoz et al. 2009). CRACC-deficient NK cells had a compromised ability to kill and produce IFN-γ in response to target cells expressing CRACC, in keeping with the notion that CRACC is a self-ligand. Interestingly, their aptitude to react to CRACC-negative targets was also reduced, although not as severely as the response to CRACC-positive targets. Hence, it would appear that CRACC–CRACC interactions promote NK-cell activation in the context of heterotypic

NK-cell–target-cell interactions and homotypic NK-cell–NK-cell interactions.

Ly108-deficient mice were observed to have a defect in TCR-mediated IL-4 secretion and compromised neutrophil bactericidal activity (Howie et al. 2005). Likewise, they showed reduced NK-T cell development, which was further compromised by combining Ly108 deficiency with SLAM deficiency (Griewank et al. 2007). The effect of Ly108 and SLAM deficiency on NK-T cell development was reported to be caused by compromised homotypic interactions among double-positive thymocytes, which contain the precursors of NK-T cells. Last, mice with a mutation of the Ly-9-encoding gene had compromised TCR-induced proliferation and IL-2 secretion (Graham et al. 2006).

These gene ablation experiments provided the best direct indication of the importance of the SLAM family in immune regulation. Moreover, they exemplified the idea that the functions of SLAM-family receptors can be triggered by heterotypic or homotypic interactions among immune cells. However, it is noteworthy that these data also failed to confirm several of the functions of SLAM-family receptors expected based on antibody stimulation experiments. This situation offers one additional example that the results of antibody stimulation experiments need to be interpreted with caution, as antibodies can be agonistic or antagonistic, or, at times, trigger nonphysiological effects.

Nonetheless, there are also some caveats to the gene ablation experiments. First, they may underestimate the functions of individual SLAM-family receptors. This is especially important for SLAM-family receptors because a given immune cell type typically expresses multiple family members. For example, T cells express SLAM, NTB-A, Ly-9, CD84, and, occasionally, CRACC and 2B4. Likewise, NK cells bear 2B4, NTB-A, CRACC, Ly-9, and CD84. The creation of mice simultaneously carrying mutations in multiple SLAM-family members will be critical to ascertain this possibility, although it is technically complicated by the close juxtaposition of *Slam*-related genes on mouse chromosome 1.

Second, it is possible that variations in genetic background underlie or mask some of the phenotypes observed in gene-ablated mice (Veillette et al. 2006). As pointed out earlier, in the mouse, the *Slam* locus corresponds to the *Sle1b* and *Nkt1* loci, which confer enhanced susceptibility to autoimmunity and are associated with altered T-cell and B-cell functions. The *Slam* locus is known to show marked structural and functional differences between C57BL/6 and 129Sv mice (Wandstrat et al. 2004). As the various genes of the *Slam* locus cosegregate during genomic recombination, it is plausible that some of the phenotypes observed in mutant mice analyzed in a mixed C57BL/6-129Sv background were caused by, or were masked by, transfer of the *Slam* locus from 129Sv to C57BL/6. Mice with a mixed background were used for mutations of SLAM, Ly108, and Ly-9 (Howie et al. 2005; Wang et al. 2004; Davidson et al. 2004; Graham et al. 2006). In the light of this, it will be important to confirm the phenotypes, or lack of phenotypes, in the mutant animals using a pure genetic background.

Interactions with SAP-Family Adaptors

The most striking, albeit only indirect, substantiation of the involvement of SLAM-family receptors in immunity was the finding that they associate with SAP-family adaptors (Schwartzberg et al. 2009; Ma et al. 2007; Veillette et al. 2007; Veillette 2006b; Veillette 2006a). The prototype of the SAP family, that is SAP, is mutated in a human immunodeficiency, X-linked lymphoproliferative (XLP) disease. Moreover, mice lacking one or more members of the SAP family have marked immune phenotypes. Thus, such findings further implicated the SLAM family in immune regulation. The SAP family will now be reviewed.

THE SAP FAMILY OF ADAPTORS

Properties

The SAP family of adaptors includes three members most commonly known as SAP (also

Table 2. SAP-family adaptors.

Adaptor	Chromosomal localization	Expression pattern	Signaling mechanism	Phenotypes knockout mice	Genetic background knockout mice	Human disease
SAP (SH2D1A)	X (H, M)	T, ?B, NK-T, NK, plat.	Fyn, ?PIX, ?Nck (arginine 78)	T_H2, T_{FH}, B (?indirect), NK-T, NK, plat.	Mixed 129-B6	XLP
EAT-2 (SH2D1B1)	1 (H, M)	NK, DC, Mφ	?PLC-γ1 (carboxy-terminal tyrosines)	NK	129	none
ERT (SH2D1B2)	1 (H, M; pseudo-gene in H)	NK	(?carboxy-terminal tyrosines)	NK	129	none

The salient features of SAP-family adaptors are depicted. H, human; M, mouse; T, T cells; B, B cells; NK-T, natural killer-T cells; NK, natural killer cells; plat., platelets; DC, dendritic cells, Mφ, macrophages; PLC, phospholipase C; C, carboxyl; T_H2, T helper 2 cells; T_{FH}, follicular helper T cells; B6, C57BL/6; XLP, X-linked lymphoproliferative disease.

named SH2D1A), Ewing's sarcoma-associated transcript-2 (EAT-2; also named SH2D1B1), and EAT-2-related transducer (ERT; also named SH2D1B2) (Table 2) (Schwartzberg et al. 2009; Ma et al. 2007; Veillette et al. 2007; Veillette 2006b; Veillette 2006a). These intracellular molecules are composed of a Src homology 2 (SH2) domain, in addition to a short carboxy-terminal region. They appear to be solely expressed in immune cells. Although SAP is present in T cells, NK cells, NK-T cells, and, perhaps, some B cells, EAT-2 is found in NK cells, DCs, and macrophages. ERT is only observed in NK cells, at least in the mouse. The SAP-encoding gene is located on the X chromosome in humans and mice. In contrast, the genes coding for EAT-2 and ERT are positioned in tandem on chromosome 1, in the two species (Calpe et al. 2006). They are located approximately 1500 kb and 1300 kb away from the *SLAM* locus in humans and mice, respectively. In humans, the ERT-encoding gene is a pseudogene and is nonfunctional.

Through their SH2 domain, SAP-family adaptors associate with the tyrosine-based motif TI/VYxxV/I (in which T is threonine, I is isoleucine, V is valine, Y is tyrosine and x is any residue) in the cytoplasmic domain of SLAM-related receptors (Sayos et al. 1998; Schwartzberg et al. 2009; Ma et al. 2007; Veillette et al. 2007; Veillette 2006b; Veillette 2006a). These interactions are of high affinity and specificity. Current evidence suggests that all SAP-family adaptors can interact with all SLAM-family receptors, with the clear exception of CRACC that associates with EAT-2 and presumably ERT, but not SAP (Tassi and Colonna 2005; Bouchon et al. 2001; Cruz-Munoz et al. 2009). Compared with other SH2 domain-mediated associations, the interactions between SAP-related adaptors and SLAM-family receptors have unusual structural characteristics. These properties and their potential significance are reviewed elsewhere (Schwartzberg et al. 2009; Ma et al. 2007; Veillette et al. 2007; Veillette 2006b; Veillette 2006a).

SAP Is Mutated in X-Linked Lymphoproliferative Disease

The SAP-encoding gene (*SH2D1A*) is mutated in 60%–70% of cases of XLP, an X-linked human immunodeficiency characterized by a disproportionate and ineffective response to Epstein-Barr virus (EBV) infection (Coffey et al. 1998; Sayos et al. 1998; Nichols et al. 1998). All SAP mutations found in XLP are null mutations or loss-of-function mutations.

Although readers are referred elsewhere for in-depth discussions of the clinical and immunological features of XLP (Schwartzberg et al. 2009; Ma et al. 2007; Veillette et al. 2007; Veillette 2006b; Veillette 2006a), it is worth mentioning that, on EBV infection, SAP-deficient patients typically show a massive lymphoproliferative illness, with marked accumulation of immune cells in lymphoid and nonlymphoid tissues. They also have a heightened frequency of malignant lymphomas. Moreover, at the immunological level, they show defects in the functions or development of several immune cell types, including $CD4^+$ T cells, $CD8^+$ T cells, NK-T cells, B cells, and NK cells (Table 2). It is believed that, combined, these alterations explain the abnormal immune response to EBV.

Mutations in genes other than *SH2D1A* were recently identified in some patients with XLP or XLP-like illnesses. The gene encoding the X-linked inhibitor of apoptosis (XIAP) was found to be mutated in 5%–10% of cases of XLP (Rigaud et al. 2006). Furthermore, one group observed that the protein tyrosine kinase (PTK) Itk was mutated in two female relatives having an XLP-like immunodeficiency (Huck et al. 2009). Whether or not XIAP and Itk are part of a common signaling pathway with SAP remains to be elucidated.

Immune Defects in Mice Lacking SAP-Family Adaptors

The involvement of SAP-family adaptors in immune regulation was also shown by the generation of mice carrying inactivating mutations in their genes (Table 2). Mutations of the SAP-encoding gene largely recreated the phenotypes of SAP-deficient humans (Schwartzberg et al. 2009; Ma et al. 2007; Veillette et al. 2007; Veillette 2006b; Veillette 2006a). Defects in $CD4^+$ T-cell functions, including reduced T-helper 2 (T_H2) cytokine production and defective follicular T-helper (T_{FH}) cell functions, were observed. This T_{FH} defect leads to severely compromised germinal center formation and antibody production by B cells, although a T cell-independent intrinsic B-cell defect may coexist (Veillette et al. 2008; Qi et al. 2008; Cannons et al. 2006). Alterations of $CD8^+$ T cells, including reduced cytotoxicity and impaired activation-induced cell death, were also noted (Chen et al. 2007; Chen et al. 2005). Moreover, a near absence of NK-T cells, caused by an early block in NK-T cell development, was found (Chung et al. 2005; Nichols et al. 2005; Pasquier et al. 2005). Likewise, reduced NK-cell-mediated cytotoxicity and IFN-γ secretion were observed (Dong et al. 2009; Bloch-Queyrat et al. 2005).

Mice lacking EAT-2 or ERT were shown to show complex alterations in NK-cell functions (Roncagalli et al. 2005). Unexpectedly, an increase, rather than a decrease, in cytotoxicity and IFN-γ production was observed in response to stimulation of various activating receptors. This finding suggested that EAT-2 and ERT are negative regulators of NK-cell activation. However, other analyses, including studies of CRACC, which binds EAT-2 but not SAP, indicated that EAT-2 and ERT can also be positive regulators of NK-cell activation (Cruz-Munoz et al. 2009). Although the basis for this dual activity still needs to be clarified, it appears that EAT-2 (and ERT) are activating when SLAM-family receptors are triggered as part of the NK-cell stimulation by target cells (Dong et al. 2009). In contrast, they are inhibitory when SLAM-family receptors are not triggered by target cells.

Recently, mice simultaneously lacking all three SAP-family adaptors were reported (Dong et al. 2009). It was observed that NK cells devoid of all SAP-family adaptors had much more severely compromised NK-cell functions than mice lacking individual SAP-family adaptors. This observation implied that the three members of the SAP family cooperate in a mechanism promoting NK-cell activation. Interestingly, NK cells lacking all SAP-family adaptors had a nearly abolished capacity to mediate cytotoxicity and produce IFN-γ in response to hematopoietic, but not nonhematopoietic, target cells. This was because of the fact that hematopoietic, but not nonhematopoietic, cells express ligands for SLAM-family receptors. As was reported for NK cells lacking EAT-2 or ERT alone, NK cells devoid of all SAP-related adaptors also showed enhanced, instead of diminished, activation in response to nonhematopoietic target cells.

Thus, studies of SAP-deficient humans and of mice lacking one or more SAP family members constituted undeniable evidence of the importance of these molecules in immune regulation. Because SAP-family adaptors interact primarily, if not exclusively, with SLAM-family receptors, these analyses also helped to implicate SLAM-related receptors in normal immunity.

Signaling Mechanisms of SAP-Family Adaptors

Given that SAP-family adaptors are almost exclusively made of an SH2 domain, it was proposed at first that they may be "natural blockers" of associations with other SH2 domain-containing molecules (Sayos et al. 1998). In keeping with this idea, it was observed that SAP and EAT-2 could prevent the binding of SLAM-family members to inhibitory signaling molecules such as SH2 domain-containing phosphatase (SHP)-2 and SHP-1, SH2 domain-containing 5' inositol phosphatase (SHIP)-1 and Csk, both in vitro and in overexpression studies (Schwartzberg et al. 2009; Ma et al. 2007; Veillette et al. 2007; Veillette 2006b; Veillette 2006a). Such a mechanism was postulated to prohibit SLAM-family receptors from mediating inhibitory signals, thereby favoring cell activation. However, this model was not validated in experiments using normal immune cells.

Subsequent studies unraveled that SAP-family adaptors possess sequences that can also couple them to specific biochemical signals (Fig. 2). SAP was found to associate with the Src-related PTK Fyn, by way of a short motif located within the SAP SH2 domain (Latour et al. 2003; Latour et al. 2001; Chan et al. 2003). This Fyn-binding motif is centered on SAP arginine 78 (R78), which binds directly to the Src homology 3 (SH3) region of Fyn. Such a sequence is not found in EAT-2 and ERT. Because the SAP-Fyn interaction occurs simultaneously with binding of SAP to SLAM-family receptors, it permits SAP to link SLAM-family receptors to protein tyrosine phosphorylation signals. This association appears to be critical for some, but not all, functions of SAP (Nunez-Cruz et al. 2008; Davidson et al. 2004; Cannons et al. 2004; Qi et al. 2008; Cannons et al. 2006). For instance, it is essential for the ability of SAP to promote T_H2 cytokine production, but has little or no role in the aptitude of SAP to augment T_{FH} functions. An intermediate requirement exists for NK-T cell development. It should be pointed out that, in addition to binding Fyn, the R78-based motif of SAP can interact with other SH3 domain-containing proteins such as PAK-interacting exchange factor (PIX) and the adaptor Nck (Gu et al. 2006; Li et al. 2009). The physiological relevance of these other associations is not clarified.

EAT-2 and ERT harbor sites of tyrosine phosphorylation in their carboxy-terminal tail (Roncagalli et al. 2005). These sites are not present in SAP. Two tyrosines exist in mouse EAT-2 and ERT (tyrosines 120 and 127), whereas only one is located in human EAT-2 (tyrosine 127). These tyrosines are positioned in classical SH2 domain-binding motifs. Evidence is accumulating that the carboxy-terminal tyrosines are critical for the activating and inhibitory effects of EAT-2 (and presumably ERT) in NK cells (Roncagalli et al. 2005; Cruz-Munoz et al. 2009; Clarkson and Brown 2009). Although it is presumed that the phosphorylated tyrosines of EAT-2 and ERT enable to couple SLAM-family receptors to other SH2 domain-containing downstream effectors, the identity of these effectors is not clearly established. Nonetheless, recent data suggested that phospholipase C (PLC)-γ may be recruited to tyrosine 127 in human EAT-2 (Clarkson and Brown 2009).

Collectively, these observations implied that SAP, EAT-2, and ERT are active signaling molecules that couple SLAM-related receptors to specific biochemical signals, although the nature of these signals still needs to be fully elucidated. It is conceivable that, under certain circumstances, SAP-family adaptors also function by blocking the association of SLAM-related with other SH2 domain-containing molecules. However, as suggested in the next section, this

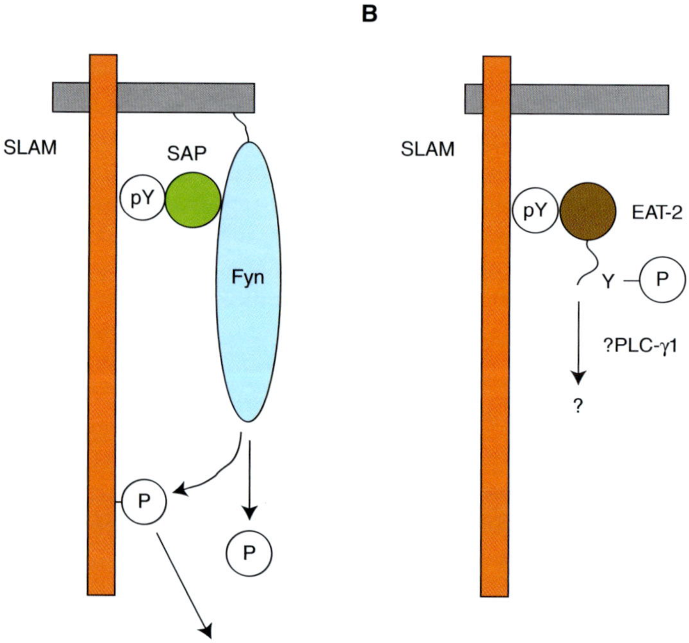

Figure 2. SAP-related adaptors mediate active biochemical signals. Through intrinsic mechanisms, SAP-family adaptors can couple SLAM-family receptors to active signals. (*A*) SAP. Via its Src homology 2 (SH2) region, SAP associates with a phosphorylated tyrosine-based motif (pY) in SLAM-related receptors (SLAM). By way of the arginine 78-based motif in its SH2 domain, SAP simultaneously binds to the Src family protein tyrosine kinase Fyn, through the Fyn Src homology 3 (SH3) domain. This allows SAP to couple SLAM-related receptors to protein tyrosine phosphorylation signals. (*B*) EAT-2. Like SAP, EAT-2 (and probably ERT; not shown) associates with a phosphorylated tyrosine-based motif (pY) in SLAM-related receptors (SLAM), through the EAT-2 SH2 domain. Via one or two phosphorylated tyrosines located in its carboxy-terminal tail, EAT-2 concomitantly interacts with its effectors. Although the identity of these effectors is not well defined, recent data suggested that phospholipase C (PLC)-γ may be involved.

may not be through a classical natural blocker model.

THE FUNCTIONS OF THE SLAM FAMILY ARE DEFINED BY THE PRESENCE OR THE ABSENCE OF THE SAP FAMILY

Differential Phenotypes of Deficiencies in SLAM Family or SAP Family

Considering the specificity of the interaction between SLAM-family receptors and SAP-family adaptors, it seems logical to speculate that SLAM-related receptors are responsible for the functions mediated by SAP-related adaptors. However, full acceptance of this notion has been hampered by the difficulty to recreate the severe phenotypes, which are caused by SAP family deficiencies, with mutations of the SLAM family. The one possible exception is the block of NK-T development caused by SAP deficiency that can be recapitulated to a considerable, albeit not full, extent by combined mutations of Ly108 and SLAM (Griewank et al. 2007).

Several possibilities may explain this discrepancy. First, as suggested earlier, it is likely that some functional redundancy exists between SLAM-family members. In this context, ablation of multiple members of the SLAM family may be needed to mimic the impact of SAP family deficiencies. Experiments to test this idea are underway in several laboratories. Second, it is possible that the functions of SLAM-family receptors are qualitatively altered, rather than eliminated, by the absence of SAP-related adaptors. Consequently, there would be distinct phenotypes for SAP family deficiencies compared with SLAM-family deficiencies. Recent analyses of NK-cell functions in mice lacking all three SAP-family adaptors provided clear support for this proposition, as will be presented in the upcoming section. And third, it is conceivable that SAP-family adaptors also mediate SLAM family independent functions. Whereas this idea is supported by the finding of associations between SAP and receptors such as CD22 and FcγRIIB, the physiological relevance of these associations is not clear (Ostrakhovitch et al. 2009; Li et al. 2008). Nonetheless, this possibility still deserves consideration.

"Switch-of-Function" of SLAM-Family Receptors in the Absence of the SAP Family

Recently, it was found that, in the absence of all SAP-family adaptors, SLAM-family receptors on mouse NK cells, including 2B4, Ly108, CRACC, Ly-9, and CD84, not only lose their activating function, but also are converted into potent inhibitory receptors (Fig. 3) (Dong et al. 2009). This inhibitory activity enables SLAM-family receptors to suppress the function of other activating NK-cell receptors such as natural-killer group 2, member D (NKG2D), CD16 and, probably, DNAX accessory molecule-1 (DNAM-1), thereby leading to a marked suppression of NK-cell responsiveness toward hematopoietic target cells. A related switch-of-function from activating to inhibitory was also described for 2B4 and NTB-A in SAP-deficient human NK cells (Bottino et al. 2001; Bottino et al. 2000). Likewise, it was noted for 2B4 in normal immature and uterine human NK cells, which express low amounts of SAP (Vacca et al. 2006; Sivori et al. 2002).

Taking into consideration these findings, it is tempting to speculate that other manifestations of SAP family deficiencies may also occur through a switch-of-function mechanism. In this manner, the T_{FH} and T_H2 defects, the block in NK-T cell development, and the $CD8^+$ T cell dysfunctions caused by SAP deficiency may be

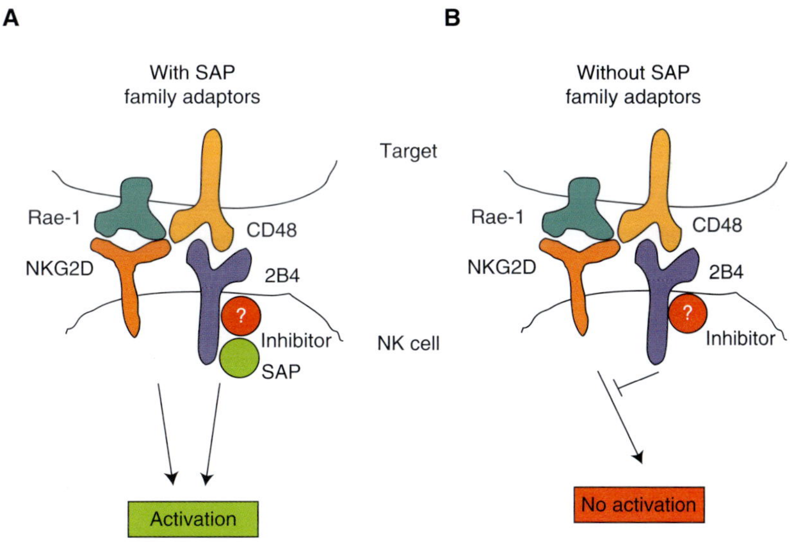

Figure 3. "Switch-of-function" of SLAM family receptors in the absence of SAP-family adaptors. (*A*) In the presence of SAP-family adaptors, SLAM-family receptors (2B4 is depicted here) mediate positive signals that promote natural killer cell (NK) activation by target cells. This is because SAP family adaptor-mediated signals dominate over signals mediated by inhibitory effectors, which are proposed to bind a tyrosine-based motif distinct from the one interacting with SAP-related adaptors. (*B*) In the absence of SAP-family adaptors, however, SLAM-family receptors mediate negative signals that inhibit natural killer cell (NK) activation by target cells. This is because inhibitory signals dominate.

due at least in part to a change, rather than a loss, in the function of SLAM-related receptors. A comparison of the phenotypes caused by the absence of the SAP family with those induced by simultaneous loss of all SLAM-related receptors will help address this possibility.

Mechanism of Switch-of-Function by SAP-Family Adaptors

A simple way to explain the switch-of-function of SLAM-related receptors in the absence of the SAP family is based on the natural-blocker model (Sayos et al. 1998). Accordingly, in the absence of SAP-related adaptors, SLAM-family receptors would preferentially associate with inhibitory effectors such as SHP-1, SHP-2, SHIP-1, or Csk, and, thereby, mediate inhibitory effects. Even though data in transfected cells supported this hypothesis, these findings were not substantiated by analyses in normal immune cells (Schwartzberg et al. 2009; Ma et al. 2007; Veillette et al. 2007; Veillette 2006b; Veillette 2006a).

Nonetheless, a recent report showed that, in the absence of SAP-family adaptors, triggering of 2B4 on mouse NK cells resulted in enhanced tyrosine phosphorylation of the inositol phosphatase SHIP-1 (Dong et al. 2009). This was accompanied by reduced activation of calcium fluxes, Erk, and Akt on coengagement of CD16. Because calcium fluxes, Erk and Akt are known to be inhibited by SHIP-1. Such findings implied that, in the absence of SAP-family adaptors, 2B4 and, possibly, other SLAM-related receptors are more efficiently coupled to inhibitory effectors like SHIP-1.

However, this phenomenon does not appear to reflect direct physical competition between SAP-family adaptors and inhibitory effectors for binding to SLAM-family receptors. Along these lines, studies in mouse NK cells showed that distinct tyrosines in the cytoplasmic domain of CRACC are responsible for CRACC-mediated activation, which occurs in the presence of EAT-2, and for CRACC-mediated inhibition, which takes place in the absence of EAT-2 (Cruz-Munoz et al. 2009). This observation implied that SAP-family adaptors and inhibitory effectors do not compete for binding to the same tyrosine. Other models such as "functional competition," "trans-regulation," or "steric hindrance" should be considered, as detailed elsewhere (Veillette et al. 2009). Interestingly, these investigations also revealed that SHIP-1, SHP-1, SHP-2, and Csk did not bind directly to the tyrosine-based motif responsible for inhibition, suggesting that they are indirectly recruited to the inhibitory tyrosine.

Altogether, these data supported the view that SHIP-1 and, perhaps, SHP-1, SHP-2, and Csk, participate in the inhibition seen in the absence of SAP-family adaptors. Obviously, genetic data supporting this view remain to be obtained. Importantly, these results also indicated that the switch-of-function mediated by the SAP family is not simply because of direct competitive binding between SAP-related adaptors and inhibitory effectors to SLAM-related receptors. However, the mechanism responsible still needs to be elucidated.

CONCLUSION

The data reviewed herein provided clear evidence that the SLAM and SAP families play crucial roles in normal immunity. These functions take place in multiple cell types, including T cells, NK-T cells, B cells, NK cells, macrophages and, probably, others. Because most SLAM-family receptors have the unusual property of being self-ligands, these activities can be triggered by homotypic or heterotypic cell–cell interactions.

The findings obtained so far also unraveled that SAP-family adaptors enable SLAM-family receptors to mediate positive signals, which usually promote immune-cell activation and/or differentiation. In the absence of SAP-family adaptors, however, SLAM-family receptors switch their function and mediate other signals, which are typically inhibitory in nature. Such effects are clearly observed in immune cells lacking SAP-family adaptors, as seen in cases of XLP in humans and in genetically engineered mutant mouse strains. Moreover, they may occur under physiological circumstances, in immune-cell types expressing little or no

SAP-family adaptors, such as B cells, mast cells, and neutrophils (Veillette et al. 2009).

In summary, SLAM-family receptors are key regulators that can mediate opposite functions in immune cells. These functions are determined by the presence or the absence of SAP-family adaptors. Although the critical importance of the SLAM and SAP families in immune regulation has been undoubtedly established, much work still remains to be performed to understand fully the functions and mechanisms of action of these families in immunity.

ACKNOWLEDGMENTS

Work in my laboratory is supported by grants from the Canadian Institutes of Health Research (CIHR), the Canadian Cancer Society (CCS), and the Howard Hughes Medical Institute (HHMI). I hold the Canada Research Chair in Signaling in the Immune System and am an HHMI International Scholar.

REFERENCES

Bleharski JR, Niazi KR, Sieling PA, Cheng G, Modlin RL. 2001. Signaling lymphocytic activation molecule is expressed on CD40 ligand-activated dendritic cells and directly augments production of inflammatory cytokines. *J Immunol* **167:** 3174–3181.

Bloch-Queyrat C, Fondaneche MC, Chen R, Yin L, Relouzat F, Veillette A, Fischer A, Latour S. 2005. Regulation of natural cytotoxicity by the adaptor SAP and the Src-related kinase Fyn. *J Exp Med* **202:** 181–192.

Bottino C, Augugliaro R, Castriconi R, Nanni M, Biassoni R, Moretta L, Moretta A. 2000. Analysis of the molecular mechanism involved in 2B4-mediated NK cell activation: Evidence that human 2B4 is physically and functionally associated with the linker for activation of T cells. *Eur J Immunol* **30:** 3718–3722.

Bottino C, Falco M, Parolini S, Marcenaro E, Augugliaro R, Sivori S, Landi E, Biassoni R, Notarangelo LD, Moretta L, et al. 2001. NTB-A [correction of GNTB-A], a novel SH2D1A-associated surface molecule contributing to the inability of natural killer cells to kill Epstein-Barr virus-infected B cells in X-linked lymphoproliferative disease. *J Exp Med* **194:** 235–246.

Bouchon A, Cella M, Grierson HL, Cohen JI, Colonna M. 2001. Cutting edge: Activation of NK cell-mediated cytotoxicity by a SAP-independent receptor of the CD2 family. *J Immunol* **167:** 5517–5521.

Brown MH, Boles K, van der Merwe PA, Kumar V, Mathew PA, Barclay AN. 1998. 2B4, the natural killer and T cell immunoglobulin superfamily surface protein, is a ligand for CD48. *J Exp Med* **188:** 2083–2090.

Calpe S, Erdos E, Liao G, Wang N, Rietdijk S, Simarro M, Scholtz B, Mooney J, Lee CH, Shin MS, et al. 2006. Identification and characterization of two related murine genes, Eat2a and Eat2b, encoding single SH2-domain adapters. *Immunogenetics* **58:** 15–25.

Cambier JC, Pleiman CM, Clark MR. 1994. Signal transduction by the B cell antigen receptor and its coreceptors. *Annu Rev Immunol* **12:** 457–486.

Cannons JL, Yu LJ, Hill B, Mijares LA, Dombroski D, Nichols KE, Antonellis A, Koretzky GA, Gardner K, Schwartzberg PL. 2004. SAP regulates T(H)2 differentiation and PKC-theta-mediated activation of NF-kappaB1. *Immunity* **21:** 693–706.

Cannons JL, Yu LJ, Jankovic D, Crotty S, Horai R, Kirby M, Anderson S, Cheever AW, Sher A, Schwartzberg PL. 2006. SAP regulates T cell-mediated help for humoral immunity by a mechanism distinct from cytokine regulation. *J Exp Med* **203:** 1551–1565.

Cao E, Ramagopal UA, Fedorov A, Fedorov E, Yan Q, Lary JW, Cole JL, Nathenson SG, Almo SC. 2006. NTB-A receptor crystal structure: Insights into homophilic interactions in the signaling lymphocytic activation molecule receptor family. *Immunity* **25:** 559–570.

Castro AG, Hauser TM, Cocks BG, Abrams J, Zurawski S, Churakova T, Zonin F, Robinson D, Tangye SG, Aversa G, et al. 1999. Molecular and functional characterization of mouse signaling lymphocytic activation molecule (SLAM): Differential expression and responsiveness in Th1 and Th2 cells. *J Immunol* **163:** 5860–5870.

Chan B, Lanyi A, Song HK, Griesbach J, Simarro-Grande M, Poy F, Howie D, Sumegi J, Terhorst C, Eck MJ. 2003. SAP couples Fyn to SLAM immune receptors. *Nat Cell Biol* **5:** 155–160.

Chen G, Tai AK, Lin M, Chang F, Terhorst C, Huber BT. 2005. Signaling lymphocyte activation molecule-associated protein is a negative regulator of the CD8 T cell response in mice. *J Immunol* **175:** 2212–2218.

Chen G, Tai AK, Lin M, Chang F, Terhorst C, Huber BT. 2007. Increased proliferation of CD8+ T cells in SAP-deficient mice is associated with impaired activation-induced cell death. *Eur J Immunol* **37:** 663–674.

Chung B, Aoukaty A, Dutz J, Terhorst C, Tan R. 2005. Signaling lymphocytic activation molecule-associated protein controls NKT cell functions. *J Immunol* **174:** 3153–3157.

Clarkson NG, Brown MH. 2009. Inhibition and activation by CD244 depends on CD2 and phospholipase C-γ1. *J Biol Chem* **284:** 24725–24734.

Cocks BG, Chang CC, Carballido JM, Yssel H, de Vries JE, Aversa G. 1995. A novel receptor involved in T-cell activation. *Nature* **376:** 260–263.

Coffey AJ, Brooksbank RA, Brandau O, Oohashi T, Howell GR, Bye JM, Cahn AP, Durham J, Heath P, Wray P, et al. 1998. Host response to EBV infection in X-linked lymphoproliferative disease results from mutations in an SH2-domain encoding gene. *Nat Genet* **20:** 129–135.

Cruz-Munoz ME, Dong Z, Shi X, Zhang S, Veillette A. 2009. Influence of CRACC, a SLAM family receptor coupled to the adaptor EAT-2, on natural killer cell function. *Nat Immunol* **10:** 297–305.

Cunninghame Graham DS, Vyse TJ, Fortin PR, Montpetit A, Cai YC, Lim S, McKenzie T, Farwell L, Rhodes B,

Chad L, et al. 2008. Association of LY9 in UK and Canadian SLE families. *Genes Immun* **9:** 93–102.

Davidson D, Shi X, Zhang S, Wang H, Nemer M, Ono N, Ohno S, Yanagi Y, Veillette A. 2004. Genetic evidence linking SAP, the X-linked lymphoproliferative gene product, to Src-related kinase FynT in T(H)2 cytokine regulation. *Immunity* **21:** 707–717.

Dong Z, Cruz-Munoz M, Zhong MC, Chen R, Latour S, Veillette A. 2009. Essential role of SAP family adaptors in natural killer cell surveillance of hematopoietic cells. *Nat Immunol* **10:** 973–980.

Garni-Wagner BA, Purohit A, Mathew PA, Bennett M, Kumar V. 1993. A novel function-associated molecule related to non-MHC-restricted cytotoxicity mediated by activated natural killer cells and T cells. *J Immunol* **151:** 60–70.

Graham DB, Bell MP, McCausland MM, Huntoon CJ, van Deursen J, Faubion WA, Crotty S, McKean DJ. 2006. Ly9 (CD229)-Deficient Mice Exhibit T Cell Defects yet Do Not Share Several Phenotypic Characteristics Associated with SLAM- and SAP-Deficient Mice. *J Immunol* **176:** 291–300.

Griewank K, Borowski C, Rietdijk S, Wang N, Julien A, Wei DG, Mamchak AA, Terhorst C, Bendelac A. 2007. Homotypic interactions mediated by slamf1 and slamf6 receptors control NKT cell lineage development. *Immunity* **27:** 751–762.

Gu C, Tangye SG, Sun X, Luo Y, Lin Z, Wu J. 2006. The X-linked lymphoproliferative disease gene product SAP associates with PAK-interacting exchange factor and participates in T cell activation. *Proc Natl Acad Sci* **103:** 14447–14452.

Henning G, Kraft MS, Derfuss T, Pirzer R, Saint-Basile G, Aversa G, Fleckenstein B, Meinl E. 2001. Signaling lymphocytic activation molecule (SLAM) regulates T cellular cytotoxicity. *Eur J Immunol* **31:** 2741–2750.

Howie D, Laroux FS, Morra M, Satoskar AR, Rosas LE, Faubion WA, Julien A, Rietdijk S, Coyle AJ, Fraser C et al. 2005. Cutting Edge: The SLAM Family Receptor Ly108 Controls T Cell and Neutrophil Functions. *J Immunol* **174:** 5931–5935.

Huck K, Feyen O, Niehues T, Ruschendorf F, Hubner N, Laws HJ, Telieps T, Knapp S, Wacker HH, Meindl A, et al. 2009. Girls homozygous for an IL-2-inducible T cell kinase mutation that leads to protein deficiency develop fatal EBV-associated lymphoproliferation. *J Clin Invest* **119:** 1350–1358.

Jordan MA, Fletcher JM, Pellicci D, Baxter AG. 2007. Slamf1, the NKT cell control gene Nkt1. *J Immunol* **178:** 1618–1627.

Kruse M, Meinl E, Henning G, Kuhnt C, Berchtold S, Berger T, Schuler G, Steinkasserer A. 2001. Signaling lymphocytic activation molecule is expressed on mature cd83(+) dendritic cells and is up-regulated by il-1β. *J Immunol* **167:** 1989–1995.

Kumar KR, Li L, Yan M, Bhaskarabhatla M, Mobley AB, Nguyen C, Mooney JM, Schatzle JD, Wakeland EK, Mohan C. 2006. Regulation of B cell tolerance by the lupus susceptibility gene Ly108. *Science* **312:** 1665–1669.

Latchman Y, McKay PF, Reiser H. 1998. Identification of the 2B4 molecule as a counter-receptor for CD48. *J Immunol* **161:** 5809–5812.

Latour S, Gish G, Helgason CD, Humphries RK, Pawson T, Veillette A. 2001. Regulation of SLAM-mediated signal transduction by SAP, the X-linked lymphoproliferative gene product. *Nat Immunol* **2:** 681–690.

Latour S, Roncagalli R, Chen R, Bakinowski M, Shi X, Schwartzberg PL, Davidson D, Veillette A. 2003. Binding of SAP SH2 domain to FynT SH3 domain reveals a novel mechanism of receptor signalling in immune regulation. *Nat Cell Biol* **5:** 149–154.

Lee KM, McNerney ME, Stepp SE, Mathew PA, Schatzle JD, Bennett M, Kumar V. 2004. 2B4 Acts As a Non-Major Histocompatibility Complex Binding Inhibitory Receptor on Mouse Natural Killer Cells. *J Exp Med* **199:** 1245–1254.

Li C, Schibli D, Li SS. 2009. The XLP syndrome protein SAP interacts with SH3 proteins to regulate T cell signaling and proliferation. *Cell Signal* **21:** 111–119.

Li C, Chung B, Tao J, Iosef C, Aoukaty A, Wang Y, Tan R, Li SS. 2008. The X-linked lymphoproliferative syndrome gene product SAP regulates B cell function through the FcγRIIB receptor. *Cell Signal* **20:** 1960–1967.

Ma CS, Nichols KE, Tangye SG. 2007. Regulation of cellular and humoral immune responses by the SLAM and SAP families of molecules. *Annu Rev Immunol* **25:** 337–379.

McNerney ME, Guzior D, Kumar V. 2005. 2B4 (CD244)-CD48 interactions provide a novel MHC class I-independent system for NK-cell self-tolerance in mice. *Blood* **106:** 1337–1340.

Mooney JM, Klem J, Wulfing C, Mijares LA, Schwartzberg PL, Bennett M, Schatzle JD. 2004. The murine NK receptor 2B4 (CD244) exhibits inhibitory function independent of signaling lymphocytic activation molecule-associated protein expression. *J Immunol* **173:** 3953–3961.

Morra M, Howie D, Grande MS, Sayos J, Wang N, Wu C, Engel P, Terhorst C. 2001. X-linked lymphoproliferative disease: A progressive immunodeficiency. *Annu Rev Immunol* **19:** 657–682.

Nakajima H, Cella M, Langen H, Friedlein A, Colonna M. 1999. Activating interactions in human NK cell recognition: The role of 2B4-CD48. *Eur J Immunol* **29:** 1676–1683.

Nanda N, Andre P, Bao M, Clauser K, Deguzman F, Howie D, Conley PB, Terhorst C, Phillips DR. 2005. Platelet aggregation induces platelet aggregate stability via SLAM family receptor signaling. *Blood* **106:** 3028–3034.

Nichols KE, Harkin DP, Levitz S, Krainer M, Kolquist KA, Genovese C, Bernard A, Ferguson M, Zuo L, Snyder E, et al. 1998. Inactivating mutations in an SH2 domain-encoding gene in X-linked lymphoproliferative syndrome. *Proc Natl Acad Sci* **95:** 13765–13770.

Nichols KE, Hom J, Gong SY, Ganguly A, Ma CS, Cannons JL, Tangye SG, Schwartzberg PL, Koretzky GA, Stein PL. 2005. Regulation of NKT cell development by SAP, the protein defective in XLP. *Nat Med* **11:** 340–345.

Nunez-Cruz S, Yeo WC, Rothman J, Ojha P, Bassiri H, Juntilla M, Davidson D, Veillette A, Koretzky GA, Nichols KE. 2008. Differential requirement for the SAP-Fyn interaction during NK T cell development and function. *J Immunol* **181:** 2311–2320.

Oliver-Vila I, Saborit-Villarroya I, Engel P, Martin M. 2008. The leukocyte receptor CD84 inhibits Fc epsilon

RI-mediated signaling through homophilic interaction in transfected RBL-2H3 cells. *Mol Immunol* **45:** 2138–2149.

Ostrakhovitch EA, Wang Y, Li SS. 2009. SAP binds to CD22 and regulates B cell inhibitory signaling and calcium flux. *Cell Signal* **21:** 540–550.

Pasquier B, Yin L, Fondaneche MC, Relouzat F, Bloch-Queyrat C, Lambert N, Fischer A, Saint-Basile G, Latour S. 2005. Defective NKT cell development in mice and humans lacking the adapter SAP, the X-linked lymphoproliferative syndrome gene product. *J Exp Med* **201:** 695–701.

Qi H, Cannons JL, Klauschen F, Schwartzberg PL, Germain RN. 2008. SAP-controlled T-B cell interactions underlie germinal centre formation. *Nature* **455:** 764–769.

Rethi B, Gogolak P, Szatmari I, Veres A, Erdos E, Nagy L, Rajnavolgyi E, Terhorst C, Lanyi A. 2006. SLAM/SLAM interactions inhibit CD40-induced production of inflammatory cytokines in monocyte-derived dendritic cells. *Blood* **107:** 2821–2829.

Rigaud S, Fondaneche MC, Lambert N, Pasquier B, Mateo V, Soulas P, Galicier L, Le DF, Rieux-Laucat F, Revy P, et al. 2006. XIAP deficiency in humans causes an X-linked lymphoproliferative syndrome. *Nature* **444:** 110–114.

Roncagalli R, Taylor JE, Zhang S, Shi X, Chen R, Cruz-Munoz ME, Yin L, Latour S, Veillette A. 2005. Negative regulation of natural killer cell function by EAT-2, a SAP-related adaptor. *Nat Immunol* **6:** 1002–1010.

Sayos J, Wu C, Morra M, Wang N, Zhang X, Allen D, van Schaik S, Notarangelo L, Geha R, Roncarolo MG, et al. 1998. The X-linked lymphoproliferative-disease gene product SAP regulates signals induced through the co-receptor SLAM. *Nature* **395:** 462–469.

Schwartzberg PL, Mueller KL, Qi H, Cannons JL. 2009. SLAM receptors and SAP influence lymphocyte interactions, development and function. *Nat Rev Immunol* **9:** 39–46.

Sidorenko SP, Clark EA. 1993. Characterization of a cell surface glycoprotein IPO-3, expressed on activated human B and T lymphocytes. *J Immunol* **151:** 4614–4624.

Sivori S, Falco M, Marcenaro E, Parolini S, Biassoni R, Bottino C, Moretta L, Moretta A. 2002. Early expression of triggering receptors and regulatory role of 2B4 in human natural killer cell precursors undergoing in vitro differentiation. *Proc Natl Acad Sci* **99:** 4526–4531.

Suzuki A, Yamada R, Kochi Y, Sawada T, Okada Y, Matsuda K, Kamatani Y, Mori M, Shimane K, Hirabayashi Y, et al. 2008. Functional SNPs in CD244 increase the risk of rheumatoid arthritis in a Japanese population. *Nat Genet* **40:** 1224–1229.

Tassi I, Colonna M. 2005. The cytotoxicity receptor CRACC (CS-1) recruits EAT-2 and activates the PI3K and phospholipase Cγ signaling pathways in human NK cells. *J Immunol* **175:** 7996–8002.

Tatsuo H, Ono N, Tanaka K, Yanagi Y. 2000. SLAM (CDw150) is a cellular receptor for measles virus. *Nature* **406:** 893–897.

Vacca P, Pietra G, Falco M, Romeo E, Bottino C, Bellora F, Prefumo F, Fulcheri E, Venturini PL, Costa M, et al. 2006. Analysis of natural killer cells isolated from human decidua: Evidence that 2B4 (CD244) functions as an inhibitory receptor and blocks NK-cell function. *Blood* **108:** 4078–4085.

Vaidya SV, Stepp SE, McNerney ME, Lee JK, Bennett M, Lee KM, Stewart CL, Kumar V, Mathew PA. 2005. Targeted Disruption of the 2B4 Gene in Mice Reveals an In Vivo Role of 2B4 (CD244) in the Rejection of B16 Melanoma Cells. *J Immunol* **174:** 800–807.

Veillette A. 2006b. Immune regulation by SLAM family receptors and SAP-related adaptors. *Nat Rev Immunol* **6:** 56–66.

Veillette A. 2006a. NK cell regulation by SLAM family receptors and SAP-related adapters. *Immunol Rev* **214:** 22–34.

Veillette A, Cruz-Munoz ME, Zhong MC. 2006. SLAM family receptors and SAP-related adaptors: Matters arising. *Trends Immunol* **27:** 228–234.

Veillette A, Dong Z, Latour S. 2007. Consequence of the SLAM-SAP Signaling Pathway in Innate-like and Conventional Lymphocytes. *Immunity* **27:** 698–710.

Veillette A, Dong Z, Pérez-Quintero L-A, Zhong MC, Cruz-Munoz M. 2009. Importance and mechanism of 'switch' function of SAP family adapters. *Immunol Rev* **232:** 229–239.

Veillette A, Zhang S, Shi X, Dong Z, Davidson D, Zhong MC. 2008. SAP expression in T cells, not in B cells, is required for humoral immunity. *Proc Natl Acad Sci* **105:** 1273–1278.

Velikovsky CA, Deng L, Chlewicki LK, Fernandez MM, Kumar V, Mariuzza RA. 2007. Structure of Natural Killer Cell Receptor 2B4 (CD244) Bound to its Ligand, CD48. *Immunity* **27:** 572–584.

Wandstrat AE, Nguyen C, Limaye N, Chan AY, Subramanian S, Tian XH, Yim YS, Pertsemlidis A, Garner HRJr, Morel L et al. 2004. Association of extensive polymorphisms in the SLAM/CD2 gene cluster with murine lupus. *Immunity* **21:** 769–780.

Wang N, Satoskar A, Faubion W, Howie D, Okamoto S, Feske S, Gullo C, Clarke K, Sosa MR, Sharpe AH, et al. 2004. The Cell Surface Receptor SLAM Controls T Cell and Macrophage Functions. *J Exp Med* **199:** 1255–1264.

Yan Q, Malashkevich VN, Fedorov A, Fedorov E, Cao E, Lary JW, Cole JL, Nathenson SG, Almor SC. 2007. Structure of CD84 provides insight into SLAM family function. *Proc Natl Acad Sci* **104:** 10583–10588.

The Coordination of T-Cell Function by Serine/Threonine Kinases

David Finlay and Doreen Cantrell

Division of Cell Biology and Immunology, University of Dundee, Dundee, United Kingdom

Correspondence: d.a.cantrell@dundee.ac.uk

The function of T-lymphocytes during adaptive immune responses is directed by antigen receptors, costimulatory molecules, and cytokines. These extrinsic stimuli are coupled to a network of serine/threonine kinases that control the epigenetic, transcriptional, and metabolic programs that determine T-cell function. It is increasingly recognized that serine/threonine kinases, notably those that are controlled by lipid second messengers such as polyunsaturated diacylglycerols (DAG) and phosphatidylinositol-(3,4,5)-trisphosphate (PIP_3), are at the core of T-cell signal transduction. In the present review the object will be to discuss some important examples of how pathways of serine/threonine phosphorylation control molecular functions of proteins and control protein localization to coordinate T-cell function in adaptive immune responses.

The magnitude and complexity of the T-cell serine/threonine phospho-proteome is beginning to be appreciated. In this context, major technological advances in mass spectrometry have allowed large-scale quantitative phospho-proteomic analyses of T-cells. For example, recent studies have identified approximately 10,000 phosphorylations in the T-leukemic cell line Jurkat of which more than 600 were regulated in response to triggering of the T-cell antigen receptor (Carrascal et al. 2008; Mayya et al. 2009; Nguyen et al. 2009). The complexity and diversity of this network of protein phosphorylation in activated T cells is striking and raises many questions not least what is the role of the individual protein phosphorylations for T-cell biology. A simplistic view of protein phosphorylation is that it can act as an on/off switch to control protein structure/function; the other key impact of protein phosphorylation is to promote protein interactions that can modify protein activity and subcellular localization. The other important consideration is that kinases are enzymes and function to amplify signal transduction pathways. The present review will not exhaustively list the pathways of serine/threonine phosphorylation in T cells but rather discuss core elements and key issues about the network of serine/threonine kinases that control T-cell function.

PROTEIN KINASES AS ON/OFF SWITCHES IN T CELLS

At a basic level serine/threonine kinases can work as a series of on/off switches that are tightly coupled to triggering of antigen and cytokine

 doi: 10.1101/cshperspect-a002261
Cite this article as *Cold Spring Harb Perspect Biol* 2011;3:a002261

receptors. There are numerous examples of kinases activated by protein phosphorylation that fit this concept in T-cells (Fig. 1). For example, T-cell antigen receptor (TCR) triggering stimulates phospholipase C (PLC)-mediated hydrolysis of phosphatidylinositol-(4,5)-bisphosphate (PIP_2), which generates inositol-(1,4,5)-trisphosphate (IP_3) that initiates a sustained increase in intracellular Ca^{2+} levels ($[Ca^{2+}]_i$) that is critical for cytokine production. One mediator of Ca^{2+}signals in T cells are the Ca^{2+}/calmodulin-dependent protein kinase kinases (CaMKKs) that phosphorylate and activate CaM-kinase IV and hence regulate cytokine genes and link Ca^{2+} signals to microtubule dynamics (Matthews and Cantrell 2009). The CaMKKs also phosphorylate and hence activate the AMP-activated protein kinase (AMPK), a protein kinase with a crucial role in maintaining cellular energy homeostasis (Tamás et al. 2006).

The hydrolysis of PIP_2 by phospholipase C also produces diacyglycerol (DAG) that drives the activation of serine/threonine kinases of the protein kinase C (PKC) and protein kinase D (PKD) families (Spitaler and Cantrell 2004; Spitaler et al. 2006) (Fig. 1). DAG binding to these kinases is required for catalytic activitation but also mediates the redistribution of PKC and PKD to the membrane. However, the key molecular switch for PKD activation is the PKC-mediated phosphorylation of core serine residues within the PKD catalytic domain. In T lymphocytes, DAG binding proteins also include the Ras guanyl releasing protein (GRP) family of nucleotide exchange factors, which activate Ras GTPases (Stone 2006). In quiescent T lymphocytes, Ras GTPases exist predominantly in an inactive GDP bound state but switch rapidly to an active GTP bound form following engagement of the TCR and regulation of intracellular diacylglycerol levels (Downward

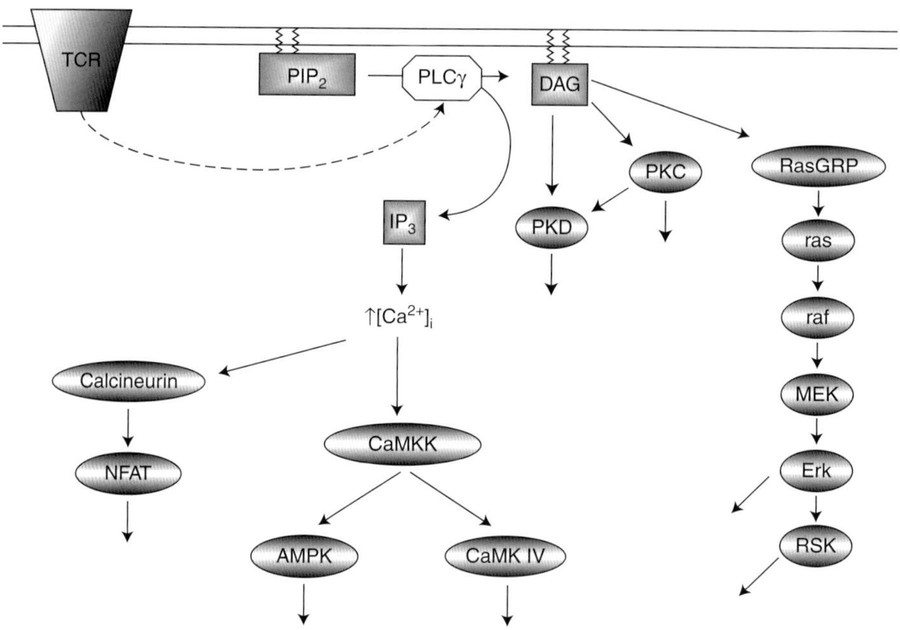

Figure 1. TCR activation of Ca^{2+}/DAG regulated kinases. T-cell receptor (TCR) ligation activates phopholipase Cγ (PLCγ), which cleaves phosphotidylinositol-(3,4)-bisphosphate (PIP_2) generating two key second messenger molecules, diacylglycerol (DAG) and inositol-(1,4,5)-trisphosphate (IP_3). IP_3 releases intracellular Ca^{2+} stores into the cytosol activating; the Ca^{2+} dependent phosphatase calcineurin inducing the expression of NFAT target genes, and the Ca^{2+} dependent kinase CaMKK which phosphorylates and activates AMPK and CaMK IV. The accumulation of DAG in the plasma membrane mediates the activation PKC and PKD protein kinases. DAG is also required for the activation of the Erk1/2 and RSK through activation of the Ras GTPase.

et al. 1990). In their active GTP bound state Ras proteins bind to the serine threonine kinase Raf-1 and activate the mitogen activated kinase MEK1 that phosphorylates and activates the MAP kinases Erk1 and Erk2 (Matthews and Cantrell 2009). The importance of the Erks for T cells is well documented and they are a classic example of kinases whose activity is switched on by protein phosphorylation and whose activity is terminated by dephosphorylation.

One other well-characterized serine/threonine kinase whose activity is tightly regulated by phosphorylation during the immune activation of lymphocytes is protein kinase B (PKB, also called Akt) (Astoul et al. 2003; Fabre et al. 2005; Finlay et al. 2009; Waugh et al. 2009). The activation of PKB requires phosphorylation of threonine 308 within the catalytic domain and serine 473 in the carboxyl terminus of the molecule (Pearce et al. 2010). In this context, PKB/Akt activity is coupled to triggering of antigen and cytokine receptors by the lipid phosphatidylinositol-(3,4,5)-trisphosphate (PIP_3) (Sinclair et al. 2008) (Fig. 2). Cellular levels of PIP_3 are determined by antigen and cytokine receptors that control the activity of class I phosphoinositide 3-kinases (PI3Ks) that phosphorylate the 3′-OH position of the inositol ring of phosphatidylinositol-(4,5)-bisphosphate (Costello et al. 2002; Sinclair et al. 2008). Extracellular stimuli can also control PIP_3 levels and activate PKB in T cells by regulating lipid phosphatases particularily PTEN (phosphatase and tensin homolog deleted on chromosome 10) (Finlay et al. 2009). PIP_3 is an important signaling molecule because it binds to the pleckstrin homology (PH) domain of PKB/Akt (Fig. 2). The rate limiting step for PKB activation is the phosphorylation of Thr308 within the PKB catalytic domain by 3-phosphoinositide-dependent protein kinase-1 (PDK1) (Pearce et al. 2010). The phosphorylation of serine 473 in the carboxyl terminus of

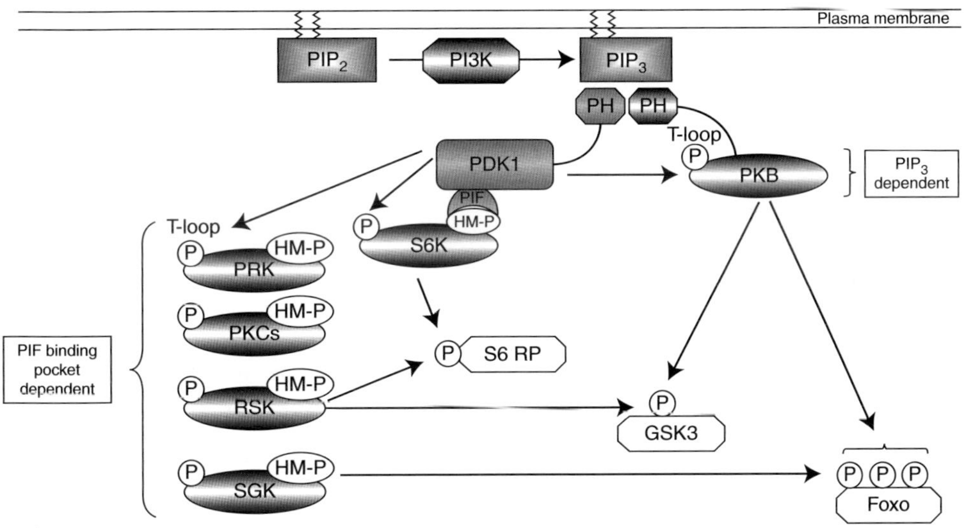

Figure 2. PDK1 signaling. PDK1 acts as a "master kinase" activating multiple AGC kinases through phosphorylation of a conserved Threonine residue ("T-loop" site) within the catalytic domain. PDK1 activation of PKB requires the second messenger molecule PIP_3 that accumulates in the plasma membrane because of the action of the lipid kinase PI3K. PDK1 activation of other AGC kinases relies on prior phosphorylation of a Ser/Thr residue within a hydrophobic motif (HM). This phosphorylated motif creates a docking site for PDK1 to bind through its PIF binding pocket, thus allowing PDK1-mediated phosphorylation of the T-loop site and kinase activation. Significant redundancy exists in PDK1 signaling as many downstream substrates, such as S6 ribosomal protein, GSK3, and FoxO transcription factors, can be phosphorylated by multiple PDK1-regulated AGC kinases.

PKB by the mammalian target of rapamycin (mTOR) complex-2 is also required for optimal kinase activity (Yang et al. 2002; Sarbassov et al. 2005). The link between PIP_3 and PKB activation is that phosphorylation of PKB Thr308 is triggered by increases in cellular PIP_3 levels that allow PIP_3 binding to the PH domain of PKB. This is essential to allow PDK1 to phosphorylate Thr308 (Calleja et al. 2007). PDK1 also has a PH domain that binds PIP_3 with high-affinity but it is important to note that PIP_3 binding to the PDK1 PH domain is not essential for PDK1 catalytic activity (Bayascas et al. 2008) or the activation of PKB in T cells. However, PDK1 binding to PIP_3 at the plasma membrane is required for efficient induction of high levels of PKB activity, probably by ensuring colocalization of the enzymes (Bayascas et al. 2008; Waugh et al. 2009).

CONSTITUTIVELY ACTIVE SERINE/ THREONINE KINASES IN "QUIESCENT" T CELLS

The discussions earlier illustrate the examples of kinases that function as on/off switches in T cells but there are equally important examples of serine/threonine kinases that have high levels of basal activity in resting T cells. For example, the rate limiting step in the triggering of PKB activity is not the activation of PDK1, because this kinase has vey high basal activity, but rather conformational/spatial restraints on the PDK1/Akt interaction (Pearce et al. 2010). In this context, the role of PDK1 extends beyond its ability to link PIP_3 to PKB and includes its ability to phosphorylate and activate p70 ribosomal S6 kinases (S6Ks), p90 ribosomal S6 kinase (RSK), serum and glucocorticoid regulated protein kinases (SGKs), protein kinase Cs, and the protein kinase C-related protein kinases (PRKs) (Williams et al. 2000; Pearce et al. 2010) (Fig. 2). The phosphorylation of S6K1, SGKs, and RSK by PDK1 is regulated by extrinsic stimuli because a key step in the activation of these kinases is their phosphorylation on hydrophobic motifs that create docking sites for the PIF binding pocket in PDK1. For S6K1 the hydrophobic motif kinase is the mTORC1 complex that phosphorylates S6K1 on Thr389. PDK1 can then dock with S6K1 and phosphorylate the kinase at its T-loop site (T229) within the catalytic domain. For SGK1, the hydrophobic motif kinase is the mTOR complex 2 (mTORC2) (Garcia-Martinez and Alessi 2008) whereas RSK activation by PDK1 is controlled by the MAP kinases Erk1/2 (Pearce et al. 2010). The initiating step for RSK activation is thus Erk1/2-mediated phosphorylation of Ser369, Thr365, and Thr577 in the carboxy-terminal catalytic domain of the kinase. The activated carboxy-terminal catalytic domain of RSK then phosphorylates Ser386 in RSK to create a docking site for PDK1, which then phosphorylates Serine 227 in the amino-terminal RSK kinase domain thereby activating the enzyme. PDK1 is thus a constitutively active kinase that integrates multiple signaling pathways initiated by antigen receptors and cytokine receptors to control the activity of other AGC kinases (Pearce et al. 2010).

Is the example of PDK1 constitutive activity unique? No there are other very important instances of constitutively active kinases in T cells and in some cases these kinases act to restrain T-cell activation. For example, members of the NFAT (nuclear factor of activated T cells) family of transcription factors (NFAT1–4) are key regulators of cytokine gene expression in T cells. NFATs are phosphorylated in quiescent cells and, as a consequence, are excluded from the nucleus and maintained in an inactive state in the cytosol. NFAT kinases are all constitutively active and include the DYRKs (dual-specificity tyrosine-phosphorylation regulated kinases), casein kinase 1 (CK1), and glycogen synthase kinase 3 (GSK3). The high basal activity of these NFAT kinases in quiescent T cells keeps NFATs phosphorylated and inactive until triggering of the T-cell antigen receptor raises intracellular free calcium levels and activates the calmodulin-dependent phosphatase calcineurin that then dephosphorylates the NFATs. The dephosphorylation of NFAT allows their translocation to the nucleus and promotes their binding to target genes. The rate limiting step for NFAT activation is thus the activation of a phosphatase: the kinases act as negative regulators of NFAT induced gene transcription (Oh-hora and Rao 2008).

REDUNDANCY BETWEEN SERINE/THREONINE KINASES IN T CELLS

Studies of the function of serine/threonine kinases in T cells have been complicated by the fact that many cells coexpress multiple isoforms of a kinase making it difficult to assess the function of these molecules with simple single-gene deletion strategies. For example, there are three isoforms of PKB: These may have unique roles but are also quite redundant with each other. Hence, in T-cell progenitors in the thymus there is very little impact of deleting a single PKB isoform; the deletion of two isoforms PKBα and PKBβ inhibits cell proliferation whereas the combined deletion of all three PKB isoforms inhibits the T-cell survival response (Fayard et al. 2007; Juntilla et al. 2007; Mao et al. 2007). There can even be redundancy among different members of a family of kinases. For example, the p90 ribosomal S6 kinase (RSK) can substitute for p70 ribosomal S6 kinases (S6Ks) and converge to phosphorylate the S6 ribosomal subunit, a key component of the translational machinery in eukaryotic cells that is essential for ribosome biogenesis (Salmond et al. 2009). Additionally, both PKB and SGK can phosphorylate and inactivate FoxO transcription factors (Brunet et al. 2001; Burgering 2008) (Fig. 2).

One way to evade issues of kinase redundancy and gain an understanding of the functional importance of a kinase family is to delete a rate limiting upstream regulator that will simultaneously inactivate multiple members of a kinase family. In this context, PDK1 phosphorylates and activates multiple members of the AGC kinase family including PKB/Akt, RSK, S6K, and PRKs. The deletion of PDK1 can thus eliminate the activity of multiple AGC family kinases (Lawlor et al. 2002). Accordingly, studies that explored the impact of PDK1 deletion in T-cell progenitors in the thymus have been much more informative about the role of AGC kinases in T cells than studies of T cell lacking individual members of this kinase family (Hinton et al. 2004). Hence the differentiation and proliferative expansion of pre-T cells in the thymus is dependent on signaling pathways mediated by PDK1. However, the substitution of wild-type PDK1 with a PDK1 L155E mutant that supports full PKB activation in T cells is not sufficient to restore normal thymus function (Kelly et al. 2007). The substitution of leucine (L) 155 in PDK1 with glutamate (E) disrupts the integrity of an important PDK1 domain termed the PIF binding pocket. This domain is not required for PKB phosphorylation but is necessary for PDK1 to interact with phosphorylated carboxy-terminal hydrophobic motifs in S6Ks and RSK and is necessary for PDK1 to activate these AGC kinases by phosphorylating their catalytic domains (Collins et al. 2003). A comparison of the function of wild type and the PDK1 L155E mutant thus reveals how multiple AGC serine kinases and not just PKB are important in T-cells (Kelly et al. 2007).

The ability of PDK1 to control the activity of many AGC kinases has generated the concept that some kinases act as "master" kinases. It is not yet clear how general this concept is for T cells but there is at least one more example that should be considered. Hence, LKB1 (or serine/threonine kinase 11-STK11) is a kinase that phosphorylates critical activating residues in the catalytic domains of multiple members of the AMPK family including the α1 and α2 isoforms of AMPK and NUAK1-2, BRSK1-2, QIK, QSK, SIK, MELK, and MARK1-4 kinases (Alessi et al. 2006) (Fig. 3). These kinases are thought to be essential to control cell polarity and to control cellular responses to energy stress. However, elimination of single members of the AMPK family of kinases such as deletion of AMPK α1 or loss of single MARK kinases has only had a relatively minor effect on peripheral T cells and causes no problems for T-cell development in the thymus (Mayer et al. 2008; Lin et al. 2009). This could indicate that AMPK family kinases are physiologically irrelevant for T-cell biology but the data are equally consistent with the idea that there is redundancy among different AMPK family members for both energy metabolism and the control of T-cell polarity. In this respect, T-lineage specific

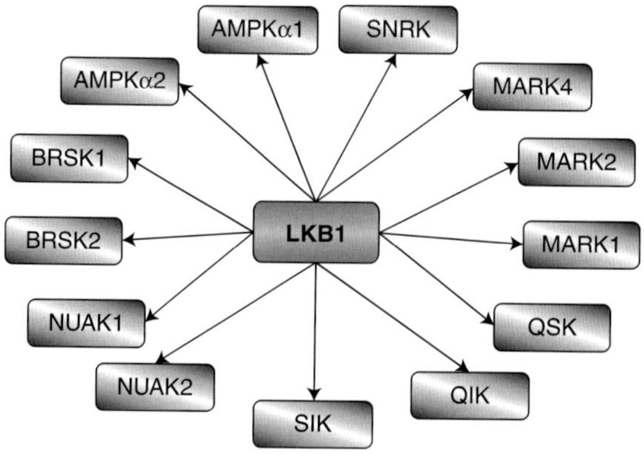

Figure 3. LKB1 is a "master kinase." LKB1 phosphorylates and activates 13 kinases within the AMPK family of protein kinases.

deletion of LKB1 shows that LKB1 is essential for the survival and development of TCRβ selected T-cell progenitors in the thymus and is also required for survival and proliferation of peripheral CD8 T cells (Cao et al.; Tamás et al. 2010). Importantly, LKB1 appears to be required for several key metabolic processes in T cells. For example, LKB1 controls expression of CD98, a key subunit of the L-system amino acid transporter in T cells and is also required to sustain the phosphorylation of the ribosomal S6 subunit (Tamás et al. 2010). As discussed, the role of LKB1 is to phosphorylate and activate multiple AMPK family members. To date there are no reports that the loss of a single AMPK isoform phenocopies the loss of LKB1 (Mayer et al. 2008). Hence it appears that there is in vivo redundancy among different AMPK family members for the control of T-cell metabolism; the elimination of LKB1 circumvents this redundancy by simultaneously eliminating activation of multiple AMPK family members (Tamás et al. 2010).

PROTEIN PHOSPHORYLATION IN T CELLS: A LINK BETWEEN THE PLASMA MEMBRANE AND THE NUCLEUS

One of the first examples of how a pathway of protein phosphorylation could couple signals generated at the plasma membrane to events in the cell nucleus came from studies of NFATs that are cytosolic and inactive when phosphorylated and only localized in the nucleus and able to regulate their target genes when they are dephosphorylated (Hogan et al. 2003) (Fig. 4). NFAT dephosphorylation is regulated by the TCR that controls intracellular calcium levels and the activity of the calcium phosphatatase calcineurin (Oh-hora and Rao 2008). In the case of NFAT protein, phosphorylation is a negative regulator of transcriptional activity. However, there are also examples of pathways of protein phosphorylation that positively regulate gene transcription notably from studies of the MAP kinases. Hence, Erk1/2 phosphorylate and activate the ternary complex factor (TCF) subfamily of ETS-domain transcription factors Elk-1, SAP1, and SAP2 that control immediate early gene expression in T cells (e.g., c-fos, Egr1, and EgR3) (Costello et al. 2004). The TCFs also control the transcription of dual specificity (Thr/Tyr) MAPK phosphatases that initiate a negative regulatory feedback pathway that acts to limit the duration and magnitude of ERK activity (Owens and Keyse 2007).

The consideration of NFAT regulation illustrates that one key way in which protein phosphorylation is important to control cell function is that it can promote protein interactions that can modify protein activity and subcellular localization (Fig. 4). In this context, 14-3-3 proteins (14-3-3s) are highly conserved dimeric molecules that bind to phosphorylated

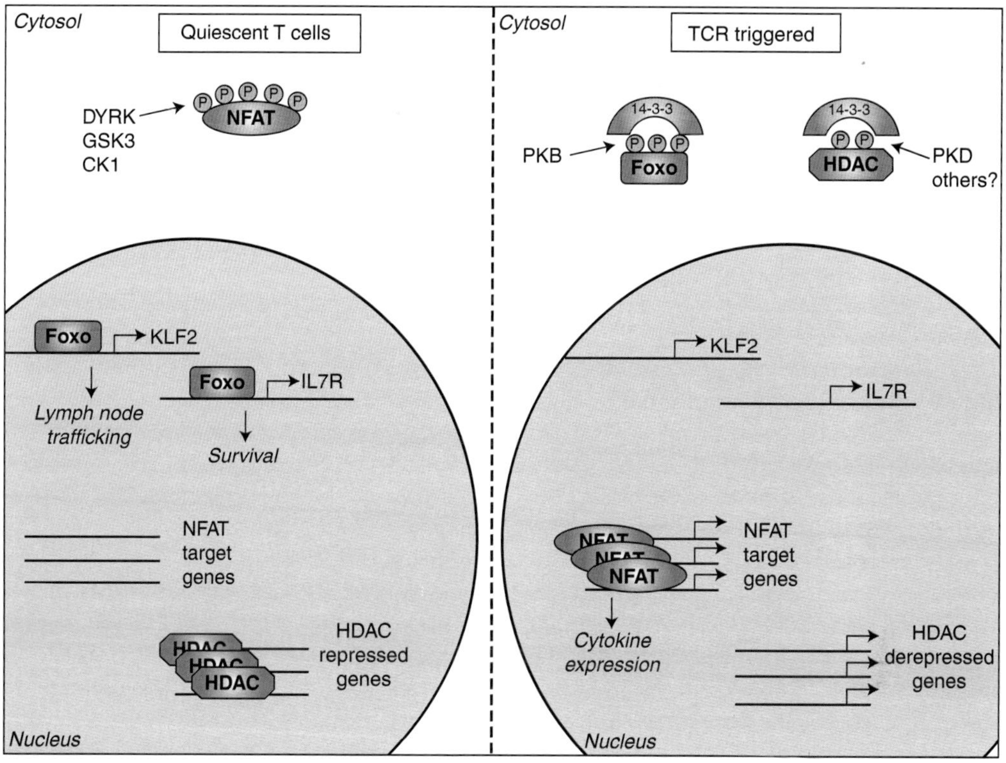

Figure 4. Phosphorylation determined intracellular localization. In quiescent T cells constitutively active kinases (DYRK, GSK3, and CK1) maintain NFAT transcription factors in a phosphorylated state excluding them from the nucleus. In contrast, the kinases that regulate FoxO transcription factors and histone deacetylases (HDACs) are inactive in quiescent cells allowing FoxOs to access the nucleus and control genes involved in T-cell trafficking and survival whereas HDACs repress gene expression through modulating chromatin structure. Following triggering of the T-cell receptor (TCR) FoxO transcription factors and HDACs are phosphorylated, resulting in their cytosolic sequestration by 14-3-3 proteins, thus preventing their nuclear actions. Meanwhile, the Ca^{2+} activated phosphatase, calcineurin, dephosphorylates NFAT transcription factors allowing nuclear entry and the expression of key cytokine genes.

serine and threonine sites in a wide range of proteins (Johnson et al.; Dubois et al. 2009; Morrison 2009). 14-3-3s target proteins include transcription factors, chromatin regulators, enzymes, and structural components of vesicle and protein trafficking. In lymphocytes, one example of how protein phosphorylation and 14-3-3 proteins can coordinate programs of gene expression is in the context of class II histone deacetylases (HDACs) (Martin et al. 2007). These control the acetylation/deacetylation of nucleosomal histones, regulate chromatin accessibility and also control gene activity by repressing the activity of MEF2 transcription factors. The suppressive effects of class II HDACs on gene transcription can be relieved by triggering of the antigen receptor that leads to the phosphorylation of serine residues within the amino-termini of the HDACs (Matthews et al. 2006; Martin et al. 2007). The proposed HDAC kinase is the diacyglycerol-regulated kinase protein kinase D. Phosphorylation at the regulatory sites promotes the redistribution of class II HDACs from the nucleus to the cytosol allowing them to bind to 14-3-3 complexes thus disrupting repressive interactions between HDACs and histones and/or transcription factors and allowing gene transcription to occur (Matthews et al. 2006). The regulated protein phosphorylation of class II HDACs is thus a key epigenetic control mechanism that determines gene expression in lymphocytes and HDAC-regulated genes in T cells are extensive and include genes that control both positive and negative selection of thymocytes (Kasler and Verdin 2007) (Fig. 4).

Another evolutionarily conserved set of transcription factors controlled by protein phosphorylation and 14-3-3s in lymphocytes are the Forkhead box O (FoxO) family of transcription factors FoxO 1, 3a, and 4 (Burgering 2008). Nonphosphorylated FoxO 1, 3a, and 4 are found in the nuclei of quiescent T cells where they drive transcription of genes encoding cytokine receptor and transcription factors (Coffer and Burgering 2004; Fabre et al. 2008; Kerdiles et al. 2009). For example, FoxOs drive expression of the receptor for interleukin 7, an essential homeostatic cytokine for T cells (Kerdiles et al. 2009; Ouyang et al. 2009). FoxO activity in naïve and memory T cells also controls the expression of the transcription factor KLF2 that directly regulates the gene transcription of key adhesion molecules and chemokine receptors that control T-cell entry and egress from secondary lymphoid tissues (Carlson et al. 2006; Finlay et al. 2009; Kerdiles et al. 2009; Waugh et al. 2009) (Fig. 4). FoxO 1 and 3a are substrates for PKB and their phosphorylation results in their nuclear export; phosphorylated FoxO transcription factors are then retained in the cytosol in a complex with 14-3-3 proteins ensuring the termination of FoxO-mediated gene transcription (Carlson et al. 2006; Burgering 2008) (Fig. 4).

The importance of PKB-mediated phosphorylation and inactivation of FoxOs is easily illustrated. First, the deletion of FoxOs in naïve T cells disrupts T-cell homing and survival (Kerdiles et al. 2009; Ouyang et al. 2009). It is also clear that much of the T-cell pathology caused by deletion of the tumor suppressor PTEN (phosphatase and tensin homolog deleted on chromosome 10) can be ascribed to FoxO deregulation. PTEN is a lipid phosphatase with specificity for the 3′ position of PIP_3 and its deletion causes T cells to accumulate high levels of PIP_3 and hence drives PDK1-mediated phosphorylation and activation of PKB with the resultant phosphorylation and inactivation of the FoxO1, 3a, and 4 (Finlay et al. 2009). This loss of FoxO transcriptional activity immediately causes loss of KLF2 expression and loss of expression of the KLF2 targets L-selectin/CD62L and CCR7 (Finlay et al. 2009). CD62L/L-selectin plays a fundamental role controlling the transmigration of T cells across high endothelial venules (Arbones et al. 1994) and the chemokine receptor CCR7 controls lymphocyte entry and retention in lymphoid tissues (Forster et al. 2008). Consequently the loss of FoxO activity in PTEN null cells changes the expression of key trafficking receptors in T cells and prevents these from cells undergoing their normal migration from blood into secondary lymphoid tissue (Finlay et al. 2009). These trafficking defects are not observed following deletion of PDK1, which causes loss of FoxO phosphorylation and restores FoxO transcriptional activity (Finlay et al. 2009). It is also striking that the activation of PDK1-mediated protein phosphorylation in PTEN null cells is key for T lymphomagenesis. Hence the tissue specific deletion of PTEN in thymocytes using Cre-loxP strategies results in T leukemogenesis or lymphomagenesis but PTEN null T-cell progenitors cannot transform or develop into invasive and fatal T lymphoma without PDK1 (Finlay et al. 2009).

CONCLUDING REMARKS

Recent studies have begun to reveal how protein phosphorylation reversibly regulates key processes in T lymphocytes. However, we have only uncovered a fraction of the relevant information and progress is constantly hampered by technical difficulties. It is still relatively difficult to study protein phosphorylation and the rationale for focusing on particular pathways of protein phosphorylation frequently stems from reagent availability rather than any intellectual deliberation. For example, the MAP kinases Erk1/2 are particularly well studied in T cells for pragmatic reasons: There are high quality phosphospecific monoclonal antibodies that selectively recognize phosphorylated and activated Erks and that can be used for single-cell analysis of ERK activity using flow cytometry. It is much harder to study many other kinase pathways using analogous reagents. However, as advances in technology reveal the magnitude and complexity of the T-cell phospho-proteome it is becoming clear that there are unique functions in T cells for ubiquitously expressed serine/threonine kinases. It is also clear that studies of serine/threonine kinases can generate a molecular understanding of how signal transduction pathways control T-cell function. Such studies have the added advantage that they provide a clear indication of the molecules that might be useful targets for manipulation of the immune response.

REFERENCES

Alessi DR, Sakamoto K, Bayascas JR. 2006. LKB1-dependent signaling pathways. *Annu Rev Biochem* **75:** 137–163.

Arbones ML, Ord DC, Ley K, Ratech H, Maynard-Curry C, Otten G, Capon DJ, Tedder TF. 1994. Lymphocyte homing and leukocyte rolling and migration are impaired in L-selectin-deficient mice. *Immunity* **1:** 247–260.

Astoul E, Laurence AD, Totty N, Beer S, Alexander DR, Cantrell DA. 2003. Approaches to define antigen receptor-induced serine kinase signal transduction pathways. *J Biol Chem* **278:** 9267–9275.

Bayascas JR, Wullschleger S, Sakamoto K, Garcia-Martinez JM, Clacher C, Komander D, van Aalten DM, Boini KM, Lang F, Lipina C, et al. 2008. Mutation of the PDK1 PH domain inhibits protein kinase B/Akt, leading to small size and insulin resistance. *Mol Cell Biol* **28:** 3258–3272.

Brunet A, Park J, Tran H, Hu LS, Hemmings BA, Greenberg ME. 2001. Protein kinase SGK mediates survival signals by phosphorylating the forkhead transcription factor FKHRL1 (FOXO3a). *Mol Cell Biol* **21:** 952–965.

Burgering BM. 2008. A brief introduction to FOXOlogy. *Oncogene* **27:** 2258–2262.

Calleja V, Alcor D, Laguerre M, Park J, Vojnovic B, Hemmings BA, Downward J, Parker PJ, Larijani B. 2007. Intramolecular and intermolecular interactions of protein kinase B define its activation in vivo. *PLoS Biol* **5:** e95.

Cao Y, Li H, Liu H, Zheng C, Ji H, Liu X. 2010. The serine/threonine kinase LKB1 controls thymocyte survival through regulation of AMPK activation and Bcl-XL expression. *Cell Res* **20:** 99–108.

Carlson CM, Endrizzi BT, Wu J, Ding X, Weinreich MA, Walsh ER, Wani MA, Lingrel JB, Hogquist KA, Jameson SC. 2006. Kruppel-like factor 2 regulates thymocyte and T-cell migration. *Nature* **442:** 299–302.

Carrascal M, Ovelleiro D, Casas V, Gay M, Abian J. 2008. Phosphorylation analysis of primary human T-lymphocytes using sequential IMAC and titanium oxide enrichment. *Jf Proteome Res* **7:** 5167–5176.

Coffer PJ, Burgering BM. 2004. Forkhead-box transcription factors and their role in the immune system. *Nat Rev Immunol* **4:** 889–899.

Collins BJ, Deak M, Arthur JS, Armit LJ, Alessi DR. 2003. In vivo role of the PIF-binding docking site of PDK1 defined by knock-in mutation. *Embo J* **22:** 4202–4211.

Costello PS, Gallagher M, Cantrell DA. 2002. Sustained and dynamic inositol lipid metabolism inside and outside the immunological synapse. *Nat Immunol* **3:** 1082–1089.

Costello PS, Nicolas RH, Watanabe Y, Rosewell I, Treisman R. 2004. Ternary complex factor SAP-1 is required for Erk-mediated thymocyte positive selection. *Nat Immunol* **5:** 289–298.

Downward J, Graves JD, Warne PH, Rayter S, Cantrell DA. 1990. Stimulation of p21ras upon T-cell activation. *Nature* **346:** 719–723.

Dubois F, Vandermoere F, Gernez A, Murphy J, Toth R, Chen S, Geraghty KM, Morrice NA, MacKintosh C. 2009. Differential 14–3–3 affinity capture reveals new downstream targets of phosphatidylinositol 3-kinase signaling. *Mol Cell Proteomics* **8:** 2487–2499.

Fabre S, Carrette F, Chen J, Lang V, Semichon M, Denoyelle C, Lazar V, Cagnard N, Dubart-Kupperschmitt A, Mangeney M, et al. 2008. FOXO1 regulates L-Selectin and a network of human T-cell homing molecules downstream of phosphatidylinositol 3-kinase. *J Immunol* **181:** 2980–2989.

Fabre S, Lang V, Harriague J, Jobart A, Unterman TG, Trautmann A, Bismuth G. 2005. Stable activation of phosphatidylinositol 3-kinase in the T-cell immunological synapse stimulates Akt signaling to FoxO1 nuclear exclusion and cell growth control. *J Immunol* **174:** 4161–4171.

Fayard E, Gill J, Paolino M, Hynx D, Hollander GA, Hemmings BA. 2007. Deletion of PKBalpha/Akt1 Affects Thymic Development. *PLoS ONE* **2:** e992.

Finlay DK, Sinclair LV, Feijoo C, Waugh CM, Hagenbeek TJ, Spits H, Cantrell DA. 2009. Phosphoinositide-dependent kinase 1 controls migration and malignant transformation but not cell growth and proliferation in PTEN-null lymphocytes. *J Exp Med* **206:** 2441–2454.

Forster R, Davalos-Misslitz AC, Rot A. 2008. CCR7 and its ligands: Balancing immunity and tolerance. *Nat Rev Immunol* **8:** 362–371.

Garcia-Martinez JM, Alessi DR. 2008. mTOR complex 2 (mTORC2) controls hydrophobic motif phosphorylation and activation of serum- and glucocorticoid-induced protein kinase 1 (SGK1). *Biochem J* **416:** 375–385.

Hinton HJ, Alessi DR, Cantrell DA. 2004. The serine kinase phosphoinositide-dependent kinase 1 (PDK1) regulates T-cell development. *Nat Immunol* **5:** 539–545.

Hogan PG, Chen L, Nardone J, Rao A. 2003. Transcriptional regulation by calcium, calcineurin, and NFAT. *Genes Dev* **17:** 2205–2232.

Johnson C, Crowther S, Stafford MJ, Campbell DG, Toth R, MacKintosh C. Bioinformatic and experimental survey of 14–3–3-binding sites. *Biochem J* **427:** 69–78.

Juntilla MM, Wofford JA, Birnbaum MJ, Rathmell JC, Koretzky GA. 2007. Akt1 and Akt2 are required for αβ thymocyte survival and differentiation. *Proc Natl Acad Sci* **104:** 12105–12110.

Kasler HG, Verdin E. 2007. Histone deacetylase 7 functions as a key regulator of genes involved in both positive and negative selection of thymocytes. *Mol Cell Biol* **27:** 5184–5200.

Kelly AP, Finlay DK, Hinton HJ, Clarke RG, Fiorini E, Radtke F, Cantrell DA. 2007. Notch-induced T-cell development requires phosphoinositide-dependent kinase 1. *Embo J* **26:** 3441–3450.

Kerdiles YM, Beisner DR, Tinoco R, Dejean AS, Castrillon DH, DePinho RA, Hedrick SM. 2009. FoxO1 links homing and survival of naive T-cells by regulating L-selectin, CCR7 and interleukin 7 receptor. *Nat Immunol* **10:** 176–184.

Lawlor MA, Mora A, Ashby PR, Williams MR, Murray-Tait V, Malone L, Prescott AR, Lucocq JM, Alessi DR. 2002. Essential role of PDK1 in regulating cell size and development in mice. *Embo J* **21:** 3728–3738.

Lin J, Hou KK, Piwnica-Worms H, Shaw AS. 2009. The polarity protein Par1b/EMK/MARK2 regulates T-cell receptor-induced microtubule-organizing center polarization. *J Immunol* **183:** 1215–1221.

Mao C, Tili EG, Dose M, Haks MC, Bear SE, Maroulakou I, Horie K, Gaitanaris GA, Fidanza V, Ludwig T, et al. 2007. Unequal contribution of Akt isoforms in the double-negative to double-positive thymocyte transition. *J Immunol* **178:** 5443–5453.

Martin M, Kettmann R, Dequiedt F. 2007. Class IIa histone deacetylases: regulating the regulators. *Oncogene* **26:** 5450–5467.

Matthews SA, Cantrell DA. 2009. New insights into the regulation and function of serine/threonine kinases in T lymphocytes. *Immunol Rev* **228:** 241–252.

Matthews SA, Liu P, Spitaler M, Olson EN, McKinsey TA, Cantrell DA, Scharenberg AM. 2006. Essential role for protein kinase D family kinases in the regulation of class II histone deacetylases in B lymphocytes. *Mol Cell Biol* **26:** 1569–1577.

Mayer A, Denanglaire S, Viollet B, Leo O, Andris F. 2008. AMP-activated protein kinase regulates lymphocyte responses to metabolic stress but is largely dispensable for immune cell development and function. *Eur J Immunol* **38:** 948–956.

Mayya V, Lundgren DH, Hwang SI, Rezaul K, Wu L, Eng JK, Rodionov V, Han DK. 2009. Quantitative phosphoproteomic analysis of T-cell receptor signaling reveals system-wide modulation of protein-protein interactions. *Sci Signaling* **2:** ra46.

Morrison DK. 2009. The 14–3–3 proteins: integrators of diverse signaling cues that impact cell fate and cancer development. *Trends Cell Biol* **19:** 16–23.

Nguyen V, Cao L, Lin JT, Hung N, Ritz A, Yu K, Jianu R, Ulin SP, Raphael BJ, Laidlaw DH, et al. 2009. A new approach for quantitative phosphoproteomic dissection of signaling pathways applied to T-cell receptor activation. *Mol Cell Proteomics* **8:** 2418–2431.

Oh-hora M, Rao A. 2008. Calcium signaling in lymphocytes. *Current Opin Immunol* **20:** 250–258.

Ouyang W, Beckett O, Flavell RA, Li MO. 2009. An essential role of the Forkhead-box transcription factor FoxO1 in control of T-cell homeostasis and tolerance. *Immunity* **30:** 358–371.

Owens DM, Keyse SM. 2007. Differential regulation of MAP kinase signalling by dual-specificity protein phosphatases. *Oncogene* **26:** 3203–3213.

Pearce LR, Komander D, Alessi DR. 2010. The nuts and bolts of AGC protein kinases. *Nat Rev Mol Cell Biol* **11:** 9–22.

Salmond RJ, Emery J, Okkenhaug K, Zamoyska R. 2009. MAPK, phosphatidylinositol 3-kinase, and mammalian target of rapamycin pathways converge at the level of ribosomal protein S6 phosphorylation to control metabolic signaling in CD8 T-cells. *J Immunol* **183:** 7388–7397.

Sarbassov DD, Guertin DA, Ali SM, Sabatini DM. 2005. Phosphorylation and regulation of Akt/PKB by the rictor-mTOR complex. *Science* **307:** 1098–1101.

Sinclair LV, Finlay D, Feijoo C, Cornish GH, Gray A, Ager A, Okkenhaug K, Hagenbeek TJ, Spits H, Cantrell DA. 2008. Phosphatidylinositol-3-OH kinase and nutrient-sensing mTOR pathways control T-lymphocyte trafficking. *Nat Immunol* **9:** 513–521.

Spitaler M, Cantrell DA. 2004. Protein kinase C and beyond. *Nat Immunol* **5:** 785–790.

Spitaler M, Emslie E, Wood CD, Cantrell D. 2006. Diacylglycerol and protein kinase D localization during T-lymphocyte activation. *Immunity* **24:** 535–546.

Stone JC. 2006. Regulation of Ras in lymphocytes: get a GRP. *Biochem Soc Trans* **34:** 858–861.

Tamás P, Hawley SA, Clarke RG, Mustard KJ, Green K, Hardie D, Cantrell DA. 2006. Regulation of the energy sensor AMP-activated protein kinase by antigen receptor and Ca2+ in T-lymphocytes. *J Exp Med* **203:** 1665–1670.

Tamás P, Macintyre A, Finlay D, Clarke R, Feijoo-Carnero C, Ashworth A, Cantrell D. 2010. LKB1 is essential for the proliferation of T-cell progenitors and mature peripheral T-cells. *Eur J Immunol* **40:** 242–253.

Waugh C, Sinclair L, Finlay D, Bayascas J, Cantrell D. 2009. PIP3 binding to Phosphoinositide dependent kinase 1 regulates a protein kinase B/Akt signalling threshold that dictates T-cell migration not proliferation. *Mol Cell Biol* **29:** 5952–5962.

Williams MR, Arthur JS, Balendran A, van der Kaay J, Poli V, Cohen P, Alessi DR. 2000. The role of 3-phosphoinositide-dependent protein kinase 1 in activating AGC kinases defined in embryonic stem cells. *Curr Biol* **10:** 439–448.

Yang J, Cron P, Thompson V, Good V, Hess D, Hemmings B, Barford D. 2002. Molecular mechanism for the regulation of protein kinase B/Akt by hydrophobic motif phosphorylation. *Mol Cell* **9:** 1227–1240.

Lipid Signaling in T-Cell Development and Function

Yina H. Huang[1] and Karsten Sauer[2]

[1]Department of Pathology, Washington University School of Medicine, St. Louis, Missouri 63110

[2]Department of Immunology and Microbial Science, The Scripps Research Institute, La Jolla, California 92037

Correspondence: ksauer@scripps.edu

Second messenger molecules relay, amplify, and diversify cell surface receptor signals. Two important examples are phosphorylated D-*myo*-inositol derivatives, such as phosphoinositide lipids within cellular membranes, and soluble inositol phosphates. Here, we review how phosphoinositide metabolism generates multiple second messengers with important roles in T-cell development and function. They include soluble inositol(1,4,5)trisphosphate, long known for its Ca^{2+}-mobilizing function, and phosphatidylinositol(3,4,5)trisphosphate, whose generation by phosphoinositide 3-kinase and turnover by the phosphatases PTEN and SHIP control a key "hub" of TCR signaling. More recent studies unveiled important second messenger functions for diacylglycerol, phosphatidic acid, and soluble inositol(1,3,4,5)tetrakisphosphate (IP_4) in immune cells. Inositol(1,3,4,5)tetrakisphosphate acts as a soluble phosphatidylinositol(3,4,5)trisphosphate analog to control protein membrane recruitment. We propose that phosphoinositide lipids and soluble inositol phosphates (IPs) can act as complementary partners whose interplay could have broadly important roles in cellular signaling.

Second messenger molecules relay, amplify, and diversify signals perceived by cell surface receptors. One important second messenger family is comprised of the phosphorylated derivatives of the small cyclic poly-alcohol D-*myo*-inositol. They include phosphoinositide lipids within cellular membranes, and soluble inositol phosphates, here termed IPs. In most stimulatory cells, the plasma membrane phosphoinositide, phosphatidylinositol(4,5)bisphosphate (here termed PIP_2 for improved readability, although several different PIP_2 isomers exist) is a key precursor for both other phosphoinositides and soluble IPs. Many of these regulate distinct and overlapping downstream effectors (Irvine and Schell 2001; Alcazar-Roman and Wente 2008; Resnick and Saiardi 2008; Sauer et al. 2009; Shears 2009; Sauer and Cooke 2010). In particular, class I phosphoinositide 3-kinases (PI3K) phosphorylate PIP_2 at the 3-position of its inositol-ring into phosphatidylinositol(3,4,5)trisphosphate (here termed PIP_3) (Fig. 1) after receptor stimulation (Vanhaesebroeck et al. 2005; Juntilla and Koretzky 2008; Buitenhuis and Coffer 2009; Fruman and Bismuth 2009). Receptor-induced

 doi: 10.1101/cshperspect.a002428
Cite this article as *Cold Spring Harb Perspect Biol* 2010;2:a002428

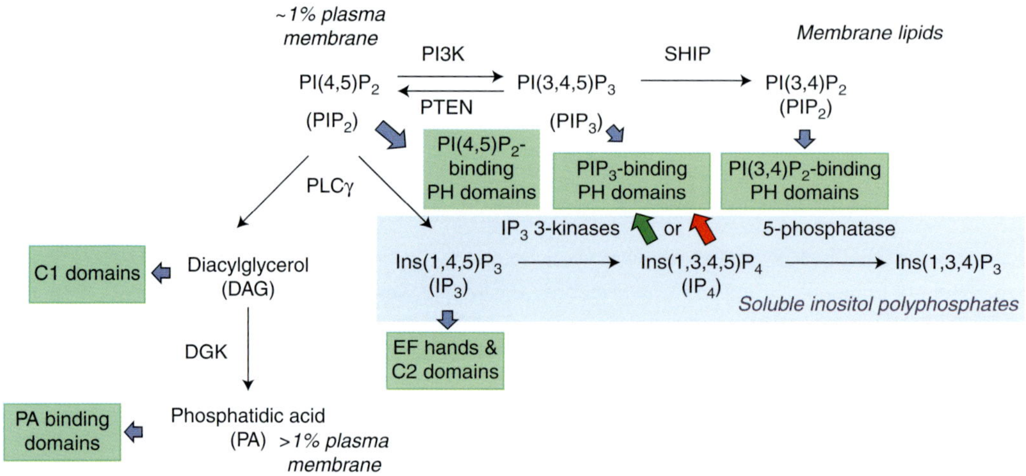

Figure 1. The phosphoinositide $PI(4,5)P_2$ is a key node of second messenger metabolism in lymphocytes. $PI(4,5)P_2$-phosphorylation by PI3-kinases and $PI(4,5)P_2$-hydrolysis by PLCγ initiate several second messenger cascades within cellular membranes and soluble compartments. These second messengers can have various functions, including the recruitment and/or activation of effector proteins by binding to specific domains within them (blue arrows and green boxes). In particular, the phosphatases SHIP1/2 or PTEN limit or down-regulate PIP_3 levels and the functions of PIP_3 and its effectors by hydrolyzing PIP_3 into $PI(3,4)P_2$ or $PI(4,5)P_2$, respectively. For group 1D, PH domains such as that of Akt, $PI(3,4)P_2$ binding sustains effector activity (Cozier et al. 2004; DiNitto and Lambright 2006; Lemmon 2008). $PI(4,5)P_2$ recruits proteins such as $RASA3/GAP1^{IP4BP}$ by binding to their PH domains (Lockyer et al. 1997; Cozier et al. 2000a; Cozier et al. 2000b). Hence, both PIP_2s and PIP_3 can have partially overlapping functions depending on the lipid-binding protein domain involved. PLCγ hydrolyzes $PI(4,5)P_2$ into protein C1 domain binding DAG and into IP_3, which binds EF and C2 domain containing proteins and mobilizes Ca^{2+}. IP_3 3-kinases convert IP_3 into IP_4, which acts as a soluble PIP_3 analog and controls the abilities of certain PH domains to bind to PIP_3 either positively (green arrow) or negatively (red arrow). An unknown 5-phosphatase metabolizes IP_4 into $I(1,3,4)P_3$, a precursor for higher-order IPs. In vitro, SHIP1 and PTEN can dephosphorylate IP_4 at the 5- or 3-positions, respectively (Erneux et al. 1998; Maehama and Dixon 1998; Pesesse et al. 1998; Caffrey et al. 2001). Whether this occurs physiologically is unknown. Finally, DAG kinases (DGKs) down-regulate DAG function by phosphorylating it into PA, itself a ligand for certain proteins. Further metabolism of all these second messengers results in the generation of many lipid and soluble metabolites. Several of these have important signaling functions (Irvine 2001; Irvine and Schell 2001; York and Hunter 2004; York 2006; Alcazar-Roman and Wente 2008; Huang et al. 2008; Jia et al. 2008b; Miller et al. 2008; Burton et al. 2009; Shears 2009; Sauer and Cooke 2010; Schell 2010). Their functions in immunocytes are unknown.

PIP_2-hydrolysis by phospholipases such as PLCγ1/2 in lymphocytes generates the lipid diacylglycerol (DAG) and the soluble IP inositol(1,4,5)trisphosphate (IP_3). PIP_3, DAG, and IP_3 have essential second messenger functions in many cells, including lymphocytes. Here, we review the importance of phosphoinositide signaling in T cells and highlight the importance of a recently identified, intriguing molecular interplay between second messenger lipids and their soluble IP counterparts.

PHOSPHOINOSITIDES CONTROL SIGNALING BY BINDING TO SPECIFIC PROTEIN DOMAINS

All phosphoinositides contain a hydrophobic membrane-embedded diacylglyceride and a hydrophilic solvent-exposed IP moiety. The inositol ring hydroxyl groups can be stereo-specifically phosphorylated by phosphoinositide-kinases. Most phosphatidylinositol-bisphosphate in the plasma membrane of unstimulated cells is phosphorylated at the inositol 4- and 5-positions. This PIP_2 comprises $<1\%$ of the inner leaflet of the plasma membrane, contrasting with much more abundant phospholipids such as phosphatidylserine and phosphatidic acid (PA) (Lemmon 2008). Despite its low abundance, PIP_2 is an important second messenger. It recruits and regulates multiple signaling proteins by binding to their pleckstrin homology (PH), epsin N-terminal homology (ENTH), 4.1, ezrin, radixin, moesin (FERM), Tubby, or Phox-homology (PX) domains with varying affinities and specificities (McLaughlin et al. 2002). Due to the constitutive PIP_2 availability in resting cells, these PIP_2-associated proteins likely maintain signaling pathways in a preactivation state. Evidence supporting this notion in T cells comes from the inhibitory effects of PIP_2 binding to the guanine nucleotide exchange factor, Vav (Han et al. 1998), and the guanine nucleotide activating factor, ArhGAP9, which maintains Rho-family GTPases in an inactive state (Ang et al. 2007; Ceccarelli et al. 2007).

PHOSPHOINOSITIDE 3-KINASES CONVERT PIP_2 INTO PHOSPHATIDYLINOSITOL(3,4,5) TRISPHOSPHATE

Despite the importance of PIP2, much greater attention has been given to the products of PIP_2 phosphorylation or PIP_2 hydrolysis that are induced following receptor activation. PIP_2 phosphorylation is mediated by PI3Ks. PI3Ks fall into four classes based upon their substrate specificity. Each class comprises one or more regulatory and catalytic subunits with partially distinct and overlapping functions (Fig. 2). Class II and III PI3Ks phosphorylate phosphatidylinositol (PI) into PI(3)P and $PI(3,4)P_2$. Their functions in lymphocytes are largely unexplored. Class I PI3Ks phosphorylate PIP_2 into PIP_3, an inducible ligand that mediates recruitment and activation of various important signaling proteins. Class IA PI3Ks are activated by most stimulatory receptors on lymphocytes including T- and B-cell antigen receptors (TCR, BCR), and co-stimulatory, Toll-like, and cytokine receptors (Vanhaesebroeck et al. 2005; Buitenhuis and Coffer 2009; Fruman and Bismuth 2009). Five regulatory (p50α, p55α, p55γ, p85α, p85β) and three catalytic (p110α, β, and δ) subunits participate in Class IA PI3K signaling. Recruitment of Class IA regulatory subunits via binding of their SH2 domains to receptor-induced tyrosine-phosphorylated ITAM motifs leads to colocalization of PI3K catalytic subunits with their substrate, membrane PIP_2. In contrast, Class IB PI3Ks are activated downstream of G-protein-coupled receptors, including chemokine receptors. Class IB comprises two regulatory subunits (p101, p84/87) and one catalytic (p110γ) subunit (Stephens et al. 1994; Stoyanov et al. 1995; Stephens et al. 1997; Suire et al. 2005). Following GPCR engagement, class IB PI3K regulatory subunits bind through their C-terminal domains to G$\beta\gamma$ subunits. Both class IA (Okkenhaug et al. 2002; Okkenhaug et al. 2006; Patton et al. 2006; Matheu et al. 2007; Liu et al. 2009a; Liu and Uzonna 2010; Soond et al. 2010) and class IB (Sasaki et al. 2000; Swat et al. 2006; Alcazar et al. 2007) PI3Ks have important functions in T-cell development and function. The T-cell phenotypes of mice with individual or combined PI3K subunit deficiencies are summarized in Table 1.

THE SECOND MESSENGER PIP_3 RECRUITS PROTEINS TO MEMBRANES BY BINDING TO THEIR PH DOMAINS

Taken together, immunoreceptor-induced PIP_3 generation is important for lymphocyte proliferation and differentiation (for reviews, see Juntilla and Koretzky 2008; Buitenhuis and Coffer 2009; Fruman and Bismuth 2009). PIP_3 mediates the cellular effects of PI3K activation by recruiting effector proteins that stereospecifically bind to PIP_3 through a subclass of pleckstrin homology or PH domains—protein modules of approximately 120 amino acids that were originally identified as a repeated domain in Pleckstrin (Haslam et al. 1993; Mayer et al. 1993). The mammalian genome contains approximately 250 PH domain proteins. Although there is little amino acid sequence conservation among PH domains, they share a common β-barrel fold comprised of two β

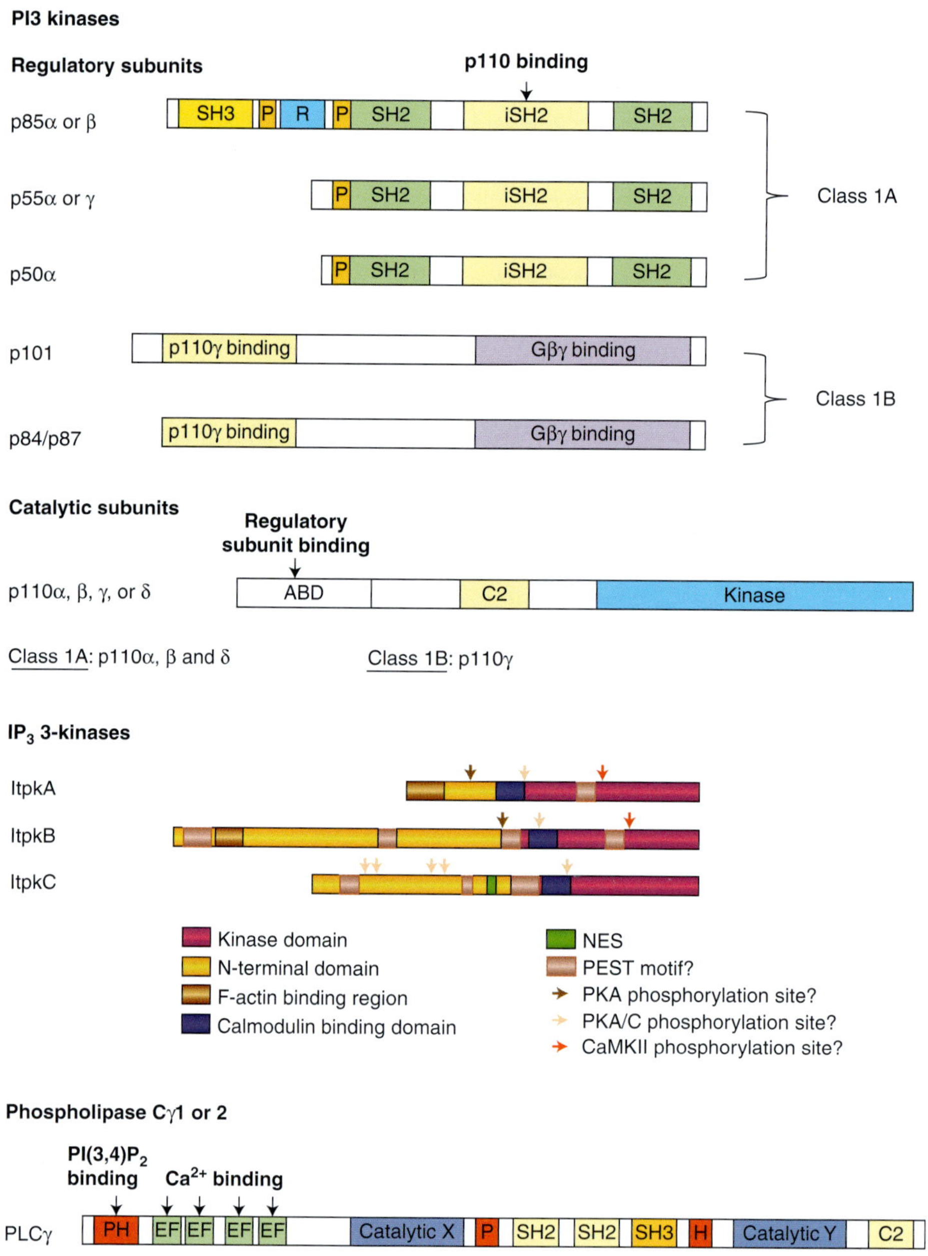

Figure 2. Schematic structures of key PIP_2 or IP_3 kinases and PIP_2 lipases in lymphocytes. For details, see text.

sheets and a C-terminal α-helical cap (Downing et al. 1994; Ferguson et al. 1994; Macias et al. 1994; Timm et al. 1994; Yoon et al. 1994). In approximately 60 PH domains, the N-terminal portion contains a pocket that binds phosphoinositols (Cozier et al. 2004; Hirata et al. 1998). These PH domains are divided into two classes. Class I PH domains bind with high affinity and specificity to $PI(4,5)P_2$, $PI(3,4)P_2$, PIP_3, and/or IP_4. Class II PH domains bind PIs more promiscuously and with lower affinity.

Different positively charged amino acid side chains in the PI-binding pocket of a PH

Table 1. Phenotypic consequences of deficiencies in phosphoinositol lipid metabolizing enzymes in lymphocytes

Deficient or functionally impaired protein(s)	T-cell phenotype	References
PI3 Kinase Regulatory Subunits		
p85α- (*pik3r1*) deficient mice	Reduced T-cell development, reduced migrational velocity, normal mature T-cell function.	(Matheu et al. 2007; Shiroki et al. 2007)
p85β- (*pik3r2*) deficient mice	Normal T-cell development, reduced cell velocity. Elevated anti-CD3/ IL-2-induced proliferation in one study. Defective peripheral T cell recall responses and CD28 signaling in another study.	(Deane et al. 2004; Matheu et al. 2007; Alcazar et al. 2009)
p85α, p55α, p50α (*pik3r1*) multi-deficient mice	None reported. Perinatal lethality.	(Fruman et al. 2000)
p85α, p55α, p50α (*pik3r1*) & p85β (*pik3r2*) multi-deficient T cells	Largely normal T-cell development. Impaired T cell Ca^{2+} signaling, proliferation and cytokines, defective Treg function and Th2 responses but normal Th1 responses, reduced cell velocity and polarization. Sjoegren's like autoimmune disease.	(Oak et al. 2006; Deane et al. 2007; Matheu et al. 2007; Fruman and Bismuth 2009)
PI3 Kinase Catalytic Subunits		
p110$\alpha^{-/-}$ mice or p110$\beta^{-/-}$ mice	Not determined. Embryonic lethal.	(Buitenhuis and Coffer 2009)
p110$\delta^{-/-}$ mice	No major abnormalities.	(Clayton et al. 2002)
p110δ^{D910A} catalytically inactive mutant expressing mice	Impaired TCR-induced Ca^{2+} signaling, proliferation, migration and cytokine production, reduced Th1, Th2 and Treg function. Protection from type 2 cytokine responses like airway hyperresponsiveness. Colitis, widespread autoimmunity.	(Okkenhaug et al. 2002; Okkenhaug et al. 2006; Patton et al. 2006; Nashed et al. 2007; Jarmin et al. 2008)
p110$\gamma^{-/-}$ mice	Some controversies between groups. Impaired T-cell development and activation, proliferation, actin polymerization. Normal activation but impaired migration in another study. Similar cell velocity but with more turning. Reduced $CD4^+$ memory cell survival, autoantibody production, glomerulonephritis, and systemic lupus.	(Sasaki et al. 2000; Rodriguez-Borlado et al. 2003; Barber et al. 2006; Alcazar et al. 2007; Martin et al. 2008; Thomas et al. 2008; Fruman and Bismuth 2009)
p110$\gamma^{-/-}$p110$\delta^{-/-}$ or p110$\gamma^{-/-}$p110δ^{D910A} mice	Impaired T-cell development, proliferation and cell survival, exaggerated Th2 over Th1 responses, high serum IgE levels.	(Webb et al. 2005; Swat et al. 2006; Ji et al. 2007)

Continued

Table 1. *Continued*

Deficient or functionally impaired protein(s)	T-cell phenotype	References
IP_3 3-Kinases		
ItpkA$^{-/-}$ mice	None reported.	(Jun et al. 1998; Kim et al. 2009)
ItpkB$^{-/-}$ mice	Blocked T-cell development and thymocyte positive selection. Normal IP_3-accumulation and Ca^{2+} signaling, but defective DAG production, Itk recruitment, and activation in thymocytes.	(Pouillon et al. 2003; Wen et al. 2004; Chamberlain et al. 2005; Huang et al. 2006; Huang et al. 2007; Jia et al. 2007; Marechal et al. 2007; Miller et al. 2007b; Huang et al. 2008; Jia et al. 2008a; Jia et al. 2008b; Miller et al. 2008; Miller et al. 2009)
ItpkC$^{-/-}$ mice	No thymic phenotype. Normal IP_3 3-kinase activity in thymocytes.	(Pouillon et al. 2003)
ItpkB$^{-/-}$ItpkC$^{-/-}$ mice	Same T-cell developmental block as *ItpkB*$^{-/-}$ mice.	(Pouillon et al. 2003)
ITPKC loss-of-function polymorphism, humans	Associated with Kawasaki disease susceptibility.	(Onouchi et al. 2008)
ITPKC knockdown, Jurkat cells	Hyperactivation.	(Onouchi et al. 2008)
ITPKC overexpression, Jurkat cells	Reduced activation.	(Onouchi et al. 2008)
IPMK$^{-/-}$ mice	None reported. Embryonic lethal.	(Frederick et al. 2005)
PIP_3/IP_4 Phosphatases		
SHIP$^{-/-}$ mice	Altered Th1/Th2 ratio, reduced CD8 cytotoxicity. Shortened lifespan, splenomegaly, autoimmunity.	(Helgason et al. 1998; Tarasenko et al. 2007)
PTEN$^{+/-}$ mice	Hyperproliferative T cells. Increased tumor incidence, autoimmunity.	(Di Cristofano et al. 1999; Di Cristofano and Pandolfi 2000)
Conditional PTEN knockout in T cells in mice	Impaired T-cell development, hyperresponsive to suboptimal TCR signals, increased cytokine production, Th2 bias, autoimmune pathology.	(Suzuki et al. 2001; Hagenbeek et al. 2004; Buckler et al. 2008)
PLCγ1$^{-/-}$ mice	Conditional knockout: Impaired thymocyte positive and negative selection, regulatory T_{reg} cell development and function, TCR-induced peripheral T-cell proliferation and cytokine production. Autoimmune disease. Conventional knockout is embryonic lethal. Chimeric mice: Severe defects in early hematopoiesis.	(Ji et al. 1997; Shirane et al. 2001; Fu et al. 2010)
PLCγ2$^{-/-}$ mice	No T-cell defect reported. Defects in mast cells, dendritic cells, osteoclasts, and neutrophils.	(Wang et al. 2000; Graham et al. 2007; Cremasco et al. 2008; Epple et al. 2008; Cremasco et al. 2010)

Continued

Table 1. *Continued*

Deficient or functionally impaired protein(s)	T-cell phenotype	References
DGKα$^{-/-}$ mice	Hyper-responsive T cells.	(Outram et al. 2002; Olenchock et al. 2006)
DGKξ$^{-/-}$ mice	Hyper-responsive T cells.	(Zhong et al. 2003; Olenchock et al. 2006)
DGKα$^{-/-}$DGKξ$^{-/-}$ mice	Lymphopenia due to a partial block in T-cell development. Reduced T_{regs}, peripheral T cell resistant to anergy-induction.	(Olenchock et al. 2006; Zha et al. 2006; Guo et al. 2008)

domain determine its ligand specificity by interacting with specific negatively charged phosphate groups within the inositol ring. The Akt (Protein kinase B) and Tec protein kinase families are part of a small but immunologically important subset of PH-domain-containing proteins whose activities are regulated by PIP_3 binding, which colocalizes the kinases with their upstream activators (August et al. 1997; Heyeck et al. 1997; Stokoe et al. 1997). In addition, the Akt PH domain can also bind to $PI(3,4)P_2$ (Cozier et al. 2004; DiNitto and Lambright 2006; James et al. 1996; Lemmon 2008).

PIP_3 CONTROLS AKT FUNCTION IN LYMPHOCYTES

Akt and its upstream activator PDK1 (Phosphoinositide-dependent kinase 1) are nonreceptor serine/threonine kinases of the AGC family. They regulate survival and proliferation of many cell types (Manning and Cantley 2007). Both kinase families are structurally relatively simple—containing only a kinase domain, a PH domain, and in the case of Akt1, a short hydrophobic motif. Before activation, the Akt and PDK1 PH domains seem to function in an inhibitory manner that is released upon membrane PIP_3 binding (Stokoe et al. 1997; Gao and Harris 2006). Full PDK1 activation requires PIP_3 binding to the PDK1 PH domain, which allows trans-phosphorylation of a C-terminal residue (Gao and Harris 2006). In the absence of lipid ligands, the Akt1 PH domain occludes access to the activation site, prohibiting its phosphorylation by PDK1 (Stokoe et al. 1997; Calleja et al. 2007). PH-domain-mediated PDK1 and Akt recruitment to membrane PIP_3 leads to phosphorylation of the catalytic T-loop of Akt by PDK1, activating Akt (Stephens et al. 1998). A second kinase of uncertain identity phosphorylates the Akt C-terminal hydrophobic motif, leading to optimal Akt kinase activity.

There are three Akt paralogs in mammals (Akt1/2/3). All are expressed in lymphocytes and have partially redundant functions. Akt promotes cell survival primarily through inhibition of apoptosis. Akt can directly or indirectly inhibit the function of proapoptotic Bcl-2 family members by phosphorylation (del Peso et al. 1997; Gardai et al. 2004; Qi et al. 2006; Hubner et al. 2008). Akt also activates the transcription factor NF-kB (Madrid et al. 2000) and induces expression of antiapoptotic Bcl-2, FLIP, and IAP proteins (Wang et al. 1998; Panka et al. 2001). Akt, moreover, promotes cell proliferation by increasing cell metabolism (Frauwirth et al. 2002; Rathmell et al. 2003; Jacobs et al. 2008), facilitating protein translation (Dufner et al. 1999), and cell cycle progression (Liang et al. 2002; Shin et al. 2002; Viglietto et al. 2002). Consequently, genetic alterations that enhance Akt activity either directly or indirectly via amplification of PI3K activity often lead to cellular transformation (Liu et al. 2009b).

Akt1 and Akt2 are particularly important during early T-cell development (Juntilla et al. 2007; Juntilla and Koretzky 2008). CD4 and CD8 T cells develop from a common thymic progenitor through a series of developmental stages that can be partially characterized by differential cell-surface expression of the CD4 and CD8 coreceptors. At the early, $CD4^-CD8^-$ double-negative (DN) stage, productive TCRβ gene rearrangement results in a pre-TCR signal that suppresses rearrangement of the second TCRβ allele and triggers proliferative expansion and progression to the $CD4^+CD8^+$ double-positive (DP) stage. During this process, the cells rearrange their TCRα chain genes. Successful TCRα protein expression results in pre-TCR replacement by a mature TCR composed of α and β chains. Akt1 or Akt2 deficiency results in a two-fold reduction in thymic cellularity due to a partial block between DN and DP cells. In contrast, Akt1 and Akt2 double-deficiency results in a 45-fold decrease in thymic cellularity (Juntilla et al. 2007). This highlights the importance of Akt kinases during the highly proliferative stage of thymic development.

TEC FAMILY PROTEIN TYROSINE KINASES ARE KEY PIP_3 EFFECTORS IN LYMPHOCYTES

The expression of certain Tec family non-receptor protein–tyrosine–kinases (TFKs) is restricted to specific immune cell types (Readinger et al. 2009). Itk and Rlk are largely restricted to T cells and NK cells. Btk is not expressed in these, but enriched in B cells, mast cells, and myeloid cells. Tec is broadly expressed in hematopoietic cells. Despite some functional redundancies, deficiency in individual TFKs causes profound functional defects (for a review, see Readinger et al. 2009). Itk deficiency causes defects in T-cell development and in certain peripheral T-cell functions (Liao and Littman 1995; Fowell et al. 1999; Schaeffer et al. 1999; Schaeffer et al. 2000; Schaeffer et al. 2001; Tomlinson et al. 2004; Finkelstein et al. 2005; Atherly et al. 2006; Broussard et al. 2006; Horai et al. 2007; Lucas et al. 2007).

At the DP stage, thymocytes are evaluated for the ability of their TCR to interact with self MHC/peptide complexes on the surfaces of thymic antigen-presenting cells. DP cells whose TCRs do not interact with sufficient strength or duration die by "neglect." DP cells whose TCRs interact moderately with self-MHC class I or II are positively selected to mature into CD8 or CD4 single positive T cells, respectively. DP cells carrying potentially autoreactive TCRs that interact too strongly or for an inappropriate duration with self-MHC are induced to die by "negative selection." A major unresolved question in immunological research is how TCR engagement can have such different outcomes.

Positively and negatively selecting TCR signals appear to activate the same proximal signaling molecules, including Itk. In Itk-deficient mice, the efficiencies of positive and negative selection are dramatically decreased (Liao and Littman 1995; Schaeffer et al. 2000). In contrast, Itk deficiency favors the accumulation of a nonconventional thymic T cell subset with an effector/memory or "innate" phenotype, particularly in the CD8 lineage (Atherly et al. 2006; Broussard et al. 2006; Horai et al. 2007). In addition, TCR signals that normally induce negative selection can, under some circumstances, promote positive selection in Itk-deficient mice (Schaeffer et al. 2000). In mature T cells, Itk appears to be required to sustain T_H2 responses (Fowell et al. 1999; Schaeffer et al. 2001).

TFKs have a relatively complex domain structure with an N-terminal PH domain, followed by a proline-rich Tec homology domain, Src homology 3 (SH3) and SH2 domains, and a C-terminal kinase domain. In Rlk, a missing N-terminal PH domain is functionally substituted by palmitoylation, resulting in constitutive membrane association (Debnath et al. 1999). Membrane recruitment of the other TFKs predominantly depends on interactions between their PH domains and membrane PIP_3 (Ching et al. 1999; Shan et al. 2000), together with interactions between their SH2 domains and adaptor proteins such as SLP-76 in T cells (Ching et al. 1999; Su et al. 1999). However, recent data may suggest that Btk membrane recruitment can also occur in a PH-domain-independent manner (Fruman and Bismuth 2009). Activation-induced association with cytosolic SLP-76 and the membrane adaptor LAT colocalizes TFKs with their upstream activators, Src kinases, and their downstream target, PLCγ1

(Wange 2000; Qi and August 2007). Mechanistically, the consequences of TFK deficiency have primarily been attributed to defective PLCγ phosphorylation and activation (Liu et al. 1998). In addition, TFKs appear required for Vav-dependent cytoskeletal organization and cell adhesion (Gomez-Rodriguez et al. 2007). However, TFK promotion of Vav function appears to involve TFK adaptor functions, rather than their kinase activities (Grasis et al. 2003; Dombroski et al. 2005).

PHOSPHOLIPASE-Cγ HYDROLYZES PIP_2 INTO THE SECOND MESSENGERS DIACYLGLYCEROL AND INOSITOL(1,4,5)TRISPHOSPHATE

$PI(4,5)P_2$ is a substrate for another immunologically important enzyme, phosphatidylinositol-specific phospholipase-Cγ (PLCγ). PLCγ hydrolyzes PIP_2 into its hydrophobic and hydrophilic components, the membrane-lipid diacylglycerol (DAG) and soluble inositol(1,4,5)trisphosphate (IP_3). Both are second messengers that regulate proteins through specific binding domains.

The two mammalian PLCγ isoforms, PLCγ1 and 2, have partially overlapping expression patterns and functions (Wilde and Watson 2001). T cells exclusively express PLCγ1. T-cell-specific PLCγ1-deletion impaired thymocyte positive and negative selection, regulatory T_{reg} cell development and function, TCR-induced peripheral T-cell proliferation, and cytokine production (Fu et al. 2010). Defective TCR activation of the MAP kinases Erk and Jnk, and of the transcription factors NFAT, AP-1, and NF-κB indicates the broad importance of PLCγ1 in TCR signaling through several pathways. Autoimmune disease symptoms show the physiological importance of PLCγ1 in T cells (Fu et al. 2010). Severe blocks in T-cell development and late onset autoimmunity in mice expressing a PLCγ1-binding-deficient LAT allele indicate the importance of LAT interactions for PLCγ1 function (Sommers et al. 2002; Sommers et al. 2005). Finally, severe defects in early hematopoiesis in chimeric mice generated with PLCγ1-deficient embryonic stem cells suggest important PLCγ1 functions in hematopoietic stem or progenitor cells (Shirane et al. 2001). In contrast, PLCγ2-deficient mice are viable with specific defects in B cells, mast cells, dendritic cells, osteoclasts, and neutrophils (Wang et al. 2000; Graham et al. 2007; Cremasco et al. 2008; Epple et al. 2008; Cremasco et al. 2010).

PLCγ1/2 have complex domain structures with an N-terminal PH domain, followed by a number of EF hands, a catalytic domain that is split by an internal regulatory domain, and a C-terminal Ca^{2+}-binding-C2 domain (Fig. 2). The regulatory domain contains a core of two tandem SH2 domains and an SH3 domain that is flanked by a split PH domain. This split PH domain can self-associate to form a functional PH domain, or associate in *trans* with a split PH domain in TRPC-family Ca^{2+} channels (van Rossum et al. 2005). Either one or both PH domains contribute to PLCγ1 membrane association by binding to PIP_3 (Bae et al. 1998; Falasca et al. 1998). Additional interactions are provided by the other modular domains. Once PLCγ1 is at the membrane, phosphorylation of tyrosine residues within its regulatory domain by TFKs induces PLCγ1 activation. Recent studies suggest that this requires phosphotyrosine-independent PLCγ1 SH2 domain binding to a noncanonical ligand motif in Itk (Joseph et al. 2007; Min et al. 2009).

DIACYLGLYCEROL CONTROLS Ras AND PKC ACTIVATION IN LYMPHOCYTES

The membrane second messenger, DAG, propagates signals via membrane recruitment of cytosolic signaling proteins by binding to their C1 domains, cysteine-rich domains of approximately 50 amino acids. Two β-sheets harbor the DAG-binding cavity.

Several well-characterized DAG-effector families include Ras guanine-nucleotide-exchange-factors/releasing proteins (RasGRPs), protein kinase C-related kinases (PKCs, PKD), chimaerin Rho/Rac-GTPase-activating proteins (Yang and Kazanietz 2007), Munc13 proteins (Betz et al. 1998), and diacylglycerol kinases (DGKs). There is some effector selectivity for different DAG species that differ in their subcellular localization.

For example, RasGRPs preferentially bind to DAG in Golgi membranes (Carrasco and Merida 2004). PKCs preferentially bind to DAG in the plasma membrane (Spitaler et al. 2006).

RasGRP membrane recruitment by DAG colocalizes these Ras activators with their substrate, inducing release of Ras-bound GDP, GTP binding, and Ras activation. Ras then activates the kinase Raf, which activates the downstream Erk cascade. The four mammalian RasGRPs1–4 have partially overlapping expression patterns and partially redundant functions. T cells predominantly express RasGRP1, the main mediator of TCR-induced Ras/Erk activation (Dower et al. 2000; Priatel et al. 2002). RasGRP1 deficiency causes a significant block of T-cell development with strong defects in positive, and some defects in negative selection (Dower et al. 2000; Priatel et al. 2002). However, this developmental block is incomplete and can be partially rescued by strong TCR activation (Priatel et al. 2006). Nevertheless, a moderate lymphopenia occurs due to T-cell exhaustion (Priatel et al. 2007). Interestingly, T_{reg} cells accumulate in the periphery of RasGRP1-deficient mice, despite perturbed T_{reg} development (Chen et al. 2008).

In contrast to RasGRPs, DAG-mediated membrane recruitment allosterically induces PKC activitation by abrogating an autoinhibitory association between the PKC pseudo-substrate and substrate-binding domains (Rosse et al. 2010). DAG promotes activation of classic (PKCα, PKCβI, PKCβII, PKCγ), novel (PKCδ, PKCε, PKCη, PKCθ), and atypical PKCs (PKCξ, λ/ι). However, classic and novel PKCs also require Ca^{2+} binding to their C2 domains (Rosse et al. 2010). Multiple studies have shown essential PKC roles in lymphocyte development and function (for reviews, see Isakov and Altman 2002; Barouch-Bentov and Altman 2006; Manicassamy et al. 2006). In particular, PKCθ is important for TCR signaling and required for thymocyte-positive selection (Morley et al. 2008). Incomplete developmental defects likely reflect redundancy among thymocyte-expressed PKCs.

Several recent publications suggest important functions for chimaerins in TCR signaling, T-cell adhesion, and chemotaxis that involve their ability to inactivate Rac (Siliceo et al. 2006; Caloca et al. 2008; Siliceo and Merida 2009). No munc13 protein roles in the immune system have been reported.

DIACYLGLYCEROL KINASES CONVERT DAG INTO PHOSPHATIDIC ACID

Aside from their production by PLCγ in lymphocytes, DAG levels are also regulated through their phosphorylation into phosphatidic acid (PA) by DAG kinases (DGKs). In T cells, this down-regulates PKC and RasGRP functions and TCR-induced-Erk activation (reviewed in Zhong et al. 2008). However, in many cell types, receptor-induced PA generation activates a series of PA-effector proteins with various functions, including vesicular trafficking, cell survival, and proliferation (Wang et al. 2006).

The ten mammalian DGKs form five groups based on their domain structure (Fig. 3). All DGKs have two to three C1 domains and a kinase domain. However, these C1 domains do not necessarily participate in DAG binding. Instead, the domains that distinguish the DGK types direct differential localization or activation requirements. T cells express at least three DAG kinases: DGKα (type I), δ (type II), and ξ (type IV). DGKα has an N-terminal RVH domain and two EF hands. Ca^{2+} binding to these three domains induces DGKα activation. DGKδ has an N-terminal PH domain, two EF hands, and a C-terminal SAM domain. The PH domain facilitates DAG binding. The SAM domain mediates DGKδ oligomerization and ER targeting. DGKξ has a central MARCKS domain, multiple ankyrin repeats, and a C-terminal PDZ-BM domain. The MARCKs domain contains a nuclear localization sequence. The PDZ-BM domain of DGKξ regulates Rac activity and membrane ruffling.

DGKα and DGKξ cooperatively down-regulate T-cell activation and facilitate T-cell tolerance (Guo et al. 2008). Different T-cell developmental and activation stages express different DGKα and DGKξ splice variants (Macian et al. 2002). DGKα- (Outram et al.

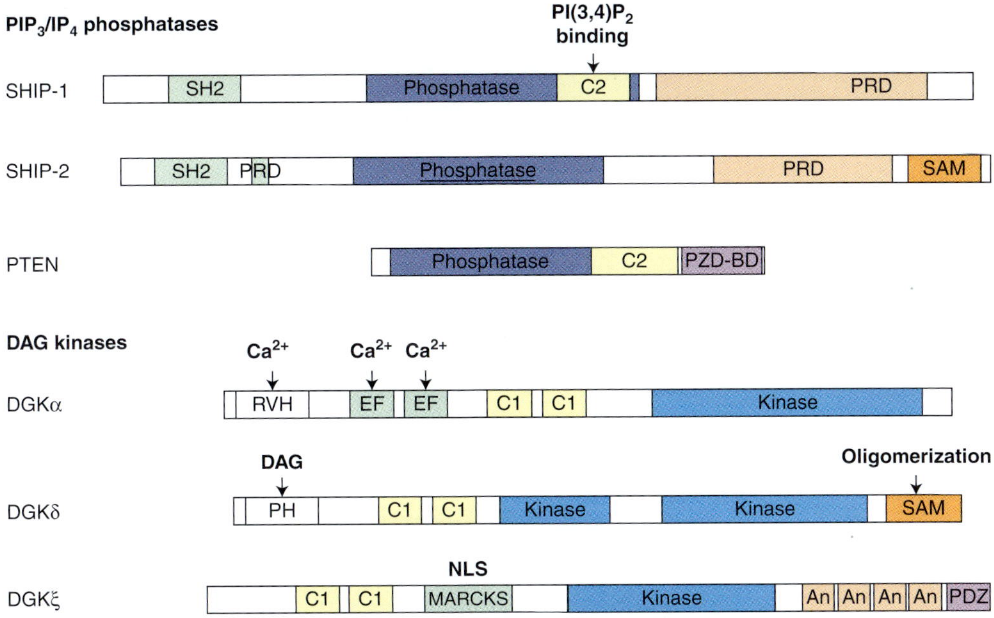

Figure 3. Schematic structures of key PIP_3 or IP_4 phosphatases and DAG kinases in lymphocytes. For details, see text.

2002; Olenchock et al. 2006) or DGKξ- (Zhong et al. 2003; Olenchock et al. 2006) deficient mice have increased DAG-dependent signaling and produce hyper-responsive T cells. DGKα/ξ double deficiency causes lymphopenia due to a partial block in T-cell development (Guo et al. 2008). The peripheral T cells contain reduced T_{reg} proportions and are resistant to anergy induction (Olenchock et al. 2006; Zha et al. 2006). Therefore, limiting DAG function in T cells is of great physiological importance, although defective PA production may contribute to the phenotype (Zhong et al. 2008).

IP_3 MEDIATES ANTIGEN-RECEPTOR-INDUCED Ca^{2+} MOBILIZATION AND CAN ACT AS A PRECURSOR FOR HIGHER-ORDER INOSITOL PHOSPHATES IN LYMPHOCYTES

Ca^{2+} is a rapid and robust, soluble second messenger with multiple downstream effectors (Feske 2007; Oh-hora and Rao 2008; Vig and Kinet 2009). mM extracellular Ca^{2+} concentrations contrast with nM cytosolic Ca^{2+} concentrations that are tightly maintained by pumps, which sequester Ca^{2+} into the extracellular space or intracellular storage compartments (Feske 2007; Oh-hora and Rao 2008). Following TCR engagement, Ca^{2+} signaling is initiated by the soluble product of PIP_2 hydrolysis, Ins(1,4,5)P_3 (here IP_3). IP_3 binding to receptors in the ER membrane releases the stored Ca^{2+} into the cytosol (Taylor et al. 2009). Subsequent sensing of the store depletion by STIM proteins induces STIM translocation to the plasma membrane, where STIM-induced opening of ORAI-family Ca^{2+} channels triggers a store-operated Ca^{2+} entry (SOCE). This causes a sustained increase of the cytoplasmic Ca^{2+} concentration that is essential for T-cell activation (Feske 2007; Oh-hora and Rao 2008; Vig and Kinet 2009).

Cytosolic Ca^{2+} binds to proteins with C2 domains and EF hands, inducing conformational changes that alter activity (Niki et al. 1996). Important examples with EF hands are the cytosolic adapter Calmodulin and STIM proteins (Feske 2007; Oh-hora and Rao 2008; Vig and Kinet 2009).

IP_3 CONVERSION INTO INOSITOL(1,3,4,5)TETRAKISPHOSPHATE BY IP_3 3-KINASES CONTROLS LYMPHOCYTE DEVELOPMENT AND FUNCTION

Aside from mediating receptor-induced Ca^{2+} mobilization, IP_3 can also act as a precursor for other IPs. A key step in their generation is IP_3 phosphorylation at its 3-position into IP_4 by IP_3 3-kinases (IP3Ks). Mammals have four IP3Ks: the closely related ITPK-family members ItpkA/B/C and the unrelated IP multikinase (IPMK).

Genetic studies in yeast and biochemical data have suggested many potential IP_4 functions (Irvine 2001; Irvine and Schell 2001; Nalaskowski and Mayr 2004; Pattni and Banting 2004; Irvine 2005; Irvine et al. 2006; York 2006; Otto et al. 2007; Huang et al. 2008; Jia et al. 2008b; Miller et al. 2008; Sauer and Cooke 2010; Schell 2010). Much research has explored potential IP3K roles in regulating Ca^{2+} mobilization either by controlling IP_3 turnover, or through distinct functions of IP_4 or IP_4-derived other IPs. The results caused a long-standing controversy that could be attributed to differences in cell type, stimulation regimen, or experimental conditions. Studies with human Jurkat T cells suggest that IP_4 can sustain TCR-induced IP_3 accumulation and Ca^{2+} mobilization through competitive inhibition of an inositol 5-phosphatase (Hermosura et al. 2000). Moreover, a recent study suggested that ITPKC knockdown in Jurkat cells promotes TCR-induced activation of the Ca^{2+}-effector NFAT and IL-2 production (Onouchi et al. 2008). These data are consistent with one IP3K role in controlling IP_3 levels or function. The precise mechanism, identity of the 5-phosphatase, and physiological relevance remain unclear. Normal anti-TCR antibody-induced IP_3 accumulation and Ca^{2+} mobilization in ItpkB-deficient thymocytes suggest that ItpkB may not regulate Ca^{2+} mobilization in murine thymocytes (Pouillon et al. 2003; Wen et al. 2004). However, IP_4 levels and total IP3K activity were only approximately 50% reduced. Thus, other IP3Ks likely contribute to IP_4 production in thymocytes, and the consequences of complete IP3K/IP_4 deficiency on Ca^{2+} mobilization still needs to be deter-mined.

IP_4 chemically resembles the PH-domain-binding headgroup of PIP_3, and less well those of $PI(4,5)P_2$ and $PI(3,4)P_2$. Not surprisingly, IP_4 can bind to some of the protein domains that bind to these phosphoinositides. This property is often used to facilitate PH domain crystallization for X-ray structure determination (Lemmon 2008). Biochemical data suggest that IP_4 binding to the PH domains of Ras GTPase-activating proteins (RASA2/$GAP1^{m}$, RASA3/$GAP1^{IP4BP}$) (Cullen et al. 1995; Cozier et al. 2000b), BTK, AKT, or certain regulators of vesicular trafficking (synaptotagmins, centaurin-α1/$p42^{IP4}$), might inhibit their membrane recruitment, activation, or protein interactions. However, until recently, the physiological relevance of these findings was unknown.

Lymphocytes predominantly express two Itpks, ItpkB and ItpkC (Vanweyenberg et al. 1995; Wen et al. 2004). ItpkC is broadly expressed by many tissues. ItpkB expression appears restricted to hematopoietic cells and the brain (Wen et al. 2004; Jia et al. 2008a). ItpkB-deficient mice recently unveiled the physiological importance of ItpkB in lymphocyte development and activation (Table 1) (Pouillon et al. 2003; Wen et al. 2004). These mice are profoundly immunocompromised due to severe peripheral T-cell deficiency (Pouillon et al. 2003; Wen et al. 2004), reduced numbers and defective activation of B cells (Marechal et al. 2007; Miller et al. 2007a; Miller et al. 2009), and increased numbers of functionally perturbed neutrophils (Jia et al. 2007). Several recent reviews describe current models for ItpkB function in B cells and neutrophils (Huang et al. 2008; Jia et al. 2008b; Miller et al. 2008; Sauer and Cooke 2010). Here, we highlight the function of IP_4 in T cells.

ItpkB-deficient mice have a block of T-cell development due to defective positive selection (Pouillon et al. 2003; Wen et al. 2004). Whether negative selection is affected is still unclear (Sauer and Cooke 2010). Selected effectors of both IP_3 (Ca^{2+}) and DAG are specifically required for positive selection. Deficiency in Calcineurin B, a Ca^{2+}-dependent phosphatase that activates NFAT (Neilson et al. 2004; Gallo et al. 2007), or in RasGRP1, partially blocks positive selection (Dower et al. 2000; Priatel et al. 2002). Intriguingly, ItpkB deficiency blocks positive selection much more profoundly in a manner that depends on the ability of ItpkB to generate IP_4 (Huang et al. 2007). Normal IP_3 levels and Ca^{2+} mobilization in ItpkB-deficient thymocytes argue against significant contributions by these messengers. Instead, the block results from an inability of ItpkB-deficient thymocytes to produce sufficient DAG amounts in response to mild, positively selecting TCR stimuli (Huang et al. 2007). Consequently, mild TCR engagement induced Ras/Erk activation is impaired in ItpkB-deficient DP cells. This likely contributes to defective positive selection (Starr et al. 2003; McGargill et al. 2009). Addition of the DAG analog, PMA, restored Erk activation and allowed ItpkB-deficient CD4 and CD8 T-cell maturation (Huang et al. 2007). Thus, IP_4 is required for DAG-induced Erk activation in response to positively selecting TCR stimuli. Yet, how does IP_4 promote DAG accumulation?

SOLUBLE IP_4 CONTROLS ITK-PH-DOMAIN BINDING TO PIP_3 IN THYMOCYTES

Mechanistic analyses revealed that IP_4 is required for membrane recruitment and subsequent activation of the TFK Itk following mild TCR stimulation (Huang et al. 2007). This was surprising since generation of PIP_3, the established membrane ligand for the Itk PH domain, was thought to be the limiting factor for Itk membrane localization. In vitro analyses then showed that low µM IP_4 concentrations, as previously found in TCR-stimulated T cells (Imboden and Pattison 1987), promoted Itk or Itk-PH-domain binding to PIP_3 (Huang et al. 2007). Very high, likely super-physiological IP_4 concentrations, competed with PIP_3 binding. The ability of the Itk-PH domain to oligomerize (Huang et al. 2007) then resulted in a model where IP_4 binding to one Itk-PH domain in a multimeric Itk complex induced conformational changes in all PH domains within the complex that allosterically increase their affinities for PIP_3 and IP_4. This promotes PIP_3 binding, as long as IP_4 is not in excess over PIP_3. Here, IP_4 catalyzes PIP_3 binding at low, but competes with it at very high concentrations (Sauer and Cooke 2010). These findings identified Itk as the first physiological IP_4 "receptor" and unveiled the first physiological IP_4 function as a second messenger that controls Itk membrane recruitment and activation downstream of the TCR (Irvine 2007). They suggest that IP_4 and PIP_3 act as partners where soluble IP_4 controls the ability of its membrane lipid counterpart to interact with the Itk-PH domain. It is exciting to hypothesize that this new principle of regulating PH-domain function through an interplay of a soluble IP and its phosphoinositide-lipid partner has broader relevance in biological signaling. Indeed, IP_4 could have similar bimodal effects on PIP_3 binding to the Tec- and RASA3-PH domains, but whether this is physiologically relevant still needs to be determined (Huang et al. 2007). In neutrophils, IP_4 has recently been suggested to inhibit Akt-PH domain interactions with membrane PIP_3 or $PI(3,4)P_2$ (Jia et al. 2007). Loss of this function may contribute to an accumulation of granulocyte-macrophage progenitors and functionally-impaired neutrophils in $ItpkB^{-/-}$ mice (Jia et al. 2007; Jia et al. 2008a; Jia et al. 2008b). On the other hand, IP_4 did not affect PLCγ1 binding to PIP_3, which is mediated through its classic- and split-PH domains (Fig. 2) (Huang et al. 2007). Thus, not all PIP_3-binding PH domains may be subject to IP_4 control. One possible explanation is that the affinities of a given PH domain for PIP_3 and IP_4 can differ, with some PH domains preferring IP_4 or PIP_3, possibly due to differential contributions of the 1-phosphate or the lipid environment to the binding free energy (DiNitto and Lambright 2006).

Finally, it is important to keep in mind that IP_4 can be a precursor for multiple soluble IPs in mammalian cells. Some of these have been found in lymphocytes (for reviews, see Sauer et al. 2009; Sauer and Cooke 2010). Hence, some aspects of the $ItpkB^{-/-}$ phenotype might reflect deficiencies in other IPs.

Due to the blocked T-cell development in ItpkB-deficient mice, analyses of specific ItpkB functions in peripheral T cells will require conditionally ItpkB-deficient mice. However, studies linking ItpkC loss-of-function to augmented Jurkat cell activation and human Kawasaki disease support an IP_4 role in regulating peripheral T-cell activation (Onouchi et al. 2008).

PHOSPHATASES CONTROL PIP_3 AND IP_4 TURNOVER

PIP_3 levels are down-regulated by lipid phosphatases, including PTEN and SHIP1/2 in lymphocytes (Figs. 1 and 3; Table 1) (Harris et al. 2008). PTEN reverses the PI3K reaction and is generally thought to limit PI3K function. SHIP1/2 remove the PIP_3 5-phosphate, generating $PI(3,4)P_2$ and also down-modulating PIP_3 levels (Damen et al. 1996; Kavanaugh et al. 1996; Lioubin et al. 1996). However, effectors such as Akt that predominantly bind to the inositol 3/4-phosphates may continue to be recruited and activated by $PI(3,4)P_2$. Thus, differential use of PTEN and SHIP1/2 can selectively alter the activities of a subset of PIP_3 effectors, qualitatively changing PI3K responses.

Complete PTEN deficiency in mice results in embryonic lethality (Di Cristofano et al. 1998). PTEN hemizygosity, or conditional deficiency caused T-cell hyperproliferation, autoimmunity, and increased tumor incidence (Di Cristofano et al. 1999; Di Cristofano and Pandolfi 2000). PTEN deletion in T cells revealed the importance of precise PIP_3 regulation for T-cell development through defective positive and negative selection (Suzuki et al. 2001). PTEN-deficient peripheral T cells were hyperresponsive to suboptimal TCR stimulation and less dependent on co-stimulation, resulting in autoimmune pathology (Suzuki et al. 2001). Activated T cells secreted increased cytokine amounts (Suzuki et al. 2001) and showed a bias toward the T_H2 lineage (Buckler et al. 2008). Thus, PIP_3 down-regulation by PTEN is required for limiting T-cell responses and maintaining self-tolerance.

PIP_3 dephosphorylation at the 5-position by SHIP1/2 is also important for T-cell tolerance. However, while the effects of SHIP deficiency partially overlap with those of PTEN deficiency, the additional importance of generating $PI(3,4)P_2$ is evident from the additional shortened lifespan of SHIP-deficient mice (Helgason et al. 1998). SHIP-deficient CD4 T cells have an altered Th1/Th2 cell ratio. SHIP-deficient CD8 T cells are less cytotoxic (Tarasenko et al. 2007). Thus far, these phenotypes have been attributed to defective PIP_3 conversion into $PI(3,4)P_2$.

However, in vitro, both SHIPs and PTEN can dephosphorylate IP_4 into $I(1,3,4)P_3$ or $Ins(1,4,5)P_3$, respectively (Fig. 1) (Erneux et al. 1998; Maehama and Dixon 1998; Pesesse et al. 1998; Caffrey et al. 2001). Whether this is relevant in vivo is unknown. If it were, then the phenotypes of $PTEN^{-/-}$ or $SHIP^{-/-}$ mice might need reinterpretation to consider potential contributions of perturbed IP_4 down-regulation. On the other hand, the presence of an IP_4-inhibited 5-phosphatase in Jurkat cells that lack SHIP and PTEN (and also show hyperactivation) suggests that other phosphatases might also regulate IP_4 levels in T cells (Hermosura et al. 2000).

CONCLUDING REMARKS

We have highlighted the importance of phosphoinositol signaling in lymphocytes, focusing on T cells. The importance of phosphoinositide lipids and of the Ca^{2+}-mobilizing soluble $I(1,4,5)P_3$ in lymphocyte signaling have long been established, although many open questions remain. For example, what influence do the specific fatty acids in a phosphoinositide have? There is evidence that many different fatty acids can be found in a given phosphoinositide isomer. They might control where it is located or other aspects of its function. Is PIP_3 really a universe of diverse molecules that differ in their fatty acid moieties and, possibly, functions?

More recent data unveiled important functions for the lipids DAG and PA, and for soluble IP_4 in lymphocyte development and function. Some insight into the molecular mechanisms through which these novel messengers act has been gained, but our understanding of their molecular functions is far from complete. In particular, the roles of additional DAG effectors other than RasGRPs and PKCs in T cells need assessment, and the molecular mechanisms through which PA contributes to DGK function in lymphocytes are only beginning to be understood (Zhong et al. 2008).

One of the most intriguing, possibly broadly relevant, recent findings is the ability of IP_4 to control PIP_3 interactions with its effectors. Because PIP_3 is the key mediator of PI3K, PTEN, and SHIP function, IP_4 might ultimately act to control the functions of these important enzymes. Biochemically, PI3K and IP3Ks catalyze the very same reaction: phosphorylation of the inositol 3-position. Physiologically, they create the Tessa Virtue and Scott Moir, respectively, of a couple of second messengers whose dance in the cellular ice ring is reminiscent of the intricacy and elegance of its 2010 Winter Olympics counterpart. In DP thymocytes, the dance establishes a feedback loop of Itk-membrane recruitment and PLCγ1 activation that is essential to allow mild TCR stimuli to cause sufficient DAG production, such that Ras/Erk can be activated and trigger positive selection (Huang et al. 2007; Sauer and Cooke 2010). It will be important to explore how broadly relevant PIP_3 control through soluble IP_4 is in other cell types and downstream of other receptors. Moreover, clarifying what determines whether a PIP_3-binding protein domain is subject to IP_4 control remains a challenging and important question. Finally, it will be exciting to determine the functions of the other higher-order IPs found in lymphocytes (Sauer et al. 2009; Sauer and Cooke 2010), and to determine which ones are IP_4-derived and might, thus, mediate aspects of IP_4 function. Clearly, exploring phosphoinositide-lipid signaling and the functions of the soluble IP code in lymphocytes will remain an exciting and important research area for years to come.

ACKNOWLEDGMENTS

We thank Claire Conche and Sabine Siegemund for critical reading of the manuscript and valuable comments. K.S. is supported by NIH grants AI070845 and GM088647, and by Scholar Award 1440-11 from The Leukemia & Lymphoma Society. The authors declare no competing financial interests.

REFERENCES

Alcazar I, Cortes I, Zaballos A, Hernandez C, Fruman DA, Barber DF, Carrera AC. 2009. p85beta phosphoinositide 3-kinase regulates CD28 coreceptor function. *Blood* **113:** 3198–3208.

Alcazar I, Marques M, Kumar A, Hirsch E, Wymann M, Carrera AC, Barber DF. 2007. Phosphoinositide 3-kinase gamma participates in T cell receptor-induced T cell activation. *J Exp Med* **204:** 2977–2987.

Alcazar-Roman AR, Wente SR. 2008. Inositol polyphosphates: a new frontier for regulating gene expression. *Chromosoma* **117:** 1–13.

Ang BK, Lim CY, Koh SS, Sivakumar N, Taib S, Lim KB, Ahmed S, Rajagopal G, Ong SH. 2007. ArhGAP9, a novel MAP kinase docking protein, inhibits Erk and p38 activation through WW domain binding. *J Mol Signal* **2:** 1.

Atherly LO, Lucas JA, Felices M, Yin CC, Reiner SL, Berg LJ. 2006. The Tec family tyrosine kinases Itk and Rlk regulate the development of conventional $CD8^+$ T cells. *Immunity* **25:** 79–91.

August A, Sadra A, Dupont B, Hanafusa H. 1997. Src-induced activation of inducible T cell kinase (ITK) requires phosphatidylinositol 3-kinase activity and the Pleckstrin homology domain of inducible T cell kinase. *Proc Natl Acad Sci U S A* **94:** 11227–11232.

Bae YS, Cantley LG, Chen CS, Kim SR, Kwon KS, Rhee SG. 1998. Activation of phospholipase C-gamma by phosphatidylinositol 3,4,5-trisphosphate. *J Biol Chem* **273:** 4465–4469.

Barber DF, Bartolome A, Hernandez C, Flores JM, Fernandez-Arias C, Rodriguez-Borlado L, Hirsch E, Wymann M, Balomenos D, Carrera AC. 2006. Class IB-phosphatidylinositol 3-kinase (PI3K) deficiency ameliorates IA-PI3K-induced systemic lupus but not T cell invasion. *J Immunol* **176:** 589–593.

Barouch-Bentov R, Altman A. 2006. Protein Kinase C-Theta (PKCtheta): New Perspectives on Its Functions in T-Cell Biology. In *Advances in Experimental Medicine and Biology* (ed. CD Tsoukas), pp. 1–14. Springer, New York, NY.

Betz A, Ashery U, Rickmann M, Augustin I, Neher E, Sudhof TC, Rettig J, Brose N. 1998. Munc13–1 is a presynaptic phorbol ester receptor that enhances neurotransmitter release. *Neuron* **21:** 123–136.

Broussard C, Fleischacker C, Horai R, Chetana M, Venegas AM, Sharp LL, Hedrick SM, Fowlkes BJ, Schwartzberg PL. 2006. Altered development of $CD8^+$ T cell lineages in mice deficient for the Tec kinases Itk and Rlk. *Immunity* **25:** 93–104.

Buckler JL, Liu X, Turka LA. 2008. Regulation of T-cell responses by PTEN. *Immunol Rev* **224:** 239–248.

Buitenhuis M, Coffer PJ. 2009. The role of the PI3K-PKB signaling module in regulation of hematopoiesis. *Cell Cycle* **8:** 560–566.

Burton A, Hu X, Saiardi A. 2009. Are inositol pyrophosphates signalling molecules? *J Cell Physiol* **220:** 8–15.

Caffrey JJ, Darden T, Wenk MR, Shears SB. 2001. Expanding coincident signaling by PTEN through its inositol 1,3,4,5,6-pentakisphosphate 3-phosphatase activity. *FEBS Lett* **499:** 6–10.

Calleja V, Alcor D, Laguerre M, Park J, Vojnovic B, Hemmings BA, Downward J, Parker PJ, Larijani B. 2007. Intramolecular and intermolecular interactions of protein kinase B define its activation in vivo. *PLoS Biol* **5:** e95.

Caloca MJ, Delgado P, Alarcon B, Bustelo XR. 2008. Role of chimaerins, a group of Rac-specific GTPase activating proteins, in T-cell receptor signaling. *Cell Signal* **20:** 758–770.

Carrasco S, Merida I. 2004. Diacylglycerol-dependent binding recruits PKCtheta and RasGRP1 C1 domains to specific subcellular localizations in living T lymphocytes. *Mol Biol Cell* **15:** 2932–2942.

Ceccarelli DF, Blasutig IM, Goudreault M, Li Z, Ruston J, Pawson T, Sicheri F. 2007. Non-canonical interaction of phosphoinositides with pleckstrin homology domains of Tiam1 and ArhGAP9. *J Biol Chem* **282:** 13864–13874.

Chamberlain PP, Sandberg ML, Sauer K, Cooke MP, Lesley SA, Spraggon G. 2005. Structural insights into enzyme regulation for inositol 1,4,5-trisphosphate 3-kinase B. *Biochem* **44:** 14486–14493.

Chen X, Priatel JJ, Chow MT, Teh HS. 2008. Preferential development of CD4 and CD8 T regulatory cells in RasGRP1-deficient mice. *J Immunol* **180:** 5973–5982.

Ching KA, Kawakami Y, Kawakami T, Tsoukas CD. 1999. Emt/Itk associates with activated TCR complexes: role of the pleckstrin homology domain. *J Immunol* **163:** 6006–6013.

Clayton E, Bardi G, Bell SE, Chantry D, Downes CP, Gray A, Humphries LA, Rawlings D, Reynolds H, Vigorito E, et al. 2002. A crucial role for the p110delta subunit of phosphatidylinositol 3-kinase in B cell development and activation. *J Exp Med* **196:** 753–763.

Cozier G, Sessions R, Bottomley JR, Reynolds JS, Cullen PJ. 2000a. Molecular modelling and site-directed mutagenesis of the inositol 1,3,4,5-tetrakisphosphate-binding pleckstrin homology domain from the Ras GTPase-activating protein GAP1IP4BP. *Biochem J* **349:** 333–342.

Cozier GE, Carlton J, Bouyoucef D, Cullen PJ. 2004. Membrane targeting by pleckstrin homology domains. *Curr Top Microbiol Immunol* **282:** 49–88.

Cozier GE, Lockyer PJ, Reynolds JS, Kupzig S, Bottomley JR, Millard TH, Banting G, Cullen PJ. 2000b. GAP1IP4BP contains a novel group I pleckstrin homology domain that directs constitutive plasma membrane association. *J Biol Chem* **275:** 28261–28268.

Cremasco V, Benasciutti E, Cella M, Kisseleva M, Croke M, Faccio R. 2010. Phospholipase C gamma 2 is critical for development of a murine model of inflammatory arthritis by affecting actin dynamics in dendritic cells. *PLoS One* **5:** e8909.

Cremasco V, Graham DB, Novack DV, Swat W, Faccio R. 2008. Vav/Phospholipase Cgamma2-mediated control of a neutrophil-dependent murine model of rheumatoid arthritis. *Arthritis Rheum* **58:** 2712–2722.

Cullen PJ, Hsuan JJ, Truong O, Letcher AJ, Jackson TR, Dawson AP, Irvine RF. 1995. Identification of a specific Ins(1,3,4,5)P4-binding protein as a member of the GAP1 family. *Nature* **376:** 527–530.

Damen JE, Liu L, Rosten P, Humphries RK, Jefferson AB, Majerus PW, Krystal G. 1996. The 145-kDa protein induced to associate with Shc by multiple cytokines is an inositol tetraphosphate and phosphatidylinositol 3,4,5-triphosphate 5-phosphatase. *Proc Natl Acad Sci U S A* **93:** 1689–1693.

Deane JA, Kharas MG, Oak JS, Stiles LN, Luo J, Moore TI, Ji H, Rommel C, Cantley LC, Lane TE, et al. 2007. T-cell function is partially maintained in the absence of class IA phosphoinositide 3-kinase signaling. *Blood* **109:** 2894–2902.

Deane JA, Trifilo MJ, Yballe CM, Choi S, Lane TE, Fruman DA. 2004. Enhanced T cell proliferation in mice lacking the p85beta subunit of phosphoinositide 3-kinase. *J Immunol* **172:** 6615–6625.

Debnath J, Chamorro M, Czar MJ, Schaeffer EM, Lenardo MJ, Varmus HE, Schwartzberg PL. 1999. rlk/TXK encodes two forms of a novel cysteine string tyrosine kinase activated by Src family kinases. *Mol Cell Biol* **19:** 1498–1507.

del Peso L, Gonzalez-Garcia M, Page C, Herrera R, Nunez G. 1997. Interleukin-3-induced phosphorylation of BAD through the protein kinase Akt. *Science* **278:** 687–689.

Di Cristofano A, Kotsi P, Peng YF, Cordon-Cardo C, Elkon KB, Pandolfi PP. 1999. Impaired Fas response and autoimmunity in $Pten^{+/-}$ mice. *Science* **285:** 2122–2125.

Di Cristofano A, Pandolfi PP. 2000. The multiple roles of PTEN in tumor suppression. *Cell* **100:** 387–390.

Di Cristofano A, Pesce B, Cordon-Cardo C, Pandolfi PP. 1998. Pten is essential for embryonic development and tumour suppression. *Nat Genet* **19:** 348–355.

DiNitto JP, Lambright DG. 2006. Membrane and juxtamembrane targeting by PH and PTB domains. *Biochim Biophys Acta* **1761:** 850–867.

Dombroski D, Houghtling RA, Labno CM, Precht P, Takesono A, Caplen NJ, Billadeau DD, Wange RL, Burkhardt JK, Schwartzberg PL. 2005. Kinase-independent functions for Itk in TCR-induced regulation of Vav and the actin cytoskeleton. *J Immunol* **174:** 1385–1392.

Dower NA, Stang SL, Bottorff DA, Ebinu JO, Dickie P, Ostergaard HL, Stone JC. 2000. RasGRP is essential for mouse thymocyte differentiation and TCR signaling. *Nat Immunol* **1:** 317–321.

Downing AK, Driscoll PC, Gout I, Salim K, Zvelebil MJ, Waterfield MD. 1994. Three-dimensional solution structure of the pleckstrin homology domain from dynamin. *Curr Biol* **4:** 884–891.

Dufner A, Andjelkovic M, Burgering BM, Hemmings BA, Thomas G. 1999. Protein kinase B localization and activation differentially affect S6 kinase 1 activity and eukaryotic translation initiation factor 4E-binding protein 1 phosphorylation. *Mol Cell Biol* **19:** 4525–4534.

Epple H, Cremasco V, Zhang K, Mao D, Longmore GD, Faccio R. 2008. Phospholipase Cgamma2 modulates integrin signaling in the osteoclast by affecting the localization and activation of Src kinase. *Mol Cell Biol* **28:** 3610–3622.

Erneux C, Govaerts C, Communi D, Pesesse X. 1998. The diversity and possible functions of the inositol polyphosphate 5-phosphatases. *Biochimica et Biophysica Acta (BBA) - Molecular and Cell Biology of Lipids* **1436:** 185–199.

Falasca M, Logan SK, Lehto VP, Baccante G, Lemmon MA, Schlessinger J. 1998. Activation of phospholipase C gamma by PI 3-kinase-induced PH domain-mediated membrane targeting. *Embo J* **17:** 414–422.

Ferguson KM, Lemmon MA, Schlessinger J, Sigler PB. 1994. Crystal structure at 2.2 A resolution of the pleckstrin homology domain from human dynamin. *Cell* **79:** 199–209.

Feske S. 2007. Calcium signalling in lymphocyte activation and disease. *Nat Rev Immunol* **7:** 690–702.

Finkelstein LD, Shimizu Y, Schwartzberg PL. 2005. Tec kinases regulate TCR-mediated recruitment of signaling molecules and integrin-dependent cell adhesion. *J Immunol* **175:** 5923–5930.

Fowell DJ, Shinkai K, Liao XC, Beebe AM, Coffman RL, Littman DR, Locksley RM. 1999. Impaired NFATc translocation and failure of Th2 development in Itk-deficient $CD4^+$ T cells. *Immunity* **11:** 399–409.

Frauwirth KA, Riley JL, Harris MH, Parry RV, Rathmell JC, Plas DR, Elstrom RL, June CH, Thompson CB. 2002. The CD28 signaling pathway regulates glucose metabolism. *Immunity* **16:** 769–777.

Frederick JP, Mattiske D, Wofford JA, Megosh LC, Drake LY, Chiou ST, Hogan BL, York JD. 2005. An essential role for an inositol polyphosphate multikinase, Ipk2, in mouse embryogenesis and second messenger production. *Proc Natl Acad Sci U S A* **102:** 8454–8459.

Fruman DA, Bismuth G. 2009. Fine tuning the immune response with PI3K. *Immunol Rev* **228:** 253–272.

Fruman DA, Mauvais-Jarvis F, Pollard DA, Yballe CM, Brazil D, Bronson RT, Kahn CR, Cantley LC. 2000. Hypoglycaemia, liver necrosis and perinatal death in mice lacking all isoforms of phosphoinositide 3-kinase p85 alpha. *Nat Genet* **26:** 379–382.

Fu G, Chen Y, Yu M, Podd A, Schuman J, He Y, Di L, Yassai M, Haribhai D, North PE, et al. 2010. Phospholipase C{gamma}1 is essential for T cell development, activation, and tolerance. *J Exp Med* **207:** 309–318.

Gallo EM, Winslow MM, Cante-Barrett K, Radermacher AN, Ho L, McGinnis L, Iritani B, Neilson JR, Crabtree GR. 2007. Calcineurin sets the bandwidth for discrimination of signals during thymocyte development. *Nature* **450:** 731–735.

Gao X, Harris TK. 2006. Role of the PH domain in regulating in vitro autophosphorylation events required for reconstitution of PDK1 catalytic activity. *Bioorg Chem* **34:** 200–223.

Gardai SJ, Hildeman DA, Frankel SK, Whitlock BB, Frasch SC, Borregaard N, Marrack P, Bratton DL, Henson PM. 2004. Phosphorylation of Bax Ser184 by Akt regulates its activity and apoptosis in neutrophils. *J Biol Chem* **279:** 21085–21095.

Gomez-Rodriguez J, Readinger JA, Viorritto IC, Mueller KL, Houghtling RA, Schwartzberg PL. 2007. Tec kinases, actin, and cell adhesion. *Immunol Rev* **218:** 45–64.

Graham DB, Robertson CM, Bautista J, Mascarenhas F, Diacovo MJ, Montgrain V, Lam SK, Cremasco V, Dunne WM, Faccio R, et al. 2007. Neutrophil-mediated oxidative burst and host defense are controlled by a Vav-PLCgamma2 signaling axis in mice. *J Clin Invest* **117:** 3445–3452.

Grasis JA, Browne CD, Tsoukas CD. 2003. Inducible T cell tyrosine kinase regulates actin-dependent cytoskeletal events induced by the T cell antigen receptor. *J Immunol* **170:** 3971–3976.

Guo R, Wan CK, Carpenter JH, Mousallem T, Boustany RM, Kuan CT, Burks AW, Zhong XP. 2008. Synergistic control of T cell development and tumor suppression by diacylglycerol kinase alpha and zeta. *Proc Natl Acad Sci U S A* **105:** 11909–11914.

Hagenbeek TJ, Naspetti M, Malergue F, Garcon F, Nunes JA, Cleutjens KBJM, Trapman J, Krimpenfort P, Spits H. 2004. The Loss of PTEN Allows TCR {alpha}{beta} Lineage Thymocytes to Bypass IL-7 and Pre-TCR-mediated Signaling. *J Exp Med* **200:** 883–894.

Han J, Luby-Phelps K, Das B, Shu X, Xia Y, Mosteller RD, Krishna UM, Falck JR, White MA, Broek D. 1998. Role of substrates and products of PI 3-kinase in regulating activation of Rac-related guanosine triphosphatases by Vav. *Science* **279:** 558–560.

Harris SJ, Parry RV, Westwick J, Ward SG. 2008. Phosphoinositide lipid phosphatases: natural regulators of phosphoinositide 3-kinase signaling in T lymphocytes. *J Biol Chem* **283:** 2465–2469.

Haslam RJ, Koide HB, Hemmings BA. 1993. Pleckstrin domain homology. *Nature* **363:** 309–310.

Helgason CD, Damen JE, Rosten P, Grewal R, Sorensen P, Chappel SM, Borowski A, Jirik F, Krystal G, Humphries RK. 1998. Targeted disruption of SHIP leads to hemopoietic perturbations, lung pathology, and a shortened life span. *Genes Dev* **12:** 1610–1620.

Hermosura MC, Takeuchi H, Fleig A, Riley AM, Potter BV, Hirata M, Penner R. 2000. InsP4 facilitates store-operated calcium influx by inhibition of InsP3 5-phosphatase. *Nature* **408:** 735–740.

Heyeck SD, Wilcox HM, Bunnell SC, Berg LJ. 1997. Lck phosphorylates the activation loop tyrosine of the Itk kinase domain and activates Itk kinase activity. *J Biol Chem* **272:** 25401–25408.

Hirata M, Kanematsu T, Takeuchi H, Yagisawa H. 1998. Pleckstrin homology domain as an inositol compound binding module. *Jpn J Pharmacol* **76:** 255–263.

Horai R, Mueller KL, Handon RA, Cannons JL, Anderson SM, Kirby MR, Schwartzberg PL. 2007. Requirements for selection of conventional and innate T lymphocyte lineages. *Immunity* **27:** 775–785.

Huang YH, Barouch-Bentov R, Herman A, Walker J, Sauer K. 2006. Integrating traditional and postgenomic approaches to investigate lymphocyte development and function. *Adv Exp Med Biol* **584:** 245–276.

Huang YH, Grasis JA, Miller AT, Xu R, Soonthornvacharin S, Andreotti AH, Tsoukas CD, Cooke MP, Sauer K. 2007. Positive regulation of Itk PH domain function by soluble IP4. *Science* **316:** 886–889.

Huang YH, Hoebe K, Sauer K. 2008. New therapeutic targets in immune disorders: ItpkB, Orai1 and UNC93B. *Expert Opin Ther Targets* **12:** 391–413.

Hubner A, Barrett T, Flavell RA, Davis RJ. 2008. Multisite phosphorylation regulates Bim stability and apoptotic activity. *Mol Cell* **30:** 415–425.

Imboden JB, Pattison G. 1987. Regulation of inositol 1,4,5-trisphosphate kinase activity after stimulation of human T cell antigen receptor. *J Clin Invest* **79:** 1538–1541.

Irvine R. 2001. Inositol phosphates: Does IP(4) run a protection racket? *Curr Biol* **11:** R172–174.

Irvine R. 2007. Cell signaling. The art of the soluble. *Science* **316:** 845–846.

Irvine RF. 2005. Inositide evolution - towards turtle domination? *J Physiol* **566:** 295–300.

Irvine RF, Lloyd-Burton SM, Yu JC, Letcher AJ, Schell MJ. 2006. The regulation and function of inositol 1,4,5-trisphosphate 3-kinases. *Adv Enzyme Regul* **46:** 314–323.

Irvine RF, Schell MJ. 2001. Back in the water: the return of the inositol phosphates. *Nat Rev Mol Cell Biol* **2:** 327–338.

Isakov N, Altman A. 2002. Protein kinase C(theta) in T cell activation. *Annu Rev Immunol* **20:** 761–794.

Jacobs SR, Herman CE, Maciver NJ, Wofford JA, Wieman HL, Hammen JJ, Rathmell JC. 2008. Glucose uptake is limiting in T cell activation and requires CD28-mediated Akt-dependent and independent pathways. *J Immunol* **180:** 4476–4486.

James SR, Downes CP, Gigg R, Grove SJ, Holmes AB, Alessi DR. 1996. Specific binding of the Akt-1 protein kinase to phosphatidylinositol 3,4,5-trisphosphate without subsequent activation. *Biochem J* **315:** 709–713.

Jarmin SJ, David R, Ma L, Chai JG, Dewchand H, Takesono A, Ridley AJ, Okkenhaug K, Marelli-Berg FM. 2008. T cell receptor-induced phosphoinositide-3-kinase p110delta activity is required for T cell localization to antigenic tissue in mice. *J Clin Invest* **118:** 1154–1164.

Ji H, Rintelen F, Waltzinger C, Bertschy Meier D, Bilancio A, Pearce W, Hirsch E, Wymann MP, Ruckle T, Camps M, et al. 2007. Inactivation of PI3Kgamma and PI3Kdelta distorts T-cell development and causes multiple organ inflammation. *Blood* **110:** 2940–2947.

Ji QS, Winnier GE, Niswender KD, Horstman D, Wisdom R, Magnuson MA, Carpenter G. 1997. Essential role of the tyrosine kinase substrate phospholipase C-gamma1 in mammalian growth and development. *Proc Natl Acad Sci U S A* **94:** 2999–3003.

Jia Y, Loison F, Hattori H, Li Y, Erneux C, Park SY, Gao C, Chai L, Silberstein LE, Schurmans S, et al. 2008a. Inositol trisphosphate 3-kinase B (InsP3KB) as a physiological modulator of myelopoiesis. *Proc Natl Acad Sci U S A* **105:** 4739–4744.

Jia Y, Schurmans S, Luo HR. 2008b. Regulation of innate immunity by inositol 1,3,4,5-tetrakisphosphate. *Cell Cycle* **7:** 2803–2808.

Jia Y, Subramanian KK, Erneux C, Pouillon V, Hattori H, Jo H, You J, Zhu D, Schurmans S, Luo HR. 2007. Inositol 1,3,4,5-tetrakisphosphate negatively regulates phosphatidylinositol-3,4,5-trisphosphate signaling in neutrophils. *Immunity* **27:** 453–467.

Joseph RE, Min L, Xu R, Musselman ED, Andreotti AH. 2007. A remote substrate docking mechanism for the tec family tyrosine kinases. *Biochem* **46:** 5595–5603.

Jun K, Choi G, Yang SG, Choi KY, Kim H, Chan GC, Storm DR, Albert C, Mayr GW, Lee CJ, et al. 1998. Enhanced hippocampal CA1 LTP but normal spatial learning in inositol 1,4,5-trisphosphate 3-kinase(A)-deficient mice. *Learn Mem* **5:** 317–330.

Juntilla MM, Koretzky GA. 2008. Critical roles of the PI3K/Akt signaling pathway in T cell development. *Immunol Lett* **116:** 104–110.

Juntilla MM, Wofford JA, Birnbaum MJ, Rathmell JC, Koretzky GA. 2007. Akt1 and Akt2 are required for alphabeta thymocyte survival and differentiation. *Proc Natl Acad Sci U S A* **104:** 12105–12110.

Kavanaugh WM, Pot DA, Chin SM, Deuter-Reinhard M, Jefferson AB, Norris FA, Masiarz FR, Cousens LS, Majerus PW, Williams LT. 1996. Multiple forms of an inositol polyphosphate 5-phosphatase form signaling complexes with Shc and Grb2. *Curr Biol* **6:** 438–445.

Kim IH, Park SK, Hong ST, Jo YS, Kim EJ, Park EH, Han SB, Shin HS, Sun W, Kim HT, et al. 2009. Inositol 1,4,5-trisphosphate 3-kinase a functions as a scaffold for synaptic Rac signaling. *J Neurosci* **29:** 14039–14049.

Lemmon MA. 2008. Membrane recognition by phospholipid-binding domains. *Nat Rev Mol Cell Biol* **9:** 99–111.

Liang J, Zubovitz J, Petrocelli T, Kotchetkov R, Connor MK, Han K, Lee JH, Ciarallo S, Catzavelos C, Beniston R, et al. 2002. PKB/Akt phosphorylates p27, impairs nuclear import of p27 and opposes p27-mediated G1 arrest. *Nat Med* **8:** 1153–1160.

Liao XC, Littman DR. 1995. Altered T cell receptor signaling and disrupted T cell development in mice lacking Itk. *Immunity* **3:** 757–769.

Lioubin MN, Algate PA, Tsai S, Carlberg K, Aebersold A, Rohrschneider LR. 1996. p150Ship, a signal transduction molecule with inositol polyphosphate-5-phosphatase activity. *Genes Dev* **10:** 1084–1095.

Liu D, Uzonna JE. 2010. The p110{delta} Isoform of Phosphatidylinositol 3-Kinase Controls the Quality of Secondary Anti-Leishmania Immunity by Regulating Expansion and Effector Function of Memory T Cell Subsets. *J Immunol* **184:** 3098–3105.

Liu D, Zhang T, Marshall AJ, Okkenhaug K, Vanhaesebroeck B, Uzonna JE. 2009a. The p110delta isoform of phosphatidylinositol 3-kinase controls susceptibility to Leishmania major by regulating expansion and tissue homing of regulatory T cells. *J Immunol* **183:** 1921–1933.

Liu KQ, Bunnell SC, Gurniak CB, Berg LJ. 1998. T cell receptor-initiated calcium release is uncoupled from capacitative calcium entry in Itk-deficient T cells. *J Exp Med* **187:** 1721–1727.

Liu P, Cheng H, Roberts TM, Zhao JJ. 2009b. Targeting the phosphoinositide 3-kinase pathway in cancer. *Nat Rev Drug Discov* **8:** 627–644.

Lockyer PJ, Bottomley JR, Reynolds JS, McNulty TJ, Venkateswarlu K, Potter BV, Dempsey CE, Cullen PJ. 1997. Distinct subcellular localisations of the putative inositol 1,3,4,5-tetrakisphosphate receptors GAP1IP4BP and GAP1m result from the GAP1IP4BP PH domain directing plasma membrane targeting. *Curr Biol* **7:** 1007–1010.

Lucas JA, Felices M, Evans JW, Berg LJ. 2007. Subtle defects in pre-TCR signaling in the absence of the Tec kinase Itk. *J Immunol* **179:** 7561–7567.

Macian F, Garcia-Cozar F, Im SH, Horton HF, Byrne MC, Rao A. 2002. Transcriptional mechanisms underlying lymphocyte tolerance. *Cell* **109:** 719–731.

Macias MJ, Musacchio A, Ponstingl H, Nilges M, Saraste M, Oschkinat H. 1994. Structure of the pleckstrin homology domain from beta-spectrin. *Nature* **369:** 675–677.

Madrid LV, Wang CY, Guttridge DC, Schottelius AJ, Baldwin AS Jr, Mayo MW. 2000. Akt suppresses apoptosis by stimulating the transactivation potential of the RelA/p65 subunit of NF-kappaB. *Mol Cell Biol* **20:** 1626–1638.

Maehama T, Dixon JE. 1998. The tumor suppressor, PTEN/MMAC1, dephosphorylates the lipid second messenger, phosphatidylinositol 3,4,5-trisphosphate. *J Biol Chem* **273:** 13375–13378.

Manicassamy S, Gupta S, Sun Z. 2006. Selective function of PKC-theta in T cells. *Cell Mol Immunol* **3:** 263–270.

Manning BD, Cantley LC. 2007. AKT/PKB signaling: navigating downstream. *Cell* **129:** 1261–1274.

Marechal Y, Pesesse X, Jia Y, Pouillon V, Perez-Morga D, Daniel J, Izui S, Cullen PJ, Leo O, Luo HR, et al. 2007. Inositol 1,3,4,5-tetrakisphosphate controls proapoptotic Bim gene expression and survival in B cells. *Proc Natl Acad Sci U S A* **104:** 13978–13983.

Martin AL, Schwartz MD, Jameson SC, Shimizu Y. 2008. Selective regulation of CD8 effector T cell migration by the p110 gamma isoform of phosphatidylinositol 3-kinase. *J Immunol* **180:** 2081–2088.

Matheu MP, Deane JA, Parker I, Fruman DA, Cahalan MD. 2007. Class IA phosphoinositide 3-kinase modulates basal lymphocyte motility in the lymph node. *J Immunol* **179:** 2261–2269.

Mayer BJ, Ren R, Clark KL, Baltimore D. 1993. A putative modular domain present in diverse signaling proteins. *Cell* **73:** 629–630.

McGargill MA, Ch'en IL, Katayama CD, Pages G, Pouyssegur J, Hedrick SM. 2009. Cutting Edge: Extracellular Signal-Related Kinase Is Not Required for Negative Selection of Developing T Cells. *J Immunol* **183:** 4838–4842.

McLaughlin S, Wang J, Gambhir A, Murray D. 2002. PIP(2) and proteins: interactions, organization, and information flow. *Annu Rev Biophys Biomol Struct* **31:** 151–175.

Michell RH, Heath VL, Lemmon MA, Dove SK. 2006. Phosphatidylinositol 3,5-bisphosphate: metabolism and cellular functions. *Trends Biochem Sci* **31:** 52–63.

Miller AT, Beisner DR, Liu D, Cooke MP. 2009. Inositol 1,4,5-trisphosphate 3-kinase B is a negative regulator of BCR signaling that controls B cell selection and tolerance induction. *J Immunol* **182:** 4696–4704.

Miller AT, Chamberlain PP, Cooke MP. 2008. Beyond IP3: roles for higher order inositol phosphates in immune cell signaling. *Cell Cycle* **7:** 463–467.

Miller AT, Sandberg M, Huang YH, Young M, Sutton S, Sauer K, Cooke MP. 2007a. Production of Ins(1,3,4,5)P4 mediated by the kinase Itpkb inhibits store-operated calcium channels and regulates B cell selection and activation. *Nat Immunol* **8:** 514–521.

Miller AT, Sandberg M, Huang YH, Young M, Sutton S, Sauer K, Cooke MP. 2007b. Production of Ins(1,3,4,5)P(4) mediated by the kinase Itpkb inhibits store-operated calcium channels and regulates B cell selection and activation. *Nat Immunol* **8:** 514–521.

Min L, Joseph RE, Fulton DB, Andreotti AH. 2009. Itk tyrosine kinase substrate docking is mediated by a nonclassical SH2 domain surface of PLCgamma1. *Proc Natl Acad Sci U S A* **106:** 21143–21148.

Morley SC, Weber KS, Kao H, Allen PM. 2008. Protein kinase C-theta is required for efficient positive selection. *J Immunol* **181:** 4696–4708.

Nalaskowski MM, Mayr GW. 2004. The families of kinases removing the Ca^{2+} releasing second messenger Ins(1,4,5)P3. *Curr Mol Med* **4:** 277–290.

Nashed BF, Zhang T, Al-Alwan M, Srinivasan G, Halayko AJ, Okkenhaug K, Vanhaesebroeck B, Hayglass KT, Marshall AJ. 2007. Role of the phosphoinositide 3-kinase p110delta in generation of type 2 cytokine responses and allergic airway inflammation. *Eur J Immunol* **37:** 416–424.

Neilson JR, Winslow MM, Hur EM, Crabtree GR. 2004. Calcineurin B1 is essential for positive but not negative selection during thymocyte development. *Immunity* **20:** 255–266.

Niki I, Yokokura H, Sudo T, Kato M, Hidaka H. 1996. Ca2+ signaling and intracellular Ca2+ binding proteins. *J Biochem* **120:** 685–698.

Oak JS, Deane JA, Kharas MG, Luo J, Lane TE, Cantley LC, Fruman DA. 2006. Sjogren's syndrome-like disease in mice with T cells lacking class 1A phosphoinositide-3-kinase. *Proc Natl Acad Sci U S A* **103:** 16882–16887.

Oh-hora M, Rao A. 2008. Calcium signaling in lymphocytes. *Curr Opin Immunol* **20:** 250–258.

Okkenhaug K, Bilancio A, Farjot G, Priddle H, Sancho S, Peskett E, Pearce W, Meek SE, Salpekar A, Waterfield MD, et al. 2002. Impaired B and T cell antigen receptor signaling in p110delta PI 3-kinase mutant mice. *Science* **297:** 1031–1034.

Okkenhaug K, Patton DT, Bilancio A, Garcon F, Rowan WC, Vanhaesebroeck B. 2006. The p110delta isoform of phosphoinositide 3-kinase controls clonal expansion and differentiation of Th cells. *J Immunol* **177:** 5122–5128.

Olenchock BA, Guo R, Carpenter JH, Jordan M, Topham MK, Koretzky GA, Zhong XP. 2006. Disruption of diacylglycerol metabolism impairs the induction of T cell anergy. *Nat Immunol* **7:** 1174–1181.

Onouchi Y, Gunji T, Burns JC, Shimizu C, Newburger JW, Yashiro M, Nakamura Y, Yanagawa H, Wakui K, Fukushima Y, et al. 2008. ITPKC functional polymorphism associated with Kawasaki disease susceptibility and formation of coronary artery aneurysms. *Nat Genet* **40:** 35–42.

Otto JC, Mulugu S, Fridy PC, Chiou ST, Armbruster BN, Ribeiro AA, York JD. 2007. Biochemical analysis of inositol phosphate kinases. *Methods Enzymol* **434:** 171–185.

Outram SV, Crompton T, Merida I, Varas A, Martinez AC. 2002. Diacylglycerol kinase alpha activity promotes survival of $CD4^+ 8^+$ double positive cells during thymocyte development. *Immunology* **105:** 391–398.

Panka DJ, Mano T, Suhara T, Walsh K, Mier JW. 2001. Phosphatidylinositol 3-kinase/Akt activity regulates c-FLIP expression in tumor cells. *J Biol Chem* **276:** 6893–6896.

Pattni K, Banting G. 2004. Ins(1,4,5)P3 metabolism and the family of IP3–3Kinases. *Cell Signal* **16:** 643–654.

Patton DT, Garden OA, Pearce WP, Clough LE, Monk CR, Leung E, Rowan WC, Sancho S, Walker LS,

Vanhaesebroeck B, et al. 2006. Cutting edge: the phosphoinositide 3-kinase p110 delta is critical for the function of CD4$^+$ CD25$^+$ Foxp3$^+$ regulatory T cells. *J Immunol* **177:** 6598–6602.

Pesesse X, Moreau C, Drayer AL, Woscholski R, Parker P, Erneux C. 1998. The SH2 domain containing inositol 5-phosphatase SHIP2 displays phosphatidylinositol 3,4,5-trisphosphate and inositol 1,3,4,5-tetrakisphosphate 5-phosphatase activity. *FEBS Lett* **437:** 301–303.

Pouillon V, Hascakova-Bartova R, Pajak B, Adam E, Bex F, Dewaste V, Van Lint C, Leo O, Erneux C, Schurmans S. 2003. Inositol 1,3,4,5-tetrakisphosphate is essential for T lymphocyte development. *Nat Immunol* **4:** 1136–1143.

Priatel JJ, Chen X, Dhanji S, Abraham N, Teh HS. 2006. RasGRP1 transmits prodifferentiation TCR signaling that is crucial for CD4 T cell development. *J Immunol* **177:** 1470–1480.

Priatel JJ, Chen X, Zenewicz LA, Shen H, Harder KW, Horwitz MS, Teh HS. 2007. Chronic immunodeficiency in mice lacking RasGRP1 results in CD4 T cell immune activation and exhaustion. *J Immunol* **179:** 2143–2152.

Priatel JJ, Teh SJ, Dower NA, Stone JC, Teh HS. 2002. RasGRP1 transduces low-grade TCR signals which are critical for T cell development, homeostasis, and differentiation. *Immunity* **17:** 617–627.

Qi Q, August A. 2007. Keeping the (kinase) party going: SLP-76 and ITK dance to the beat. *Sci STKE* **2007:** e39.

Qi XJ, Wildey GM, Howe PH. 2006. Evidence that Ser87 of BimEL is phosphorylated by Akt and regulates BimEL apoptotic function. *J Biol Chem* **281:** 813–823.

Rathmell JC, Elstrom RL, Cinalli RM, Thompson CB. 2003. Activated Akt promotes increased resting T cell size, CD28-independent T cell growth, and development of autoimmunity and lymphoma. *Eur J Immunol* **33:** 2223–2232.

Readinger JA, Mueller KL, Venegas AM, Horai R, Schwartzberg PL. 2009. Tec kinases regulate T-lymphocyte development and function: new insights into the roles of Itk and Rlk/Txk. *Immunol Rev* **228:** 93–114.

Resnick AC, Saiardi A. 2008. Inositol polyphosphate multikinase: metabolic architect of nuclear inositides. *Front Biosci* **13:** 856–866.

Rodriguez-Borlado L, Barber DF, Hernandez C, Rodriguez-Marcos MA, Sanchez A, Hirsch E, Wymann M, Martinez AC, Carrera AC. 2003. Phosphatidylinositol 3-kinase regulates the CD4/CD8 T cell differentiation ratio. *J Immunol* **170:** 4475–4482.

Rosse C, Linch M, Kermorgant S, Cameron AJ, Boeckeler K, Parker PJ. 2010. PKC and the control of localized signal dynamics. *Nat Rev Mol Cell Biol* **11:** 103–112.

Sasaki T, Irie-Sasaki J, Jones RG, Oliveira-dos-Santos AJ, Stanford WL, Bolon B, Wakeham A, Itie A, Bouchard D, Kozieradzki I, et al. 2000. Function of PI3Kgamma in thymocyte development, T cell activation, and neutrophil migration. *Science* **287:** 1040–1046.

Sauer K, Cooke MP. 2010. Regulation of immune cell development through soluble inositol(1,3,4,5)tetrakisphosphate. *Nat Rev Immunol* **10**, 257–271.

Sauer K, Huang YH, Ying H, Sandberg M, Mayr GW. 2009. Phosphoinositide Analysis in Lymphocyte Activation. *Curr Protoc Immunol* **87:** 11.11.11–11.11.46.

Schaeffer EM, Broussard C, Debnath J, Anderson S, McVicar DW, Schwartzberg PL. 2000. Tec family kinases modulate thresholds for thymocyte development and selection. *J Exp Med* **192:** 987–1000.

Schaeffer EM, Debnath J, Yap G, McVicar D, Liao XC, Littman DR, Sher A, Varmus HE, Lenardo MJ, Schwartzberg PL. 1999. Requirement for Tec kinases Rlk and Itk in T cell receptor signaling and immunity. *Science* **284:** 638–641.

Schaeffer EM, Yap GS, Lewis CM, Czar MJ, McVicar DW, Cheever AW, Sher A, Schwartzberg PL. 2001. Mutation of Tec family kinases alters T helper cell differentiation. *Nat Immunol* **2:** 1183–1188.

Schell MJ. 2010. Inositol trisphosphate 3-kinases: focus on immune and neuronal signaling. *Cell Mol Life Sci* **67:** 1755–1778.

Shan X, Czar MJ, Bunnell SC, Liu P, Liu Y, Schwartzberg PL, Wange RL. 2000. Deficiency of PTEN in Jurkat T cells causes constitutive localization of Itk to the plasma membrane and hyperresponsiveness to CD3 stimulation. *Mol Cell Biol* **20:** 6945–6957.

Shears SB. 2009. Molecular basis for the integration of inositol phosphate signaling pathways via human ITPK1. *Adv Enzyme Regul* **49:** 87–96.

Shin I, Yakes FM, Rojo F, Shin NY, Bakin AV, Baselga J, Arteaga CL. 2002. PKB/Akt mediates cell-cycle progression by phosphorylation of p27(Kip1) at threonine 157 and modulation of its cellular localization. *Nat Med* **8:** 1145–1152.

Shirane M, Sawa H, Kobayashi Y, Nakano T, Kitajima K, Shinkai Y, Nagashima K, Negishi I. 2001. Deficiency of phospholipase C-gamma1 impairs renal development and hematopoiesis. *Development* **128:** 5173–5180.

Shiroki F, Matsuda S, Doi T, Fujiwara M, Mochizuki Y, Kadowaki T, Suzuki H, Koyasu S. 2007. The p85alpha regulatory subunit of class IA phosphoinositide 3-kinase regulates beta-selection in thymocyte development. *J Immunol* **178:** 1349–1356.

Siliceo M, Garcia-Bernal D, Carrasco S, Diaz-Flores E, Coluccio Leskow F, Teixido J, Kazanietz MG, Merida I. 2006. Beta2-chimaerin provides a diacylglycerol-dependent mechanism for regulation of adhesion and chemotaxis of T cells. *J Cell Sci* **119:** 141–152.

Siliceo M, Merida I. 2009. T cell receptor-dependent tyrosine phosphorylation of beta2-chimaerin modulates its Rac-GAP function in T cells. *J Biol Chem* **284:** 11354–11363.

Sommers CL, Lee J, Steiner KL, Gurson JM, Depersis CL, El-Khoury D, Fuller CL, Shores EW, Love PE, Samelson LE. 2005. Mutation of the phospholipase C-gamma1-binding site of LAT affects both positive and negative thymocyte selection. *J Exp Med* **201:** 1125–1134.

Sommers CL, Park CS, Lee J, Feng C, Fuller CL, Grinberg A, Hildebrand JA, Lacana E, Menon RK, Shores EW, et al. 2002. A LAT mutation that inhibits T cell development yet induces lymphoproliferation. *Science* **296:** 2040–2043.

Soond DR, Bjorgo E, Moltu K, Dale VQ, Patton DT, Torgersen KM, Galleway F, Twomey B, Clark J, Gaston JH, et al. 2010. PI3K p110{delta} regulates T cell cytokine production during primary and secondary immune responses in mice and humans. *Blood* **115:** 2203–2213.

Spitaler M, Emslie E, Wood CD, Cantrell D. 2006. Diacylglycerol and protein kinase D localization during T lymphocyte activation. *Immunity* **24:** 535–546.

Starr TK, Jameson SC, Hogquist KA. 2003. Positive and negative selection of T cells. *Annu Rev Immunol* **21:** 139–176.

Stephens L, Anderson K, Stokoe D, Erdjument-Bromage H, Painter GF, Holmes AB, Gaffney PR, Reese CB, McCormick F, Tempst P, et al. 1998. Protein kinase B kinases that mediate phosphatidylinositol 3,4,5-trisphosphate-dependent activation of protein kinase B. *Science* **279:** 710–714.

Stephens L, Smrcka A, Cooke FT, Jackson TR, Sternweis PC, Hawkins PT. 1994. A novel phosphoinositide 3 kinase activity in myeloid-derived cells is activated by G protein beta gamma subunits. *Cell* **77:** 83–93.

Stephens LR, Eguinoa A, Erdjument-Bromage H, Lui M, Cooke F, Coadwell J, Smrcka AS, Thelen M, Cadwallader K, Tempst P, et al. 1997. The G beta gamma sensitivity of a PI3K is dependent upon a tightly associated adaptor, p101. *Cell* **89:** 105–114.

Stokoe D, Stephens LR, Copeland T, Gaffney PR, Reese CB, Painter GF, Holmes AB, McCormick F, Hawkins PT. 1997. Dual role of phosphatidylinositol-3,4,5-trisphosphate in the activation of protein kinase B. *Science* **277:** 567–570.

Stoyanov B, Volinia S, Hanck T, Rubio I, Loubtchenkov M, Malek D, Stoyanova S, Vanhaesebroeck B, Dhand R, Nurnberg B, et al. 1995. Cloning and characterization of a G protein-activated human phosphoinositide-3 kinase. *Science* **269:** 690–693.

Su YW, Zhang Y, Schweikert J, Koretzky GA, Reth M, Wienands J. 1999. Interaction of SLP adaptors with the SH2 domain of Tec family kinases. *Eur J Immunol* **29:** 3702–3711.

Suire S, Coadwell J, Ferguson GJ, Davidson K, Hawkins P, Stephens L. 2005. p84, a new Gbetagamma-activated regulatory subunit of the type IB phosphoinositide 3-kinase p110gamma. *Curr Biol* **15:** 566–570.

Suzuki A, Yamaguchi MT, Ohteki T, Sasaki T, Kaisho T, Kimura Y, Yoshida R, Wakeham A, Higuchi T, Fukumoto M, et al. 2001. T cell-specific loss of Pten leads to defects in central and peripheral tolerance. *Immunity* **14:** 523–534.

Swat W, Montgrain V, Doggett TA, Douangpanya J, Puri K, Vermi W, Diacovo TG. 2006. Essential role of PI3Kdelta and PI3Kgamma in thymocyte survival. *Blood* **107:** 2415–2422.

Tarasenko T, Kole HK, Chi AW, Mentink-Kane MM, Wynn TA, Bolland S. 2007. T cell-specific deletion of the inositol phosphatase SHIP reveals its role in regulating Th1/Th2 and cytotoxic responses. *Proc Natl Acad Sci U S A* **104:** 11382–11387.

Taylor CW, Rahman T, Tovey SC, Dedos SG, Taylor EJ, Velamakanni S. 2009. IP3 receptors: some lessons from DT40 cells. *Immunol Rev* **231:** 23–44.

Thomas MS, Mitchell JS, DeNucci CC, Martin AL, Shimizu Y. 2008. The p110gamma isoform of phosphatidylinositol 3-kinase regulates migration of effector CD4 T lymphocytes into peripheral inflammatory sites. *J Leukoc Biol* **84:** 814–823.

Timm D, Salim K, Gout I, Guruprasad L, Waterfield M, Blundell T. 1994. Crystal structure of the pleckstrin homology domain from dynamin. *Nat Struct Biol* **1:** 782–788.

Tomlinson MG, Kane LP, Su J, Kadlecek TA, Mollenauer MN, Weiss A. 2004. Expression and function of Tec, Itk, and Btk in lymphocytes: evidence for a unique role for Tec. *Mol Cell Biol* **24:** 2455–2466.

van Rossum DB, Patterson RL, Sharma S, Barrow RK, Kornberg M, Gill DL, Snyder SH. 2005. Phospholipase Cgamma1 controls surface expression of TRPC3 through an intermolecular PH domain. *Nature* **434:** 99–104.

Vanhaesebroeck B, Ali K, Bilancio A, Geering B, Foukas LC. 2005. Signalling by PI3K isoforms: insights from gene-targeted mice. *Trends Biochem Sci* **30:** 194–204.

Vanweyenberg V, Communi D, D'Santos CS, Erneux C. 1995. Tissue- and cell-specific expression of Ins(1,4,5)P3 3-kinase isoenzymes. *Biochem J* **306:** 429–435.

Vig M, Kinet JP. 2009. Calcium signaling in immune cells. *Nat Immunol* **10:** 21–27.

Viglietto G, Motti ML, Bruni P, Melillo RM, D'Alessio A, Califano D, Vinci F, Chiappetta G, Tsichlis P, Bellacosa A, et al. 2002. Cytoplasmic relocalization and inhibition of the cyclin-dependent kinase inhibitor p27(Kip1) by PKB/Akt-mediated phosphorylation in breast cancer. *Nat Med* **8:** 1136–1144.

Wang CY, Mayo MW, Korneluk RG, Goeddel DV, Baldwin ASJr. 1998. NF-kappaB antiapoptosis: induction of TRAF1 and TRAF2 and c-IAP1 and c-IAP2 to suppress caspase-8 activation. *Science* **281:** 1680–1683.

Wang D, Feng J, Wen R, Marine JC, Sangster MY, Parganas E, Hoffmeyer A, Jackson CW, Cleveland JL, Murray PJ, et al. 2000. Phospholipase Cgamma2 is essential in the functions of B cell and several Fc receptors. *Immunity* **13:** 25–35.

Wang X, Devaiah SP, Zhang W, Welti R. 2006. Signaling functions of phosphatidic acid. *Prog Lipid Res* **45:** 250–278.

Wange RL. 2000. LAT, the linker for activation of T cells: a bridge between T cell-specific and general signaling pathways. *Sci STKE* **2000:** re1.

Webb LM, Vigorito E, Wymann MP, Hirsch E, Turner M. 2005. Cutting edge: T cell development requires the combined activities of the p110gamma and p110delta catalytic isoforms of phosphatidylinositol 3-kinase. *J Immunol* **175:** 2783–2787.

Wen BG, Pletcher MT, Warashina M, Choe SH, Ziaee N, Wiltshire T, Sauer K, Cooke MP. 2004. Inositol (1,4,5) trisphosphate 3 kinase B controls positive selection of T cells and modulates Erk activity. *Proc Natl Acad Sci U S A* **101:** 5604–5609.

Wilde JI, Watson SP. 2001. Regulation of phospholipase C gamma isoforms in haematopoietic cells: why one, not the other? *Cell Signal* **13:** 691–701.

Yang C, Kazanietz MG. 2007. Chimaerins: GAPs that bridge diacylglycerol signalling and the small G-protein Rac. *Biochem J* **403:** 1–12.

Yoon HS, Hajduk PJ, Petros AM, Olejniczak ET, Meadows RP, Fesik SW. 1994. Solution structure of a pleckstrin-homology domain. *Nature* **369:** 672–675.

York JD. 2006. Regulation of nuclear processes by inositol polyphosphates. *Biochim Biophys Acta* **1761:** 552–559.

York JD, Hunter T. 2004. Signal transduction. Unexpected mediators of protein phosphorylation. *Science* **306:** 2053–2055.

Zha Y, Marks R, Ho AW, Peterson AC, Janardhan S, Brown I, Praveen K, Stang S, Stone JC, Gajewski TF. 2006. T cell anergy is reversed by active Ras and is regulated by diacylglycerol kinase-alpha. *Nat Immunol* **7:** 1166–1173.

Zhong XP, Guo R, Zhou H, Liu C, Wan CK. 2008. Diacylglycerol kinases in immune cell function and self-tolerance. *Immunol Rev* **224:** 249–264.

Zhong XP, Hainey EA, Olenchock BA, Jordan MS, Maltzman JS, Nichols KE, Shen H, Koretzky GA. 2003. Enhanced T cell responses due to diacylglycerol kinase zeta deficiency. *Nat Immunol* **4:** 882–890.

Antigen Receptor Signaling to NF-κB via CARMA1, BCL10, and MALT1

Margot Thome, Jean Enno Charton, Christiane Pelzer, and Stephan Hailfinger

Department of Biochemistry, University of Lausanne, Chemin des Boveresses 155, CH-1066 Epalinges, Switzerland

Correspondence: margot.thomemiazza@unil.ch

The signaling pathway controlling antigen receptor-induced regulation of the transcription factor NF-κB plays a key role in lymphocyte activation and development and the generation of lymphomas. Work of the past decade has led to dramatic progress in the identification and characterization of new players in the pathway. Moreover, novel enzymatic activities relevant for this pathway have been discovered, which represent interesting drug targets for immuno-suppression or lymphoma treatment. Here, we summarize these findings and give an outlook on interesting open issues that need to be addressed in the future.

Lymphocytes play a major role in the defense against infection and tumors through their capacity to detect specific disease-associated antigens. Triggering of lymphocyte cell-surface antigen receptors (AgR) leads to the initiation of signaling pathways that regulate the activation, proliferation and survival of the stimulated lymphocyte. One such signaling pathway that has recently gained a lot of interest is the AgR-dependent activation of the nuclear factor-κB (NF-κB) transcription factor family. Genetic defects in this pathway are linked to immune deficiencies, whereas aberrant constitutive NF-κB activation is associated with the development of autoimmune diseases and neoplastic disorders (Karin et al. 2002; Li and Verma 2002). The precise mechanism by which AgR triggering controls NF-κB activation in lymphocytes is therefore a focus of intense investigation.

NF-κB designates a family of heterodimeric transcription factors that share a Rel homology domain (RHD), required for DNA binding and homo- or hetero-dimerization. Transcriptionally active NF-κB dimers are composed of a member with a RHD (p50 or p52) and another member with a transcription activation domain in addition to the RHD (RelA, RelB, or c-Rel). NF-κB family members can be activated by either the classical (also called canonical) pathway, which depends upon p50, RelA, and c-Rel, or by the alternative (also called noncanonical) pathway, which is p52- and RelB-dependent (Bonizzi and Karin 2004). We will focus on the classical pathway, as the phenotype of mice deficient for functional p50, RelA or c-Rel provide strong evidence for engagement of this pathway in AgR-dependent lymphocyte activation (Li and Verma 2002).

 doi: 10.1101/cshperspect.a003004
Cite this article as *Cold Spring Harb Perspect Biol* 2010;2:a003004

Before activation of the classical pathway, NF-κB function is suppressed through interaction with the inhibitor of κB (IκB) family of cytoplasmic inhibitors, which block nuclear translocation of NF-κB family members. As a consequence, engagement of the classical pathway requires IκB degradation before NF-κB can enter the nucleus and drive transcription. This is achieved by the stimulus-dependent activation of the IκB kinase (IKK) complex, which phosphorylates IκBα on S32 and S36 within a conserved D**S**GLD**S** motif, thereby targeting the protein for rapid ubiquitination-dependent degradation by the proteasome (Li and Verma 2002; Bonizzi and Karin 2004). The IκB family members IκBβ and IκBε harbor similar motifs, suggesting these are also regulated by IKK-dependent mechanisms (Li and Verma 2002). Although all stimuli leading to classical NF-κB activation appear to converge on IKK-mediated IκB phosphorylation, the upstream events controlling IKK activation are distinct and specific to the individual type of NF-κB-activating stimulus. TNFα-regulated IKK activation, for example, depends upon the TRAF2 ubiquitin ligase and RIP-1 kinase, whereas lipopolysaccharide-regulated NF-κB activation requires the adaptor protein MyD88 and IRAK family kinases. Recent studies revealed that AgR-regulated IKK activation specifically requires the signaling protein CARMA1 and its binding partners BCL10 and MALT1 (Ruland et al. 2001; Egawa et al. 2003; Hara et al. 2003; Jun et al. 2003; Newton and Dixit 2003; Ruefli-Brasse et al. 2003; Ruland et al. 2003). Here, we will summarize the current state of knowledge regarding the molecular and biological function of the CBM (CARMA1, BCL10, and MALT1) proteins during lymphocyte activation, with a particular focus on T-cell activation.

STRUCTURAL FEATURES OF CBM PROTEINS

The scaffold protein CARMA1 (CARD-MAGUK1, also called CARD11 or Bimp3) is characterized by the presence of a caspase recruitment domain (CARD) and its homology to proteins of the MAGUK (membrane-associated guanylate kinase) family (Bertin et al. 2001; Gaide et al. 2001; McAllister-Lucas et al. 2001; Thome 2004). CARMA1 shares with MAGUK proteins a number of family-specific protein–protein interaction domains. These domains support the function of MAGUKs as scaffolding proteins at sites of cell–cell contact, such as for example the neuronal synapse or tight junctions (Thome 2004; Funke et al. 2005; Feng and Zhang 2009). MAGUK family members typically contain one to three PDZ domains (named after the domain-containing PSD-95, Dlg and ZO-1 proteins), followed by SH3 (src homology 3) and GUK (guanylate kinase) domains (Funke et al. 2005) (Fig. 1). PDZ domains target proteins to the plasma membrane by binding to a four amino-acid motif present in the cytoplasmic carboxyl terminus of transmembrane proteins (Ponting et al. 1997). In the case of MAGUK proteins, such interactions are thought to contribute to the structure and dynamics of synaptic cell structures (Feng and Zhang 2009). For the MAGUK protein PSD-95, the SH3 domain undergoes an unusual intramolecular interaction with the GUK domain that does not involve a typical Pro-rich SH3 target motif (McGee et al. 2001; Tavares et al. 2001). This intramolecular interaction may negatively regulate MAGUK function by keeping the protein in a closed, inactive conformation. Interaction with a regulatory ligand may be necessary for the conformational change that enables dimerization or oligomerization by intermolecular SH3-GUK interactions (Yaffe 2001; Feng and Zhang 2009). Whether such a mechanism regulates CARMA1 function is currently unclear.

In addition to its carboxy-terminal MAGUK-like features, CARMA1 contains a CARD motif and a coiled coil domain that are functionally crucial (Fig. 1). CARD motifs are protein–protein interaction domains that can mediate homotypic CARD/CARD interactions between two, or possibly even three CARD-containing binding partners (Hofmann et al. 1997; Weber and Vincenz 2001). The CARD of CARMA1 mediates homotypic interaction with the adaptor BCL10 (Bertin et al. 2001; Gaide et al. 2001; Gaide et al. 2002), which contains an amino-terminal CARD motif and a

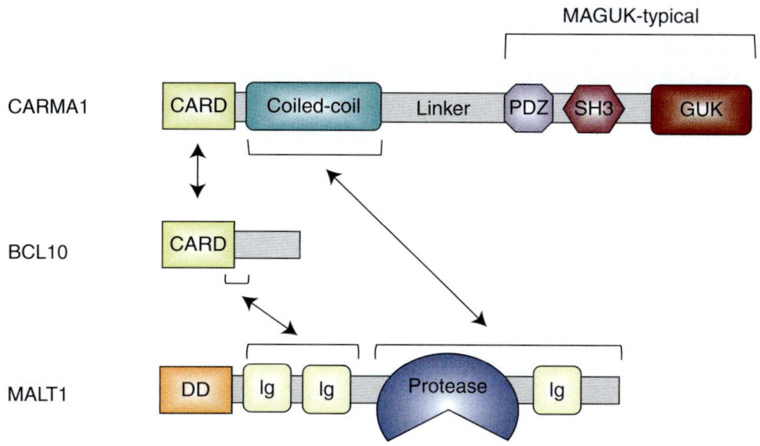

Figure 1. Molecular structure of CARMA1, BCL10, and MALT1. CARMA1 has an amino-terminal caspase-recruitment domain (CARD) that is followed by a coiled coil domain and a linker region. The carboxy-terminal region of CARMA1 contains a PDZ domain (a motif present in the Psd-95, Dlg and ZO-1 proteins), a Src homology-3 (SH3) domain and a guanylate kinase (GUK) motif characteristic of proteins of the membrane-associated guanylate kinase (MAGUK) family (Thome and Tschopp 2003). BCL10 harbors an amino-terminal CARD motif and a carboxy-terminal extension of unknown structure. The carboxy-terminal part of the BCL10 CARD and a short stretch of 13 amino acids following the CARD are required for constitutive binding to MALT1 (Lucas et al. 2001; Langel et al. 2008). MALT1 contains an amino-terminal death domain (DD) that is followed by two immunoglobulin (Ig)-like domains required for BCL10 binding (Lucas et al. 2001). The carboxy-terminal part of MALT1 contains the protease domain, a third Ig domain, and an extension of unknown structure (Uren et al. 2000).

Ser/Thr-rich carboxyl terminus of unknown structure (Costanzo et al. 1999; Koseki et al. 1999; Srinivasula et al. 1999; Thome et al. 1999; Willis et al. 1999; Yan et al. 1999; Zhang et al. 1999) (Fig. 1). BCL10 constitutively forms a complex with MALT1 (Uren et al. 2000; Lucas et al. 2001); this interaction may stabilize these proteins, as knockout of BCL10 or MALT1 reduces MALT1 or BCL10 levels, respectively (Klemm et al. 2006). BCL10 interaction with MALT1 depends upon both a short stretch of amino acids following the CARD motif of BCL10 (aa 107-119), and an additional site within the CARD (Bonizzi and Karin 2004; Lucas et al. 2007).

MALT1 (also called paracaspase) contains several interesting structural features that are suggestive of its molecular functions (Fig. 1). The most striking feature of MALT1 is its protease domain, which shares homology with proteases of the caspase and metacaspase family (Uren et al. 2000; Vercammen et al. 2007). In addition to this catalytic domain, MALT1 contains an amino-terminal death domain (DD) predicted to associate with an unidentified DD-containing protein, two immunoglobulin-like domains required for BCL10 interaction (Uren et al. 2000; Lucas et al. 2007), as well as binding motifs for the ubiquitin ligase TNF receptor-associated factor-6 (TRAF6) (Sun et al. 2004; Noels et al. 2007). In the following discussion, we will outline the present understanding of the molecular function of CARMA1, BCL10, and MALT1 in lymphocyte activation, with a particular focus on the individual domain structures that enable function.

MOLECULAR FUNCTION OF CARMA1

CARMA1 contains a number of defined protein domains (Fig. 1), most of which are required for mediating activation of the NF-κB pathway. Reconstitution experiments in CARMA1-silenced cells have demonstrated that the

CARD, coiled coil, SH3 and GUK domains are all essential for the NF-κB-activating function of CARMA1, whereas the PDZ domain appears dispensable (Pomerantz et al. 2002). Consistent with the known function of the CARMA1 CARD motif in mediating BCL10 interaction (Bertin et al. 2001; Gaide et al. 2001), lymphocytes expressing CARMA1 constructs with a CARD point mutation or deletion show impaired NF-κB and JNK responses (Gaide et al. 2002; Pomerantz et al. 2002; Newton and Dixit 2003). Moreover, mice expressing a CARD-deficient CARMA1 mutation are immunodeficient, probably as a consequence of impaired proliferation of lymphocytes in response to AgR or PMA/ionomycin stimulation (Newton and Dixit 2003). The CARMA1 coiled coil domain appears to mediate oligomerization (Rawlings et al. 2006; Tanner et al. 2007), a contention supported by the dominant negative activity of coiled coil-only CARMA1 mutants that inhibit NF-κB activation and IL-2 production in stimulated Jurkat T cells (Tanner et al. 2007). As the coiled coil motif bound MALT1 in a two-hybrid assay (Che et al. 2004), this domain may also contribute to the recruitment of the preformed MALT1-BCL10 complexes, CBM complex stabilization and/or to the oligomerization-dependent MALT1 activation (see later). The specific contributions of the SH3 and GUK domains to CARMA1 function are less well understood. For other MAGUK proteins such as PSD-95, it has been proposed that the SH3 and GUK domains appear to mediate auto-repression via intramolecular interaction, whereas intermolecular SH3-GUK interactions may mediate signal-induced oligomerization (Tavares et al. 2001; Yaffe 2001; Thome 2004). It remains unclear whether such mechanisms regulate CARMA1 activity. The phenotype of a fortuitous CARMA1 mutation, L808P, suggests the SH3 domain of CARMA1 regulates membrane localization and protein kinase C-θ (PKCθ), BCL10 and IKK complex recruitment to the immunological synapse (Wang et al. 2004). When overexpressed, the GUK domain binds to the Ser/Thr kinase PDK1, and PDK1 silencing affected TCR-induced NF-κB activation (Lee et al. 2005). However, these findings were contradicted in a later study, in which conditional PDK1 knockout in chicken DT40 cells did not affect BCR-induced IKK activation (Shinohara et al. 2007). Thus, it remains unclear whether PDK1 participates in AgR-stimulated IKK activation via CARMA1.

REGULATION OF CARMA1 FUNCTION BY PHOSPHORYLATION

The molecular function of CARMA1 appears to be regulated by a complex array of phosphorylation events modulating CARMA1 association with BCL10 and MALT1 in activated lymphocytes (Matsumoto et al. 2005; Sommer et al. 2005; Shinohara et al. 2007) (Table 1; Fig. 2). Although the kinases involved and the order and priority of individual phosphorylation events remain somewhat unclear, data obtained with CARMA1 point mutants, phosphorylation-specific antibodies and mass spectrometric approaches suggest the following general model for phosphorylation-dependent CARMA1 regulation. For the sake of clarity, we will refer later to human CARMA1 residue positions; the corresponding positions for mouse and chicken are described in Table 1.

AgR stimulation rapidly induces phosphorylation of CARMA1 within a region linking the coiled coil and the PDZ domains (Rueda and Thome 2005) (Fig. 2). Within this linker region, S552 and S645 were identified as targets of PKCβ or PKCθ in two independent studies (Matsumoto et al. 2005; Sommer et al. 2005), while phosphorylation of S555 by IKKβ was proposed in a subsequent study (Shinohara et al. 2007). S565 was identified as an in vivo phosphorylation site in activated T cells by mass spectrometry, and mutation of this site resulted in impaired NF-κB activation (Matsumoto et al. 2005). The kinase responsible for S565 phosphorylation has not yet been determined. Another site, S551, was targeted for phosphorylation by hematopoietic progenitor kinase-1 (HPK1) in vitro, consistent with the observed requirement for CARMA1 in HPK1-mediated NF-κB activation in T cells (Brenner et al. 2009). Collectively, these phosphorylation

Table 1. Overview on CARMA1 phosphorylation sites and functions

Human CARMA1	S109	T110	S551	S552	S555	S565	S608	S637	S645
Mouse CARMA1	S116	T117	S563	S564	S567	S577	S620	S649	S657
Chicken CARMA1	S118	T119	–	S575	S578	S588	S631	S660	S668
Proposed kinase(s)	CaMKII	PKCβ/θ ?	HPK1	PKCβ/θ	IKKβ	unknown	CK1α	PKCβ/θ nPKC	PKCβ/θ
Determined by:									
- in vitro kinase assay	+	–	+	+	+	–	–	+ (PKCβ/θ)	+
- mass spectrometry	–	–	–	+	–	–	–	–	+
- P-specific antibodies	–	+	–	–	+	–	+	+	+
Effect of Ala mutation:									
- NF-κB activation	Impaired	Impaired	Impaired	Impaired	Impaired	Impaired	Increased	normal or increased	Impaired
- JNK activation	n.d.	impaired	n.d.	impaired	impaired	n.d.	normal ?	n.d.	impaired
Effect on association with BCL10 & MALT1:	S/D increased	T/A impaired	n.d.		S/A impaired		S/A normal		S/A impaired
References	(Ishiguro et al. 2007)	(Shinohara et al. 2007)	(Brenner et al. 2009)	(Matsumoto et al. 2005; Sommer et al. 2005; Shinohara et al. 2007)	(Matsumoto et al. 2005; Shinohara et al. 2007)	(Matsumoto et al. 2005)	(Shinohara et al. 2007; Bidere et al. 2009)	(Matsumoto et al. 2005; Sommer et al. 2005; Moreno-Garcia et al. 2009)	(Matsumoto et al. 2005; Sommer et al. 2005; Shinohara et al. 2007)

The table summarizes proposed Serine (S) and Threonine (T) phosphorylation sites present in human, mouse, or chicken CARMA1, and the functional effects of corresponding CARMA1 point mutations. Abbreviations: n.d., not determined; nPKC, novel PKC; S/A or T/A, mutation of Serine or Threonine into Alanine; S/D, mutation of Serine into Aspartic acid; P-specific, phosphorylation site-specific.

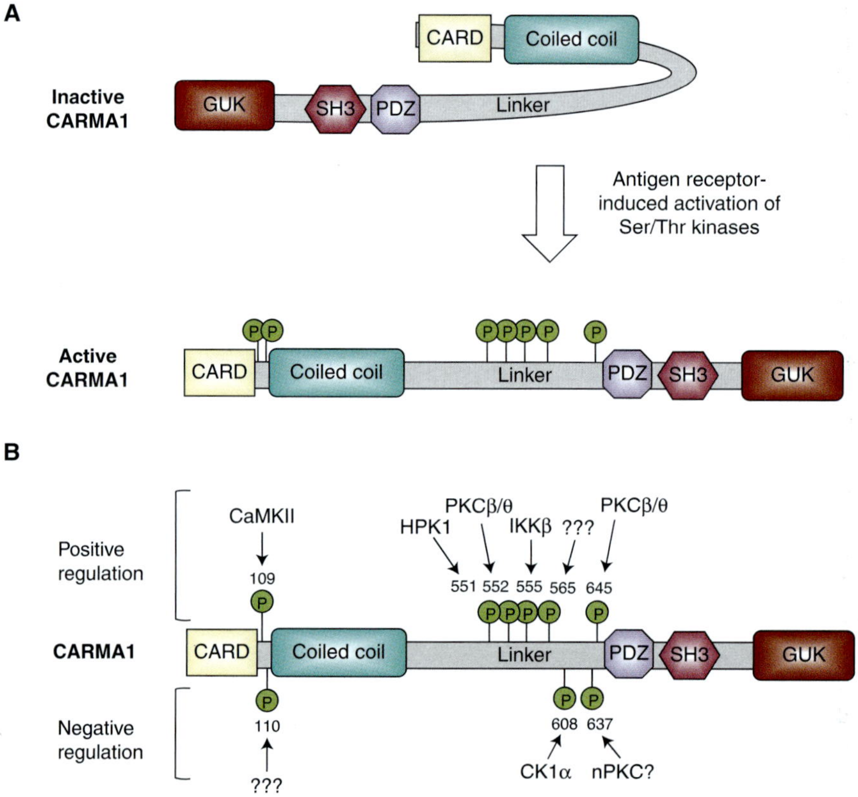

Figure 2. Phosphorylation events affecting CARMA1 function. (*A*) In resting cells, CARMA1 adopts an inactive conformation that is stabilized by an intramolecular interaction between the CARD and the linker region. Antigen receptor stimulation triggers the phosphorylation of the CARMA1 linker motif and the region between the CARD and coiled coil motifs. This is thought to induce a conformational change, possibly by electrostatic repulsion, which unfolds the CARMA1 conformation, allowing the CARD to bind to the CARD motif of BCL10. (*B*) Summary of proposed phosphorylation events proposed to positively or negatively regulate CARMA1 function. See Table 1 and text for further details.

events appear to trigger a conformational change in CARMA1, allowing BCL10-MALT1 recruitment (Fig. 2A). In the inactive, closed conformation, the CARMA1 linker region may bind back to the CARD motif, thereby blocking CARD-dependent protein-protein interactions. Stimulation-induced phosphorylation of the S552 and S645 linker residues may weaken this interaction, inducing an open conformation of CARMA1. In this conformation, the CARD motif and potentially the coiled coil domain are accessible for interaction with BCL10 and MALT1, respectively (Che et al. 2004; Sommer et al. 2005). This model is mainly suggested by in vitro binding assays using recombinant constructs containing the linker and CARD motifs of CARMA1 (Sommer et al. 2005). In these assays, recombinant active PKCθ suppressed the in vitro association of the CARD with a wild-type, but not a S552/S645 double mutant linker construct (Sommer et al. 2005).

Another phosphorylation cluster at positions 109 and 110, carboxy-terminal of the CARD domain, appears to regulate CARMA1 function (Ishiguro et al. 2006; Shinohara et al. 2007). Gene silencing and inhibitor experiments suggested that calmodulin-dependent kinase II (CaMKII) contributes to TCR-regulated NF-κB activation through CARMA1 S109 phosphorylation (Ishiguro et al. 2006).

In support of such a model, the CARMA1 mutant S109D, which mimics phosphorylation at S109, drives strong constitutive NF-κB activation and BCL10 interaction. However, further experimental evidence is required to confirm S109 as a *bona fide* phosphorylation site.

Collectively, the data discussed above suggests that initial phosphorylation at S551, S552, S555, and S645, and possibly at S109, promotes CARMA1 activation. PKCβ/θ, IKKβ, CaMKII, and HPK1 were identified as responsible for phosphorylation-dependent CARMA1 activation, but other kinases may also regulate CARMA1 function. For example, PDK1 and Akt were reported to interact with CARMA1 and their regulation of NF-κB activity was CARMA1-dependent (Lee et al. 2005; Narayan et al. 2006). Future studies are required to establish whether these kinases modulate NF-κB activity by phosphorylating CARMA1.

Recent findings suggest that phosphorylation events may also suppress CARMA1 activity, possibly during negative feedback regulation. S637, for example, is phosphorylated during the later stages of AgR signaling, and may serve to turn off CARMA1 signaling (Moreno-Garcia et al. 2009). In support of this, an S637A mutant of CARMA1 potentiated NF-κB activation. Although PKC inhibitors blocked S637 phosphorylation, this event occurred independently of PKCβ, suggesting that a member of the novel PKC family is responsible for S637 phosphorylation (Moreno-Garcia et al. 2009). The mechanism by which S637 phosphorylation inhibits CARMA1 activity remains unclear.

Another site, S608, that is likely to be a phosphorylation target of Casein kinase 1α, may inhibit CARMA1-dependent NF-κB activation (Bidere et al. 2009). The carboxyl terminus of CK1α binds to CARMA1 coiled coil and linker regions (Bidere et al. 2009), and mutation of S608 abolished CK1α-mediated phosphorylation of such a recombinant CARMA1 protein fragment in 293T cells (Bidere et al. 2009). Whether S608 is indeed phosphorylated by CK1α in vivo is still unknown, however, CARMA1 S608A mutation clearly augmented its NF-κB activating potency (Bidere et al. 2009).

Finally, CARMA1 T110 was clearly identified as a stimulation-induced phosphorylation site (Shinohara et al. 2007). In B cells, phosphorylation at T110 still occurred in the absence of IKKβ or PKCβ, suggesting that another kinase might be responsible (Shinohara et al. 2007). Interestingly, expression of a T110A mutant potentiated basal IKK activation, suggesting that phosphorylation at this site negatively regulates CARMA1 function. Similarly, an S883A mutant showed impaired inducible IKK activation, but high constitutive IKK activity (Shinohara et al. 2007). In contrast to T110, there is no evidence for in vivo phosphorylation of S883, so such findings should be interpreted with caution.

In conclusion, the available data support a model in which CARMA1 phosphorylation at multiple sites by the S/T kinases PKC β/θ, IKKβ, HPK1 and CaMKII trigger CARMA1 activation by regulating the accessibility of the effector CARD motif. Phosphorylation at residues T110, S608, and S637 may mediate autorepression, as mutation of these sites augmented the constitutive or inducible CARMA1-mediated IKK activity; the inhibitory mechanisms responsible for such an effect remain to be explored.

MULTIPLE FUNCTIONS OF BCL10 IN LYMPHOCYTE ACTIVATION

Genetic evidence suggests that the adaptor protein BCL10 performs non-redundant functions in a number of disparate biological pathways, such as normal development of B1 and marginal zone B cells (Ruland et al. 2001; Xue et al. 2003; Li et al. 2009), AgR-dependent activation of naïve lymphocytes (Ruland et al. 2001), and neural tube closure and normal brain development in mice (Ruland et al. 2001). Some of these BCL10 functions appear to be independent of association with CARMA1 and MALT1 (Thome and Weil 2007) (Fig. 3). For example, AgR triggering results in rapid BCL10 phosphorylation at S138 and another, unidentified residue (Rueda et al. 2007). These phosphorylation events are required for TCR-induced actin polymerization, cellular spreading

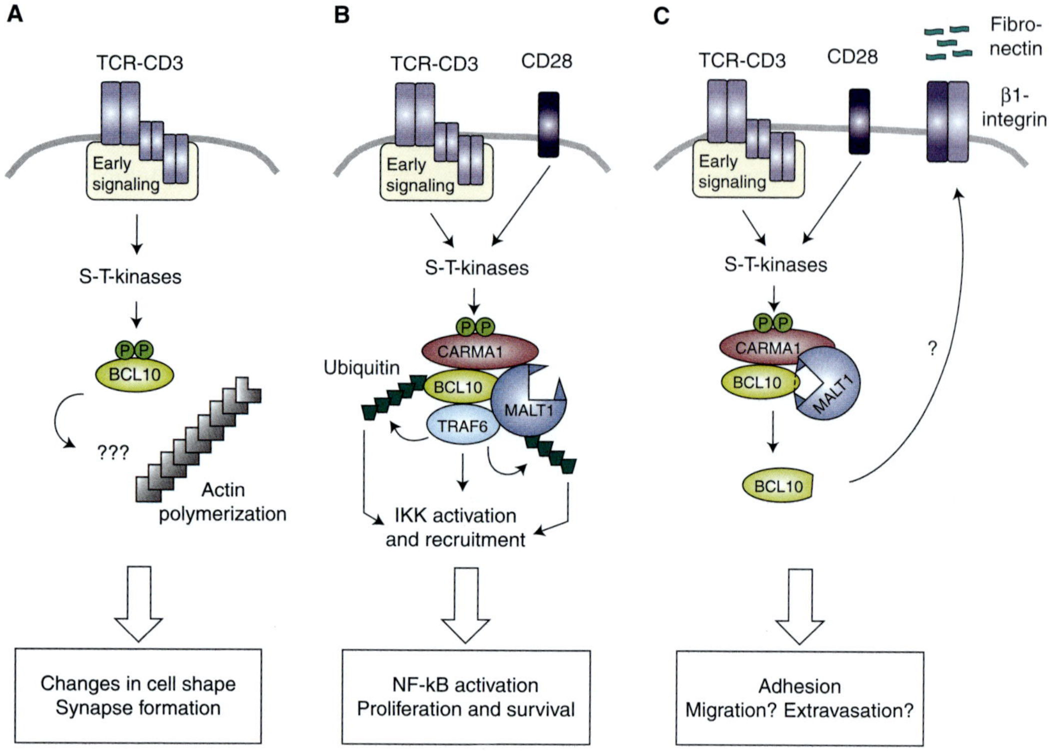

Figure 3. Overview of the multiple functions of BCL10 and its binding partners. (*A*) T-cell receptor (TCR) triggering leads to the rapid phosphorylation of BCL10, which is likely to control TCR-induced actin polymerization in a CARMA1- and MALT1-independent manner. (*B*) TCR/CD28 costimulation induces formation of the CBM complex, which controls IKK mediated NF-κB activation by promoting the activation of the IKK complex. Within this complex, MALT1-bound TRAF6 is thought to mediate K63-linked ubiquitination of BCL10 and MALT1, promoting recruitment of the IKK complex via the ubiquitin-binding IKKγ subunit. Additionally, TRAF6-mediated ubiquitination of IKKγ may contribute to IKK activation. (*C*) CBM complex formation is required for the activation of MALT1 protease function. BCL10 serves as a substrate of MALT1; its cleavage may regulate β1 integrin-mediated T-cell adhesion.

and the stabilization of the immunological synapse, but occur independently of CARMA1 and MALT1 (Rueda et al. 2007) (Fig. 3A). It is also likely that BCL10 performs additional, CARMA1- and MALT1- independent functions in neuronal development, which is normal in mice deficient for CARMA1 or MALT1 (Egawa et al. 2003; Hara et al. 2003; Ruefli-Brasse et al. 2003; Ruland et al. 2003). Cooperation of all three proteins is required for AgR-dependent engagement of the NF-κB and JNK signaling pathways in lymphocytes (Fig. 3B and C). While the relevance of AgR-dependent JNK activation is not yet precisely understood (Rincon and Pedraza-Alva 2003; Blonska et al. 2006), numerous studies support the critical function of activated NF-κB in the adaptive immune response (Li and Verma 2002; Thome 2004), and several recent studies document the contribution of CBM proteins to NF-κB activation (Rawlings et al. 2006; Schulze-Luehrmann and Ghosh 2006; Thome and Weil 2007; Thome 2008). Molecular events directing BCL10 and MALT1 activation are still somewhat unclear, but several lines of evidence suggest that T cell triggering via both, the TCR/CD3 complex and CD28 mobilizes the pool of BCL10 constitutively associated with MALT1 (Uren et al. 2000; Gaide et al. 2002; Shinohara et al. 2005; Su et al. 2005; Wegener et al. 2006;

Oeckinghaus et al. 2007). This is likely to require a phosphorylation-regulated conformational change in CARMA1 that exposes its CARD motif and allows interaction with the BCL10 CARD motif (Matsumoto et al. 2005; Rueda and Thome 2005; Sommer et al. 2005). The resulting CARMA1-BCL10-MALT1 (CBM) complex may be stabilized by interaction between the CARMA1 coiled coil motif and a C-terminal MALT1 region that lacks the DD and first two Ig domains (Che et al. 2004). Mechanisms by which CBM-associated BCL10 supports lymphocyte activation are not completely understood. Ubiquitination of BCL10 by the MALT1-binding ubiquitin ligase TRAF6 may regulate this function (Wu and Ashwell 2008). Ubiquitinated BCL10 may recruit the IKK complex via the ubiquitin-binding domain of the IKK subunit IKKγ, and this may promote further IKK activation by TRAF6-dependent ubiquitination (Sun et al. 2004; Wu and Ashwell 2008) (Fig. 3B). Additional BCL10 ubiquitination events may target BCL10 for lysosomal or proteasomal degradation upon its IKK mediated phosphorylation (Scharschmidt et al. 2004; Lobry et al. 2007). Together with CARMA1, BCL10 is required for the activation of the protease activity of MALT1 (Rebeaud et al. 2008), which contributes to NF-κB activation in an IKK-independent manner (see below). Finally, work from our laboratory demonstrated that MALT1-dependent cleavage of BCL10 affects β1 integrin-dependent T-cell adhesion by an unknown mechanism (Rebeaud et al. 2008) (Fig. 3C). This suggests that CBM proteins may participate not only in modulating NF-κB-dependent gene regulation, but also in efficient immune synapse formation and lymphocyte retention, positioning, migration and extravasation (Sixt et al. 2006).

REGULATION OF LYMPHOCYTE ACTIVATION BY MALT1

MALT1 is a scaffold protein that activates the NF-κB-regulating IKK complex by physically bridging activated CARMA1 with BCL10 and TRAF6 (Thome 2008). The first evidence indicating a role for MALT1 in NF-κB activation came from two studies showing MALT1 and BCL10 synergistically activated NF-κB in 293T cell reporter assays (Uren et al. 2000; Lucas et al. 2001). Studies with MALT1 knock-out mice revealed that MALT1 controls AgR-dependent NF-κB activation in B- and T cells (Ruefli-Brasse et al. 2003; Ruland et al. 2003), as well as NF-κB activity downstream of other receptors with immunoreceptor tyrosine-based activation motifs, such as Fc and NK receptors and of some G-protein coupled receptors (Wegener and Krappmann 2007; Thome 2008). Moreover, MALT1 plays an important pathogenic role in the development of specific types of B-cell lymphomas (see below). The molecular mechanism engaged by MALT1 to control NF-κB activation in normal and malignant lymphocytes is thus a matter of considerable interest. Recent studies suggest that MALT1-dependent NF-κB activation is mediated by both scaffold and a protease functions (Thome 2008). The scaffold function of MALT1 depends upon its multiple protein-protein interaction domains (Uren et al. 2000; Lucas et al. 2001; Sun et al. 2004; Noels et al. 2007) (Figs. 1 and 3). MALT1 contains a central caspase-like domain, together with a death domain (DD), three immunoglobulin domains and a C-terminal extension. The MALT1 C-terminal Ig domain and extension contain two binding motifs for TRAF6 (Sun et al. 2004). Presently no binding partner of the DD has been identified, however, the two immunoglobulin (Ig)-like domains following the DD are required for constitutive binding to BCL10 (Lucas et al. 2001) (Fig. 1). The C-terminal Ig domain and the remaining C-terminal extension contain two binding motifs for TRAF6 (Sun et al. 2004; Noels et al. 2007). A third TRAF6 binding site within the second immunoglobulin-like domain of MALT1 was recently identified in a MALT1 fusion protein present in MALT lymphoma (Noels et al. 2007). TRAF6 is a ubiquitin ligase that plays a central role in the IKK-dependent canonical NF-κB pathway (Deng et al. 2000; Kobayashi et al. 2004; Sun et al. 2004). It may activate NF-κB signaling compounds by K63-linked (non degradative) polyubiquitination (Deng et al. 2000; Kobayashi et al. 2004). Interestingly, the MALT1 region C-terminal to its protease domain contains multiple ubiquitination sites that can be ubiquitinated by TRAF6 and thereby contribute to the physical recruitment of the IKK complex (Oeckinghaus et al. 2007) (Fig. 3B). A two-hybrid approach demonstrated MALT1 directly binds to the coiled-coil domain of CARMA1 (Che et al. 2004), suggesting such an interaction aids in formation or stabilization of the CBM complex. Within the frame of the CBM complex, MALT1 also bound to caspase-8 and FADD (Su et al. 2005; Misra et al. 2007), but the exact nature and functional relevance of such interactions remains poorly understood (Hailfinger et al. 2009b).

MALT1 PROTEASE ACTIVITY AND NF-κB ACTIVATION

MALT1 was originally identified through a caspase homology search, and called paracaspase because of its homology to mammalian caspases and to an additional family of caspase-like proteins known as metacaspases that are found in yeast, plants and parasites (Uren et al. 2000). MALT1 shares with these protease families the presence of conserved Cys and His residues predicted to be important for catalytic activity (Uren et al. 2000). Attempts to demonstrate MALT1 protease activity were initially unsuccessful (Uren et al. 2000; Snipas et al. 2004), but such activity was recently demonstrated using MALT1 oligomerization-inducing conditions and optimized substrates (Coornaert et al. 2008; Rebeaud et al. 2008). These studies established that, in contrast to caspases, which cleave their substrates after a negatively-charged Asp residue, MALT1 cleaves protein or peptide substrates after a positively-charged Arg residue. In this respect, MALT1 more closely resembles members of the metacaspase family, that cleave substrates after Arg or Lys residues (Vercammen et al. 2007). In keeping with this, MALT1 shares with metacaspases the presence of amino-acid residues directing Arg or Lys specificity (Vercammen et al. 2007).

The mechanism by which MALT1 protease activity regulates lymphocyte activation is a matter of intense investigation. Initial overexpression experiments in 293T cells showed that mutation in the predicted MALT1 active site Cys 464 reduced its ability to activate an NF-κB reporter (Uren et al. 2000; Lucas et al. 2001), suggesting that MALT1-dependent cleavage of unknown substrate(s) may be required for optimal NF-κB activation. Surprisingly however, treatment with a cell-permeable MALT1 protease inhibitor did not affect IκB phosphorylation by the IKK complex (Rebeaud et al. 2008; Duwel et al. 2009). This suggests that the protease activity of MALT1 potentiates NF-κB activation in an IKK-independent manner, at a crucial step downstream of, or parallel to, the IKK complex. The two known MALT1 substrates, BCL10 and A20, are unlikely to mediate this effect. BCL10 cleavage regulated T-cell adhesion but was largely dispensable for NF-κB activation (see above) (Rebeaud et al. 2008). The exact role of A20 cleavage is not well understood. A20 inhibits TCR-mediated IKK activation (Duwel et al. 2009), most likely via its ability to remove K63-linked polyubiquitination from the ubiquitin ligase TRAF6, the IKKγ subunit and MALT1 itself (Heyninck and Beyaert 2005; Duwel et al. 2009). MALT1 activity was required for the cleavage of a small fraction of total A20, but dispensable for down-modulation of total A20 levels in activated T cells (Coornaert et al. 2008; Duwel et al. 2009). Thus, both the predominant role of A20 in IKK-dependent NF-κB activation and the small proportion of A20 cleavage argue against a major role for A20 in the IKK-independent control of NF-κB activity by MALT1 protease activity.

Therefore, the substrate responsible for MALT1 protease-dependent NF-κB activation remains to be identified. It is possible that this substrate specifically affects c-Rel but not RelA activation, as a recent study using MALT1 deficient B cells demonstrated impaired c-Rel activation in B cells (Ferch et al. 2007). Surprisingly, this correlated with reduced IκBβ but not IκBα degradation (Ferch et al. 2007), implying a role for MALT1 in IκBβ-dependent c-Rel activation, but the exact role of the MALT1 protease activity in this process remains to be tested.

In conclusion, the presently available experimental evidence suggests that the protease activity of MALT1 controls NF-κB activation in an IKK-independent manner, either by affecting IκBβ stability, or by controlling other events required for the nuclear translocation or transcriptional activity of NF-κB1 complexes (Fig. 4).

CBM PROTEINS AND LYMPHOMAS

Given the critical role for CBM proteins and the NF-κB pathway in the control of lymphocyte survival and proliferation, it is not surprising to find altered expression and/or function of CBM proteins in specific types of B-cell lymphomas (Isaacson and Du 2004; Zhou et al. 2005; Noels et al. 2007; Sagaert et al. 2007; Coornaert et al. 2008; Lenz et al. 2008; Compagno et al. 2009; Ferch et al. 2009; Hailfinger et al.

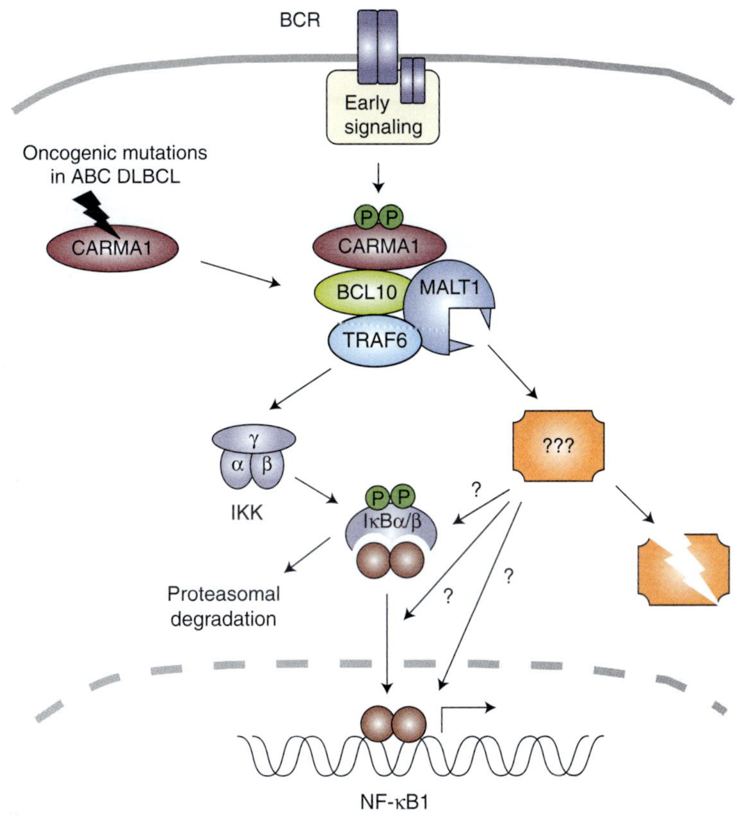

Figure 4. Proposed model for MALT1 function in NF-κB activation. B-cell receptor (BCR) stimulation induces formation of the CBM complex, which controls NF-κB activation through at least two different mechanisms. The scaffold function of BCL10 and MALT1 are required for TRAF6 recruitment and ubiquitination-dependent activation of the IKK complex, which phosphorylates IκB proteins. The protease activity of MALT1 also contributes to NF-κB activation in an IKK-independent manner, through cleavage of an unknown substrate (???). This might control IκBβ degradation, which occurs with slower kinetics than IκBα degradation and is sustained (Whiteside and Israel 1997), possibly because it depends on additional signaling events. Alternatively, MALT1 protease activity might control NF-κB1 activation at the level of its nuclear translocation or transcriptional activity. The presence of oncogenic mutations in CARMA1 or its upstream regulators may render this pathway constitutively active in diffuse large B-cell lymphoma of the activated B-cell type (ABC DLBCL) (see text for details).

2009a). Lymphomas of the mucosa-associated lymphoid tissue (MALT lymphomas) are derived from extranodal marginal zone B cells and arise typically in mucosal or non-mucosal sites as a result of chronic infection or inflammation (Sagaert et al. 2007). Such lymphomas were associated with chromosomal translocations of the genes encoding BCL10 or MALT1. These are thought to result in overexpression of BCL10 or MALT1 or in the generation of fusion proteins containing the C-terminal part of MALT1 and the N-terminal part of the cellular inhibitor of apoptosis protein-2 (cIAP2) (Isaacson and Du 2004; Sagaert et al. 2007). Interestingly, the resulting cIAP2-MALT1 fusion proteins showed both constitutive MALT1 protease activity and NF-κB inducing capacity (Uren et al. 2000; Lucas et al. 2007; Coornaert et al. 2008), suggesting aberrant MALT1 activity drives the growth of these lymphomas. Altered CBM functions were also identified in another type of B-cell lymphoma, the diffuse large B-cell lymphoma (DLBCL) (Lenz et al. 2008; Compagno et al. 2009; Ferch et al. 2009; Hailfinger et al. 2009a). Constitutive activation of NF-κB is characteristic of the activated B-cell (ABC) subtype of DLBCL, which depends on NF-κB signals for survival and/or proliferation (Staudt and Dave 2005). Silencing of the expression of CARMA1, BCL10 or MALT1 led to reduced growth and induction of apoptosis in cell lines derived from ABC DLBCL (Ngo et al. 2006). Recently, two studies identified oncogenic CARMA1 mutations in patients with ABC DLBCL (Lenz et al. 2008; Compagno et al. 2009) and, less frequently, DLBCL of the germinal center B-cell type (Lenz et al. 2008; Compagno et al. 2009). We and others showed constitutive MALT1 activity in cell lines derived from ABC DLBCL (Ferch et al. 2009; Hailfinger et al. 2009a). Treatment of these cell lines with a MALT1 inhibitor specifically affected cellular growth by blocking NF-κB activity (Ferch et al. 2009; Hailfinger et al. 2009a), and in particular the nuclear translocation of c- Rel (Ferch et al. 2009). This suggests that constitutive CBM activity, induced by activating mutations in CARMA1 or its upstream regulators, drives constitutive NF-κB activity in these lymphomas (Fig. 4). Thus, the CBM signaling pathway, and in particular the protease activity of MALT1, reveals to be an interesting target for the development of specific anti-lymphoma drugs.

CONCLUSIONS

The past decade has witnessed dramatic progress in the characterization of signaling mechanisms controlling AgR-induced NF-κB signaling and downstream lymphocyte activation, development and tumorigenesis. A number of open questions remain and will be fertile ground for future studies. For example, the identification and characterization of additional MALT1 substrates, and further delineation of mechanisms directing MALT1 activity are currently hot topics. Further dissection of mechanisms that positively or negatively regulate pathway signaling components, and the precise regulation and function of individual NF-κB family members present other interesting aspects. Ultimately, novel enzymatic activities relevant to this pathway will require validation as new drug targets for immuno-suppression or treatment of lymphomas.

ACKNOWLEDGMENTS

The authors would like to thank Kate Schroder for critical review of the manuscript. Research in the lab is supported by grants from the Swiss National Science Foundation, the Swiss Cancer League and the foundations Leenaards, Pierre Mercier and Emma Muschamp. J.E.C. is supported by a PhD fellowship from the Faculty of Biology and Medicine of the University of Lausanne.

REFERENCES

Bertin J, Wang L, Guo Y, Jacobson MD, Poyet JL, Srinivasula SM, Merriam S, DiStefano PS, Alnemri ES. 2001. CARD11 and CARD14 are novel caspase recruitment domain (CARD)/membrane- associated guanylate kinase (MAGUK) family members that interact with BCL10 and activate NF-kappa B. *J Biol Chem* **276:** 11877–11882.

Bidere N, Ngo VN, Lee J, Collins C, Zheng L, Wan F, Davis RE, Lenz G, Anderson DE, Arnoult D, et al. 2009. Casein kinase 1alpha governs antigen-receptor-induced NF-kappaB activation and human lymphoma cell survival. *Nature* **458:** 92–96.

Blonska M, Pappu BP, Matsumoto R, Li H, Su B, Wang D, Lin X. 2006. The CARMA1-Bcl10 Signaling Complex Selectively Regulates JNK2 Kinase in the T Cell Receptor-Signaling Pathway. *Immunity* **26:** 55–66.

Bonizzi G, Karin M. 2004. The two NF-kappaB activation pathways and their role in innate and adaptive immunity. *Trends Immunol* **25:** 280–288.

Brenner D, Brechmann M, Rohling S, Tapernoux M, Mock T, Winter D, Lehmann WD, Kiefer F, Thome M, Krammer PH, et al. 2009. Phosphorylation of CARMA1 by HPK1 is critical for NF-kappaB activation in T cells. *Proc Natl Acad Sci U S A* **106:** 14508–14513.

Che T, You Y, Wang D, Tanner MJ, Dixit VM, Lin X. 2004. MALT1/paracaspase is a signaling component downstream of CARMA1 and mediates T cell receptor-induced NF-kappaB activation. *J Biol Chem* **279:** 15870–15876.

Compagno M, Lim WK, Grunn A, Nandula SV, Brahmachary M, Shen Q, Bertoni F, Ponzoni M, Scandurra M, Califano A, et al. 2009. Mutations of multiple genes cause deregulation of NF-kappaB in diffuse large B-cell lymphoma. *Nature* **459:** 717–721.

Coornaert B, Baens M, Heyninck K, Bekaert T, Haegman M, Staal J, Sun L, Chen ZJ, Marynen P, Beyaert R. 2008. T cell antigen receptor stimulation induces MALT1 paracaspase-mediated cleavage of the NF-kappaB inhibitor A20. *Nature Immunol* **9:** 263–271.

Costanzo A, Guiet C, Vito P. 1999. c-E10 is a caspase-recruiting domain-containing protein that interacts with components of death receptors signaling pathway and activates nuclear factor-kappaB. *J Biol Chem* **274:** 20127–20132.

Deng L, Wang C, Spencer E, Yang L, Braun A, You J, Slaughter C, Pickart C, Chen ZJ. 2000. Activation of the IkappaB kinase complex by TRAF6 requires a dimeric ubiquitin-conjugating enzyme complex and a unique polyubiquitin chain. *Cell* **103:** 351–361.

Duwel M, Welteke V, Oeckinghaus A, Baens M, Kloo B, Ferch U, Darnay BG, Ruland J, Marynen P, Krappmann D. 2009. A20 negatively regulates T cell receptor signaling to NF-kappaB by cleaving Malt1 ubiquitin chains. *J Immunol* **182:** 7718–7728.

Egawa T, Albrecht B, Favier B, Sunshine MJ, Mirchandani K, O'Brien W, Thome M, Littman DR. 2003. Requirement for CARMA1 in antigen receptor-induced NF-kappa B activation and lymphocyte proliferation. *Curr Biol* **13:** 1252–1258.

Feng W, Zhang M. 2009. Organization and dynamics of PDZ-domain-related supramodules in the postsynaptic density. *Nat Rev Neurosci* **10:** 87–99.

Ferch U, Kloo B, Gewies A, Pfander V, Duwel M, Peschel C, Krappmann D, Ruland J. 2009. Inhibition of MALT1 protease activity is selectively toxic for activated B cell-like diffuse large B cell lymphoma cells. *J Exp Med* **206:** 2313–2320.

Ferch U, zum Buschenfelde CM, Gewies A, Wegener E, Rauser S, Peschel C, Krappmann D, Ruland J. 2007. MALT1 directs B cell receptor-induced canonical nuclear factor-kappaB signaling selectively to the c-Rel subunit. *Nature Immunol* **8:** 4–991.

Funke L, Dakoji S, Bredt DS. 2005. Membrane-associated guanylate kinases regulate adhesion and plasticity at cell junctions. *Annual review of biochemistry* **74:** 219–245.

Gaide O, Favier B, Legler DF, Bonnet D, Brissoni B, Valitutti S, Bron C, Tschopp J, Thome M. 2002. CARMA1 is a critical lipid raft-associated regulator of TCR-induced NF-kappa B activation. *Nat Immunol* **3:** 836–843.

Gaide O, Martinon F, Micheau O, Bonnet D, Thome M, Tschopp J. 2001. Carma1, a CARD-containing binding partner of Bcl10, induces Bcl10 phosphorylation and NF-kappaB activation. *FEBS Lett* **496:** 121–127.

Hailfinger S, Lenz G, Ngo V, Posvitz-Fejfar A, Rebeaud F, Guzzardi M, Penas EM, Dierlamm J, Chan WC, Staudt LM, et al. 2009a. Essential role of MALT1 protease activity in activated B cell-like diffuse large B-cell lymphoma. *Proc Natl Acad Sci U S A* **106:** 19946–19951.

Hailfinger S, Rebeaud F, Thome M. 2009b. Adapter and enzymatic functions of proteases in T-cell activation. *Immunol Rev* **232:** 334–347.

Hara H, Wada T, Bakal C, Kozieradzki I, Suzuki S, Suzuki N, Nghiem M, Griffiths EK, Krawczyk C, Bauer B, et al. 2003. The MAGUK family protein CARD11 is essential for lymphocyte activation. *Immunity* **18:** 763–775.

Heyninck K, Beyaert R. 2005. A20 inhibits NF-kappaB activation by dual ubiquitin-editing functions. *Trends in biochemical sciences* **30:** 1–4.

Hofmann K, Bucher P, Tschopp J. 1997. The CARD domain: a new apoptotic signalling motif. *Trends in biochemical sciences* **22:** 155–156.

Isaacson PG, Du MQ. 2004. MALT lymphoma: from morphology to molecules. *Nat Rev Cancer* **4:** 644–653.

Ishiguro K, Ando T, Goto H, Xavier R. 2007. Bcl10 is phosphorylated on Ser138 by Ca(2+)/calmodulin-dependent protein kinase II. *Mol Immunol* **44:** 2105–2110.

Ishiguro K, Green T, Rapley J, Wachtel H, Giallourakis C, Landry A, Cao Z, Lu N, Takafumi A, Goto H, et al. 2006. Ca2+/calmodulin-dependent protein kinase II is a modulator of CARMA1-mediated NF-kappaB activation. *Mol Cell Biol* **26:** 5497–5508.

Jun JE, Wilson LE, Vinuesa CG, Lesage S, Blery M, Miosge LA, Cook MC, Kucharska EM, Hara H, Penninger JM, et al. 2003. Identifying the MAGUK protein Carma-1 as a central regulator of humoral immune responses and atopy by genome-wide mouse mutagenesis. *Immunity* **18:** 751–762.

Karin M, Cao Y, Greten FR, Li ZW. 2002. NF-kappaB in cancer: from innocent bystander to major culprit. *Nat Rev Cancer* **2:** 301–310.

Klemm S, Gutermuth J, Hultner L, Sparwasser T, Behrendt H, Peschel C, Mak TW, Jakob T, Ruland J. 2006. The Bcl10-Malt1 complex segregates Fc epsilon RI-mediated nuclear factor kappa B activation and cytokine production from mast cell degranulation. *J Exp Med* **203:** 337–347.

Kobayashi T, Walsh MC, Choi Y. 2004. The role of TRAF6 in signal transduction and the immune response. *Microbes Infect* **6:** 1333–1338.

Koseki T, Inohara N, Chen S, Carrio R, Merino J, Hottiger MO, Nabel GJ, Nunez G. 1999. CIPER, a novel NF kappaB-activating protein containing a caspase recruitment domain with homology to Herpesvirus-2 protein E10. *J Biol Chem* **274:** 9955–9961.

Langel FD, Jain NA, Rossman JS, Kingeter LM, Kashyap AK, Schaefer BC. 2008. Multiple protein domains mediate interaction between Bcl10 and MALT1. *J Biol Chem* **283:** 32419–32431.

Lee KY, D'Acquisto F, Hayden MS, Shim JH, Ghosh S. 2005. PDK1 nucleates T cell receptor-induced signaling complex for NF-kappaB activation. *Science* **308:** 114–118.

Lenz G, Davis RE, Ngo VN, Lam L, George TC, Wright GW, Dave SS, Zhao H, Xu W, Rosenwald A, et al. 2008. Oncogenic CARD11 Mutations in Human Diffuse Large B Cell Lymphoma. *Science.*

Li Q, Verma IM. 2002. NF-kappaB regulation in the immune system. *Nat Rev Immunol* **2:** 725–734.

Li Z, Wang H, Xue L, Shin DM, Roopenian D, Xu W, Qi CF, Sangster MY, Orihuela CJ, Tuomanen E, et al. 2009. Emu-BCL10 mice exhibit constitutive activation of both canonical and noncanonical NF-kappaB pathways generating marginal zone (MZ) B-cell expansion as a precursor to splenic MZ lymphoma. *Blood* **114:** 4158–4168.

Lobry C, Lopez T, Israël A, Weil R. 2007. Negative feed back loop in T-cell activation through IkappaB kinase-induced phosphorylation and degradation of Bcl10. *Proc Natl Acad Sci U S A* **104:** 908–913.

Lucas PC, Kuffa P, Gu S, Kohrt D, Kim DS, Siu K, Jin X, Swenson J, McAllister-Lucas LM. 2007. A dual role for the API2 moiety in API2-MALT1-dependent NF-kappaB activation: heterotypic oligomerization and TRAF2 recruitment. *Oncogene* **26:** 5643–5654.

Lucas PC, Yonezumi M, Inohara N, McAllister-Lucas LM, Abazeed ME, Chen FF, Yamaoka S, Seto M, Nunez G. 2001. Bcl10 and MALT1, independent targets of chromosomal translocation in malt lymphoma, cooperate in a novel NF-kappa B signaling pathway. *J Biol Chem* **276:** 19012–19019.

Matsumoto R, Wang D, Blonska M, Li H, Kobayashi M, Pappu B, Chen Y, Lin X. 2005. Phosphorylation of CARMA1 plays a critical role in T Cell receptor-mediated NF-kappaB activation. *Immunity* **23:** 575–585.

McAllister-Lucas LM, Inohara N, Lucas PC, Ruland J, Benito A, Li Q, Chen S, Chen FF, Yamaoka S, Verma IM, et al. 2001. Bimp1, a MAGUK family member linking protein kinase C activation to Bcl10-mediated NF-kappaB induction. *J Biol Chem* **276:** 30589–30597.

McGee AW, Dakoji SR, Olsen O, Bredt DS, Lim WA, Prehoda KE. 2001. Structure of the SH3-guanylate kinase module from PSD-95 suggests a mechanism for regulated assembly of MAGUK scaffolding proteins. *Mol Cell* **8:** 1291–1301.

Misra RS, Russell JQ, Koenig A, Hinshaw-Makepeace JA, Wen R, Wang D, Huo H, Littman DR, Ferch U, Ruland J, et al. 2007. Caspase-8 and c-FLIPL associate in lipid rafts with NF-kappaB adaptors during T cell activation. *J Biol Chem* **282:** 19365–19374.

Moreno-Garcia ME, Sommer K, Haftmann C, Sontheimer C, Andrews SF, Rawlings DJ. 2009. Serine 649 phosphorylation within the protein kinase C-regulated domain down-regulates CARMA1 activity in lymphocytes. *J Immunol* **183:** 7362–7370.

Narayan P, Holt B, Tosti R, Kane LP. 2006. CARMA1 is required for Akt-mediated NF-kappaB activation in T cells. *Mol Cell Biol* **26:** 2327–2336.

Newton K, Dixit VM. 2003. Mice lacking the CARD of CARMA1 exhibit defective B lymphocyte development and impaired proliferation of their B and T lymphocytes. *Curr Biol* **13:** 1247–1251.

Ngo VN, Davis RE, Lamy L, Yu X, Zhao H, Lenz G, Lam LT, Dave S, Yang L, Powell J, et al. 2006. A loss-of-function RNA interference screen for molecular targets in cancer. *Nature* **441:** 106–110.

Noels H, van Loo G, Hagens S, Broeckx V, Beyaert R, Marynen P, Baens M. 2007. A Novel TRAF6 Binding Site in MALT1 Defines Distinct Mechanisms of NF-kappaB Activation by API2-MALT1 Fusions. *J Biol Chem* **282:** 10180–10189.

Oeckinghaus A, Wegener E, Welteke V, Ferch U, Arslan SC, Ruland J, Scheidereit C, Krappmann D. 2007. Malt1 ubiquitination triggers NF-kappaB signaling upon T-cell activation. *EMBO J* **26:** 4634–4645.

Pomerantz JL, Denny EM, Baltimore D. 2002. CARD11 mediates factor-specific activation of NF-kappaB by the T cell receptor complex. *EMBO J* **21:** 5184–5194.

Ponting CP, Phillips C, Davies KE, Blake DJ. 1997. PDZ domains: targeting signalling molecules to submembranous sites. *Bioessays* **19:** 469–479.

Rawlings DJ, Sommer K, Moreno-Garcia ME. 2006. The CARMA1 signalosome links the signalling machinery of adaptive and innate immunity in lymphocytes. *Nature Rev Immunol* **6:** 799–812.

Rebeaud F, Hailfinger S, Posevitz-Fejfar A, Tapernoux M, Moser R, Rueda D, Gaide O, Guzzardi M, Iancu EM, Rufer N, et al. 2008. The proteolytic activity of the paracaspase MALT1 is key in T cell activation. *Nature Immunol* **9:** 272–281.

Rincon M, Pedraza-Alva G. 2003. JNK and p38 MAP kinases in CD4+ and CD8+ T cells. *Immunol Rev* **192:** 131–142.

Rueda D, Gaide O, Ho L, Lewkowicz E, Niedergang F, Hailfinger S, Rebeaud F, Guzzardi M, Conne B, Thelen M, et al. 2007. Bcl10 controls TCR- and FcgammaR-induced actin polymerization. *J Immunol* **178:** 4373–4384.

Rueda D, Thome M. 2005. Phosphorylation of CARMA1: the link(er) to NF-kappaB activation. *Immunity* **23:** 551–553.

Ruefli-Brasse AA, French DM, Dixit VM. 2003. Regulation of NF-kappaB-dependent lymphocyte activation and development by paracaspase. *Science* **302:** 1581–1584.

Ruland J, Duncan GS, Elia A, del Barco Barrantes I, Nguyen L, Plyte S, Millar DG, Bouchard D, Wakeham A, Ohashi PS, et al. 2001. Bcl10 is a positive regulator of antigen receptor-induced activation of NF-kappaB and neural tube closure. *Cell* **104:** 33–42.

Ruland J, Duncan GS, Wakeham A, Mak TW. 2003. Differential requirement for Malt1 in T and B cell antigen receptor signaling. *Immunity* **19:** 749–758.

Sagaert X, De Wolf-Peeters C, Noels H, Baens M. 2007. The pathogenesis of MALT lymphomas: where do we stand? *Leukemia* **21:** 389–396.

Scharschmidt E, Wegener E, Heissmeyer V, Rao A, Krappmann D. 2004. Degradation of Bcl10 induced by T-cell activation negatively regulates NF-kappa B signaling. *Mol Cell Biol* **24:** 3860–3873.

Schulze-Luehrmann J, Ghosh S. 2006. Antigen-receptor signaling to nuclear factor kappa B. *Immunity* **25:** 701–715.

Shinohara H, Maeda S, Watarai H, Kurosaki T. 2007. IkappaB kinase beta-induced phosphorylation of CARMA1 contributes to CARMA1 Bcl10 MALT1 complex formation in B cells. *J Exp Med* **204:** 3285–3293.

Shinohara H, Yasuda T, Aiba Y, Sanjo H, Hamadate M, Watarai H, Sakurai H, Kurosaki T. 2005. PKC beta regulates BCR-mediated IKK activation by facilitating the interaction between TAK1 and CARMA1. *J Exp Med* **202:** 1423–1431.

Sixt M, Bauer M, Lammermann T, Fassler R. 2006. Beta1 integrins: zip codes and signaling relay for blood cells. *Curr Opin Cell Biol* **18:** 482–490.

Snipas SJ, Wildfang E, Nazif T, Christensen L, Boatright KM, Bogyo M, Stennicke HR, Salvesen GS. 2004. Characteristics of the caspase-like catalytic domain of human paracaspase. *Biol Chem* **385:** 1093–1098.

Sommer K, Guo B, Pomerantz JL, Bandaranayake AD, Moreno-Garcia ME, Ovechkina YL, Rawlings DJ. 2005. Phosphorylation of the CARMA1 linker controls NF-kappaB activation. *Immunity* **23:** 561–574.

Srinivasula SM, Ahmad M, Lin JH, Poyet JL, Fernandes-Alnemri T, Tsichlis PN, Alnemri ES. 1999. CLAP, a novel caspase recruitment domain-containing protein in the tumor necrosis factor receptor pathway, regulates NF-kappaB activation and apoptosis. *J Biol Chem* **274:** 17946–17954.

Staudt LM, Dave S. 2005. The biology of human lymphoid malignancies revealed by gene expression profiling. *Adv Immunol* **87:** 163–208.

Su H, Bidere N, Zheng L, Cubre A, Sakai K, Dale J, Salmena L, Hakem R, Straus S, Lenardo M. 2005. Requirement for caspase-8 in NF-kappaB activation by antigen receptor. *Science* **307:** 1465–1468.

Sun L, Deng L, Ea CK, Xia ZP, Chen ZJ. 2004. The TRAF6 ubiquitin ligase and TAK1 kinase mediate IKK activation by BCL10 and MALT1 in T lymphocytes. *Mol Cell* **14:** 289–301.

Tanner MJ, Hanel W, Gaffen SL, Lin X. 2007. CARMA1 Coiled-coil domain is involved in the oligomerization and subcellular localization of CARMA1, and is required for T cell receptor-induced NF-kappa B activation. *J Biol Chem* **282:** 17141–17147.

Tavares GA, Panepucci EH, Brunger AT. 2001. Structural characterization of the intramolecular interaction between the SH3 and guanylate kinase domains of PSD-95. *Mol Cell* **8:** 1313–1325.

Thome M. 2004. CARMA1, BCL-10 and MALT1 in lymphocyte development and activation. *Nature reviews* **4:** 348–359.

Thome M. 2008. Multifunctional roles for MALT1 in T-cell activation. *Nature Rev Immunol* **8:** 495–500.

Thome M, Martinon F, Hofmann K, Rubio V, Steiner V, Schneider P, Mattmann C, Tschopp J. 1999. Equine herpesvirus-2 E10 gene product, but not its cellular homologue, activates NF-kappaB transcription factor and c-Jun N-terminal kinase. *J Biol Chem* **274:** 9962–9968.

Thome M, Tschopp J. 2003. TCR-induced NF-kB activation: a crucial role for Carma1, Bcl10 and MALT1. *Trends Immunol* (in press).

Thome M, Weil R. 2007. Post-translational modifications regulate distinct functions of CARMA1 and BCL10. *Trends Immunol* **28:** 281–288.

Uren AG, O'Rourke K, Aravind LA, Pisabarro MT, Seshagiri S, Koonin EV, Dixit VM. 2000. Identification of paracaspases and metacaspases: two ancient families of caspase-like proteins, one of which plays a key role in MALT lymphoma. *Mol Cell* **6:** 961–967.

Vercammen D, Declercq W, Vandenabeele P, Van Breusegem F. 2007. Are metacaspases caspases? *J Cell Biol* **179:** 375–380.

Wang D, Matsumoto R, You Y, Che T, Lin XY, Gaffen SL, Lin X. 2004. CD3/CD28 costimulation-induced NF-kappaB activation is mediated by recruitment of protein kinase C-theta, Bcl10, and IkappaB kinase beta to the immunological synapse through CARMA1. *Mol Cell Biol* **24:** 164–171.

Weber CH, Vincenz C. 2001. The death domain superfamily: a tale of two interfaces? *Trends in biochemical sciences* **26:** 475–481.

Wegener E, Krappmann D. 2007. CARD-Bcl10-Malt1 signalosomes: missing link to NF-kappaB. *Sci STKE* **2007:** e21.

Wegener E, Oeckinghaus A, Papadopoulou N, Lavitas L, Schmidt-Supprian M, Ferch U, Mak TW, Ruland J, Heissmeyer V, Krappmann D. 2006. Essential role for IkappaB kinase beta in remodeling Carma1-Bcl10-Malt1 complexes upon T cell activation. *Mol Cell* **23:** 13–23.

Wertz IE, O'Rourke KM, Zhou H, Eby M, Aravind L, Seshagiri S, Wu P, Wiesmann C, Baker R, Boone DL, et al. 2004. De-ubiquitination and ubiquitin ligase domains of A20 downregulate NF-kappaB signalling. *Nature* **430:** 694–699.

Whiteside ST, Israel A. 1997. I kappa B proteins: structure, function and regulation. *Semin Cancer Biol* **8:** 75–82.

Willis TG, Jadayel DM, Du MQ, Peng H, Perry AR, Abdul-Rauf M, Price H, Karran L, Majekodunmi O, Wlodarska I, et al. 1999. Bcl10 is involved in t(1;14)(p22;q32) of MALT B cell lymphoma and mutated in multiple tumor types [see comments]. *Cell* **96:** 35–45.

Wu CJ, Ashwell JD. 2008. NEMO recognition of ubiquitinated Bcl10 is required for T cell receptor-mediated NF-{kappa}B activation. *Proc Natl Acad Sci U S A* **105:** 3023–3028.

Xue L, Morris SW, Orihuela C, Tuomanen E, Cui X, Wen R, Wang D. 2003. Defective development and function of Bcl10-deficient follicular, marginal zone and B1 B cells. *Nat Immunol* **4:** 857–865.

Yaffe MB. 2001. MAGUK SH3 Domains—Swapped and Stranded by Their Kinases? *Structure* **10:** 3–9.

Yan M, Lee J, Schilbach S, Goddard A, Dixit V. 1999. mE10, a novel caspase recruitment domain-containing pro-apoptotic molecule. *J Biol Chem* **274:** 10287–10292.

Zhang Q, Siebert R, Yan M, Hinzmann B, Cui X, Xue L, Rakestraw KM, Naeve CW, Beckmann G, Weisenburger DD, et al. 1999. Inactivating mutations and overexpression of BCL10, a caspase recruitment domain-containing gene, in MALT lymphoma with t(1;14)(p22;q32). *Nat Genet* **22:** 63–68.

Zhou H, Du MQ, Dixit VM. 2005. Constitutive NF-kappaB activation by the t(11;18)(q21;q21) product in MALT lymphoma is linked to deregulated ubiquitin ligase activity. *Cancer Cell* **7:** 425–431.

Mediation of T-Cell Activation by Actin Meshworks

Peter Beemiller and Matthew F. Krummel

Department of Pathology, University of California, San Francisco, San Francisco, California 94143-0511

Correspondence: matthew.krummel@ucsf.edu

Although the actin cytoskeleton and T-cell receptor (TCR) signaling complexes are seemingly distinct molecular structures, they are tightly integrated in T cells. The signaling pathways initiated by TCRs binding to peptide MHC complexes are extensively influenced by the actin cytoskeletal activities of the motile phase before TCR signaling, the signalosome scaffolding function of the cytoskeleton, and the translocation of signaling clusters that precedes the termination of signaling at these complexes. As these three successive phases constitute essentially all the steps consequent to immune synapse formation, it has become clear that the substantial physical forces and signaling interactions generated by the actin cytoskeleton dominate the signaling life cycle of TCR signalosomes. We discuss the contributions of the actin cytoskeleton to TCR signaling phases and model some remaining questions about how specific cytoskeletal factors regulate TCR signaling outcomes.

The activation of T cells is controlled primarily by T-cell receptors (TCRs) interacting with peptide-loaded major histocompatibility complexes (pMHCs) as T cells scan the surface of antigen presenting cells (APCs). Because T cells are continuously motile cells that transit through lymph nodes in their surveillance, it is clear that TCR triggering must occur within the context of physical forces that might rapidly separate TCRs from agonist pMHCs. Moreover, crawling T cells do not truly come to rest at the surfaces of APCs following TCR engagement. Instead, they continuously extend protrusions over APCs and move along the surface of their partner (Gunzer et al. 2000). In their initial encounters with antigen-bearing dendritic cells (DCs), T cells also often rapidly couple and uncouple on the order of minutes, rather than dwelling for extended periods of time on single DCs (Gunzer et al. 2000; Mempel et al. 2004). This dynamic coupling allows T cells to quickly sample a large proportion of the total APC membrane pool in search of their cognate antigen. Still, these transient contacts are productive—they induce calcium fluxes and the expression of markers of activated T cells—indicating that TCR signalosome outputs can be initiated in mere minutes and survive the dissolution of contacts, even under the mechanical stress of cytoskeletal remodeling.

TCR signaling requires the dynamic recruitment of a macromolecular complex of kinases, scaffolding molecules, and other signaling effectors to a triggered TCR. Assembly of this macromolecular signaling complex must be very sensitive and occur rapidly, or there is a

 doi: 10.1101/cshperspect.a002444
Cite this article as *Cold Spring Harb Perspect Biol* 2010;2:a002444

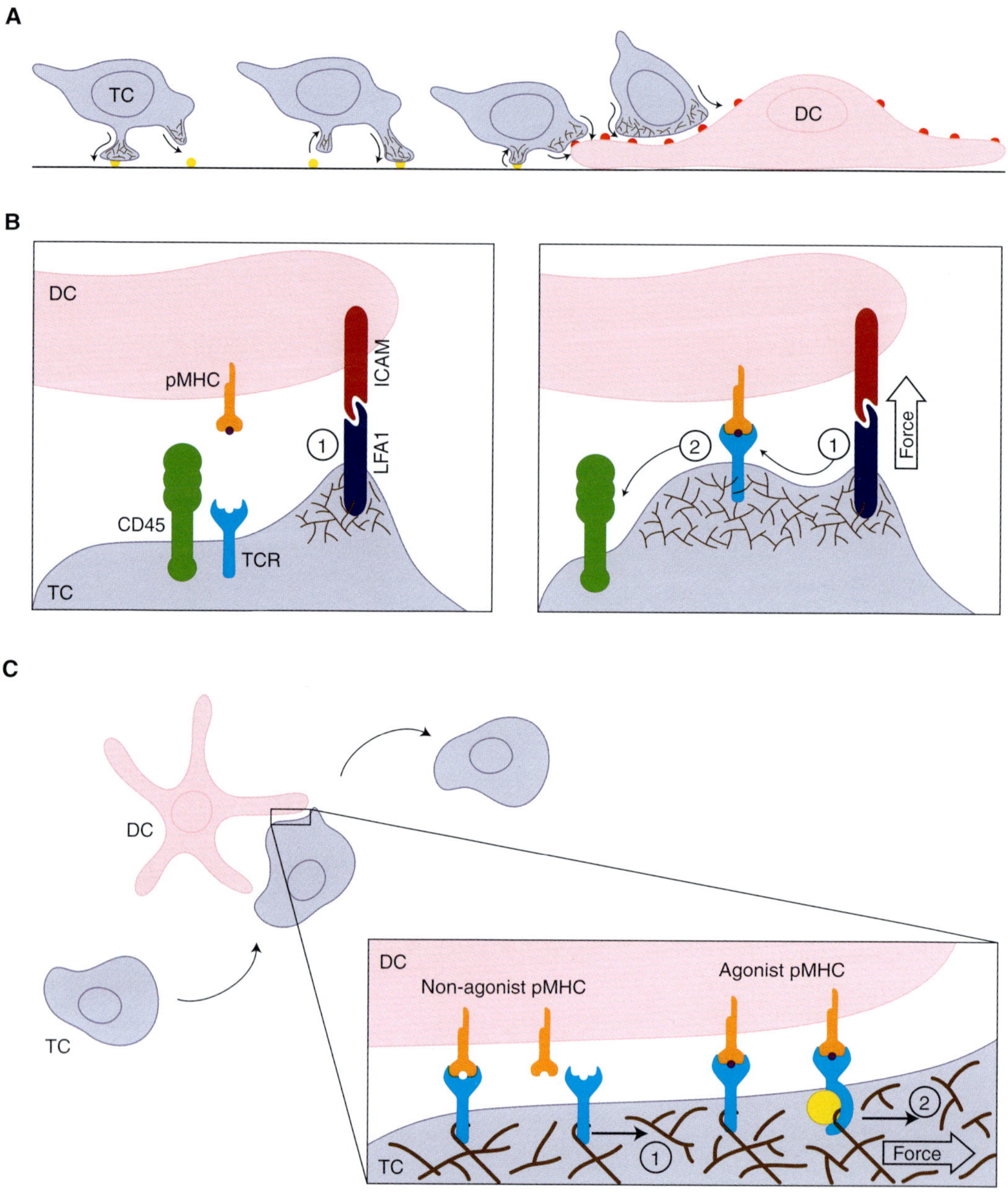

Figure 1. The actin meshwork of the motile T cell establishes a mechanical regime of TCR triggering. (*A*) T cells (TC) display various modes of motility depending on the adhesive properties of their microenvironment. To achieve high velocities, T cells create series of small, short-lived contacts with substrates. The pseudopod sequences are generated by cycles of actin polymerization followed by myosin generated contractility. When T cells encounter the adhesion promoting surface of an APC, such as a dendritic cell (DC), integrin ligation causes the T cells to spread more extensively, maximizing surface interactions between the cells. Cell body motion does not completely stop, though, and T cells continue to drift over the surface of the APC as they scan for their cognate pMHC. (*B*) The actin cytoskeletal remodeling induced by integrin ligation during T-APC coupling assists TCR-pMHC interactions. Branched actin network formation (brown), initiated from ligated LFA1 (purple), presses the cell surfaces together (1) to a length scale that allows pMHC (gold) binding to TCRs (blue). (*See facing page for legend.*)

risk that the TCR will release the pMHC ligand, and the T cell will fail to register the antigen hit. Conversely, the signalosome assembly mechanism needs to discriminate against TCRs interacting transiently with a vast array of pMHCs presenting nonagonist peptides. Viewed in this manner, a scheme that rapidly dissociates TCRs from MHCs loaded with endogenous peptide, freeing them to rebind and test other MHCs, is desirable. It is notable that several TCR signaling factors carry binding sites for actin binding proteins or actin itself (Rozdzial et al. 1995; Zhang et al. 1999; Zeng et al. 2003; Phee et al. 2005; Gomez et al. 2006). Through these actin-associated factors, agonist-triggered TCRs rapidly assemble stabilized signaling platforms that survive mechanical disruption.

In concert with adhesive integrin interactions and costimulatory receptor signaling, TCRs orchestrate a reorganization of the T-cell plasma membrane that may begin with a handful of receptors and eventually encompasses the entire contact face with the APC (some 50–100 μm^2). TCRs first aggregate into micron scale clusters of TCRs, then flow to the center of the contact face, generating the central supramolecular activating complex (cSMAC) of the immune synapse (Monks et al. 1998; Grakoui et al. 1999; Krummel et al. 2000). Underscoring the importance of the cytoskeleton, the actin depolymerizing toxins latrunculin A and cytochalasin D are potent inhibitors of T-cell activation and block both TCR microcluster formation and cSMAC coalescence (Wulfing et al. 1998; Grakoui et al. 1999; Krummel et al. 2000; Varma et al. 2006). Ultimately, it is the coordination of the local interactions between receptors and effectors with the cell morphological level rearrangements that determines the nature and magnitude of T-cell responses to pathogens. Regulation of TCR signaling lifecycles and T-cell responses, therefore, falls squarely on the actin cytoskeleton.

TRIGGERING

TCR triggering occurs on interstitially motile T cells undergoing constant morphological remodeling and crawling at speeds of approximately 10 $\mu m/min$ (Miller et al. 2002). In vitro, the actinomyosin cytoskeleton reacts to local integrin levels to allow T cells to adopt one of two crawling modes (Jacobelli et al. 2009). In the faster "walking" mode, T cells create relatively small contacts to grab surfaces and bound from one contact to the next (Fig. 1A). In a mesenchymal mode, T cells place large portions of their membrane into contact with the surfaces they are surveying. These large, contiguous contacts translate as the T cell moves forward, giving rise to a "moving walkway" of T-cell membrane that enables extensive scanning of the underlying surface. Although these modes are only presumed to occur in vivo, recent evidence indicates that T cells employ a "millipede" type motility to move along the blood vessel endothelium that is similar to the walking mode seen in vitro (Shulman et al. 2009). The need for the fast mode may be most pronounced when navigating between loose adherence zones on follicular reticular cell fibers in the lymph nodes, which T cells appear to use to help guide their circuit from afferent to efferent vessels (Bajenoff et al. 2008). Furthermore, transitions between modes are hinted at by the variety of cell shapes and speeds observed within the T-cell zone (Miller et al. 2002). The actin cytoskeletal mechanics of these motility modes are an essential parameter for TCR triggering, because TCR-pMHC ligation logically

Figure 1. (*Continued*) By pressing the cell membranes into close proximity, actin polymerization drives bulky, inhibitory phosphatases, such as CD45 (green), from the zone of pMHC-TCR ligation (2). (*C*) T-cell motility places continuous stress of molecules interaction across the T cell-APC junction. In the receptor deformation model, actin cytoskeletal movements introduce shear force that can break TCR-pMHC interactions. When a perpendicular force is applied (1), TCRs bound to pMHCs that are not bearing the T cells agonist ligand (*left*) rupture, freeing the TCR to scan another pMHC. When a TCR binds a pMHC bearing the agonist ligand (purple, at *right*), shear force (2) induces a conformational change, triggering the receptor and recruitment of signalosome factors (yellow).

requires tight juxtaposition between domains on the T-cell surface that bear TCRs and membrane regions on the APC that display agonist pMHC complexes. By establishing the rate of T-cell scanning across APC membranes and the extent of the T cell-APC contact area, the actin cytoskeleton dictates the time window in which a TCR will bind a specific pMHC complex.

In addition to establishing the transience of membrane-membrane interactions, the cytoskeleton also facilitates T cells' exceptional responsiveness to TCR engagement. From the host perspective, it is undesirable for T cells to fail to activate following encounters with antigens, no matter how limiting the dose. TCRs, therefore, must be very sensitive—and there is no doubt that they are sensitive: CD8-positive T cells are potentially capable of responding to a single agonist peptide (Sykulev et al. 1996), whereas 1–10 pMHCs can activate CD4 T cells thanks to their coreceptor (Irvine et al. 2002). At a naive glance, the ideal mechanism to achieve such sensitivity might seem obvious: A receptor should combine a very fast association rate with a very slow dissociation rate for agonist ligands. Studying the measured binding and unbinding kinetics of some model TCRs quickly shows that not to be the case for TCRs (Davis et al. 2003). It appears that T cells instead take advantage of rapid unbinding of pMHCs to sequentially trigger multiple TCRs, (Valitutti et al. 1995; Lanzavecchia et al. 1999). In effect, the successive ligation of TCRs serves as a gain, turning each pMHC into many activated receptors. The development of mathematical models has helped to explain how T cells have adapted to TCR biochemical properties to achieve both fidelity and sensitivity (McKeithan 1995; Coombs et al. 2002; Wedagedera and Burroughs 2006).

But what sort of biomechanical processes can we imagine that would facilitate this mechanism of triggering TCRs? Conformational change models of TCR triggering fell into disfavor because of an apparent lack of structural data to support large scale structural alterations following pMHC binding (Ding et al. 1999; Willcox et al. 1999; Degano et al. 2000). However, evidence for rearrangements between members of the TCR complex or within various TCR subunits has been described (Krogsgaard et al. 2003; Kim et al. 2009). These studies provide clues to how information about the TCR binding partner is transmitted across the T-cell membrane, but don't immediately address how the TCR translates this into a signaling output. Under the kinetic-segregation model, T cells use the intercellular spacing to exclude inhibitory molecules from the domain of triggered TCRs (van der Merwe et al. 2000). By creating zones of close contact in which TCRs can bind to pMHCs, integral membrane phosphatases with large ectodomains that antagonize TCR phosphorylation can be excluded. Specificity is achieved because relatively weak interactions, such as between TCRs and self-peptide MHCs, dissociate more rapidly, and the phosphorylated TCRs diffuse out of the close contact zone. This allows inhibitory phosphatase activities to deactivate the released TCRs. In contrast, relatively stable TCR-pMHC interactions would tether the TCR in the zone of tight contact longer, providing a window to assemble an activated, phosphatase-resistant signalosome. Although lipid rafts might explain how TCRs and other prosignaling factors remain tethered within the area of close contact, it is not clear how the weak adhesive strength from a limited number of pMHC-TCR interactions could maintain a closely spaced junction between the two cellular membranes at the size scale required for the kinetic-segregation model (Burroughs et al. 2006). To this end, mechanical pressure exerted by the actin cytoskeleton of the leading edge may be instrumental in pressing the apposed cell membranes together, squeezing proteins with larger ectodomains away from the tightest contact (Fig. 1B). Consistent with this idea, increasing the length of the pMHC extracellular domain impacted TCR segregation from CD45 and antagonized receptor triggering (Choudhuri et al. 2005). Kinetic-segregation might partially explain why integrin-mediated adhesion enhances TCR signaling even on bilayer systems (Grakoui et al. 1999; Porter et al. 2002; Suzuki et al. 2007), as the longer LFA1-ICAM pairs would establish zones of looser contact to accommodate bulky phosphatases.

It is important to note that one of the phosphatases associated with Lck inhibition and TCR deactivation, CD45, also appears to positively regulate Lck by dephosphorylating an inhibitory phosphotyrosine of Lck (Biffen et al. 1994; Stone et al. 1997). As a result, it would appear that CD45 must have access to TCR signalosomes near the time of triggering; otherwise, phosphorylation of the Lck inhibitory site could inactivate Lck and cause a triggering failure. Balancing the activating and inhibitory potential of CD45, then, may come down to timing access and exclusion through actin-mediated force generation. Following dephosphorylation of the inhibitory site of Lck, CD45 might be rapidly driven from the zone of close contact as integrin-mediated actin polymerization presses the cell membranes together, preventing it from dephosphorylating the active site phosphotyrosine of Lck (Thomas and Brown 1999). Spatiotemporal analysis of CD45 at T-APC interfaces indicated that CD45 was colocalized with Lck in young cSMACs and segregated from Lck at more mature interfaces (Freiberg et al. 2002). So far, though, only limited evidence for the sort of microscale exclusion of CD45 envisioned by the kinetic-segregation model has been presented (Choudhuri et al. 2005; Choudhuri et al. 2009). Furthermore, the importance of the extracellular domain to signaling is not obvious, and, in fact, smaller extracellular domains of CD45 appear to positively correlate with coreceptor association and antigen receptor signaling (Leitenberg et al. 1996; Shenoi et al. 1999; Trowbridge and Thomas 2003). Therefore, a variation on this apposition-induced triggering mechanism could be that the force pressing the membranes together induces the bending of TCR complexes in TCR-pMHC pairs, revealing sites on CD3 for phosphorylation. Weaker, non-agonist pMHC-TCR interactions would release before phosphorylation sites expose, allowing the TCR to relax into the untriggered, resting conformation.

Ma and others recently proposed a very intriguing modified conformational change model (Ma et al. 2008) hypothesizing that actin cytoskeleton generated shear force parallel to the synapse, rather than a force perpendicular to TCRs, facilitates receptor triggering. In this receptor deformation model, they argue that the stress on TCR-pMHC pairs induced by spreading over an APC essentially analyzes TCR-pMHCs interactions: Strong agonists provide a sufficiently high resistance to bond rupture to induce a deformation of the TCR upon actin cytoskeletal induced stress. That deformation is interpreted as an authentic antigen encounter only if the deformation is large enough or sufficiently persistent to lead to signalosome assembly on a deformed TCR (Fig. 1C). This hypothesis is particularly interesting when viewed in the context of a mixture of integrin and TCR signaling. As the T cells crawl over an APC, LFA-1 stimulation will be temporally coincident with initial TCR-pMHC ligation. The background of integrin signaling can facilitate dynamic actin cytoskeletal polymerization that will generate the shear force needed to test recently bound TCRs. As a result, this model lends itself to a very elegant mechanical explanation of why integrin cosignaling is required to sustain TCR signaling—beyond the adhesion strength supplied to hold the T-APC couple together, it drives the shear induced bond rupture testing with actin remodeling (Fig. 1C). This fits nicely with observations from Takashi Saito's group. They found that LFA1 signaling induced an actin cloud at the center of contact sites between T cells and APCs in the absence of antigen (Suzuki et al. 2007). The formation of this cloud lowered the threshold for T-cell activation after subsequent antigen exposure.

It is not clear if the sheer-induced TCR-pMHC bond testing would require the assembly of a nascent signalosome with signaling factors providing specific binding sites for actin filaments. The cytoskeleton could exert force on a TCR without a specific molecular coupling linking receptors to filaments through general steric interactions. By first assembling a nascent actin-linked signalosome around the receptor, though, the cytoskeleton could apply force to the TCR-pMHC bond no matter what direction the filament moved relative to the TCR. If the receptor deformation testing mechanism is engaged after the assembly of an initial signalosome, a secondary proof-reading mechanism is needed to explain how the signalosomes of TCRs bound to weak agonists, which fail the sheer test, can be reverted. SHP-1 phosphatase regulates a negative feedback loop that deactivates TCR signaling initiated by weak agonists, whereas ERK can stabilize Lck modification and stabilize signalosomes (Stefanova et al. 2003). Combining this observation with the receptor deformation model, ERK may be recruited specifically to those receptors that have undergone deformation, exposing a binding site for ERK or an ERK-binding intermediate.

This model can also explain why soluble multivalent, but not monomeric, pMHCs can stimulate T cells (Boniface et al. 1998; Cochran et al. 2000), which is sometimes taken as evidence for a multimerization model of TCR triggering. For single TCRs ligated to soluble pMHCs, the added mass of the bound pMHC is unlikely to distort the TCR conformation under a pulling force applied by actin. When two pMHCs are bound to nearby TCRs, though, small deviations in the directions or magnitudes of the forces applied to the TCRs could induce conformation changes in one or both receptors. This model might even be able to explain the "pseudodimer" mechanism (Krogsgaard et al. 2005), if the added stability of the low-affinity member of the pseudodimer can increase the rupture strength of the high-affinity counterpart just enough to induce a conformational change.

An intriguing recent observation indicates that the interaction of untriggered TCRs with the actin cytoskeleton increases in response to calcium release (Dushek et al. 2008). In T cells stimulated to release calcium with ionomycin, rather than antigenic stimulation of TCRs, F-actin increased and TCR mobility decreased. Although the decrease in TCR mobility could be described by diffusion trapping, the authors argue that a model combining diffusion trapping with receptor-actin binding better explained the observed changes in TCR mobility. Even if we dismiss the actin binding component, diffusion trapping alone of untriggered TCRs has significant implications for the biology TCRs. Migrating T cells are most sensitive to TCR triggering at the leading edge (Negulescu et al. 1996), site of the filamentous actin-rich lamellipod and lamella (Pollard and Borisy 2003). Trapping TCRs might prepare them for signalosome assembly by placing them in proximity to a number of effectors of TCR signaling, or facilitate formation of the cytoskeletal linkages that will test pMHC interactions through the receptor deformation mechanism. It would appear, then, that TCR triggering mechanisms are adapted to achieve maximum sensitivity within the micro-environmental context of the actin cytoskeleton of motile T cells. This suggests an additional benefit for scanning T cells to continuously slide over APC surfaces—it maintains the dynamic actin meshwork that facilitates TCR triggering. Following triggering, it would be expected that TCR signalosome and microcluster life cycles are likewise coordinated with the morphological rearrangements of continuous motility.

SCAFFOLDING

With the cytoskeletal forces of motility potentially reducing the lifetime of TCR-pMHC interactions on crawling T cells to a few seconds or even less (Huppa et al. 2010), a system that facilitates the rapid assembly of a signaling complex around a triggered TCR can greatly enhance the sensitivity of TCRs. For technical reasons, imaging TCR signaling dynamics proves to be much easier on stimulatory surfaces, such as antiCD3ϵ-coated glass or ICAM-1/pMHC laden bilayers. When T cells contacting these surfaces are imaged, TCR microclusters appear very rapidly. Within seconds, a number of effectors are recruited to TCRs, including Lck, ZAP-70, LAT, and SLP-76 (Bunnell et al. 2002; Ehrlich et al. 2002). In a similar kinetic window, cytoskeletal remodeling factors, such as Vav1, WASP, Cdc42, and Rac, colocalize with triggered receptors (Barda-Saad et al. 2005; Tskvitaria-Fuller et al. 2006). By structurally integrating with receptor signalosomes, the actin cytoskeleton serves as a relatively stable scaffold, which can rapidly localize many signaling proteins to receptors (Fig. 2A). Additionally, in linking the

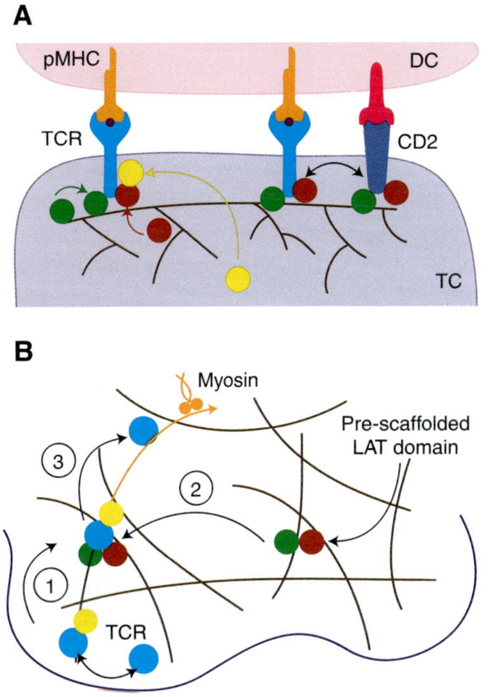

Figure 2. Actin meshwork scaffolding stabilizes signaling clusters and promotes rapid signalosome assembly. (*A*) Association of TCRs with actin filaments holds receptors in place following triggering, and places the receptors in proximity to actin-associated effectors that can be rapidly recruited to stabilize the triggered TCR on the cytoskeleton (*left*, red and green). This can increase the lifetime of intermolecular interactions to allow diffusion of soluble effectors (yellow) to signalosomes and microcluster coalescence. Additionally, costimulatory receptors signal through many of the same effectors as TCRs. CD2 has been observed to colocalize with TCRs following stimulation, and can assemble TCR signaling factors into a signalosome resembling a TCR signalosome even before TCR triggering. By sharing effectors, association of CD2 and TCR microdomains allows rapid coalescence of a signalosome around a triggered TCR. (*B*) TCRs transiently associate with the actin meshwork, either directly or through an intermediate binding partner (yellow). Functional signalosomes can be quickly assembled following receptor triggering by concatenating actin-dependent TCR membrane microdomains (1) with prescaffolded, LAT-enriched membrane microdomains (2). To terminate signaling, TCRs can be separated from signalosome factors by myosin based actin filament movement (in gold) that pulls TCRs from the signaling supportive periphery into the center of the immune synapse (3).

signaling complexes to the relatively stable actin filaments, the cytoskeleton allows signalosomes to survive pMHC unbinding, facilitating T-cell activation by even a small number of agonists. This allows a single pMHC to serially trigger a number of TCRs (Valitutti et al. 1995), without "resetting" the stimulus at each point. Again, this can help T cells to discriminate strong and weak agonists: Strong agonists ligate TCRs long enough to facilitate an actin-linked signaling assembly and once stabilized, the signaling cluster can survive pMHC release, at least temporarily.

The guanine nucleotide exchange factor Vav1 is well-established as a regulator of TCR signaling (Wu et al. 1995; Costello et al. 1999). As an activator of Rho-family GTPases, Vav1 can activate actin cytoskeletal remodeling through a number of pathways and can mediate indirect association with actin filaments through several binding partners. A few of those interactions in particular, though, help to explain how Vav1 quickly integrates TCRs into the cytoskeleton network and catalyzes signalosome assembly. Although Vav proteins typically have a predominantly cytosolic distribution, a fraction of Vav1 was found to be constitutively phosphorylated and associated with the CD3ζ chain of the TCR complex in resting cells (Valensin et al. 2002). This result was obtained using cell lines, so the constitutively phosphorylated Vav1 was possibly an artifact of a disregulated signaling pathway. TCR signaling, though, is initiated in the context of costimulatory receptor and integrin ligation, which might preactivate a pool of Vav1 that can associate with CD3ζ. Salojin et al. reported that CD28 costimulation increased Zap-70-mediated phosphorylation of Vav1, and increased LAT-Vav1 interactions (Salojin et al. 1999). CD28 and TCR cosegregation into signaling clusters would give TCRs access to costimulatory receptor activated Vav1 proteins (Yokosuka et al. 2008). Through its association with ZAP-70, then, Vav1 is tied to one of the earliest factors recruited to TCR signalosomes (Salojin et al. 1999; Bunnell et al. 2002; Yokosuka et al. 2005).

Although it is intimately associated with actin remodeling, Vav1 does not directly bind actin. A recently discovered interaction in T cells, though, explains how Vav1 can load TCRs onto actin filaments (Gomez et al. 2006). Phosphorylation of HS1 by Lck and ZAP-70 stabilizes actin filaments at the synapse and helps maintain Vav localization to synapses. By mediating a linkage between Vav1 and actin, HS1 can immobilize a multimolecular complex consisting of the TCR with CD3ζ, ZAP-70, Lck, and Vav1 on an actin filament. As noted above, the signaling fate of this complex may be controlled by conformational deformation induced when the associated microfilament shifts and stresses the TCR-pMHC bond. Somewhat surprisingly, it was recently reported that the GEF activity of Vav1 is not necessary for the transduction of TCR signaling (Miletic et al. 2009). This suggests that Vav1 mediated scaffolding onto the actin cytoskeleton is more important for initiating TCR signaling than the cytoskeletal rearrangements induced by Vav1 downstream of its exchange activity (Miletic et al. 2006).

Prescaffolding signaling molecules into membrane microdomains maintained by the actin cytoskeleton can also increase the rate of signalosome activation following TCR triggering. This can reduce signalosome assembly to a few steps, rather than a sequential integration sequence of diffusing factors. Additionally, holding signaling factors in domains exclusive from TCRs provides a safety mechanism to prevent inappropriate signaling complex formation. Work by Mark Davis' and Bridget Wilson's research groups revealed islands of membrane-associated proteins that depended on the integrity of the actin cytoskeleton (Lillemeier et al. 2006). Recently published experiments following up on this approach showed that TCRs and the critical T-cell signaling linker LAT exist in separate membrane-associated protein islands that aggregate, but do not quite mix, following TCR stimulation (Lillemeier et al. 2010). Concatenation of these domains generated functional TCR signaling complexes. This type of segregation might also account for nonmixing populations of CD3 and Lck revealed previously by single-molecule imaging (Douglass and Vale 2005). The fusion of two actin-associated domains, then, allows the rapid colocalization of the factors needed to construct a complete signalosome, in contrast to a serial diffusion-based mechanism (Fig. 2B). Costimulation receptors, not just TCRs, use actin-mediated signalosome assembly to bring together signaling factors, allowing receptors to cooperatively assemble signaling complexes. Even in the absence of TCR triggering, the CD2 coreceptor generated actin-scaffolded microdomains of TCR signaling factors that included Lck, LAT, and the TCR CD3ζ chain (Kaizuka et al. 2009). Fusion of these LAT-bearing microdomains with a triggered TCR microdomain would immediately generate a functional TCR signalosome, allowing CD2 to "prefabricate" a TCR signalosome on an actin filament for a triggered receptor (Fig. 2A).

Although the stabilization provided by actin scaffolding is critical to the assembly of TCR signalosomes and microcluster coalescence (Campi et al. 2005), at some point, microclusters of TCRs become auto-stable. Microclusters formed in T cells before treatment with Latrunculin A persist for minutes after addition of the toxin (Varma et al. 2006). The auto-stabilized components of the clusters do not appear to include all the factors necessary for signaling, though, as the Latrunculin challenge terminated calcium fluxes within a minute. This indicates that the scaffolding provided by the actin cytoskeleton is required to maintain the association between TCRs and signaling factors throughout the signaling life cycle of the receptors. The collection of TCRs into the auto-stabilized fraction of the microclusters has implications for their life cycle on entering cSMACs. TCRs are held in the cSMAC, even in the absence of continued pMHC interactions, and despite the lack of an extensive actin filament network in the cSMAC (Barda-Saad et al. 2005; Varma et al. 2006). This is probably because receptors held in large aggregates cannot significantly diffuse across the barrier imposed by the dense actin meshwork of the peripheral SMAC (pSMAC). An interesting possibility is that TCR microcluster auto-stability serves as a mechanism to keep TCRs from reentering the signaling-conducive pSMAC region and reactivating, while integrins and other receptors that travel to the edge of the cSMAC through centripetal flow (discussed later) are presumably free to recycle and continue stimulating adhesion. By imposing a strict dependence on actin cytoskeletal scaffolding for signal propagation and confining extinguished TCRs in a zone that does not support signaling, the actin cytoskeleton establishes a ceiling for TCR signal generation. This may prevent run-away signal propagation and the induction of T-cell apoptosis in response to strong pathogens.

TRANSLOCATION

In T cells transferred onto activating lipid bilayers, the organization of receptors into micron-scale signaling clusters is followed within minutes by the radial translocation of clusters to the center of the contact site. Accumulation of individually trafficked clusters to the center leads to the coalescence of cSMACs, as observed in T-APC couples, and originally reported by Kupfer. As in T-APC couples, the LFA1-ICAM pair occupies the periphery of the interface (pSMAC), although it now appears that integrins also undergo centripetal streaming to some extent (Kaizuka et al. 2007). The functional consequences of the large-scale membrane reorganization initiated by TCR signaling now appear to be the termination of signaling and the internalization of receptors (Mossman et al. 2005; Varma et al. 2006). Despite this, the existence, let alone significance, of long-lasting immune synapses with clearly defined cSMACs and pSMACs in vivo is not clear.

In addition to scaffolding the assembly of signalosomes, the actin cytoskeleton orchestrates the sorting of different receptors into the pSMAC and cSMAC domains. At the periphery of the synapse, TCR microcluster centripetal flow is coupled to actin retrograde flow (Fig. 3A), although it is not exactly clear that this cytoskeletal force extends all the way to the cSMAC. The molecular interactions that facilitate this linkage are incompletely described, but it appears to be a frictional coupling mechanism that allows some slippage of microclusters against the actin meshwork (Kaizuka et al. 2007; DeMond et al. 2008). Using high-speed video microscopy to image the dynamics of TCRs and integrins during synapse formation on bilayers, Kaizuka et al. showed that TCRs and integrins formed spatially separate microcluster domains at the earliest observed time points (Kaizuka et al. 2007). Similar to TCRs, the assembly of integrin clusters required actin cytoskeletal integrity. However, unlike mature TCRs microclusters, which achieved a level of actin-independent stability, intregrins clusters disintegrated following latrunculin treatment. The authors provide additional insight by pointing out that although integrin clusters flowed centripetally, they did not enter the actin-depleted cSMAC.

The observation that both integrin and TCR signalosomes couple with retrograde actin flow poses an interesting problem for models of receptor sorting into the cSMAC and pSMAC domains. If both receptor signaling complexes flow inward, what mechanism allows the cell to distinguish and sort signalosomes so that integrin clusters do not reach the central zone but TCR clusters do? An absolute dependence on the actin cytoskeleton for stability explains why the integrin clusters disassemble at the cSMAC, but it does not indicate which property of a cluster confers stability. This issue is particularly troublesome if, indeed, coupling of signalosomes to the actin cytoskeleton is nonspecific. The work of Jay Groves' research group, which has provided extensive insight into the mechanics of cSMAC generation, indicates that the centripetal travel of a signaling cluster is related to its size (Hartman et al. 2009). By generating cross-linked dimers and tetramers of integrins, they show that increasing the size of integrin clusters increases the extent to which clusters centralize. Tetrameric integrins clusters were able to reach the center of interfaces, with the resulting integrin distribution appearing remarkably similar to TCRs in a cSMAC. The group went on to show that when centralizing receptors reached a barrier, the clusters sorted into a size hierarchy—the largest clusters (typically TCR clusters) pressed against the barrier and forced lower total molecular weight integrin clusters to take a more peripheral position (Fig. 3A). Similarly, tetrameric integrin clusters sorted inside dimeric integrin clusters.

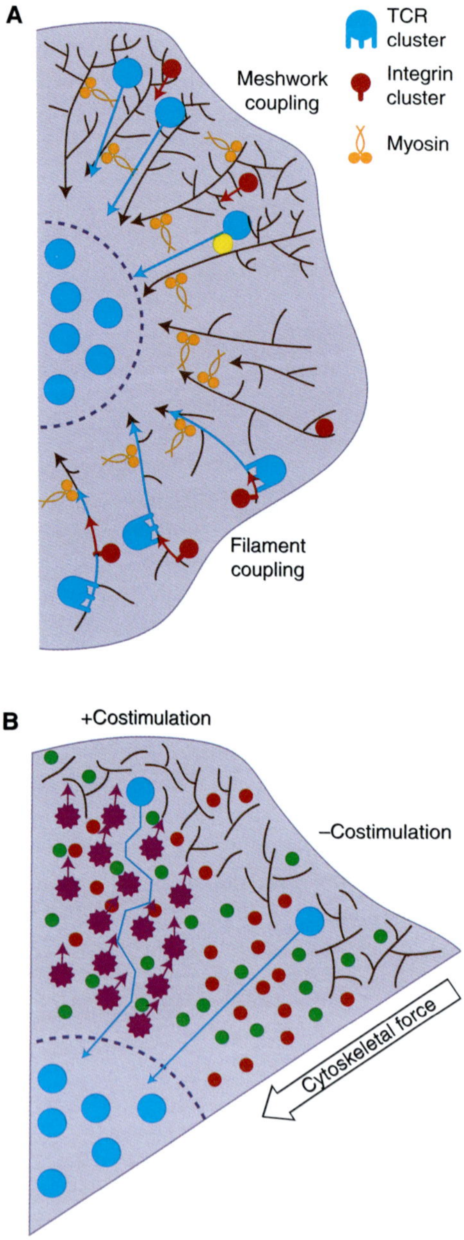

Figure 3. Friction applied by the actin meshwork regulates T-cell activation through actin retrograde flow. (*A*) TCR microclusters (cyan circles) stream centripetally toward the cSMAC boundary (indicated by the dashed purple semi-circle). Receptors couple to retrograde actin flow (brown arrows) generated by actin-associated myosin motor proteins (gold) and by actin polymerization at the periphery of the interface. Because of the density of the actin meshwork in the periphery, receptors need not travel along filaments, but rather can be forced inward nonspecifically (shown at *top*). Alternatively, differential binding to an actin-associated signaling factor (yellow) could determine the extent to which a receptor centralizes. A generic coupling mechanism is suggested by recent work that indicates that both LFA1 (red circles) and TCRs couple to the centripetal actin flow. The extent to which receptors centralize is controlled by the size of clusters, with larger TCR microclusters consistently sorting to the interior of smaller integrin microclusters (shown at *bottom*). In a "clutched" mechanism, the larger TCR microclusters (cyan) generate more links to actin filaments than smaller intergrin clusters (red). As a result, TCR signaling clusters show increased centralization. (*B*) Some costimulation receptors (purple) may function by increasing the time that TCR microclusters spend in the signaling factor enriched periphery of the synapse. By creating a series of frictional barriers (shown at *left*), ligated costimulation receptors generate forces (purple arrows) that counteract retrograde flow of the cytoskeletal meshwork, increasing the length of the path (cyan line) that microclusters follow to reach the cSMAC. This increases the residence time of the TCR microcluster in the signaling effector enriched pSMAC. In the absence of ligated coreceptors, TCRs quickly pass through the pSMAC, decreasing their residence time in the signal generating region of the synapse (shown at *right*).

This size-dependent sorting could explain how T cells modulate TCR signaling to achieve generally comparable outcomes in response to low doses or high doses of antigen: Once clusters reach a certain size (a certain number of TCRs are triggered), a signaling threshold/integral is reached and the clusters flow to the actin-poor central region to be down-regulated. On encountering an APC presenting agonist ligands at low density, microclusters would accumulate triggered TCRs slowly and spend more time in the periphery in which TCRs can interact with signaling molecules (Varma et al. 2006; Nguyen et al. 2008). Conversely, high concentrations of agonist peptides on an APC surface would trigger more TCRs, leading to increased rates of TCR microcluster growth, increased centralization rates and more rapid attenuation of signaling. Incorporating work from Stephen Bunnell's group, costimulatory receptor effects can be integrated into this model. Nguyen et al. reported that ligation of the VLA-4 integrin reduced the actin-dependent rate of microcluster centralization (Nguyen et al. 2008). They found that this VLA4-mediated effect increased the time that SLP76 remained associated with LAT in active peripheral microclusters (Fig. 3B). Therefore, a major contribution of this kind of costimulation may be to modulate the rate of TCR centralization by increasing friction in the pSMAC, creating a temporary barrier in the periphery (DeMond et al. 2008).

What is not clear yet is the exact nature of the coupling of TCRs to the retrograde actin flow and the molecular motors activities that force receptors to the cSMAC. Myosin generated contractions could pull filaments toward the cSMAC, with TCR microclusters forming stable linkages to filaments. Myosin II was recently implicated in TCR microcluster translocation (Ilani et al. 2009). However, these experiments relied on imaging conditions that can induce blebbistatin crosslinking, the inhibitor used to block myosin II activity, and are at odds with earlier work that found no role for myosin II in synapse stabilization and cSMAC coalescence (Jacobelli et al. 2004). Other motors might also function in the pSMAC to generate flow, but little is known about the distribution of these other myosins in immune synapses. Alternatively, actin polymerization can also generate retrograde flow independently of myosin motor activity (Henson et al. 1999). This could generate centripetal flow of receptors by pushing the actin meshwork and associated receptors away from the cell edge.

As a final note, it should be mentioned that the actin cytoskeleton has roles in immune synapse formation and function beyond regulating the signaling life cycles of TCRs. Centrosome polarization toward the cSMAC depends on the actin cytoskeleton and is critical for directional secretion of granules across the cell-cell interface (Stinchcombe et al. 2006). This process appears to be mediated by formin family actin nucleating factors using aspects of the actin cytoskeleton that are distinct from TCR induced F-actin structures (Gomez et al. 2007). Very recent work indicates that agonist potency and the kinetics of TCR signaling control granule recruitment and cytotoxic T lymphocyte killing activity (Beal et al. 2009; Jenkins et al. 2009). The actin cytoskeleton, therefore, can also form a structural framework to coordinate TCR signaling life cycles with effector T-cell functions.

REFERENCES

Bajenoff M, Glaichenhaus N, Germain RN. 2008. Fibroblastic reticular cells guide T lymphocyte entry into and migration within the splenic T cell zone. *J Immunol* **181:** 3947–3954.

Barda-Saad M, Braiman A, Titerence R, Bunnell SC, Barr VA, Samelson LE. 2005. Dynamic molecular interactions linking the T cell antigen receptor to the actin cytoskeleton. *Nat Immunol* **6:** 80–89.

Beal AM, Anikeeva N, Varma R, Cameron TO, Vasiliver-Shamis G, Norris PJ, Dustin ML, Sykulev Y. 2009. Kinetics of early T cell receptor signaling regulate the pathway of lytic granule delivery to the secretory domain. *Immunity* **31:** 632–642.

Biffen M, McMichael-Phillips D, Larson T, Venkitaraman A, Alexander D. 1994. The CD45 tyrosine phosphatase regulates specific pools of antigen receptor-associated p59fyn and CD4-associated p56lck tyrosine in human T-cells. *EMBO J* **13:** 1920–1929.

Boniface JJ, Rabinowitz JD, Wülfing C, Hampl J, Reich Z, Altman JD, Kantor RM, Beeson C, McConnell HM, Davis MM. 1998. Initiation of signal transduction through the T cell receptor requires the multivalent engagement of peptide/MHC ligands. *Immunity* **9:** 459–466.

Bunnell SC, Hong DI, Kardon JR, Yamazaki T, McGlade CJ, Barr VA, Samelson LE. 2002. T cell receptor ligation induces the formation of dynamically regulated signaling assemblies. *J Cell Biol* **158:** 1263–1275.

Burroughs NJ, Lazic Z, van der Merwe PA. 2006. Ligand detection and discrimination by spatial relocalization: A kinase-phosphatase segregation model of TCR activation. *Biophys J* **91:** 1619–1629.

Campi G, Varma R, Dustin ML. 2005. Actin and agonist MHC-peptide complex-dependent T cell receptor microclusters as scaffolds for signaling. *J Exp Med* **202:** 1031–1036.

Choudhuri K, Parker M, Milicic A, Cole DK, Shaw MK, Sewell AK, Stewart-Jones G, Dong T, Gould KG, van der Merwe PA. 2009. Peptide-major histocompatibility complex dimensions control proximal kinase-phosphatase balance during T cell activation. *J Biol Chem* **284:** 26096–26105.

Choudhuri K, Wiseman D, Brown MH, Gould K, van der Merwe PA. 2005. T-cell receptor triggering is critically dependent on the dimensions of its peptide-MHC ligand. *Nature* **436:** 578–582.

Cochran JR, Cameron TO, Stern LJ. 2000. The relationship of MHC-peptide binding and T cell activation probed using chemically defined MHC class II oligomers. *Immunity* **12:** 241–250.

Coombs D, Kalergis AM, Nathenson SG, Wofsy C, Goldstein B. 2002. Activated TCRs remain marked for internalization after dissociation from pMHC. *Nat Immunol* **3:** 926–931.

Costello PS, Walters AE, Mee PJ, Turner M, Reynolds LF, Prisco A, Sarner N, Zamoyska R, Tybulewicz VLJ. 1999. The Rho-family GTP exchange factor Vav is a critical transducer of T cell receptor signals to the calcium, ERK, and NF-kappa B pathways. *Proc Natl Acad Sci* **96:** 3035–3040.

Davis MM, Krogsgaard M, Huppa JB, Sumen C, Purbhoo MA, Irvine DJ, Wu LC, Ehrlich L. 2003. Dynamics of cell surface molecules during T cell recognition. *Ann Rev Biochem* **72:** 717–742.

Degano M, Garcia KC, Apostolopoulos V, Rudolph MG, Teyton L, Wilson IA. 2000. A functional hot spot for antigen recognition in a superagonist TCR/MHC complex. *Immunity* **12:** 251–261.

DeMond AL, Mossman KD, Starr T, Dustin ML, Groves JT. 2008. T cell receptor microcluster transport through molecular mazes reveals mechanism of translocation. *Biophys J* **94:** 3286–3292.

Ding Y-H, Baker BM, Garboczi DN, Biddison WE, Wiley DC. 1999. Four A6-TCR/Peptide/HLA-A2 structures that generate very different T cell signals are nearly identical. *Immunity* **11:** 45–56.

Douglass AD, Vale RD. 2005. Single-molecule microscopy reveals plasma membrane microdomains created by protein-protein networks that exclude or trap signaling molecules in T cells. *Cell* **121:** 937–950.

Dushek O, Mueller S, Soubies S, Depoil D, Caramalho I, Coombs D, Valitutti S. 2008. Effects of intracellular calcium and actin cytoskeleton on TCR mobility measured by fluorescence recovery. *PLoS ONE* **3:** e3913.

Ehrlich LIR, Ebert PJR, Krummel MF, Weiss A, Davis MM. 2002. Dynamics of p56lck translocation to the T cell immunological synapse following agonist and antagonist stimulation. *Immunity* **17:** 809–822.

Freiberg BA, Kupfer H, Maslanik W, Delli J, Kappler J, Zaller DM, Kupfer A. 2002. Staging and resetting T cell activation in SMACs. *Nat Immunol* **3:** 911–917.

Gomez TS, Kumar K, Medeiros RB, Shimizu Y, Leibson PJ, Billadeau DD. 2007. Formins Regulate the actin-related protein 2/3 complex-independent polarization of the centrosome to the immunological synapse. *Immunity* **26:** 177–190.

Gomez TS, McCarney SD, Carrizosa E, Labno CM, Comiskey EO, Nolz JC, Zhu P, Freedman BD, Clark MR, Rawlings DJ, et al. 2006. HS1 Functions as an Essential Actin-Regulatory Adaptor Protein at the Immune Synapse. *Immunity* **24:** 741–752.

Grakoui A, Bromley SK, Sumen C, Davis MM, Shaw AS, Allen PM, Dustin ML. 1999. The immunological synapse: a molecular machine controlling T cell activation. *Science* **285:** 221–227.

Gunzer M, Schäfer A, Borgmann S, Grabbe S, Zänker KS, Bröcker E-B, Kämpgen E, Friedl P. 2000. Antigen presentation in extracellular matrix: Interactions of T cells with dendritic cells are dynamic, short lived, and sequential. *Immunity* **13:** 323–332.

Hartman NaC, Nye JA, Groves JT. 2009. Cluster size regulates protein sorting in the immunological synapse. *Proc Natl Acad Sci* **106:** 12729–12734.

Henson JH, Svitkina TM, Burns AR, Hughes HE, MacPartland KJ, Nazarian R, Borisy GG. 1999. Two components of actin-based retrograde flow in sea urchin coelomocytes. *Mol Biol Cell* **10:** 4075–4090.

Huppa JB, Axmann M, Mortelmaier MA, Lillemeier BF, Newell EW, Brameshuber M, Klein LO, Schutz GJ, Davis MM. 2010. TCR-peptide-MHC interactions in situ show accelerated kinetics and increased affinity. *Nature* **463:** 963–967.

Ilani T, Vasiliver-Shamis G, Vardhana S, Bretscher A, Dustin ML. 2009. T cell antigen receptor signaling and immunological synapse stability require myosin IIA. *Nat Immunol* **10:** 531–539.

Irvine DJ, Purbhoo MA, Krogsgaard M, Davis MM. 2002. Direct observation of ligand recognition by T cells. *Nature* **419:** 845–849.

Jacobelli J, Bennett FC, Pandurangi P, Tooley AJ, Krummel MF. 2009. Myosin-IIA and ICAM-1 regulate the interchange between two distinct modes of T cell migration. *J Immunol* **182:** 2041–2050.

Jacobelli J, Chmura SA, Buxton DB, Davis MM, Krummel MF. 2004. A single class II myosin modulates T cell motility and stopping, but not synapse formation. *Nat Immunol* **5:** 531–538.

Jenkins MR, Tsun A, Stinchcombe JC, Griffiths GM. 2009. The strength of T cell receptor signal controls the polarization of cytotoxic machinery to the immunological synapse. *Immunity* **31:** 621–631.

Kaizuka Y, Douglass AD, Vardhana S, Dustin ML, Vale RD. 2009. The coreceptor CD2 uses plasma membrane microdomains to transduce signals in T cells. *J Cell Biol* **185:** 521–534.

Kaizuka Y, Douglass AD, Varma R, Dustin ML, Vale RD. 2007. Mechanisms for segregating T cell receptor and adhesion molecules during immunological synapse formation in Jurkat T cells. *Proc Natl Acad Sci* **104:** 20296–20301.

Kim ST, Takeuchi K, Sun Z-YJ, Touma M, Castro CE, Fahmy A, Lang MJ, Wagner G, Reinherz EL. 2009. The αβ T cell receptor is an anisotropic mechanosensor. *J Biol Chem* **284:** 31028–31037.

Krogsgaard M, Li Q-j, Sumen C, Huppa JB, Huse M, Davis MM. 2005. Agonist/endogenous peptide-MHC heterodimers drive T cell activation and sensitivity. *Nature* **434:** 238–243.

Krogsgaard M, Prado N, Adams EJ, He X-l, Chow D-C, Wilson DB, Garcia KC, Davis MM. 2003. Evidence that structural rearrangements and/or flexibility during TCR binding can contribute to T cell activation. *Molecular Cell* **12:** 1367–1378.

Krummel MF, Sjaastad MD, Wulfing C, Davis MM. 2000. Differential clustering of CD4 and CD3zeta during T cell recognition. *Science* **289:** 1349–1352.

Lanzavecchia A, Iezzi G, Viola A. 1999. From TCR engagement to T cell activation: a kinetic view of T cell behavior. *cell* **96:** 1–4.

Leitenberg D, Novak T, Farber D, Smith B, Bottomly K. 1996. The extracellular domain of CD45 controls association with the CD4-T cell receptor complex and the response to antigen-specific stimulation. *J Exp Med* **183:** 249–259.

Lillemeier BF, Mortelmaier MA, Forstner MB, Huppa JB, Groves JT, Davis MM. 2010. TCR and Lat are expressed on separate protein islands on T cell membranes and concatenate during activation. *Nat Immunol* **11:** 90–96.

Lillemeier BF, Pfeiffer JR, Surviladze Z, Wilson BS, Davis MM. 2006. Plasma membrane-associated proteins are clustered into islands attached to the cytoskeleton. *Proc Natl Acad Sci* **103:** 18992–18997.

Ma Z, Sharp KA, Janmey PA, Finkel TH. 2008. Surface-anchored monomeric agonist pMHCs alone trigger TCR with high sensitivity. *PLoS Biology* **6:** e43.

McKeithan TW. 1995. Kinetic proofreading in T-cell receptor signal transduction. *Proc Natl Acad Sci* **92:** 5042–5046.

Mempel TR, Henrickson SE, von Andrian UH. 2004. T-cell priming by dendriticcells in lymph nodes occurs in three distinct phases. *Nature* **427:** 154–159.

Miletic AV, Graham DB, Sakata-Sogawa K, Hiroshima M, Hamann MJ, Cemerski S, Kloeppel T, Billadeau DD, Kanagawa O, Tokunaga M, et al. 2009. Vav links the T cell antigen receptor to the actin cytoskeleton and T cell activation independently of intrinsic guanine nucleotide exchange activity. *PLoS ONE* **4:** e6599.

Miletic AV, Sakata-Sogawa K, Hiroshima M, Hamann MJ, Gomez TS, Ota N, Kloeppel T, Kanagawa O, Tokunaga M, Billadeau DD, et al. 2006. Vav1 acidic region tyrosine 174 is required for the formation of T cell receptor-induced microclusters and is essential in T cell development and activation. *J Biol Chem* **281:** 38257–38265.

Miller MJ, Wei SH, Parker I, Cahalan MD. 2002. Two-photon imaging of lymphocyte motility and antigen response in intact lymph node. *Science* **296:** 1869–1873.

Monks CRF, Freiberg BA, Kupfer H, Sciaky N, Kupfer A. 1998. Three-dimensional segregation of supramolecular activation clusters in T cells. *Nature* **395:** 82–86.

Mossman KD, Campi G, Groves JT, Dustin ML. 2005. Altered TCR signaling from geometrically repatterned immunological synapses. *Science* **310:** 1191–1193.

Negulescu PA, Krasieva TB, Khan A, Kerschbaum HH, Cahalan MD. 1996. Polarity of T cell shape, motility, and sensitivity to antigen. *Immunity* **4:** 421–430.

Nguyen K, Sylvain NR, Bunnell SC. 2008. T cell costimulation via the integrin VLA-4 inhibits the actin-dependent centralization of signaling microclusters containing the adaptor SLP-76. *Immunity* **28:** 810–821.

Phee H, Abraham RT, Weiss A. 2005. Dynamic recruitment of PAK1 to the immunological synapse is mediated by PIX independently of SLP-76 and Vav1. *Nat Immunol* **6:** 608–617.

Pollard TD, Borisy GG. 2003. Cellular motility driven by assembly and disassembly of actin filaments. *Cell* **112:** 453–465.

Porter JC, Bracke M, Smith A, Davies D, Hogg N. 2002. Signaling through integrin LFA-1 leads to filamentous actin polymerization and remodeling, resulting in enhanced T cell adhesion. *J Immunol* **168:** 6330–6335.

Rozdzial MM, Malissen B, Finkel TH. 1995. Tyrosine-phosphorylated T cell receptor zeta chain associates with the actin cytoskeleton upon activation of mature T lymphocytes. *Immunity* **3:** 623–633.

Salojin KV, Zhang J, Delovitch TL. 1999. TCR and CD28 are coupled via ZAP-70 to the activation of the Vav/Rac-1-/PAK-1/p38 MAPK signaling pathway. *J Immunol* **163:** 844–853.

Shenoi H, Seavitt J, Zheleznyak A, Thomas ML, Brown EJ. 1999. Regulation of integrin-mediated T cell adhesion by the transmembrane protein tyrosine phosphatase CD45. *J Immunol* **162:** 7120–7127.

Shulman Z, Shinder V, Klein E, Grabovsky V, Yeger O, Geron E, Montresor A, Bolomini-Vittori M, Feigelson SW, Kirchhausen T, et al. 2009. Lymphocyte crawling and transendothelial migration require chemokine triggering of high-affinity LFA-1 integrin. *Immunity* **30:** 384–396.

Stefanova I, Hemmer B, Vergelli M, Martin R, Biddison WE, Germain RN. 2003. TCR ligand discrimination is enforced by competing ERK positive and SHP-1 negative feedback pathways. *Nat Immunol* **4:** 248–254.

Stinchcombe JC, Majorovits E, Bossi G, Fuller S, Griffiths GM. 2006. Centrosome polarization delivers secretory granules to the immunological synapse. *Nature* **443:** 462–465.

Stone J, Conroy L, Byth K, Hederer R, Howlett S, Takemoto Y, Holmes N, Alexander D. 1997. Aberrant TCR-mediated signaling in CD45-null thymocytes involves dysfunctional regulation of Lck, Fyn, TCR-zeta, and ZAP-70. *J Immunol* **158:** 5773–5782.

Suzuki J-i, Yamasaki S, Wu J, Koretzky GA, Saito T. 2007. The actin cloud induced by LFA-1-mediated outside-in signals lowers the threshold for T-cell activation. *Blood* **109:** 168–175.

Sykulev Y, Joo M, Vturina I, Tsomides TJ, Eisen HN. 1996. Evidence that a single peptide MHC complex on a target cell can elicit a cytolytic T cell response. *Immunity* **4:** 565–571.

Thomas ML, Brown EJ. 1999. Positive and negative regulation of Src-family membrane kinases by CD45. *Immunol Today* **20:** 406–411.

Trowbridge IS, Thomas ML. 2003. CD45: An emerging role as a protein tyrosine phosphatase required for lymphocyte activation and development. *Ann Rev Immunol* **12:** 85–116.

Tskvitaria-Fuller I, Seth A, Mistry N, Gu H, Rosen MK, Wulfing C. 2006. Specific patterns of Cdc42 activity are related to distinct elements of T cell polarization. *J Immunol* **177:** 1708–1720.

Valensin S, Paccani Silvia R, Ulivieri C, Mercati D, Pacini S, Patrussi L, Hirst T, Lupetti P, Baldari Cosima T. 2002. F-actin dynamics control segregation of the TCR signaling cascade to clustered lipid rafts. *Eur J Immunol* **32:** 435–446.

Valitutti S, Muller S, Cella M, Padovan E, Lanzavecchia A. 1995. Serial triggering of many T-cell receptors by a few peptideÂ–MHC complexes. *Nature* **375:** 148–151.

van der Merwe PA, Davis SJ, Shaw AS, Dustin ML. 2000. Cytoskeletal polarization and redistribution of cell-surface molecules during T cell antigen recognition. *Seminars in Immunology* **12:** 5–21.

Varma R, Campi G, Yokosuka T, Saito T, Dustin ML. 2006. T cell receptor-proximal signals are sustained in peripheral microclusters and terminated in the central supramolecular activation cluster. *Immunity* **25:** 117–127.

Wedagedera JR, Burroughs NJ. 2006. T-cell activation: A queuing theory analysis at low agonist density. *Biophys J* **91:** 1604–1618.

Willcox BE, Gao GF, Wyer JR, Ladbury JE, Bell JI, Jakobsen BK, van der Merwe PA. 1999. TCR Binding to peptide-MHC stabilizes a flexible recognition interface. *Immunity* **10:** 357–365.

Wu J, Katzav S, Weiss A. 1995. A functional T-cell receptor signaling pathway is required for p95vav activity. *Mol Cell Biol* **15:** 4337–4346.

Wulfing C, Sjaastad MD, Davis MM. 1998. Visualizing the dynamics of T cell activation: Intracellular adhesion molecule 1 migrates rapidly to the T cell/B cell interface and acts to sustain calcium levels. *Proc Natl Acad Sci* **95:** 6302–6307.

Yokosuka T, Kobayashi W, Sakata-Sogawa K, Takamatsu M, Hashimoto-Tane A, Dustin ML, Tokunaga M, Saito T. 2008. Spatiotemporal regulation of T cell costimulation by TCR-CD28 microclusters and protein kinase C, Translocation. *Immunity* **29:** 589–601.

Yokosuka T, Sakata-Sogawa K, Kobayashi W, Hiroshima M, Hashimoto-Tane A, Tokunaga M, Dustin ML, Saito T. 2005. Newly generated T cell receptor microclusters initiate and sustain T cell activation by recruitment of Zap70 and SLP-76. *Nat Immunol* **6:** 1253–1262.

Zeng R, Cannon JL, Abraham RT, Way M, Billadeau DD, Bubeck-Wardenberg J, Burkhardt JK. 2003. SLP-76 Coordinates Nck-dependent Wiskott-Aldrich syndrome protein recruitment with Vav-1/Cdc42-Dependent Wiskott-Aldrich syndrome protein activation at the T Cell-APC contact site. *J Immunol* **171:** 1360–1368.

Zhang J, Shehabeldin A, da Cruz LAG, Butler J, Somani A-K, McGavin M, Kozieradzki I, dos Santos AO, Nagy A, Grinstein S, et al. 1999. Antigen receptor-induced activation and cytoskeletal rearrangement are impaired in Wiskott-Aldrich syndrome protein-deficient lymphocytes. *J Exp Med* **190:** 1329–1342.

Understanding the Structure and Function of the Immunological Synapse

Michael L. Dustin[1], Arup K. Chakraborty[2], and Andrey S. Shaw[3]

[1]Program in Molecular Pathogenesis, Skirball Institute of Biomolecular Medicine and Department of Pathology, New York University, New York, New York 10016

[2]Department of Chemical Engineering, Department of Chemistry, Division of Biological Engineering, Massachusetts Institute of Technology, Cambridge, Massachusetts 02139

[3]Department of Pathology and Immunology and Howard Hughes Medical Institute, Washington University School of Medicine, St. Louis, Missouri 63110

Correspondence: shaw@pathology.wustl.edu

The immunological synapse has been an area of very active scientific interest over the last decade. Surprisingly, much about the synapse remains unknown or is controversial. Here we review some of these current issues in the field: how the synapse is defined, its potential role in T-cell function, and our current understanding about how the synapse is formed.

T cells are activated when they recognize peptide-MHC complexes on the surface of antigen presenting cells (APC) (Babbitt et al. 1985). But the exact process regarding how antigenic pMHC complexes are recognized and transduced into signals is still incompletely understood. Naïve T cells enter secondary lymphoid organs such as the lymph node and scan dendritic cells for the presence of rare specific pMHC complexes (Miller et al. 2004). After recognizing less than 10 specific pMHC complexes, naïve T cells maintain long contacts (6–18 h) with dendritic cells before being committed to enter cell cycle and differentiate into effector T cells (Iezzi et al. 1998; Irvine et al. 2002; Mempel et al. 2004).

The immunological synapse (IS) refers to the organization of membrane proteins that occurs at the interface between the T cell and the APC during these long contacts and also during the effector phase (Grakoui et al. 1999; Monks et al. 1998). Interest in studying the IS stems from ideas that the supramolecular structures that form at the IS underlies the high sensitivity of T cell recognition and that understanding these structures will lead to better insights into how antigen recognition leads to the decision of a T cell to proliferate, differentiate, and function.

Springer first put forward the concept that receptors would segregate laterally during cell interactions (Springer 1990). Subsequently, Kupfer was the first to show that proteins in the contact area between a T cell and APC segregate laterally (Monks et al. 1998). Specifically, he noted that the integrin, LFA-1, became concentrated in an outer ring, known as the peripheral supramolecular activation complex (pSMAC) and the TCR became concentrated in the center,

 doi: 10.1101/cshperspect.a002311
Cite this article as *Cold Spring Harb Perspect Biol* 2010;2:a002311

in a zone known as the central supramolecular activation complex (cSMAC) (Monks et al. 1998)(Fig. 1). We showed that CD2 could segregate from LFA-1 and concentrate in the center of a hybrid cell-planar bilayer junction and suggested that these patterns and those described by Monks et al. (1998) provided evidence for the previously hypothesized immunological synapse (Dustin et al. 1998; Norcross 1984). The function of this receptor segregation is still not completely understood but it was initially hypothesized that formation of this pattern might be related to T-cell activation and constitute a "molecular machine" that would be formed in response to the presence of antigenic ligand and that this "molecular machine" might function to sustain signaling for long periods of time and direct subsequent T-cell differentiation (Grakoui et al. 1999).

FUNCTIONS OF THE IMMUNOLOGICAL SYNAPSE

Over the last 12 years, intensive research into the function of the IS implicates the IS in two major categories of functions: 1) priming of responses in which the T cell is receiving information and 2) effector functions in which the T cell is sending information. For example, the pSMAC has been suggested to function for cytolytic CD8+ cells as a sealing ring, preventing the leakage of cytolytic granule contents to bystander cells (Stinchcombe et al. 2006). Much work also implicates the cSMAC as a site for receptor internalization and degradation, which is important for regulation of longer-term responses of the T cell (Lee et al. 2002; Lee et al. 2003; Varma et al. 2006). There is evidence that the ubiquitination machinery is concentrated in the cSMAC and that formation of the cSMAC is required for receptor degradation (Lee et al. 2003; Lee et al. 2002; Wiedemann et al. 2005). Lastly, the application of highly sensitive microscopy techniques has revealed that sustained signaling in a mature IS is mediated by TCRs in small clusters of 5–30 molecules and that these clusters move centripetally toward the cSMAC (Varma et al. 2006; Yokosuka et al. 2005). In some cases the signaling elements assembled by the TCR display centripetal motion after separating from the TCR (Bunnell et al. 2002).

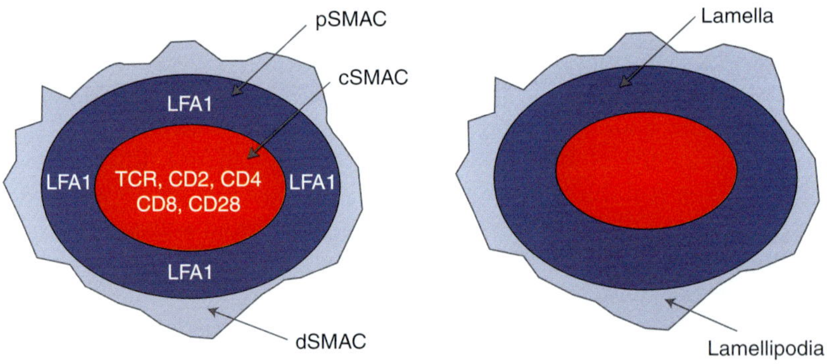

Figure 1. Structure of the immunological synapse. The basic structure of the "organized" immunological synapse with SMACs is shown (*left*). In the center is the central supramolecular activation complex or cSMAC, which contains receptors like the TCR, CD28, CD4, CD8, and CD2. Newer studies suggest that the cSMAC may be divided into an outer area containing CD28 and an inner area containing the TCR (not shown). The ring that surrounds the cSMAC is called the peripheral supramolecular activation complex or pSMAC. This domain is mainly populated by the integrin molecule LFA-1. Outside of the pSMAC is another domain known as the distal supramolecular activation complex. Originally the dSMAC was thought not be important and contain all of the molecules that are not specifically recruited to the cSMAC or pSMAC but it is increasingly becoming appreciated that the dSMAC is an area of active membrane movement. This suggests that the pSMAC and dSMAC may be analogous to the actin structures known as the lamellae and lamellipodia, respectively (*right*).

There are many situations in which mature ISs with pSMACs and cSMACs are not observed in vitro and it remains an open question how common these structures are in vivo (Barcia et al. 2006; Brossard et al. 2005; McGavern et al. 2002; Purtic et al. 2005; Reichert et al. 2001; Tseng et al. 2008). Mature IS formation is most readily observed with B cell tumors, resting B cells, and supported planar bilayers and is dependent on antigen quality (Grakoui et al. 1999; Monks et al. 1998). Furthermore, the size of the cSMAC is linearly dependent on antigen quantity in the supported planar bilayer system in which formation can be most sensitively tracked (Grakoui et al. 1999; Varma et al. 2006). The frequency of mature IS visualized during priming of naïve T cells is typically less than 50% at any time, but this may, in part, reflect the periodic symmetry breaking and migration that is interspersed with stable IS formation (Sims et al. 2007). Reflecting its potential importance in cell killing, effector CD8 T cells form synapses very efficiently with ~80% forming mature synapses on lipid bilayers (Beal et al. 2008). The applicability of lessons from the mature IS models (e.g., B = cell tumors and planar bilayers) to situations in which such highly ordered structures do not appear to form (e.g., priming of naïve T cells by DC) is a matter of controversy.

In this review, we examine current thoughts about how synapses form and the function of the IS. Advanced imaging methods have led to highly detailed images that have given new insight into how synapses form and what the basic parameters are in vivo. Computational approaches have helped us to understand what is happening at the IS and what effects synapse formation will have on T-cell activation events.

GETTING LYMPHOCYTES TO STOP!

Lymphocytes are constantly moving through the body in search of antigen. When they finally recognize antigen, an important step is they must stop and form a stable contact with the antigen-presenting cell. The antigen-induced up-regulation of LFA-1 binding to ICAM-1 to generate a strong adhesive surface is a critical step in slowing cell motility (Dustin and Springer 1989; Dustin et al. 1997; Scholer et al. 2008). Although this is probably the first step in the process of immunological synapse formation, adhesion strengthening itself is not sufficient.

Early experiments using different in vitro models made strikingly different predictions about what might occur in vivo. Methods in which T cells encountered antigen in the absence of extracellular matrix showed antigen stop signals (Dustin et al. 1997; Grakoui et al. 1999; Negulescu et al. 1996), whereas methods in which T cells encounter dendritic cells in collagen gels showed continual motility (Gunzer et al. 2000). This controversy was resolved by in vivo imaging that consistently showed T cell arrest with antigen presenting dendritic cells, but also showed that the time from first antigen recognition to arrest was variable and related to the quantity of antigen present; more antigen resulted in the faster initiation of stable contacts (Henrickson et al. 2008; Mempel et al. 2004). The duration of the stable T-DC interactions in vivo is on the order of 1 d and activation of T cells may involve multiple cycles of stable in teractions during the proliferative phase of the response (Celli et al. 2007).

Signaling through the TCR stimulates calcium entry and calcium is likely to play a key role in mediating the stop signal (Skokos et al. 2007). Analysis of T-DC interactions in mice lacking ICAM-1 showed that the LFA-1-ICAM-1 interaction is required for stable interactions and development of memory responses but is not required for cytokine production or proliferation of CD8 T cells (Scholer et al. 2008). This suggests that stable T-DC interactions may be important to induce the formation of immunological memory, but not required for other early aspects of T-cell activation.

Front and Back: Repolarizing the Cell

In addition to increased adhesion, the repolarization of the T cell is a critical event in generation of the immunological synapse. The first type of polarization occurs when the cells exit the blood and migrate into tissue. Chemokines are thought to be the initial trigger to prepare the cell for extravasation. Chemokine signaling through G-protein coupled receptors results in localized PIP3 production and subsequent polarization of the actin filament system (Hirsch et al. 2000; Li et al. 2000; Niggli and Keller 1997). Chemokines can be present in soluble gradients, but in lymphoid tissues, the chemokine is thought to decorate the surface of the reticular fiber network. Polarization of the migrating lymphocyte results in a protrusive leading forward edge, or lamellipodium, that is highly sensitive to the detection of antigen and a bulbous appendage at the rear, the uropod. (Sanchez-Madrid and Serrador 2009).

POLARITY AND CELL MIGRATION

Lymphocyte migration can be mediated by the formation of integrin-mediated adhesions (focal contacts) at the leading edge, which are translocated rearward resulting in forward motility (Smith et al. 2005). In fibroblasts and epithelial cells, the focal contacts move rearward and mature into larger structures called focal adhesions (Vogel and Sheetz 2006). In most cells, the highest concentration of focal adhesions are associated with the cellular actin structure known as the lamella, a structure found just behind the lamellopodia (Ponti et al. 2004; Ponti et al. 2005). Although the term focal adhesions is not widely used by T-cell biologists, integrin clusters are abundant within the pSMAC, a structure potentially analogous to the lamellae (Grakoui et al. 1999) (Fig. 1). The uropod contains the microtubule-organizing center (MTOC), and is also enriched in myosin IIA (Jacobelli et al. 2004; Morin et al. 2008). Myosin mediated contraction of the uropod is important for de-adhesion during motility (Jay et al. 1995). The positioning of the MTOC in the rear of the cell in lymphocytes is unusual because studies of most migrating cells, which are often much larger than lymphocytes, usually results in the positioning of the MTOC just behind the leading edge (Anderson et al. 1982; Gotlieb et al. 1981). Because the MTOC is the cytoplasmic center of the cell, its positioning to the rear of migrating lymphocytes may reflect packaging efficiency in the smaller cell and allow the lymphocyte extra flexibility in squeezing through endothelium and tightly packed tissue spaces.

POLARITY AND ANTIGEN RECOGNITION

The MTOC has long been used as a marker of changes in T-cell polarization in response to antigen stimulation. Its movement to the contact is thought to play a critical role in allowing directed secretion of cytotoxic granules by killer lymphocytes (Beal et al. 2009; Geiger et al. 1982; Kupfer et al. 1983; Stinchcombe et al. 2006). After a stable contact is formed, the uropod disassembles and the MTOC moves to a new position between the nucleus and the contact surface (Kupfer et al. 1983; Stinchcombe et al. 2006). With the MTOC in this position, the axis of polarity is now toward the antigen presenting cell surface, with the motility apparatus configured to sustain contact over many minutes to hours.

The exact mechanism of how the MTOC is reoriented is unclear but recent data suggest that one important factor is di-acyl-glycerol (DAG) (Quann et al. 2009). DAG is generated during the hydrolysis of PIP2 by the activation of phospholipase C. Because DAG is confined to the plasma membrane, and has a relatively short half-life, the localized production of DAG at the IS marks the position of activated receptors at the contact membrane. Localization of PLC-γ to the IS allows for the local production of DAG and presumably the recruitment of proteins with DAG binding C1 domains. In fact, MTOC reorientation can be triggered solely by acute generation of DAG at one pole of a T cell (Quann et al. 2009).

Movement of microtubules generally requires microtubule motors. Microtubules have a fast growing "plus" end and a slower growing "minus" end. The minus end is associated with the MTOC and the plus end is pointed at the plasma membrane. Two groups have shown that the minus end motor protein, dynein, is recruited to the immunological synapse and is required for MTOC repolarization (Combs et al. 2006; Martin-Cofreces et al. 2008). Dynein may be recruited to the IS by binding to the scaffold protein ADAP, suggesting that dynein recruitment may be linked to the activation of integrins (Combs et al. 2006; Peterson et al. 2001). Dynein may induce the reorientation of the MTOC by pulling microtubules and the MTOC toward the immunological synapse. The proximity of the MTOC to the contact surface may be critical as Griffiths and coworkers have shown that in CTLs, direct contact of the MTOC with the IS is critical for cytolytic granule release because the process appears to only depend on +end-directed motors (Stinchcombe et al. 2006).

PAR PROTEINS AND THE GENETICS OF POLARITY

The partition defective or Par proteins were originally identified based on their role in asymmetric cell division of the *Caenorhabditis elegans* oocyte (Kemphues et al. 1988). More recently, they have been shown to play important roles in directing the polarity of epithelial cells (Joberty et al. 2000; Lin et al. 2000). The role of Par proteins in regulating polarity in T lymphocytes is largely unknown. There are six distinct Par proteins spanning four structural families. Two of the proteins, Par1 and Par4 are serine/threonine kinases. Par3 and Par6 are PDZ domain containing adapter proteins. Par2 is a RING finger protein and Par 5 is a 14-3-3 protein. Another serine-threonine kinase, an atypical protein kinase C, PKC-ζ, interacts with the Par proteins and plays an important role in polarity.

Typically, the Par3 and Par6s form a complex with PKC-ζ. In the *C. elegans* oocyte, Par3/6/PKC-ζ are associated with the anterior portion of the cell whereas Par1 localizes to the posterior pole (Kemphues et al. 1988). In epithelial cells, the Par3/6/PKC-ζ complex is associated with apical junctional complexes (Joberty et al. 2000; Lin et al. 2000), whereas Par1 is associated with the basolateral surface (Bohm et al. 1997). The mutually exclusive localization of the Par3 complex and Par1 is thought to be mediated by PKC-ζ phosphorylation of Par1 resulting in 14-3-3 binding and subsequent removal of Par1 from Par3/6/PKC-ζ containing membranes (Suzuki et al. 2004).

In T cells, the role of the Par proteins is less clear. Par3 moves to the synapse whereas PKC-ζ apparently localizes to the pole of the cell distal to the synapse (Ludford-Menting et al. 2005). Par1 appears to relocate to the synapse where it is phosphorylated by PKC-ζ and resides in the cytoplasm below the synapse (Lin et al. 2009). Par1 appears to be important for polarization because expression of a dominant negative form of Par1 can block MTOC polarization (Lin et al. 2009). Interestingly, knockout mice lacking one of the Par1 isoforms, Par1b (EMK) develop autoimmunity suggesting that Par1 and cell polarity may play an important role in immunoregulation (Hurov et al. 2001).

Another important class of polarity proteins are PDZ containing adapter proteins known as Scribble (Scrb), Disc Large (Dlg), and Lethal Giant Larva (Lgl). In epithelial cells, these three molecules appear to interact with each other, are required for apical/basal polarity, and are localized to the basal/lateral membrane (Qin et al. 2005). Early in synapse formation, Scrb and Dlg are recruited to the synapse but at later time-points, they end up localized to the membrane distal to the synapse (Ludford-Menting et al. 2005). Inhibition of Scrb expression blocks motility and conjugate formation but the mechanism is not known.

Forming the Bullseye

The bullseye pattern with the integrins, forming the outer ring surrounding the TCR that is concentrated in a central spot, is a highly recognized feature of the T-cell immunological synapse (Grakoui et al. 1999; Monks et al. 1998). Although this pattern does not appear to be a universal feature of immunological synapses (Brossard et al. 2005; Tseng et al. 2008), its distinct structure has served as a central focus of researchers in the field.

Recent progress in understanding the formation of the immunological synapse has been facilitated by the use of artificial planar lipid bilayers as well as a method of imaging known as total interference reflection microscopy (TIRFM) (Campi et al. 2005; Yokosuka et al. 2005). The incorporation of mobile ligands into lipid bilayers has allowed the reorganization of surface receptors during T-cell activation to be visualized in real-time and with optimal resolution (Grakoui et al. 1999). In contrast, the reorganization of TCRs and integrins into pSMACs and cSMACs does not occur in systems using antibody coated glass (Bunnell et al. 2002; Douglass and Vale 2005). In addition, the highly sensitive TIRFM method allows individual molecules or small clusters of molecules to be imaged (Varma et al. 2006; Yokosuka et al. 2005). These studies show that new engagement of antigen by the TCR first takes place in an area outside of the pSMAC, known as the distal-SMAC or dSMAC.

MICROCLUSTERS IN THE SYNAPSE

Antigen induces the formation of small TCR microclusters on the order of 11–17 TCRs per microcluster (Varma et al. 2006). Although the TCR microclusters can also include molecules like CD2, CD4, CD8, and CD28, they exclude the tyrosine phosphatase CD45 (Varma et al. 2006; Yokosuka et al. 2008), a step that may be important for TCR triggering (Choudhuri et al. 2005). Interestingly, LFA-1 also forms microclusters in the dSMAC that are segregated from the TCR microclusters (Kaizuka et al. 2007). Although both TCR and LFA-1 microclusters then move centripetally through the pSMAC (Kaizuka et al. 2007), experiments using barriers placed in the bilayer (Mossman et al. 2005) show that the TCR microclusters are more stable than the more labile LFA-1 clusters. The TCR microclusters have a limited capacity to navigate around small obstacles in their path to the cSMAC, supporting a "frictional coupling" model between TCR and the actin cytoskeleton (DeMond et al. 2008). The TCR microclusters are then translocated into the actin poor cSMAC by a TSG101 dependent mechanism (Vardhana 2010), whereas free ICAM-1 in the bilayer appears to be poorly able to diffuse into the cSMAC, perhaps limiting accumulation of LFA-1-ICAM-1 complexes. It should be noted that TCR microclusters have yet to be visualized in the IS of bona fide T cell/APC conjugates.

The centripetal movement of TCR and LFA-1 microclusters is actin dependent and parallels the retrograde actin flow that occurs during cell spreading and migration (Kaizuka et al. 2007; Varma et al. 2006). At the extreme periphery of the contact area in the dSMAC, actin filaments grow rapidly because of the high concentration of Arp2/3 and cofilin, which increase the number of growing filaments (Sims et al. 2007). Behind the dSMAC, in the pSMAC, the actin filaments are anchored to clusters of integrins like LFA-1 through adapters such as talin (Monks et al. 1998). This anchoring impedes the free centripetal movement of filaments but contributes to the force of the rapidly growing actin filaments on the plasma membrane (Hu et al. 2007). In the pSMAC, the retrograde actin flow is linked to substrate attached integrins and the speed of cell movement is determined by the slippage between the integrin mediated anchors as well as the activity of myosin IIa, which contracts and integrates the actin network (Ilani et al. 2009; Jacobelli et al. 2004; Morin et al. 2008). The LFA1 ligand ICAM-1 can slide across the membrane of the antigen-presenting cell allowing for the formation of the rings of LFA-1 and ICAM-1 on the T cell and APC respectively (Carrasco et al. 2004).

Because of the absence of actin in the cSMAC, it was initially proposed that the ability of the TCR to accumulate in the cSMAC might be based on an unknown mechanism that was actin independent. Although the formation of both TCR and LFA-1 microclusters requires actin, TCR, and not LFA-1 microclusters become F-actin-independent suggesting a distinct transport mechanism (Bunnell et al. 2002; Kaizuka et al. 2007; Sims et al. 2007; Varma et al. 2006). In the case of the actin dependent LFA-1 clusters, one idea is that they disperse when they reach F-actin free cSMAC (Kaizuka et al. 2007). This model postulates that the cSMAC is a container that collects cargo that falls off the actinomyosin conveyor belt as it reaches the border of the pSMAC and the cSMAC. In this model, because TCR clusters are actin independent, they can persist after falling into the cSMAC. This model is supported by recent evidence showing that LFA-1 cross-linking antibodies promote the transport of LFA-1-ICAM-1 complexes into the cSMAC (Hartman et al. 2009). Although the authors interpreted this result in terms of cluster size and frictional coupling to centripetally moving F-actin, it is also possible that the cross-linking antibodies stabilize the LFA-1 clusters and allows the clusters to persist in the F-actin poor cSMAC.

It is also possible to explain the inability of LFA-1 clusters to penetrate the border of the cSMAC and the pSMAC based on the size mismatches between the TCR and LFA-1 (Springer, 1990). The TCR bound to the MHC is estimated to span a distance of about 15 nm, whereas the distance spanned by the LFA-1/ICAM pair is estimated at over 40 nm (Springer 1990). Consistent with this, LFA-1 and TCR microclusters are segregated laterally as soon as they form and the accumulation of TCR microclusters in cSMAC might form a tight surface that is not penetrable by the LFA-1/ICAM-1 pair (Kaizuka et al. 2007; Qi et al. 2001; Shaw and Dustin 1997).

Studies of the localization of PKC-θ (Monks et al. 1998) and CD28 in the synapse suggest that the cSMAC may have two distinct compartments. CD28 and PKC-θ are strongly colocalized and high-resolution imaging shows lateral segregation from most of the TCR accumulated in the core of the cSMAC (Yokosuka et al. 2008). This result splits the cSMAC into two distinct compartments. The localization of a signaling molecule in a distinct outer zone suggests that the cSMAC may be divided into two functional domains, with a signaling component localized in the area just inside the pSMAC (Yokosuka et al. 2008). These dynamic central compartments may account for signaling activity detected in the cSMAC in CD4 and CD8 T cells (Beal et al. 2009; Cemerski et al. 2008; Jenkins et al. 2009) and is discussed in more detail in the next section of the article.

Function of the Synapse

The function of the immunological synapse has been a controversial area and confused by different definitions of the immunological synapse (Davis and Dustin 2004). If the term is meant to define stable contact surface between T cell and APC (Dustin et al. 1998; Mempel et al. 2004), then the question is what is the role of sustained contact between these two cells. For others, the term refers to the reorganization of the contact area into the pSMAC and the cSMAC (Monks et al. 1998). In this case, the question refers mainly to the function of the cSMAC versus the pSMAC Lastly, the term immunological synapse sometimes refers to the contact surface between a CTL and a target cell (Stinchcombe et al. 2006). In this case, the issue is what the role of the pSMAC and cSMAC are in cytolytic killing. Regardless, each of these areas has been an active area of research.

THE SYNAPSE AND STABLE CONTACTS

The requirement for stable contacts for T-cell activation is an issue that has been largely resolved over the last 5 yr, but the biological implications remain controversial. Lanzavecchia initially proposed that naïve T cells require many hours of stable contact before becoming committed to divide (Iezzi et al. 1998). Continuous imaging of T cells interacting with dendritic cells in a three-dimensional collagen matrix showed that T cells can, however, proliferate after many transient interactions (Gunzer et al. 2000). In vivo imaging showed that T cells transition from transient interactions to stable contacts that persists for many hours (Mempel et al. 2004; Shakhar et al. 2005). Experiments with ICAM-1 deficient mice and delivery of antigen to all dendritic cells confirmed that stable contacts are not required for T-cell proliferation and early cytokine release, but suggested that they are required for T-cell memory (Scholer et al. 2008). It remains possible that stable contacts between T cells and DC will be more important when antigen is presented on more physiological subpopulations of dendritic cells (Dustin, 2008).

THE FUNCTION OF THE pSMAC AND cSMAC

What is the role of the cSMAC and pSMAC? Because the cSMAC is notable for high concentrations of TCRs, it was initially postulated to function as a site of TCR signaling by facilitating sustained recognition of low-affinity ligands for the TCR (Grakoui et al. 1999; Monks et al. 1998). But it is clear now that formation of the cSMAC is not necessary to initiate signaling, rather, as discussed previously, TCRs form small microclusters in the pSMAC that are associated with the initiation of signaling (Campi et al. 2005; Yokosuka et al. 2005). These studies also suggest that continuous microcluster formation is required for sustained signaling because microcluster signaling appears to attenuate as the microclusters reach the cSMAC (Campi et al. 2005; Yokosuka et al. 2005). One possible function of the cSMAC is that it may function to terminate signaling.

Initially, because TCRs are concentrated in the cSMAC, it was assumed that signaling by the TCR would occur mainly in the cSMAC (Grakoui et al. 1999; Monks et al. 1998). Staining the IS with phosphotyrosine antibodies suggested surprisingly that signaling was low in the cSMAC compared to the pSMAC (Lee et al. 2002; Lee et al. 2003). Further experimentation showed that this issue was complicated as the intensity of phosphotyrosine staining varied depending on the quality of the agonist. Strong peptides tended to result in less phosphotyrosine staining in the cSMAC whereas weak peptides tended to show more phosphotyrosine staining in the cSMAC (Cemerski et al. 2008). These experiments are complicated by the fact that the efficiency of cSMAC formation is also related to agonist quality, with stronger peptides generating cSMACs with better efficiency (Grakoui et al. 1999; Monks et al. 1998).

Because the eventual fate of many TCRs that are triggered by agonist ligands is degradation in lysosomal compartments (Valitutti et al. 1997), one interpretation is that the cSMAC functions as a specific compartment to facilitate ubiquitination and degradation of the TCR (Lee et al. 2003; Lee et al. 2002). Strong antigenic peptides would both enhance the formation of the cSMAC and degradation of the T-cell receptor resulting in lower phosphotyrosine detection in the cSMAC. Because weak antigenic peptides might have less capacity to stimulate receptor degradation, their stimulation results in more persistent phosphotyrosine detection in the cSMAC. The observation that the cSMAC is rich in lysobisphosphatidic acid (Varma et al. 2006), a marker of multivesicular bodies, supports a role of the cSMAC in receptor degradation for strong peptides and further suggests that the cSMAC may function actively in the sorting process leading to lysosomal targeting of membrane proteins (Williams and Urbe 2007).

Combining a specific role for the cSMAC in receptor degradation with the microcluster data suggests that continuous and sustained signaling might require the continuous formation of new TCR microclusters in the periphery of the contact (Varma et al. 2006; Yokosuka et al. 2005). This model requires that the APC also have a continuous supply of antigenic pMHC complexes available at the periphery of the contact site for many hours. This model is consistent with much of the existing data with but also with the caveat that most of this data was generated with TIRFM on artificial lipid bilayers. If the model is correct, one important question to be resolved is why signaling occurs in the peripheral microclusters and not in the TCRs clustered in the cSMAC.

This model is, however, not consistent with the presence of phosphotyrosine staining in the cSMAC seen with weak agonist stimulation. One possible explanation is the recent data suggesting that the cSMAC has at least two functional domains. The finding that CD28 and PKC-θ are strongly colocalized and segregated from most of the TCR accumulated in the core of the cSMAC (Yokosuka et al. 2008). This result splits the cSMAC into two distinct compartments, one that is more dynamic and may remain active in signaling. Similar structures were detected in multifocal T cell-DC IS, in which CD80 and PKC-θ rich compartments were seen to be continuously dependent on dynamic TCR signaling (Tseng et al. 2008). These dynamic central compartments could account for signaling activity detected in the cSMAC in CD4 and CD8 T cells (Cemerski et al. 2008; Jenkins et al. 2009).

COMPUTATIONAL APPROACHES TO UNDERSTAND SYNAPSE FUNCTION

Elucidating the signaling function of the synapse, and the cSMAC in particular, has proven to be difficult because there are many competing effects and components at play, which makes it difficult to intuit mechanisms from experimental observations of a few variables. Computational and theoretical models have been applied to complement experimental studies to help unravel the interplay of complex-competing effects in various aspects of T-cell signaling (Altan-Bonnet and Germain 2005; Cemerski et al. 2007; Chakraborty and Das 2010; Das et al. 2009; Goldstein et al. 2002; Lee et al. 2003). This is advantageous because computational models can keep track of the outcomes of different mechanistic hypotheses for each signaling component and determine whether each hypothesis yields results consistent with experimental observations. Because of the complex nature of the interactions, some hypotheses that seem plausible intuitively may be inconsistent with experimental observations, and these can be ruled out by computational modeling. The computational models can also be used to design experiments that can sensitively discriminate between competing plausible hypotheses, which are consistent with all known experimental facts. Such complementary computational and experimental investigations have aided the quest to understand the formation of and signaling in the immunological synapse.

Shortly after the first report of the low levels of active signaling molecules in the cSMAC (Lee et al. 2002), a computational model was formulated (Lee et al. 2003) to try and shed light on the mechanistic origin of this observation. The computational model placed the TCR and pMHC molecules on two apposed surfaces, and simulated their binding to each other when they were within a certain distance of each other. A threshold level of signaling was used to initiate forces that led to cSMAC formation. The model also simulated the phosphorylation and dephosphorylation of the ITAMs as a function of the binding kinetics of the TCR and pMHC, allowing for fully or partially phosphorylated ITAMs. As the model was developed, it incorporated aspects of ubiquitination and receptor down-regulation and degradation (Liu et al. 2000; Naramura et al. 2002). Because biochemical reactions are stochastic events, the computational models explicitly considered these stochastic fluctuations. Importantly, the formation of the cSMAC could be turned on or off at will in the simulations, thus allowing one to assess the potential role of cSMAC formation on TCR signaling under different conditions.

Computational studies indicated that the overall amount of signaling as well as its ability to be detected could be affected by cSMAC formation. The model suggested that for strong agonists, the long half-life of the TCR/pMHC complex would allow for efficient TCR triggering before being transported to the cSMAC, consistent with what was subsequently seen in the TCR microcluster studies (Varma et al. 2006). The model also showed that the process of concentrating receptors in the center of the contact area (cSMAC) would promote receptor degradation by helping to concentrate both phosphorylated TCRs and ubiquitination agents in the same local area. The model, however, also warned experimentalists that low levels of signaling intermediates in the cSMAC might actually reflect rapid degradation of active signaling components and not low levels of signaling. Supporting this idea, experiments using agents to slow receptor degradation resulted in increased detection of phosphotyrosine in the cSMAC (Cemerski et al. 2008).

The computational studies also suggested other implications for the effects of cSMAC formation on signaling. The calculations predicted that assuming that the cSMAC could form, it could enhance signaling induced by weak ligands. Although weak ligands would result in only partially phosphorylated receptors prior to cSMAC formation, concentrating them in the cSMAC along with other signaling components would promote more complete phosphorylation because the higher concentration of receptors and ligands would enhance receptor occupancy. Only subsequently, would the cSMAC enhance degradation. Using manuipulations that enhance cSMAC formation, it was shown that cSMAC formation could enhance signaling by weak ligands (Cemerski et al. 2007). Although the efficiency of cSMAC formation is related to antigen quality, suggesting that cSMACs do not form in the presence of weak ligands, evidence suggests that nonTCR stimuli can allow for cSMAC formation in the absence of antigen. For example, the cytokine IL-12 and the expression of stress activated ligands for NKG2D can lower the threshold for cSMAC formation. (Markiewicz et al. 2005; Somersalo et al. 2004). It therefore remains a possibility that the cSMAC could serve to enhance signaling by weak ligands.

In spite of these insights, it should be noted that the computational models have not yet incorporated some of the new details that are emerging from the high-resolution TIRF microscopy studies. For example, the different compartments within the cSMAC have not yet been considered in the computational models. Combining models for microcluster formation, transport and SMAC formation explicitly with models for signaling will provide a pathway for quantitatively testing our understanding of how these events are coordinated. Toward this end, adapting and further advancing models of the coarse-grained structures that form during immunological synapse formation may be useful (Burroughs and Wulfing 2002; Qi et al. 2001; Weikl and Lipowsky 2004). The ability of such physical models to relate patterns to functional outcomes likely point to important physical processes linking membrane bending fluctuations, chemical kinetics, molecular topology and cytoskeletal transport, processes also likely to operate at the level of microcluster formation. Because the rules for microcluster formation are poorly understood simulations should play an important role in testing hypotheses.

THE IMMUNOLOGICAL SYNAPSE AND THE KISS OF DEATH

Lastly, formation of the cSMAC and pSMAC may be important for cytolytic killing. In this case, it has been postulated that the pSMAC functions as a "sealing ring" to prevent the escape of cytolytic agents secreted into the cSMAC during cytolysis as well as allow for the polarized secretion of cytokines and cytolytic agents into the cSMAC. Griffiths and coworkers showed using fluorescence and electron microscopy that in the CTL synapse with target cell, the cSMAC is divided into two zones (Stinchcombe et al. 2001). One zone is characterized by the concentration of receptors previously shown to be in the cSMAC, whereas the other zone functions as a space for the release of cytolytic granule contents. This suggests that the formation of a specific zone that is actin free and within the actin-rich pSMAC may be important for cytolytic killing.

In a recent series of studies, Sykulev and colleagues quantified the contribution of the pSMAC and granule delivery to the killing efficiency of cytotoxic T cells (Anikeeva et al. 2005; Beal et al. 2008; Beal et al. 2009). LFA-1 makes profound contributions to T-cell sensitivity and its interactions uniquely define the pSMAC. These effects were separated out in two studies with the conclusion that an intact pSMAC increases the efficiency of killing by threefold (Beal et al. 2008). The rate of granule exocytosis to the target cell is controlled by efficient granule delivery to the cSMAC. This parameter was controlled in turn by the intensity of signaling from the synapse and appears to account for a 30-fold component of the difference between fast killing $CD8^+$ and slow killing $CD4^+$ CTL (Beal et al. 2009). Similar results were obtained with a $CD8^+$ T cell system in which altered peptide ligands with weaker signals were less able to recruit granules to the MTOC at the cSMAC (Jenkins et al. 2009). Thus, the pSMAC and cSMAC both make significant quantitative contributions to the efficiency of CTL, neither structure is absolutely required for killing of targets.

REFERENCES

Altan-Bonnet G, Germain RN. 2005. Modeling T cell antigen discrimination based on feedback control of digital ERK responses. *PLoS Biol* **3:** e356.

Anderson DC, Wible LJ, Hughes BJ, Smith CW, Brinkley BR. 1982. Cytoplasmic microtubules in polymorphonuclear leukocytes: effects of chemotactic stimulation and colchicine. *Cell* **31:** 719–729.

Anikeeva N, Somersalo K, Sims TN, Thomas VK, Dustin ML, Sykulev Y. 2005. Distinct role of lymphocyte function-associated antigen-1 in mediating effective cytolytic activity by cytotoxic T lymphocytes. *Proc Natl Acad Sci* **102:** 6437–6442.

Babbitt BP, Allen PM, Matsueda G, Haber E, Unanue ER. 1985. Binding of immunogenic peptides to Ia histocompatibility molecules. *Nature* **317:** 359–361.

Barcia C, Thomas CE, Curtin JF, King GD, Wawrowsky K, Candolfi M, Xiong WD, Liu C, Kroeger K, Boyer O, et al. 2006. In vivo mature immunological synapses forming SMACs mediate clearance of virally infected astrocytes from the brain. *J Exp Med* **203:** 2095–2107.

Beal AM, Anikeeva N, Varma R, Cameron TO, Norris PJ, Dustin ML, Sykulev Y. 2008. Protein Kinase C{theta} Regulates stability of the peripheral adhesion ring junction and contributes to the sensitivity of target cell lysis by CTL. *J Immunol* **181:** 4815–4824.

Beal AM, Anikeeva N, Varma R, Cameron TO, Vasiliver-Shamis G, Norris PJ, Dustin ML, Sykulev Y. 2009. Kinetics of early T cell receptor signaling regulate the pathway of lytic granule delivery to the secretory domain. *Immunity* **31:** 632–642.

Bohm H, Brinkmann V, Drab M, Henske A, Kurzchalia TV. 1997. Mammalian homologues of *C. elegans* PAR-1 are asymmetrically localized in epithelial cells and may influence their polarity. *Curr Biol* **7:** 603–606.

Brossard C, Feuillet V, Schmitt A, Randriamampita C, Romao M, Raposo G, Trautmann A. 2005. Multifocal structure of the T cell—dendritic cell synapse. *Eur J Immunol* **35:** 1741–1753.

Bunnell SC, Hong DI, Kardon JR, Yamazaki T, McGlade CJ, Barr VA, Samelson LE. 2002. T cell receptor ligation induces the formation of dynamically regulated signaling assemblies. *J Cell Biol* **158:** 1263–1275.

Burroughs NJ, Wulfing C. 2002. Differential segregation in a cell-cell contact interface: the dynamics of the immunological synapse. *Biophys J* **83:** 1784–1796.

Campi G, Varma R, Dustin ML. 2005. Actin and agonist MHC-peptide complex-dependent T cell receptor microclusters as scaffolds for signaling. *J Exp Med* **202:** 1031–1036.

Carrasco YR, Fleire SJ, Cameron T, Dustin ML, Batista FD. 2004. LFA-1/ICAM-1 interaction lowers the threshold of B cell activation by facilitating B cell adhesion and synapse formation. *Immunity* **20:** 589–599.

Celli S, Lemaitre F, Bousso P. 2007. Real-time manipulation of T cell-dendritic cell interactions in vivo reveals the importance of prolonged contacts for CD4+ T cell activation. *Immunity* **27:** 625–634.

Cemerski S, Das J, Giurisato E, Markiewicz MA, Allen PM, Chakraborty AK, Shaw AS. 2008. The balance between T cell receptor signaling and degradation at the center of the immunological synapse is determined by antigen quality. *Immunity* **29:** 414–422.

Cemerski S, Das J, Locasale J, Arnold P, Giurisato E, Markiewicz MA, Fremont D, Allen PM, Chakraborty AK, Shaw AS. 2007. The stimulatory potency of T cell antigens is influenced by the formation of the immunological synapse. *Immunity* **26:** 345–355.

Chakraborty AK, Das J. 2010. Pairing computation with experimentation: A powerful coupling for understanding T cell signalling. *Nat Rev Immunol* **10:** 59–71.

Choudhuri K, Wiseman D, Brown MH, Gould K, van der Merwe PA. 2005. T-cell receptor triggering is critically dependent on the dimensions of its peptide-MHC ligand. *Nature* **436:** 578–582.

Combs J, Kim SJ, Tan S, Ligon LA, Holzbaur EL, Kuhn J, Poenie M. 2006. Recruitment of dynein to the Jurkat immunological synapse. *Proc Natl Acad Sci* **103:** 14883–14888.

Das J, Kardar M, Chakraborty AK. 2009. Positive feedback regulation results in spatial clustering and fast spreading of active signaling molecules on a cell membrane. *J Chem Phys* **130:** 245102.

Davis DM, Dustin ML. 2004. What is the importance of the immunological synapse? *Trends Immunol* **25:** 323–327.

DeMond AL, Mossman KD, Starr T, Dustin ML, Groves JT. 2008. T cell receptor microcluster transport through molecular mazes reveals mechanism of translocation. *Biophys J* **94:** 3286–3292.

Douglass AD, Vale RD. 2005. Single-molecule microscopy reveals plasma membrane microdomains created by protein-protein networks that exclude or trap signaling molecules in T cells. *Cell* **121:** 937–950.

Dustin ML. 2008. T-cell activation through immunological synapses and kinapses. *Immunol Rev* **221:** 77–89.

Dustin ML, Springer TA. 1989. T-cell receptor cross-linking transiently stimulates adhesiveness through LFA-1. *Nature* **341:** 619–624.

Dustin ML, Bromley SK, Kan Z, Peterson DA, Unanue ER. 1997. Antigen receptor engagement delivers a stop signal to migrating T lymphocytes. *Proc Natl Acad Sci* **94:** 3909–3913.

Dustin ML, Olszowy MW, Holdorf AD, Li J, Bromley S, Desai N, Widder P, Rosenberger F, van der Merwe PA, Allen PM, et al. 1998. A novel adaptor protein orchestrates receptor patterning and cytoskeletal polarity in T-cell contacts. *Cell* **94:** 667–677.

Geiger B, Rosen D, Berke G. 1982. Spatial relationships of microtubule-organizing centers and the contact area of cytotoxic T lymphocytes and target cells. *J Cell Biol* **95:** 137–143.

Goldstein B, Faeder JR, Hlavacek WS, Blinov ML, Redondo A, Wofsy C. 2002. Modeling the early signaling events mediated by FcepsilonRI. *Mol Immunol* **38:** 1213–1219.

Gotlieb AI, May LM, Subrahmanyan L, Kalnins VI. 1981. Distribution of microtubule organizing centers in migrating sheets of endothelial cells. *J Cell Biol* **91:** 589–594.

Grakoui A, Bromley SK, Sumen C, Davis MM, Shaw AS, Allen PM, Dustin ML. 1999. The immunological synapse: A molecular machine controlling T cell activation. *Science* **285:** 221–227.

Gunzer M, Schafer A, Borgmann S, Grabbe S, Zanker KS, Brocker EB, Kampgen E, Friedl P. 2000. Antigen presentation in extracellular matrix: interactions of T cells with dendritic cells are dynamic, short lived, and sequential. *Immunity* **13:** 323–332.

Hartman NC, Nye JA, Groves JT. 2009. Cluster size regulates protein sorting in the immunological synapse. *Proc Natl Acad Sci* **106:** 12729–12734.

Henrickson SE, Mempel TR, Mazo IB, Liu B, Artyomov MN, Zheng H, Peixoto A, Flynn MP, Senman B, Junt T, et al. 2008. T cell sensing of antigen dose governs interactive behavior with dendritic cells and sets a threshold for T cell activation. *Nat Immunol* **9:** 282–291.

Hirsch E, Katanaev VL, Garlanda C, Azzolino O, Pirola L, Silengo L, Sozzani S, Mantovani A, Altruda F, Wymann MP. 2000. Central role for G protein-coupled phosphoinositide 3-kinase γin inflammation. *Science* **287:** 1049–1053.

Hu K, Ji L, Applegate KT, Danuser G, Waterman-Storer CM. 2007. Differential transmission of actin motion within focal adhesions. *Science* **315:** 111–115.

Hurov JB, Stappenbeck TS, Zmasek CM, White LS, Ranganath SH, Russell JH, Chan AC, Murphy KM, Piwnica-Worms H. 2001. Immune system dysfunction and autoimmune disease in mice lacking Emk (Par-1) protein kinase. *Mol Cell Biol* **21:** 3206–3219.

Huse M, Klein LO, Girvin AT, Faraj JM, Li QJ, Kuhns MS, Davis MM. 2007. Spatial and temporal dynamics of T cell receptor signaling with a photoactivatable agonist. *Immunity* **27:** 76–88.

Iezzi G, Karjalainen K, Lanzavecchia A. 1998. The duration of antigenic stimulation determines the fate of naive and effector T cells. *Immunity* **8:** 89–95.

Ilani T, Vasiliver-Shamis G, Vardhana S, Bretscher A, Dustin ML. 2009. T cell antigen receptor signaling and immunological synapse stability require myosin IIA. *Nat Immunol* **10:** 531–539.

Irvine DJ, Purbhoo MA, Krogsgaard M, Davis MM. 2002. Direct observation of ligand recognition by T cells. *Nature* **419:** 845–849.

Jacobelli J, Chmura SA, Buxton DB, Davis MM, Krummel MF. 2004. A single class II myosin modulates T cell motility and stopping, but not synapse formation. *Nat Immunol* **5:** 531–538.

Jay PY, Pham PA, Wong SA, Elson EL. 1995. A mechanical function of myosin II in cell motility. *J Cell Sci* **108:** 387–393.

Jenkins MR, Tsun A, Stinchcombe JC, Griffiths GM. 2009. The strength of T cell receptor signal controls the polarization of cytotoxic machinery to the immunological synapse. *Immunity* **31:** 621–631.

Joberty G, Petersen C, Gao L, Macara IG. 2000. The cell-polarity protein Par6 links Par3 and atypical protein kinase C to Cdc42. *Nat Cell Biol* **2:** 531–539.

Kaizuka Y, Douglass AD, Vardhana S, Dustin ML, Vale RD. 2009. The coreceptor CD2 uses plasma membrane microdomains to transduce signals in T cells. *J Cell Biol* **185:** 521–534.

Kaizuka Y, Douglass AD, Varma R, Dustin ML, Vale RD. 2007. Mechanisms for segregating T cell receptor and adhesion molecules during immunological synapse formation in Jurkat T cells. *Proc Natl Acad Sci* **104:** 20296–20301.

Kemphues KJ, Priess JR, Morton DG, Cheng NS. 1988. Identification of genes required for cytoplasmic localization in early *C. elegans* embryos. *Cell* **52:** 311–320.

Kupfer A, Dennert G, Singer SJ. 1983. Polarization of the Golgi apparatus and the microtubule-organizing center within cloned natural killer cells bound to their targets. *Proc Natl Acad Sci* **80:** 7224–7228.

Lee KH, Dinner AR, Tu C, Campi G, Raychaudhuri S, Varma R, Sims TN, Burack WR, Wu H, Wang J, et al. 2003. The immunological synapse balances T cell receptor signaling and degradation. *Science* **302:** 1218–1222.

Lee KH, Holdorf AD, Dustin ML, Chan AC, Allen PM, Shaw AS. 2002. T cell receptor signaling precedes immunological synapse formation. *Science* **295:** 1539–1542.

Li Z, Jiang H, Xie W, Zhang Z, Smrcka AV, Wu D. 2000. Roles of PLC-β2 and -β3 and PI3Kγ in chemoattractant-mediated signal transduction. *Science* **287:** 1046–1049.

Lin D, Edwards AS, Fawcett JP, Mbamalu G, Scott JD, Pawson T. 2000. A mammalian PAR-3-PAR-6 complex implicated in Cdc42/Rac1 and aPKC signalling and cell polarity. *Nat Cell Biol* **2:** 540–547.

Lin J, Hou KK, Piwnica-Worms H, Shaw AS. 2009. The polarity protein Par1b/EMK/MARK2 regulates T cell receptor-induced microtubule-organizing center polarization. *J Immunol* **183:** 1215–1221.

Liu H, Rhodes M, Wiest DL, Vignali DA. 2000. On the dynamics of TCR:CD3 complex cell surface expression and downmodulation. *Immunity* **13:** 665–675.

Ludford-Menting MJ, Oliaro J, Sacirbegovic F, Cheah ET, Pedersen N, Thomas SJ, Pasam A, Iazzolino R, Dow LE, Waterhouse NJ, et al. 2005. A network of PDZ-containing proteins regulates T cell polarity and morphology during migration and immunological synapse formation. *Immunity* **22:** 737–748.

Markiewicz MA, Carayannopoulos LN, Naidenko OV, Matsui K, Burack WR, Wise EL, Fremont DH, Allen PM, Yokoyama WM, Colonna M, et al. 2005. Costimulation through NKG2D enhances murine CD8+ CTL function: Similarities and differences between NKG2D and CD28 costimulation. *J Immunol* **175:** 2825–2833.

Martin-Cofreces NB, Robles-Valero J, Cabrero JR, Mittelbrunn M, Gordon-Alonso M, Sung CH, Alarcon B, Vazquez J, Sanchez-Madrid F. 2008. MTOC translocation modulates IS formation and controls sustained T cell signaling. *J Cell Biol* **182:** 951–962.

McGavern DB, Christen U, Oldstone MB. 2002. Molecular anatomy of antigen-specific CD8(+) T cell engagement and synapse formation in vivo. *Nat Immunol* **3:** 918–925.

Mempel TR, Henrickson SE, Von Andrian UH. 2004. T-cell priming by dendritic cells in lymph nodes occurs in three distinct phases. *Nature* **427:** 154–159.

Miller MJ, Hejazi AS, Wei SH, Cahalan MD, Parker I. 2004. T cell repertoire scanning is promoted by dynamic dendritic cell behavior and random T cell motility in the lymph node. *Proc Natl Acad Sci* **101:** 998–1003.

Monks CR, Freiberg BA, Kupfer H, Sciaky N, Kupfer A. 1998. Three-dimensional segregation of supramolecular activation clusters in T cells. *Nature* **395:** 82–86.

Morin NA, Oakes PW, Hyun YM, Lee D, Chin EY, King MR, Springer TA, Shimaoka M, Tang JX, Reichner JS, et al. 2008. Nonmuscle myosin heavy chain IIA mediates integrin LFA-1 de-adhesion during T lymphocyte migration. *J Exp Med* **205:** 195–205.

Mossman KD, Campi G, Groves JT, Dustin ML. 2005. Altered TCR signaling from geometrically repatterned immunological synapses. *Science* **310:** 1191–1193.

Naramura M, Jang IK, Kole H, Huang F, Haines D, Gu H. 2002. c-Cbl and Cbl-b regulate T cell responsiveness by promoting ligand-induced TCR down-modulation. *Nat Immunol* **3:** 1192–1199.

Negulescu PA, Krasieva TB, Khan A, Kerschbaum HH, Cahalan MD. 1996. Polarity of T cell shape, motility, and sensitivity to antigen. *Immunity* **4:** 421–430.

Niggli V, Keller H. 1997. The phosphatidylinositol 3-kinase inhibitor wortmannin markedly reduces chemotactic peptide-induced locomotion and increases in cytoskeletal actin in human neutrophils. *Eur J Pharmacol* **335:** 43–52.

Norcross MA. 1984. A synaptic basis for T-lymphocyte activation. *Ann Immunol (Paris)* **135D:** 113–134.

Peterson EJ, Woods ML, Dmowski SA, Derimanov G, Jordan MS, Wu JN, Myung PS, Liu QH, Pribila JT, Freedman BD, et al. 2001. Coupling of the TCR to integrin activation by Slap-130/Fyb. *Science* **293:** 2263–2265.

Ponti A, Machacek M, Gupton SL, Waterman-Storer CM, Danuser G. 2004. Two distinct actin networks drive the protrusion of migrating cells. *Science* **305:** 1782–1786.

Ponti A, Matov A, Adams M, Gupton S, Waterman-Storer CM, Danuser G. 2005. Periodic patterns of actin turnover in lamellipodia and lamellae of migrating epithelial cells analyzed by quantitative Fluorescent Speckle Microscopy. *Biophys J* **89:** 3456–3469.

Purtic B, Pitcher LA, van Oers NS, Wulfing C. 2005. T cell receptor (TCR) clustering in the immunological synapse integrates TCR and costimulatory signaling in selected T cells. *Proc Natl Acad Sci* **102:** 2904–2909.

Qi SY, Groves JT, Chakraborty AK. 2001. Synaptic pattern formation during cellular recognition. *Proc Natl Acad Sci* **98:** 6548–6553.

Qin Y, Capaldo C, Gumbiner BM, Macara IG. 2005. The mammalian Scribble polarity protein regulates epithelial cell adhesion and migration through E-cadherin. *J Cell Biol* **171:** 1061–1071.

Quann EJ, Merino E, Furuta T, Huse M. 2009. Localized diacylglycerol drives the polarization of the microtubule-organizing center in T cells. *Nat Immunol* **10:** 627–635.

Reichert P, Reinhardt RL, Ingulli E, Jenkins MK. 2001. Cutting edge: In vivo identification of TCR redistribution and polarized IL-2 production by naive CD4 T cells. *J Immunol* **166:** 4278–4281.

Sanchez-Madrid F, Serrador JM. 2009. Bringing up the rear: defining the roles of the uropod. *Nat Rev Mol Cell Biol* **10:** 353–359.

Scholer A, Hugues S, Boissonnas A, Fetler L, Amigorena S. 2008. Intercellular adhesion molecule-1-dependent stable interactions between T cells and dendritic cells determine CD8+ T cell memory. *Immunity* **28:** 258–270.

Shakhar G, Lindquist RL, Skokos D, Dudziak D, Huang JH, Nussenzweig MC, Dustin ML. 2005. Stable T cell-dendritic cell interactions precede the development of both tolerance and immunity in vivo. *Nat Immunol* **6:** 707–714.

Shaw AS, Dustin ML. 1997. Making the T cell receptor go the distance: A topological view of T cell activation. *Immunity* **6:** 361–369.

Sims TN, Soos TJ, Xenias HS, Dubin-Thaler B, Hofman JM, Waite JC, Cameron TO, Thomas VK, Varma R, Wiggins CH, et al. 2007. Opposing effects of PKCθ and WASp on symmetry breaking and relocation of the immunological synapse. *Cell* **129:** 773–785.

Skokos D, Shakhar G, Varma R, Waite JC, Cameron TO, Lindquist RL, Schwickert T, Nussenzweig MC, Dustin ML. 2007. Peptide-MHC potency governs dynamic interactions between T cells and dendritic cells in lymph nodes. *Nat Immunol* **8:** 835–844.

Smith A, Carrasco YR, Stanley P, Kieffer N, Batista FD, Hogg N. 2005. A talin-dependent LFA-1 focal zone is formed by rapidly migrating T lymphocytes. *J Cell Biol* **170:** 141–151.

Somersalo K, Anikeeva N, Sims TN, Thomas VK, Strong RK, Spies T, Lebedeva T, Sykulev Y, Dustin ML. 2004. Cytotoxic T lymphocytes form an antigen-independent ring junction. *J Clin Invest* **113:** 49–57.

Springer TA. 1990. Adhesion receptors of the immune system. *Nature* **346:** 425–434.

Stinchcombe JC, Bossi G, Booth S, Griffiths GM. 2001. The immunological synapse of CTL contains a secretory domain and membrane bridges. *Immunity* **15:** 751–761.

Stinchcombe JC, Majorovits E, Bossi G, Fuller S, Griffiths GM. 2006. Centrosome polarization delivers secretory granules to the immunological synapse. *Nature* **443:** 462–465.

Suzuki A, Hirata M, Kamimura K, Maniwa R, Yamanaka T, Mizuno K, Kishikawa M, Hirose H, Amano Y., Izumi N, et al. 2004. aPKC acts upstream of PAR-1b in both the establishment and maintenance of mammalian epithelial polarity. *Curr Biol* **14:** 1425–1435.

Tseng SY, Waite JC, Liu M, Vardhana S, Dustin ML. 2008. T cell-dendritic cell immunological synapses contain TCR-dependent CD28-CD80 clusters that recruit protein kinase Ctheta. *J Immunol* **181:** 4852–4863.

Valitutti S, Muller S, Salio M, Lanzavecchia A. 1997. Degradation of T cell receptor (TCR)-CD3-ζcomplexes after antigenic stimulation. *J Exp Med* **185:** 1859–1864.

Vardhana S, Choudhuri K, Varma R, Dustin ML. 2010. Essential role of ubiquitin and TSG101 protein in formation and function of the central supramolecular activation cluster. *Immunity* **32:** 531–540.

Varma R, Campi G, Yokosuka T, Saito T, Dustin ML. 2006. T cell receptor-proximal signals are sustained in peripheral microclusters and terminated in the central supramolecular activation cluster. *Immunity* **25:** 117–127.

Vogel V, Sheetz M. 2006. Local force and geometry sensing regulate cell functions. *Nat Rev Mol Cell Biol* **7:** 265–275.

Weikl TR, Lipowsky R. 2004. Pattern formation during T-cell adhesion. *Biophys J* **87:** 3665–3678.

Wiedemann A, Muller S, Favier B, Penna D, Guiraud M, Delmas C, Champagne E, Valitutti S. 2005. T-cell activation is accompanied by an ubiquitination process occurring at the immunological synapse. *Immunol Lett* **98:** 57–61.

Williams RL, Urbe S. 2007. The emerging shape of the ESCRT machinery. *Nat Rev Mol Cell Biol* **8:** 355–368.

Woolf E, Grigorova I, Sagiv A, Grabovsky V, Feigelson SW, Shulman Z, Hartmann T, Sixt M, Cyster JG, Alon R. 2007. Lymph node chemokines promote sustained T lymphocyte motility without triggering stable integrin adhesiveness in the absence of shear forces. *Nat Immunol* **8:** 1076–1085.

Yokosuka T, Kobayashi W, Sakata-Sogawa K, Takamatsu M, Hashimoto-Tane A, Dustin ML, Tokunaga M, Saito T. 2008. Spatiotemporal regulation of T cell costimulation by TCR-CD28 microclusters and protein kinase C θtranslocation. *Immunity* **29:** 589–601.

Yokosuka T, Sakata-Sogawa K, Kobayashi W, Hiroshima M, Hashimoto-Tane A, Tokunaga M, Dustin ML, Saito T. 2005. Newly generated T cell receptor microclusters initiate and sustain T cell activation by recruitment of Zap70 and SLP-76. *Nat Immunol* **6:** 1253–1262.

Perspectives for Computer Modeling in the Study of T-Cell Activation

Jesse Coward[1], Ronald N. Germain[2], and Grégoire Altan-Bonnet[1]

[1]ImmunoDynamics Group, Programs in Computational Biology and Immunology, Memorial Sloan-Kettering Cancer Center, New York, New York 10065

[2]Lymphocyte Biology Section, Laboratory of Immunology, Program in Systems Immunology and Infectious Disease Modeling (PSIIM), National Institute of Allergy and Infectious Diseases, National Institutes of Health, DHHS, Bethesda, Maryland 20892-1892

Correspondence: altanbog@mskcc.org

The T-cell receptor (TCR) is responsible for discriminating between self- and foreign-derived peptides, translating minute differences in amino-acid sequence into large differences in response. Because of the great variability in the TCR and its ligands, activation of T cells by foreign peptides is a quantitative process, dependent on a mix of upstream signals and downstream integration. Accordingly, quantitative data and computational models have shed light on many important aspects of this process: molecular noise in ligand recognition, spatial dynamics in T cell-APC (antigen presenting cell) interactions, graded versus all-or-none decision-making by the TCR apparatus, mechanisms of peptide antagonism and synergism, and the tunability and robustness of activation thresholds. Though diverse in their formalism, these studies together paint a picture of how modeling has shaped and will continue to shape understanding of T-cell immunobiology.

Deciphering the detailed basis of antigen-specific T-cell activation is classically considered to reside in the realm of cell biology and biochemistry, both "wet lab" disciplines. Molecular approaches complemented by increasingly high resolution imaging tools have identified the components and interactions that translate antigen recognition into gene activation. The application of biophysical tools, initially surface plasmon resonance and later, in situ FRET, have added a quantitative element to the pool of data available in this field (Davis et al. 1998). But these advances alone, though critical, have not provided a deep conceptual understanding of how ligand discrimination occurs or the peculiarities of the discrimination and activation processes. In particular, they have not yet yielded a framework that can reliably predict how T cells will respond when molecular components of the transduction machinery are altered in concentration or function, such as in the case of genetic polymorphism or mutation.

Over the past decade, mathematical and computational modeling has emerged as a means of integrating the experimental observations resulting from biochemical, biophysical, and imaging studies into a quantitative view of T-cell responses to antigens. In this

 doi: 10.1101/cshperspect.a005538
Cite this article as *Cold Spring Harb Perspect Biol* 2010;2:a005538

communication we review the contributions of computer modeling to our understanding of T-cell immunobiology. First, we discuss how recent technical developments have yielded more quantitative measurements of immune responses and triggered modeling efforts dealing with the dynamics of T-cell activation. We will present the computer modeling tools that have been used and their increasing accessibility to immunobiologists. In a second section, we present different conundrums related to T-cell activation and describe how they have been addressed through computer modeling (Fig. 1). Finally, we will discuss why modeling remains only peripherally relevant to experimental immunology, and how this will change through the implementation of better tools and practices.

QUANTITATIVE DATA AND APPROACHES TO MODELING

Quantitative characterizations of T lymphocyte biology are providing both the impetus and the raw materials for models correlating biophysical observables with functional outcomes. Indeed, many immunological components have been characterized, from the organ level (e.g., lymph node architecture), to the cellular level (e.g., varied states of differentiation of lymphocytes), down to the molecular level (e.g., expression of signaling proteins, number of secreted molecules). The molecular components of T-cell antigen and cytokine signaling have been isolated and purified, and studies of the isolated molecules have precisely determined many of the biophysical parameters governing ligand-receptor interactions.

A prime example of this latter category involves T-cell receptor (TCR) interaction with its ligand, a peptide presented by major histocompatibility complex (MHC) surface proteins. Most (but not all) studies have shown a correlation between minute changes in a single kinetic parameter (off-rate for the receptor/ligand complex) and large differences in signaling patterns and functional responses (Davis et al.

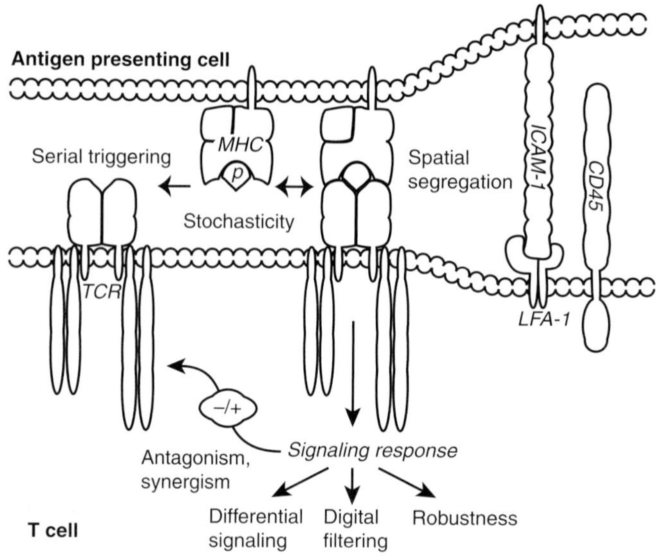

Figure 1. Aspects of T-cell signaling clarified through computational modeling. Stochasticity (molecular noise) has been incorporated into models of the TCR, operating at the level of recognition of just a few antigens. Serial triggering of multiple TCRs by the same antigenic peptide (p) presented in a major histocompatibilty complex protein (MHC) and segregation of bulky phosphatases (CD45) and adhesive molecule pairs (LFA-1 and ICAM-1) from smaller pMHC-TCR complexes have been studied in models quantifying spatial regulation of T-cell activation. Antagonism, synergism, differential signaling, digital filtering, and robustness have been examined using models describing the biochemical signaling induced by TCR ligation.

1998). As a quantitative explanation for this observation, T. McKeithan first offered a "kinetic proofreading" scheme (Hopfield 1974; Ninio 1975) to account for T-cell ligand discrimination (McKeithan 1995). In that scheme, receptors engaging stably binding ligands engender a complete phosphorylation of TCR-associated proteins, whereas receptors engaging transiently binding ligands fail to reach the endpoint required to trigger effective cellular responses (reviewed extensively in Feinerman et al. 2008a). McKeithan's seminal study emphasized that static dose-response equilibrium descriptions of this system insufficiently capture its true behavior. Building on this work, following studies emphasized how dynamic interactions within the TCR apparatus can account for the keen specificity of T-cell activation. Most of these have focused on the development of meso-scale biochemical models that typically include 10–30 components such as receptors, adapters, kinases, and phosphatases, representing limited signaling pathways and operating on short timescales (Schoeberl et al. 2002; Faeder et al. 2003; Altan-Bonnet and Germain 2005). They are often well-mixed, deterministic models that rely on classical ordinary differential equations (mass-action laws, Michaelis-Menten or others) to describe protein–protein interaction and enzymological activities.

The principal benefit of such biochemical reaction models is their ability to simulate the behavior of large numbers of molecules by dealing with concentrations rather than individual molecules. The drawback, however, is that the modeler must enumerate all molecular states and transition rates in a series of linked equations for any given signaling pathway. This is reasonably straightforward for a network of reactions in which enzymes and substrates are soluble, do not aggregate, and have a small number of reaction sites. Desktop modeling packages such as CellDesigner (Funahashi et al. 2003), JDesigner (Le Novere et al. 2009), and COPASI (Hoops et al. 2006), which allow schematically specified biochemical networks to be translated into explicit biochemical equations, have expanded the accessibility of such techniques. The behavior of these systems over time can then be simulated with classical deterministic or stochastic methods (Gillespie 2007).

The limitations of this method of model specification become apparent, however, when faced with reactions involving aggregations of components or molecules with large numbers of possible modifications. In these situations, the specification of the network becomes a challenge that has spurred the development of new approaches for defining models. As one example of why this is a central issue in modeling T-cell activation, consider the organization of the TCR complex and the earliest biochemical events induced by binding of its ligand, a complex of peptide and MHC (pMHC). The TCR complex (composed of the $\alpha\beta$ chains of the TCR as wells as the CD3/ζ chains) has 10 immunoreceptor tyrosine-based activation motifs (ITAMs) that are phosphorylated during TCR activation and provide docking sites for downstream signaling molecules. Each ITAM consists of two tyrosines, for a total of 20 tyrosines in the TCR complex. Hence, there are theoretically $2^{20} > 10^6$ different phospho-tyrosine states for the TCR complex. Explicitly enumerating all of these phosphorylated states in a model would be very tedious and time-consuming (Blinov et al. 2004). A common simplification is to limit the biochemical models to fewer ITAMs (e.g., the three ITAMs associated with one ζ chain). This approach is problematic, though, as it both underestimates the importance of a full complement of ITAMs (Holst et al. 2008; Malissen 2008) and still allows large models—20 pages in our previous work (Altan-Bonnet and Germain 2005)—that are difficult to proofread and troubleshoot or to expand when multiplexing with other signaling pathways.

Accordingly, the field is applying a new theoretical formalism—known as rule-based modeling (Faeder et al. 2003; Blinov et al. 2004; Hlavacek et al. 2006; Meier-Schellersheim et al. 2006; Lipniacki et al. 2008)—to develop *in silico* models of signaling cascades in lymphocytes. With rule-based modeling, the modeler articulates single biochemical reactions (e.g., interaction between two proteins or phosphorylation of one tyrosine) as rules; the full network of interactions, including the combinatorial complexity in a given signaling cascade, is then generated automatically using a computational algorithm.

In general, rule-based formalism has the advantage of a simple syntax that translates biochemical cartoons into quantitative models. These are thus more accessible to biologists and easier to troubleshoot. More specifically in the case of T-cell signaling, rule-based formalism simplifies the development of models where the combinatorial complexity of the signalosome is daunting (Malissen et al. 2005). Indeed, aggregation of signaling components on phosphorylated ITAMs or on LAT (linker for activation of T cells) downstream of the TCR is automatically accounted for by rules linking individual components, whereas biochemical formalism would require explicit specification of each molecular complex (Fig. 2). Furthermore, cross-talk of the TCR signaling pathway with cytokine- or costimulation-induced signaling can be easily tested: Independently constructed and validated rule-based models for signaling pathways are automatically integrated when combined in a comprehensive model.

Other avenues of exploration attempted to avoid the issue of complexity altogether. Some investigators have tried to manage the large quantity of qualitative observations and dearth of dynamic parameters in the T-cell activation literature by developing a Boolean (logic-based) representation of the T-cell signaling network (Saez-Rodriguez et al. 2007). Even without exact kinetic parameters, it is possible to extract further information from current knowledge by building a large-scale logical model, curated from observations in the literature. For example, Saez-Rodriguez et al. used their Boolean model to make a surprising prediction that they subsequently validated experimentally: Cross-linking of CD28 was sufficient to trigger JNK activation without antigen activation. However, because of their inherent on/off nature, Boolean systems are not suited to investigation of phenomena such as ligand discrimination or thymic selection that involve translation of minute differences in biophysical characteristics of pMHC-TCR interaction into discrete functional differences (Altan-Bonnet and Germain 2005; Daniels et al. 2006).

Although the dynamics of molecular pathways downstream of the TCR account for a great deal of antigen recognition, incorporating physical processes that modulate this biochemistry remains a challenge with current methods. In particular, computational time can be forbidding when modeling T-cell activation with diffusion of signaling components (Loew and Schaff 2001; Meier-Schellersheim et al. 2006) or with combinatorial complexity in the biochemical network (Hlavacek et al. 2006). More complex phenomena such as membrane deformation and cytoskeletal coupling are even harder computational problems and have been modeled only in frameworks constructed specifically for that purpose (Qi et al. 2001). Thus, there exist many ad hoc computational methods that were specifically designed to address specific biological questions. We expect the field of computational biology to converge on a *lingua franca* for systems immunology that will make the communication of models more efficient (Le Novere et al. 2009). This will benefit the computational community, as experimental parameters will be more easily shared, and model-building efforts will be cross-validated. The community of immunologists will also benefit as a "reference" model of T-cell activation will emerge from computational efforts.

QUESTIONS IN SIGNAL TRANSDUCTION AND T-CELL ANTIGEN RECEPTOR SIGNALING THAT HAVE BEEN ADDRESSED BY QUANTITATIVE MODELING

Digital Filtering and the Specificity, Sensitivity, and Speed of TCR Signaling

A striking feature of T-cell activation is the specificity of its response to pMHC ligand. Three key observations have set conditions that models of TCR signaling must satisfy (Altan-Bonnet and Germain 2005). First, T-cell activation is a very fast process that can be detected within seconds of T:APC contacts (by kinase recruitment [Bunnell et al. 2002; Huse et al. 2007], calcium influx [Delon et al. 1998], TCR

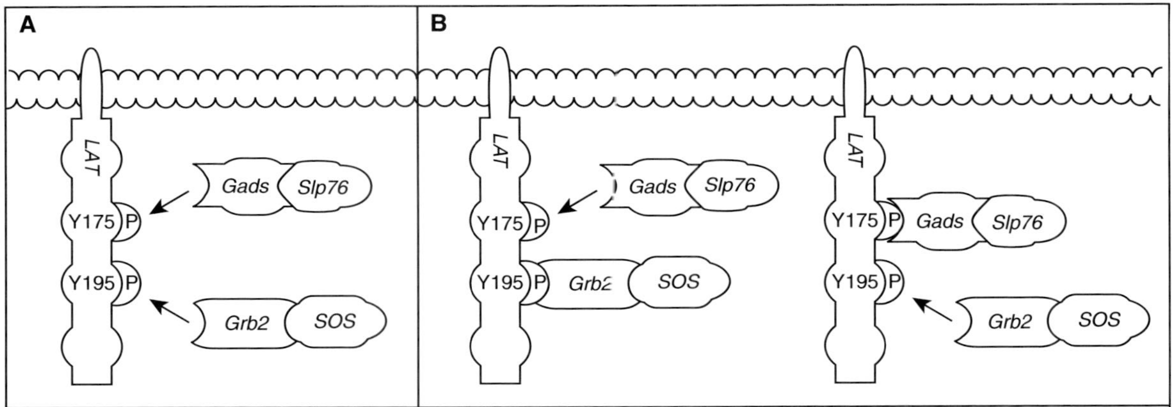

Rule-based model description formalism simplifies simulation of LAT signalosome assembly. Slp76 and SOS are recruited to the signalosome through association of constitutively-associated adaptor proteins (Gads and Grb2, respectively) with two independent phosphotyrosines.

A. The first step in model specification is similar in rule-based and differential equation formalisms; both involve descriptions of interactions between unbound LAT and each of the effectors.

Differential equations specify the relationships between concentrations of reactants and change in the concentration of products.

```
d[LAT_Y175P_Y195P_SOS]/dt    = k1_on*[LAT_Y175P_Y195P][SOS] - k1_off*[LAT_Y175P_Y195P_SOS]
d[LAT_Y175P_Slp76_Y195P]/dt = k2_on*[LAT_Y175P_Y195P][Slp76] - k2_off*[LAT_Y175P_Slp76_Y195P]
```

Rules specify interactions between specific molecular sites (BioNetGen syntax is used here).

```
LAT(Y175~P, Y195~P) + SOS(Grb2)  <-> LAT(Y175~P, Y195~P!1).SOS(Grb2!1)   k1_on k1_off
LAT(Y175~P, Y195~P) + Slp76(Gads)<-> LAT(Y175~P!1, Y195~P).Slp76(Gads!1) k2_on k2_off
```

B. The additional step of describing the interaction of one unbound effector with a complex of LAT and the opposite effector are unnecessary when using rule-based formalism, but must be specified in a differential equation system.

```
d[LAT_Y175P_Slp76_Y195P_SOS]/dt = k1_on*[Lat_Y175P_Slp76_Y195P][SOS] +
k2_on*[LAT_Y175P_Y195P_SOS][Slp76] - k1_off*[LAT_Y175P_Slp76_Y195P_SOS] -
k2_off*[LAT_Y175P_Slp76_Y195P_SOS]
```

Figure 2. Rule-based modeling versus biochemical modeling of LAT signalosome aggregation.

phosphorylation [Germain and Stefanova 1999] or others). Second, T-cell activation is very sensitive, as one or few ligands are enough to trigger a response (Sykulev et al. 1996; Irvine et al. 2002). Third, T-cell activation (beginning at timescales far below a minute) is quite specific: One amino acid variation in an antigenic peptide can ablate its stimulatory function (Davis et al. 1998). The relevance of this specificity has been challenged recently (Zehn et al. 2009), but close examination of these data revealed that minute differences in peptide sequence still yields quantitative difference in T-cell proliferation in vivo. Aside from allosteric models (Gil et al. 2002; Schamel et al. 2006; Palmer and Naeher 2009), few of the modeling attempts building primarily on McKeithan's proofreading formalism have managed to simultaneously incorporate all these key features of T-cell responses into the output of their model (as reviewed in Feinerman et al. 2008a).

One facet of T-cell signaling that has attracted a great deal of theoretical attention is the observation that the signal transduction cascade associated with the TCR translates a spectrum of input strength (quality and quantity of antigens) into all-or-none responses associated with cell differentiation or mitotic progression. Theoretical work has shown that both signaling feedback and allosteric cooperativity have the potential to impart switchlike or "ultrasensitive" behavior (Fig. 3A–C). Analyses of the TCR signaling apparatus have revealed that multiple modules within the signaling cascade can serve as ultrasensitive switches. The ERK mitogen-activated protein kinase (MAPK) pathway contains three interconnected switches, whereas regulation of NFAT provides a fourth, somewhat independent switch (Fig. 3A).

The MAPK pathway passes signals from an upstream kinase (such as Ras) to Raf, MEK, then finally ERK. MEK and ERK each have two sites of phosphorylation that are phosphorylated sequentially in two separate enzyme-substrate interactions (Ferrell and Bhatt 1997). Using an analytical model of this cascade, Huang and Ferrell (Huang and Ferrell 1996) showed that the consequence of this pattern of phosphorylation is a sharpening of the input-output dose-response curve at each step of the pathway. Hence, in the absence of feedback, activation of ERK displays ultrasensitivity to input signals from Ras or analogous kinases. This theoretical result was confirmed in vitro by titrating amounts of input signal and demonstrating ultrasensitive phosphorylation of MEK and ERK in *Xenopus* eggs (Ferrell 2002; Xiong and Ferrell 2003).

Within the TCR apparatus, structural and biochemical findings have pointed to the existence of feedback loops—positive (Das et al. 2009) and double-negative (Stefanova et al. 2003; Altan-Bonnet and Germain 2005)—which are both expected to lead to digital behavior (Ferrell 2002). Positive allosteric feedback is mediated through SOS, a Ras-activating enzyme, which is itself activated by active Ras. In the case of the TCR and B-cell receptor (BCR) signaling machinery, Das et al (2009) showed that this positive feedback resulted in bistability. Note that bistability is a hallmark of a positive feedback signaling module, but is absent in ultrasensitive switches. In the same model, the authors showed that a second Ras-activating enzyme (Ras-GRP) that lacks a positive feedback mechanism is able to turn on the MAPK pathway in a graduated, analog manner. Most interestingly, the computer model predicted that bistability in ERK activation as a function of SOS activity should be hysteretic: Once the MAPK activity reaches its high state, it is primed to regain that high state on receipt of subsequent, smaller stimuli (Fig. 3D). Hysteresis within the TCR, which the authors confirmed experimentally, may have important implications for T-cell antigen recognition in vivo, where T cell-APC contacts are often brief but repeated (Henrickson et al. 2008).

Another signaling module enabling T-cell decision-making is a double-negative feedback regulation of TCR signaling. Using computer modeling, we have shown how upstream negative feedback from TCR-induced activation of SHP-1 phosphatase prevents signals from large quantities of subthreshold ligands from initiating full downstream TCR signaling. However, we also found that a downstream feedback mediated by ERK acts to shut down the SHP-1 negative

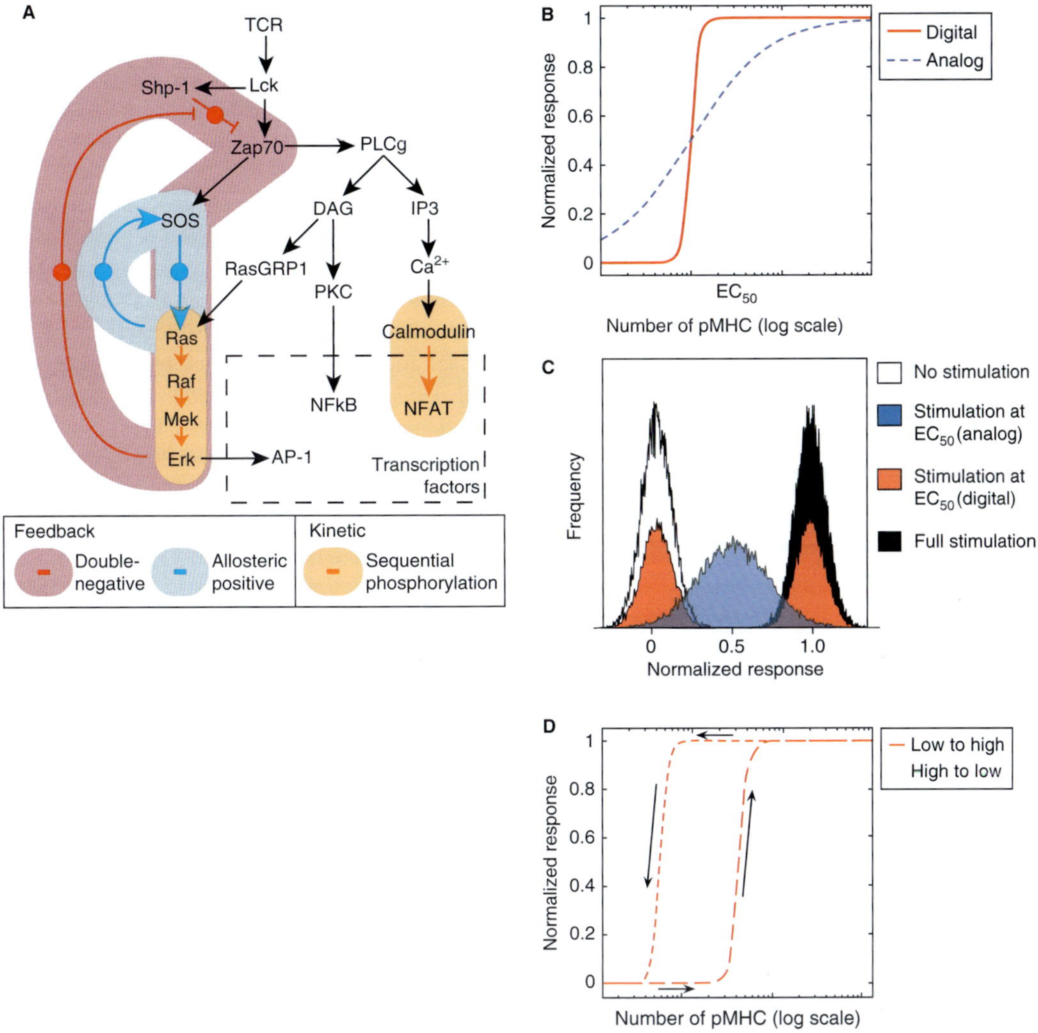

Figure 3. Digital responses in TCR signaling. *(A)* Many signaling responses in T cells are digital even on short timescales (1 min). Three signaling features contribute to this digital character: Ultrasensitivity through multiple dephosphorylation (cf. NFAT translocation to the nucleus) or through kinase cascade (cf. MAPK activation) or bistability through positive feedback (cf. SOS switch) or double-negative feedback (cf. SHP-1 regulation). *(B)* Ultrasensitivity yields a sharp dose response that is reversible. *(C)* The hallmark of digital responses is the bimodal (i.e., all-or-none) distribution of signaling responses: at EC_{50}, 50% of cells have a null response, 50% of the cells have a total response. On the other hand, an analog response yields a gradual increase with a unimodal distribution of response centered at 50% of the maximal response at EC_{50}. *(D)* Feedback regulation also generates also a sharp dose response similar to that generated by ultrasensitivity but with hysteresis.

feedback. Such a double-negative feedback drives sustained activation for above-threshold ligands, and stably limits the noise from subthreshold ligands (Feinerman et al. 2008a).

As in the case of ERK, single-cell measurements on T cells showed that nuclear factor of activated T cells (NFAT) is activated in an all-or-none response (Fiering et al. 1990; Macian 2005). Two groups (Salazar and Hofer 2003; Podtschaske et al. 2007) have modeled quantitatively the mechanism by which graded calcium influx drives a graded dephosphorylation of

NFAT at many residues, leading to all-or-none nuclear translocation and activation. The proposed mechanism for the ultrasensitive activation of NFAT is therefore a classical scheme of enzymatic activation through multiple modifications (similar to the double phosphorylation of MAPK that triggers its enzymatic activity).

It is worth noting that nuclear localization of NF-κB downstream of the TCR is *not* digital, but rather graded in proportion to stimulation strength (Podtschaske et al. 2007). Consequently, cells may commit to digital responses, such as cytotoxic release of granzymes and perforins, secretion of IL-2 or entry into mitosis, while retaining analog functions such as IFNγ secretion, CD69 expression, and others, allowing diverse immune responses to different antigens (Rabinowitz et al. 1996; Itoh and Germain 1997; Hemmer et al. 1998).

The study of analog-to-digital filtering in TCR signaling is a prime example of productive dialogue between experimental and theoretical immunology. Established theoretical groundwork provided explanations for many of the initial descriptions of all-or-none phenotypes seen in T cells and helped to illuminate how sensitivity, specificity, and speed are wired into the ligand-sensing signaling network. Further exploration of hallmarks of digital circuits through computational modeling has extended the understanding of T-cell activation: Hysteresis, characteristic of positive feedback, indicates that T cells may have short-term memory of stimulation. Double-negative feedback, well-understood to suppress noise in near-threshold systems, may also provide insights into the poorly understood phenomena of peptide antagonism and synergism (discussed below). In this way, studies of digital circuits in T cells have provided both understanding of all-or-none decision-making and hints at further phenotypes for experimental investigation.

Antagonism, Synergism

One of the counterintuitive facets of TCR signaling is that pMHC ligands that bind the TCR nearly as well as agonists *decrease* its sensitivity to agonists when copresented with otherwise stimulatory ligands, whereas peptides that are far from being agonists increase the sensitivity of T-cell responses (Krogsgaard et al. 2007). We review here how computer modeling has contributed to our understanding of antagonism and synergism in T-cell activation.

Early work to unravel the biochemical mechanism of antagonism showed that it is dependent on signals received through the TCR (Jameson et al. 1993; Racioppi et al. 1996) and is associated with the diffusible spread of the phosphatase SHP-1 (Plas et al. 1996; Dittel et al. 1999; Stefanova et al. 2003). In computational models of this process, briefly binding ligands induce minimal SHP-1 activation, whereas ligands with longer half-lives induce greater tyrosine phosphorylation of SHP-1 and its subsequent association with Lck in the TCR complex. Consequently, as ligand binding time approaches the agonist threshold, induction of negative feedback increases, directly antagonizing signals coming from neighboring stimulatory (agonist) ligands (Altan-Bonnet and Germain 2005). Thus, previous computer models automatically accounted for antagonism by implementing the suppression of substimulatory ligand signaling through SHP-1-mediated negative feedback (Altan-Bonnet and Germain 2005; Wylie et al. 2007; Lipniacki et al. 2008; Owens et al. 2009).

Synergism (coagonist function) was introduced by the Davis group, who showed that endogenous peptides can increase T-cell sensitivity to limiting densities of agonist ligands. Evidence for synergism in TCR signaling is extensive (Krogsgaard et al. 2005; Krogsgaard et al. 2007; Yachi et al. 2007) though not unchallenged (Sporri and Reis e Sousa 2002; Ma et al. 2008). One effort to explain and model synergistic effects expanded the TCR-pMHC picture to include intracellular interaction between Lck and the coreceptor (either CD4 or CD8) and extracellular interaction between the coreceptor and the MHC (Li et al. 2004), with the coreceptor acting as a bridging molecule. This model allows even momentary interaction between the TCR and self-pMHC to bring TCR ITAM residues near Lck for phosphorylation by dimerizing self-pMHC and agonist-pMHC molecules. Although this model simulated synergy well, it had a number of weaknesses: It could not account for antagonism, allowed for large numbers of endogenous ligands to trigger signaling, and was not validated beyond its ability to qualitatively show a role for endogenous pMHC-TCR complexes in enhancing signaling from agonist pMHC-TCR complexes. At a conceptual level, however, it showed that cross-linking by the coreceptor could produce lattices of pMHC-TCR complexes and hence may be the key to incorporating synergy into models of TCR triggering.

A model simultaneously accounting for both antagonism and synergism has been attempted (Wylie et al. 2007) but failed to achieve digital activation of the MAPK pathway. However, two existing models of TCR signaling may account for synergism through the "memory" inherent in their signaling cascades. Both our model (Feinerman et al. 2008a), which allows protection of previously unengaged TCR through feedback inhibition of SHP-1 by activated ERK, and the model of Das et al (2009), through hysteretic activation of Ras, may allow weak ligands to contribute in a positive manner to signaling. In these models, crossing of the digital activation threshold by the agonist could allow parallel TCR engagement by weak ligands to drive the analog aspects of the response measured in a synergism assay. Thus existing models may already account for the phenomena of antagonism and synergism, emphasizing the added value of computer modeling efforts to study T-cell activation.

Tunability of the T-Cell Response to Ligand

T lymphocytes must undergo a strict developmental program in the thymus before being released as naïve T cells to the periphery. Hematopoietic progenitors entering the thymus are tested against self pMHC to pass two selecting processes. The absence of responsiveness to self pMHC presented by epithelial cells induces death by neglect, whereas proper responsiveness green-lights further differentiation—so called positive selection. But over-responsiveness toward self pMHC induces apoptosis (negative selection). Ultimately, thymic differentiation leads to the export of mature T cells whose signaling machinery is properly "wired" as indicated by a modest, subactivating level of signaling in response to self-pMHC (Stefanova et al. 2002). T cells endowed mainly with receptors of intermediate affinity for self pMHC ligands pass the filters of positive/negative selection (Alam et al. 1996; Savage et al. 1999; Savage and Davis 2001). Beyond the role of the intrinsic affinity of the TCR for selecting ligands in the maturation process, it is also clear that T cells modify the intracellular components of the signaling apparatus to tune the amplitude of their signaling response induced by self-ligand during thymocyte development, extinguishing responses to weak ligands while not constraining responses to strong ligands (Hogquist et al. 1994; Lucas et al. 1999; Yasutomo et al. 2000; Hogquist 2001; Starr et al. 2003).

This fine-tuning of the activation threshold to diverse ligands is an appealing issue to address with computational models. There is clear evidence for changes at the level of proximal TCR signaling in this regard, based on direct analysis of ZAP-70 and ζ phosphorylation in immature versus mature thymocytes offered ligands of various binding affinities for the TCR (Lucas et al. 1999). Studies involving miR181a (Li et al. 2007) and the ERK cascade (Altan-Bonnet and Germain 2005) as well as the expression of Shp-1 (Plas et al. 1999) suggest that one aspect of the tuning involves a rebalancing of ERK activity relative to SHP-1. Immature cells have fewer ERK phosphatases and higher basal ERK activity, but lower SHP-1 levels than more mature thymocytes; our model predicts that such a state of the feedback control elements would allow weak ligands of the TCR to become stimulatory, as direct biochemical analysis indicates (Li et al. 2007). Loss of the elevated ERK state and increased SHP-1 in more mature cells would extinguish signaling in response to weak (self-) ligands but preserve responses to strong TCR binders as experiments show (Lucas et al. 1999) and our model predicts (Altan-Bonnet and Germain 2005). Examination of how these changes drive positive and negative selection during thymocyte differentiation is a challenge well suited to computer modeling.

Stochasticity

Computer modeling has also been particularly illuminating in dealing with the topic of stochasticity in T-cell activation. Early experimental measurements estimated that T cells could respond to one or few pMHC (Sykulev et al. 1996; Irvine et al. 2002; Altan-Bonnet and Germain 2005). Models have pointed out how signaling modules (e.g., MAPK cascade or Ca^{2+} response) could convert such a minute stimulation into large, all-or-none (digital) effects downstream. The small number of ligands implies that a certain degree of stochasticity (i.e., molecular noise) might be critical when studying the system's dynamics. Unfortunately, the more commonly employed deterministic models predict a fixed and reproducible threshold for activation (e.g., T cells would be activated reproducibly by a certain number of antigenic ligands).

Artyomov and colleagues studied a toy model of competition between positive and negative feedback to argue for the functional relevance of stochasticity in TCR signaling (Artyomov et al. 2007). Indeed, in their differential signaling model, the feedbacks are finely balanced, with similar characteristic times of activation. Stochastic fluctuations then generate the experimentally observed bistability, whereby the system decides between activation and repression depending on which feedback—positive or negative—is activated first. However, given the large fluctuations in signaling protein levels from T cell to T cell and the ensuing variation in signaling dynamics (cf. pp. 271–272), additional mechanisms would be necessary to explain how the system self-regulates to be constantly poised at this tipping point of activation.

More generally, a small amount of random variation in ultrasensitive systems stimulated at near-threshold levels can lead to the system occupying both active and inactive states, either on a cell-by-cell basis (two populations) or temporally switching back and forth between two states (Lipniacki et al. 2008). In such cases, biological stochasticity would generate an immune response that would be qualitatively different from that generated by a deterministic mechanism (Germain 2001). For these, computer models become a critical tool of analysis, without which experimental data may be hard to interpret.

Spatial Regulation of T-Cell Activation

Modeling the TCR signaling cascade as a biochemical network in a well-mixed environment has provided great insight into its dynamics and biological function. However, spatial organization of T-cell components may be relevant to the regulation of a receptor by its neighbors or by other membrane-signaling proteins within the plasma membrane.

For example, serial engagement of multiple adjacent receptors by a single low-affinity pMHC has been proposed as a mechanism by which few ligands may assemble a robust functional response over time (Valitutti et al. 1995; Itoh and Germain 1997; Hudrisier et al. 1998; Wofsy et al. 2001). Wofsy et al. estimated analytically the competition between ligand binding/debinding, diffusion, and internalization for a given TCR to illustrate how extensive serial engagement of a given ligand could be. One prediction of this model is that there exists an optimal dwell-time for the ligand-receptor interaction below which the T cell truncates its signaling response and above which ligands fail to serially trigger many TCRs because of excessively long engagement times. Some experimental evidence supports the existence of this optimal dwell time (Kalergis et al. 2001). However, these findings are incompatible with the high potency of pMHC engineered for extremely lengthy TCR engagement (Holler et al. 2000).

Settling this issue requires computer models of T-cell activation on longer timescales (10 min–1 h) that must also account for well-documented lateral segregation of proteins within the plasma membrane. Initial pMHC-TCR binding results in the steric exclusion of the negative regulatory phosphatase CD45 (Choudhuri et al. 2005) and the concentration of actively signaling TCRs into so-called microclusters (Bunnell et al. 2002; Campi et al. 2005) that move toward the center of the T cell-APC contact (Varma et al. 2006). This dynamic reorganization into an immunological synapse (IS) concentrates TCR-pMHC contacts in the center with a surrounding adhesive ring of LFA-1-ICAM-1 interactions. Theoretical work has shown that self-organization of the membrane components can be sufficient for formation of the IS pattern, even in the absence of active cytoskeletal forces (Qi et al. 2001). This model combined measured values of antigen-receptor interaction and physical properties of the plasma membrane to account for synapse formation. The main prediction of this model—that lower expression of TCR in thymocytes would yield smaller, more transient, multifocal synapse formation—was subsequently confirmed experimentally (Lee et al. 2003a).

Further modeling efforts have sought to illuminate how the immune synapse modulates T-cell activation by suppressing signaling from strong agonists through increased down-regulation of the TCR (Cemerski et al. 2008), as well as by enhancing signaling from less avid agonist peptides through concentration of signaling components (Chan et al. 2001; Lee et al. 2003a). Other work has probed how exclusion of the bulky CD45 phosphatase could account for TCR triggering—albeit without adequate speed in the models relative to experiment (Burroughs et al. 2006; Burroughs and van der Merwe 2007). This work accounted for experimental observations whereby genetic modification of the volume or heights of receptors modulated the separation between T cell and APC within the contact area, and resulted in varied levels of activation of T cells (Choudhuri et al. 2005). As discussed above, the coupling of local reaction-diffusion with global elastic constraints is a difficult theoretical problem that necessitates specific modeling formalism. Yet, such theoretical effort will certainly illuminate experimental observations on the interplay between TCR signaling and membrane reorganization (currently a field of intense study). Though pharmacological and genetic methods of synapse interruption often have wide-ranging effects that preclude their attribution solely to spatial organization (Lee et al. 2003a; Ilani et al. 2009), recent work using micropatterned surfaces of membrane and ligands (Mossman et al. 2005; DeMond and Groves 2007) or using photoactivatable ligands (Huse et al. 2007) enables the direct testing of the effects of geometric constraints on T-cell activation. The quantitative and qualitative data from these efforts will in turn allow the revision and validation of models of immune synapse formation and a better understanding of T-cell activation on longer timescales (>10 min).

Reconciliation of Robustness and Variability in T-Cell Activation

Robustness is a concept that was introduced to describe a salient feature of cell signaling: Although cells have fluctuating levels of the many proteins involved in signal transduction, proper physiological function requires predictable responses to extracellular stimuli (Barkai and Leibler 1997). First modeled and experimentally validated in bacteria (Alon et al. 1999; Cluzel et al. 2000; Korobkova et al. 2004), robustness is also critical to T-cell discrimination between self and nonself. Quantitative analysis of T-cell ligand sensing has shown that the calcium response to a set number of ligands does not vary as much as would be expected from variation in the levels of individual signaling proteins that determine this response (Irvine et al. 2002; Li et al. 2004; Purbhoo et al. 2004; Feinerman et al. 2008a).

To investigate the effects of this variation on the robustness of self/nonself discrimination by T cells, we have systematically investigated the effects of modulating individual species within an experimentally validated computational model of the TCR apparatus (Feinerman et al. 2008b). By varying each component within its physiological range, we characterized the effects of natural variability in signaling components in peripheral T lymphocytes. This analysis predicted that signaling components would fall into three categories: (1) noncritical signaling proteins whose variation in expression within the observed range does not measurably affect ligand response, (2) analog regulators whose level of expression fine-tunes the activation threshold in terms of number of ligands required to exceed this threshold, and (3) digital regulators whose level of expression switches a cell between responsive and unresponsive states without affecting activation thresholds in responsive cells (Feinerman et al. 2008b).

These theoretical investigations predicted that ERK-1, despite its crucial role in regulation of ligand discrimination, is a noncritical component because its average expression is in such excess that variation within the physiological range does not affect the T-cell response. Other noncritical components are Lck, ZAP70, LAT, Raf, and MEK. In contrast, the CD8 coreceptor was predicted to be an analog-positive regulator, increasing sensitivity with increased expression. SHP-1, on the other hand, was predicted to be a digital negative regulator: For high levels of SHP-1, T cells will not respond even to high doses of antigen. Consequently, our model predicted that even in a clonal population of T cells, the natural range of expression of SHP-1 and CD8 would lead to highly variable sensitivity to antigen.

We then developed an experimental assay to correlate T-cell responsiveness with the expression levels of signaling components within individual cells. This assay relies on staining of endogenous proteins (e.g., CD8 and SHP-1) and phospho-proteins (e.g., ppERK), followed by flow cytometry to achieve single-cell resolution. Our experimental measurements validated the predictions from our computer model, demonstrating variability across a population of cells, but also added insight into how the system remains robust: Correlation between expression levels of positive and negative regulators (CD8 and SHP-1, respectively) restricts the actual range of sensitivity to a smaller range than would be predicted from theoretical analysis of uncorrelated protein expression variation. This study shed light on the delicate regulation of ligand discrimination in T cells by showing how it can be robustly defined while remaining tunable and flexible in the face of irreducible cell-to-cell variation in molecular phenotype. In this case, the computational model was not the endpoint of the effort, but a tool to enable more precise selection of experimental approaches.

COMPUTATION AND THE STUDY OF T-CELL IMMUNOBIOLOGY: GOING FORWARD

In this review, we have discussed computational models that yielded insights on issues of specificity, antagonism, synergism, stochasticity, spatial regulation, and robustness in T-cell activation. The majority of these models served principally as illustrations of complex hypotheses, using quantitative data to explain observed phenomena. However, successful application of modeling requires more than accurate fitting of experimental observations or clarification of purely theoretical points. Models with the ability to generate testable predictions and extend current understanding based on a validated mathematical framework have the greatest potential to contribute to immunology (Qi et al. 2001; Lee et al. 2003b; Altan-Bonnet and Germain 2005; Feinerman et al. 2008a; Feinerman et al. 2008b). In our experience, the added value of such models is correlated to two simple criteria: model parameters that are tightly constrained by experimental measurements and model predictions that are designed to be testable (and hopefully tested) at the bench.

Proliferation of such added-value models will benefit from current trends in both experimental immunology and modeling software development. From the experimental side, the wider application of quantitative methods, including multiplexed surface plasmon resonance (Lee et al. 2006), absolute quantitative mass spectrometry (Bantscheff et al. 2007), and quantitative flow cytometry (Irish et al. 2004), will continue to restrain the set of unknown parameters, improving the statistical relevance of computer models. From the theoretical side, the rapid dissemination of standardized modeling platforms (particularly Virtual Cell [Loew and Schaff 2001], BioNetGen [Blinov et al. 2004] CellDesigner [Funahashi et al. 2003], and Simmune [Meier-Schellersheim et al. 2006]) will accelerate the exchange of models for further editing, testing, and validation. Thus, one prospect for computer modeling in immunology is more direct participation from immunologists who will both utilize and extend models of T-cell signaling in order to develop a more complete model of T-cell activation. In the long term, one can envision a time when immunologists will systematically test pre-established models of T-cell signaling to assess the significance of new experimental observations and challenge pre-existing mechanisms.

ACKNOWLEDGMENTS

This work was supported in part by funds from the Intramural Research Program, NIAID, National Institutes of Health (NIH). J.C. and G.A.-B.'s research is supported by an NSF grant #0848030 and an NIH grant #R01-AI083408. G.A.-B. is the incumbent of a Bristol-Myers Squibb / James D. Robinson III junior faculty chair at MSKCC. J.C. and G.A.-B. wish to thank Matthew Hathorn and Karen Tkach for critical reading of the manuscript.

REFERENCES

Alam SM, Travers PJ, Wung JL, Nasholds W, Redpath S, Jameson SC, Gascoigne NR. 1996. T-cell-receptor affinity and thymocyte positive selection. *Nature* **381:** 616–620.

Alon U, Surette MG, Barkai N, Leibler S. 1999. Robustness in bacterial chemotaxis. *Nature* **397:** 168–171.

Altan-Bonnet G, Germain RN. 2005. Modeling T-cell antigen discrimination based on feedback control of digital ERK responses. *PLoS Biol* **3:** e356.

Artyomov MN, Das J, Kardar M, Chakraborty AK. 2007. Purely stochastic binary decisions in cell signaling models without underlying deterministic bistabilities. *Proc Natl Acad Sci* **104:** 18958–18963.

Bantscheff M, Schirle M, Sweetman G, Rick J, Kuster B. 2007. Quantitative mass spectrometry in proteomics: A critical review. *Anal Bioanal Chem* **389:** 1017–1031.

Barkai N, Leibler S. 1997. Robustness in simple biochemical networks. *Nature* **387:** 913–917.

Blinov ML, Faeder JR, Goldstein B, Hlavacek WS. 2004. BioNetGen: Software for rule-based modeling of signal transduction based on the interactions of molecular domains. *Bioinformatics* **20:** 3289–3291.

Bunnell SC, Hong DI, Kardon JR, Yamazaki T, McGlade CJ, Barr VA, Samelson LE. 2002. T-cell receptor ligation induces the formation of dynamically regulated signaling assemblies. *J Cell Biol* **158:** 1263–1275.

Burroughs NJ, van der Merwe PA. 2007. Stochasticity and spatial heterogeneity in T-cell activation. *Immunol Rev* **216:** 69–80.

Burroughs NJ, Lazic Z, van der Merwe PA. 2006. Ligand detection and discrimination by spatial relocalization: A kinase-phosphatase segregation model of TCR activation. *Biophys J* **91:** 1619–1629.

Campi G, Varma R, Dustin ML. 2005. Actin and agonist MHC-peptide complex-dependent T-cell receptor microclusters as scaffolds for signaling. *J Exp Med* **202:** 1031–1036.

Cemerski S, Das J, Giurisato E, Markiewicz MA, Allen PM, Chakraborty AK, Shaw AS. 2008. The balance between T-cell receptor signaling and degradation at the center of the immunological synapse is determined by antigen quality. *Immunity* **29:** 414–422.

Chan C, George AJ, Stark J. 2001. Cooperative enhancement of specificity in a lattice of T-cell receptors. *Proc Natl Acad Sci* **98:** 5758–5763.

Choudhuri K, Wiseman D, Brown MH, Gould K, van der Merwe PA. 2005. T-cell receptor triggering is critically dependent on the dimensions of its peptide-MHC ligand. *Nature* **436:** 578–582.

Cluzel P, Surette M, Leibler S. 2000. An ultrasensitive bacterial motor revealed by monitoring signaling proteins in single cells. *Science* **287:** 1652–1655.

Daniels MA, Teixeiro E, Gill J, Hausmann B, Roubaty D, Holmberg K, Werlen G, Hollander GA, Gascoigne NR, Palmer E. 2006. Thymic selection threshold defined by compartmentalization of Ras/MAPK signalling. *Nature* **444:** 724–729.

Das J, Ho M, Zikherman J, Govern C, Yang M, Weiss A, Chakraborty AK, Roose JP. 2009. Digital signaling and hysteresis characterize Ras activation in lymphoid cells. *Cell* **136:** 337–351.

Davis MM, Boniface JJ, Reich Z, Lyons D, Hampl J, Arden B, Chien Y. 1998. Ligand recognition by alpha βT-cell receptors. *Ann Rev Immunol* **16:** 523–544.

Delon J, Bercovici N, Raposo G, Liblau R, Trautmann A. 1998. Antigen-dependent and -independent Ca^2+ responses triggered in T-cells by dendritic cells compared with B cells. *J Exp Med* **188:** 1473–1484.

DeMond AL, Groves JT. 2007. Interrogating the T-cell synapse with patterned surfaces and photoactivated proteins. *Curr Opin Immunol* **19:** 722–727.

Dittel BN, Stefanova I, Germain RN, Janeway CA, Jr. 1999. Cross-antagonism of a T-cell clone expressing two distinct T-cell receptors. *Immunity* **11:** 289–298.

Faeder JR, Hlavacek WS, Reischl I, Blinov ML, Metzger H, Redondo A, Wofsy C, Goldstein B. 2003. Investigation of early events in Fc xRI-mediated signaling using a detailed mathematical model. *J Immunol* **170:** 3769–3781.

Feinerman O, Germain RN, Altan-Bonnet G. 2008a. Quantitative challenges in understanding ligand discrimination by αβ T-cells. *Mol Immunol* **45:** 619–631.

Feinerman O, Veiga J, Dorfman JR, Germain RN, Altan-Bonnet G. 2008b. Variability and robustness in T-cell activation from regulated heterogeneity in protein levels. *Science* **321:** 1081–1084.

Ferrell JE. 2002. Self-perpetuating states in signal transduction: Positive feedback, double-negative feedback and bistability. *Current Opinion Cell Biol* **14:** 140–148.

Ferrell JE, Jr, Bhatt RR. 1997. Mechanistic studies of the dual phosphorylation of mitogen-activated protein kinase. *J Biol Chem* **272:** 19008–19016.

Fiering S, Northrop JP, Nolan GP, Mattila PS, Crabtree GR, Herzenberg LA. 1990. Single cell assay of a transcription factor reveals a threshold in transcription activated by signals emanating from the T-cell antigen receptor. *Genes Develop* **4:** 1823–1834.

Funahashi A, Morohashi M, Kitano H, Tanimura N. 2003. CellDesigner: A process diagram editor for gene-regulatory and biochemical networks *BioSilico* **1:** 159–162.

Germain RN. 2001. The art of the probable: System control in the adaptive immune system. *Science* **293:** 240–245.

Germain RN, Stefanova I. 1999. The dynamics of T-cell receptor signaling: Complex orchestration and the key roles of tempo and cooperation. *Ann Rev Immunol* **17:** 467–522.

Gil D, Schamel WW, Montoya M, Sanchez-Madrid F, Alarcon B. 2002. Recruitment of Nck by CD3ε reveals a ligand-induced conformational change essential for T-cell receptor signaling and synapse formation. *Cell* **109:** 901–912.

Gillespie DT. 2007. Stochastic simulation of chemical kinetics. *Annu Rev Phys Chem* **58:** 35–55.

Hemmer B, Stefanova I, Vergelli M, Germain RN, Martin R. 1998. Relationships among TCR ligand potency, thresholds for effector function elicitation, and the quality of early signaling events in human T-cells. *J Immunol* **160:** 5807–5814.

Henrickson SE, Mempel TR, Mazo IB, Liu B, Artyomov MN, Zheng H, Peixoto A, Flynn MP, Senman B, Junt T et al. 2008. T-cell sensing of antigen dose governs interactive behavior with dendritic cells and sets a threshold for T-cell activation. *Nat Immunol* **9:** 282–291.

Hlavacek WS, Faeder JR, Blinov ML, Posner RG, Hucka M, Fontana W. 2006. Rules for modeling signal-transduction systems. *Sci STKE* **2006:** re6.

Hogquist KA. 2001. Signal strength in thymic selection and lineage commitment. *Curr Opin Immunol* **13:** 225–231.

Hogquist KA, Jameson SC, Heath WR, Howard JL, Bevan MJ, Carbone FR. 1994. T-cell receptor antagonist peptides induce positive selection. *Cell* **76:** 17–27.

Holler PD, Holman PO, Shusta EV, O'Herrin S, Wittrup KD, Kranz DM. 2000. In vitro evolution of a T-cell receptor with high affinity for peptide/MHC. *Proc Natl Acad Sci***97:** 5387–5392.

Holst J, Wang H, Eder KD, Workman CJ, Boyd KL, Baquet Z, Singh H, Forbes K, Chruscinski A, Smeyne R et al. 2008. Scalable signaling mediated by T-cell antigen receptor-CD3 ITAMs ensures effective negative selection and prevents autoimmunity. *Nat Immunol* **9:** 658–666.

Hoops S, Sahle S, Gauges R, Lee C, Pahle J, Simus N, Singhal M, Xu L, Mendes P, Kummer U. 2006. COPASI–a COmplex PAthway SImulator. *Bioinformatics* **22:** 3067–3074.

Hopfield JJ. 1974. Kinetic proofreading: A new mechanism for reducing errors in biosynthetic processes requiring high specificity. *Proc Natl Acad Sci USA* **71:** 4135–4139.

Huang CY, Ferrell JE, Jr. 1996. Ultrasensitivity in the mitogen-activated protein kinase cascade. *Proc Natl Acad Sci* **93:** 10078–10083.

Hudrisier D, Kessler B, Valitutti S, Horvath C, Cerottini JC, Luescher IF. 1998. The efficiency of antigen recognition by CD8+ CTL clones is determined by the frequency of serial TCR engagement. *J Immunol* **161:** 553–562.

Huse M, Klein LO, Girvin AT, Faraj JM, Li Q-J, Kuhns MS, Davis MM. 2007. Spatial and temporal dynamics of T-cell receptor signaling with a photoactivatable agonist. *Immunity* **27:** 76–88.

Ilani T, Vasiliver-Shamis G, Vardhana S, Bretscher A, Dustin ML. 2009. T-cell antigen receptor signaling and immunological synapse stability require myosin IIA. *Nat Immunol* **10:** 531–539.

Irish JM, Hovland R, Krutzik PO, Perez OD, Bruserud O, Gjertsen BT, Nolan GP. 2004. Single cell profiling of potentiated phospho-protein networks in cancer cells. *Cell* **118:** 217–228.

Irvine DJ, Purbhoo MA, Krogsgaard M, Davis MM. 2002. Direct observation of ligand recognition by T-cells. *Nature* **419:** 845–849.

Itoh Y, Germain RN. 1997. Single cell analysis reveals regulated hierarchical T-cell antigen receptor signaling thresholds and intraclonal heterogeneity for individual cytokine responses of CD4+ T-cells. *J Exp Med* **186:** 757–766.

Jameson SC, Carbone FR, Bevan MJ. 1993. Clone-specific T-cell receptor antagonists of major histocompatibility complex class I-restricted cytotoxic T-cells. *J Exp Med* **177:** 1541–1550.

Kalergis AM, Boucheron N, Doucey MA, Palmieri E, Goyarts EC, Vegh Z, Luescher IF, Nathenson SG. 2001. Efficient T-cell activation requires an optimal dwell-time of interaction between the TCR and the pMHC complex. *Nat Immunol* **2:** 229–234.

Korobkova E, Emonet T, Vilar JM, Shimizu TS, Cluzel P. 2004. From molecular noise to behavioural variability in a single bacterium. *Nature* **428:** 574–578.

Krogsgaard M, Juang J, Davis MM. 2007. A role for "self" in T-cell activation. *Sem Immunol* **19:** 236–244.

Krogsgaard M, Li Q-j, Sumen C, Huppa JB, Huse M, Davis MM. 2005. Agonist/endogenous peptide-MHC heterodimers drive T-cell activation and sensitivity. *Nature* **434:** 238–243.

Le Novere N. 2006. Model storage, exchange and integration. *BMC Neurosci* **7:** S11.

Le Novere N, Hucka M, Mi H, Moodie S, Schreiber F, Sorokin A, Demir E, Wegner K, Aladjem MI, Wimalaratne SM et al. 2009. The systems biology graphical notation. *Nat Biotechnol* **27:** 735–741.

Lee SJ, Hori Y, Chakraborty AK. 2003b. Low T-cell receptor expression and thermal fluctuations contribute to formation of dynamic multifocal synapses in thymocytes. *Proc Natl Acad Sci* **100:** 4383–4388.

Lee HJ, Nedelkov D, Corn RM. 2006. Surface plasmon resonance imaging measurements of antibody arrays for the multiplexed detection of low molecular weight protein biomarkers. *Anal Chem* **78:** 6504–6510.

Lee K-H, Dinner AR, Tu C, Campi G, Raychaudhuri S, Varma R, Sims TN, Burack WR, Wu H, Wang J et al.

2003a. The immunological synapse balances T-cell receptor signaling and degradation. *Science* **302:** 1218–1222.

Li QJ, Chau J, Ebert PJ, Sylvester G, Min H, Liu G, Braich R, Manoharan M, Soutschek J, Skare P et al. 2007. miR-181a is an intrinsic modulator of T-cell sensitivity and selection. *Cell* **129:** 147–161.

Li Q-J, Dinner AR, Qi S, Irvine DJ, Huppa JB, Davis MM, Chakraborty AK. 2004. CD4 enhances T-cell sensitivity to antigen by coordinating Lck accumulation at the immunological synapse. *Nat Immunol* **5:** 791–799.

Lipniacki T, Hat B, Faeder JR, Hlavacek WS. 2008. Stochastic effects and bistability in T-cell receptor signaling. *Journal of Theoretical Biology* **254:** 110–122.

Loew LM, Schaff JC. 2001. The Virtual Cell: A software environment for computational cell biology. *Trends Biotechnol* **19:** 401–406.

Lucas B, Stefanova I, Yasutomo K, Dautigny N, Germain RN. 1999. Divergent changes in the sensitivity of maturing T-cells to structurally related ligands underlies formation of a useful T-cell repertoire. *Immunity* **10:** 367–376.

Ma Z, Sharp KA, Janmey PA, Finkel TH. 2008. Surface-anchored monomeric agonist pMHCs alone trigger TCR with high sensitivity. *PLoS Biol* **6:** e43.

Macian F. 2005. NFAT proteins: Key regulators of T-cell development and function. *Nat Rev Immunol* **5:** 472–484.

Malissen B. 2008. CD3 ITAMs count! *Nat Immunol* **9:** 583–584.

Malissen B, Aguado E, Malissen M. 2005. Role of the LAT adaptor in T-cell development and Th2 differentiation. *Adv Immunol* **87:** 1–25.

McKeithan TW. 1995. Kinetic proofreading in T-cell receptor signal transduction. *Proc Natl Acad Sci* **92:** 5042–5046.

Meier-Schellersheim M, Xu X, Angermann B, Kunkel EJ, Jin T, Germain RN. 2006. Key role of local regulation in chemosensing revealed by a new molecular interaction-based modeling method. *PLoS Comput Biol* **2:** e82.

Mossman KD, Campi G, Groves JT, Dustin ML. 2005. Altered TCR signaling from geometrically repatterned immunological synapses. *Science* **310:** 1191–1193.

Ninio J. 1975. Kinetic amplification of enzyme discrimination. *Biochimie* **57:** 587–595.

Owens ND, Timmis J, Greensted A, Tyrrell A. 2010. Elucidation of T-cell signalling models. *J Theor Biol* **262:** 452–470.

Palmer E, Naeher D. 2009. Affinity threshold for thymic selection through a T-cell receptor-co-receptor zipper. *Nat Rev Immunol* **9:** 207–213.

Plas DR, Johnson R, Pingel JT, Matthews RJ, Dalton M, Roy G, Chan AC, Thomas ML. 1996. Direct regulation of ZAP-70 by SHP-1 in T-cell antigen receptor signaling. *Science* **272:** 1173–1176.

Plas DR, Williams CB, Kersh GJ, White LS, White JM, Paust S, Ulyanova T, Allen PM, Thomas ML. 1999. Cutting edge: the tyrosine phosphatase SHP-1 regulates thymocyte positive selection. *J Immunol* **162:** 5680–5684.

Podtschaske M, Benary U, Zwinger S, Hofer T, Radbruch A, Baumgrass R. 2007. Digital NFATc2 activation per cell transforms graded T-cell receptor activation into an all-or-none IL-2 expression. *PLoS ONE* **2:** e935.

Purbhoo MA, Irvine DJ, Huppa JB, Davis MM. 2004. T-cell killing does not require the formation of a stable mature immunological synapse. [erratum appears in Nat Immunol. 2004 Jun;5(6):658]. *Nat Immunol* **5:** 524–530.

Qi SY, Groves JT, Chakraborty AK. 2001. Synaptic pattern formation during cellular recognition. *Proc Natl Acad Sci* **98:** 6548–6553.

Rabinowitz JD, Beeson C, Wulfing C, Tate K, Allen PM, Davis MM, McConnell HM. 1996. Altered T-cell receptor ligands trigger a subset of early T-cell signals. *Immunity* **5:** 125–135.

Racioppi L, Matarese G, D'Oro U, De Pascale M, Masci AM, Fontana S, Zappacosta S. 1996. The role of CD4-Lck in T-cell receptor antagonism: Evidence for negative signaling. *Proc Natl Acad Sci* **93:** 10360–10365.

Saez-Rodriguez J, Simeoni L, Lindquist JA, Hemenway R, Bommhardt U, Arndt B, Haus U-U, Weismantel R, Gilles ED, Klamt S et al. 2007. A logical model provides insights into T-cell receptor signaling. *PLoS Comput Biol* **3:** e163.

Salazar C, Hofer T. 2003. Allosteric regulation of the transcription factor NFAT1 by multiple phosphorylation sites: A mathematical analysis. *J Mol Biol* **327:** 31–45.

Savage PA, Davis MM. 2001. A kinetic window constricts the T-cell receptor repertoire in the thymus. *Immunity* **14:** 243–252.

Savage PA, Boniface JJ, Davis MM. 1999. A kinetic basis for T-cell receptor repertoire selection during an immune response. *Immunity* **10:** 485–492.

Schamel WW, Risueno RM, Minguet S, Ortiz AR, Alarcon B. 2006. A conformation- and avidity-based proofreading mechanism for the TCR-CD3 complex. *Trends Immunol* **27:** 176–182.

Schoeberl B, Eichler-Jonsson C, Gilles ED, Muller G. 2002. Computational modeling of the dynamics of the MAP kinase cascade activated by surface and internalized EGF receptors. *Nat Biotechnol* **20:** 370–375.

Sporri R, Reis e Sousa C. 2002. Self peptide/MHC class I complexes have a negligible effect on the response of some CD8(+) T-cells to foreign antigen. *Eur J Immunol* **32:** 3161–3170.

Starr TK, Jameson SC, Hogquist KA. 2003. Positive and negative selection of T-cells. *Annu Rev Immunol* **21:** 139–176.

Stefanova I, Dorfman JR, Germain RN. 2002. Self-recognition promotes the foreign antigen sensitivity of naive T lymphocytes. *Nature* **420:** 429–434.

Stefanova I, Hemmer B, Vergelli M, Martin R, Biddison WE, Germain RN. 2003. TCR ligand discrimination is enforced by competing ERK positive and SHP-1 negative feedback pathways. *Nat Immunol* **4:** 248–254.

Sykulev Y, Joo M, Vturina I, Tsomides TJ, Eisen HN. 1996. Evidence that a single peptide-MHC complex on a targeT-cell can elicit a cytolytic T-cell response. *Immunity* **4:** 565–571.

Valitutti S, Muller S, Cella M, Padovan E, Lanzavecchia A. 1995. Serial triggering of many T-cell receptors by a few peptide-MHC complexes.[see comment]. *Nature* **375:** 148–151.

Varma R, Campi G, Yokosuka T, Saito T, Dustin ML. 2006. T-cell receptor-proximal signals are sustained in peripheral microclusters and terminated in the central supramolecular activation cluster. *Immunity* **25:** 117–127.

Wofsy C, Coombs D, Goldstein B. 2001. Calculations show substantial serial engagement of T-cell receptors. *Biophys J* **80:** 606–612.

Wylie DC, Das J, Chakraborty AK. 2007. Sensitivity of T-cells to antigen and antagonism emerges from differential regulation of the same molecular signaling module. *Proc Natl Acad Sci* **104:** 5533–5538.

Xiong W, Ferrell JE, Jr. 2003. A positive-feedback-based bistable 'memory module' that governs a cell fate decision.[see comment]. *Nature* **426:** 460–465.

Yachi PP, Lotz C, Ampudia J, Gascoigne NR. 2007. T-cell activation enhancement by endogenous pMHC acts for both weak and strong agonists but varies with differentiation state. *J Exp Med* **204:** 2747–2757.

Yasutomo K, Lucas B, Germain RN. 2000. TCR signaling for initiation and completion of thymocyte positive selection has distinct requirements for ligand quality and presenting cell type. *J Immunol* **165:** 3015–3022.

Zehn D, Lee SY, Bevan MJ. 2009. Complete but curtailed T-cell response to very low-affinity antigen. *Nature* **458:** 211–214.

Index